Saline and Alkaline Soils in Latin America

Edith Taleisnik · Raúl S. Lavado
Editors

Saline and Alkaline Soils in Latin America

Natural Resources, Management and Productive Alternatives

Springer

Editors
Edith Taleisnik
Consejo Nacional de Investigaciones
Científicas y Técnicas (CONICET)
Buenos Aires, Argentina

Instituto de Fisiología y Recursos Genéticos
Vegetales (IFRGV), Instituto Nacional
de Tecnología Agropecuaria (INTA)
Centro de Investigaciones
Agropecuarias (CIAP)
Córdoba, Argentina

Facultad de Ciencias Agropecuarias
Universidad Católica de Córdoba
Córdoba, Argentina

Raúl S. Lavado
Consejo Nacional de Investigaciones
Científicas y Técnicas (CONICET)
Buenos Aires, Argentina

Facultad de Agronomía
Universidad de Buenos Aires
Buenos Aires, Argentina

ISBN 978-3-030-52591-0 ISBN 978-3-030-52592-7 (eBook)
https://doi.org/10.1007/978-3-030-52592-7

This Springer imprint is published by the registered company Springer Nature Switzerland AG
The registered company address is: Gewerbestrasse 11, 6330 Cham, Switzerland

Foreword

As world population continues to expand, global need for increased food production is pushing agriculture into regions with lower rainfall and salt-affected soils. This increases the pressure to develop and extend irrigation schemes and to bring marginal land into production. Soil salinization is becoming more extensive as a result of land clearing and unsustainable irrigation practices, and is a major cause of land degradation. The need for more food means more dependence on irrigation schemes which typically result in 25–50% of the irrigated land being salinized. Clearing of natural vegetation for cropping or grazing brings salt to the surface. Low-lying countries are susceptible to coastal inundation due to rising sea levels and extreme weather events caused by global warming, and a combination of salinity plus waterlogging is doubly disastrous.

There is a clear and urgent need for an evaluation of the extent of salt-affected soils, the extent to which they are expanding, and their ability to support agriculture and forestry without further land degradation. If these lands are to be cultivated, it is imperative they be managed sensitively by innovative management, new forms of agriculture, using new genetic resources. These issues apply to all continents including South America, and most importantly to the three large countries in Latin America: Mexico, Argentina and Brazil. These are the production mainstay for the world's beef and sugar, as well as for a large share of soybean, maize and several other agricultural commodities.

This book is a valuable contribution to the issues surrounding salinized soils. It is timely as global food production is under threat from climate change. The editors recognised the need for this book and have selected papers for two reasons. One is to provide a holistic coverage—an integrated perspective of the issues surrounding soil salinity, landscape management, and crop and forage production—for this has rarely been published. While soil salinity and its effects on crop production have been addressed in a large number of publications in specialised literature, integrative perspectives in any country or continent are not commonly published in a single volume.

The second rationale for this volume is that many key papers on salinity in Latin America are published in the language of that country, Spanish or Portuguese, and not readily accessed by scientists in other continents. Yet the fundamentals of managing salinized soils in Latin America are the same as in many other countries or continents particularly Australia, Asia and North America.

The editors of this work have a strong international reputation through their publications and participation at international conferences. Both editors are from Argentina: Dr. Edith Taleisnik is a plant physiologist specialising in salt-tolerant forage grasses, and Dr. Raúl Lavado is a soil scientist working also on plant nutrition and soil toxicities to crop production. Many of the contributors to this book also have a strong international presence while others more often publish in national journals or attend conferences where Spanish or Portuguese is the common language.

I recommend this volume as a valuable resource for land-use managers, irrigation engineers, agronomists, and also for crop scientists and plant breeders working to increase production on saline soils. While applying to all countries in Latin America, the principles are of equal importance to scientists from North America and other continents. This book provides a perspective that will allow the reader to extend their understanding of the issues faced by agriculture on salt-affected land, and the planning of land use in the foreseeable future.

Rana Munns
School of Agriculture and Environment and
ARC Centre of Excellence in Plant Energy Biology
University of Western Australia
Perth, Australia

CSIRO Agriculture and Food
Canberra, Australia

Preface

Latin America is a cultural entity extending from 23°38′4.2″ N in North America, to the southernmost tip of South America, at 59°29′20″ S. It is a vast area, spanning for 19.2 million km^2 and home for approximately 650 million inhabitants. Spanish and Portuguese are the most widely used languages in the region, while English, French and Dutch are also spoken in some countries. This extensive territory features not just a huge variety of climates and soils, it includes the one of the world's driest regions (the Atacama desert), the longest continental mountain range (the Andes), the Amazon jungle, one of the largest wetlands of the world (the Brazilian Pantanal), as well as extensive fertile plains (the Argentine pampas and the campos).

Soil salinity and alkalinity are concerning environmental constraints and are found in diverse environments throughout the region. In non-irrigated arid and semi-arid zones, very saline soils support naturally adapted vegetation. Saline soils also occur in irrigated areas in those zones, where intensive agriculture is practiced. The existence of sodic soils is registered mainly in humid and sub-humid regions and has spread under the effects of complementary irrigation, particularly when poor-quality water is used. On the other hand, agricultural expansion in recent years, fueled by increased world demand for soybeans, has led to extensive forest clearing and, in some areas, consequent soil salinization. Soil salinity and alkalinity have well-documented negative impacts, that stem from the susceptibility of plants in general, and specifically crop plants, to these conditions. The adverse consequences for crop yield and quality generate, in turn, social and economic penalties, immediately affecting the communities established in those environments, and, evidently, society as a whole. Challenges posed by those penalties have stimulated local research endeavors to overcome them.

While soil salinity and its effects have been addressed in a large number of publications in specialized literature, an integrated perspective in Latin America had not been presented previously. The chapters in this book aim at providing such perspective, and this has been one of the purposes of this publication. The second purpose of this book has been to provide an entry way to a large body of knowledge developed on salinity-related problems, often published locally in each country.

Thus, a number of case studies is included, supported by references to work published in Spanish or Portuguese, which is often not registered by foreign readers. Together, the 106 contributing authors to this volume provide a dynamic perspective on salinity and alkalinity problems and approaches in LA.

The book has been organized in five parts. The first provides a global overview on the salinity situation in LA, particularly in irrigated areas. Through a case study in Brazil, it addresses some of the complex social outcomes of soil and water salinity. The second part, focuses on soils, water, agriculture and management in arid environments, while the third part addresses the same topics in humid environments. The fourth part considers plant resources from saline soils and their contribution to productive and ecological sustainability. The chapter in the fifth part provides a perspective on expected changes in vulnerable environments under global climate change and potential management and research actions to mitigate them.

The information about the actual area occupied by salt affected soils in Latin America is relatively uncertain. Most data are different estimates from varied sources made along 50 years and therefore, it is not uncommon to arrive at different numbers when authors quote papers or reports showing relatively different information. The variability of estimations of the extent of saline and salinized areas reveals a weakness which has not been overcome yet. There is an urgent need to update and coordinate the data that these estimates are based upon. This highlights one of the challenges that must be addressed if concerted actions to deal with salinity in the region are initiated in the future. While salinity (and sodicity) usually translate into local social impacts, long-term, extra-regional policies are required to protect vulnerable ecosystems and their communities from expected effects of large-scale land-use changes.

The volume is intended for an academic and also a technical audience. The editors are indebted to all the authors that very generously contributed with their knowledge and expertise to this book, and responded positively to editorial comments and suggestions. We acknowledge topic cover has been involuntarily partial and extend our apologies to numerous research groups that work in the general subject of this book, and whose contributions have not been included. It is our hope that, as a reaction, this publication will stimulate efforts for increasing the visibility of all work related to salinity in Latin America.

Córdoba, Argentina Edith Taleisnik
Buenos Aires, Argentina Raúl S. Lavado
2020

Commonly Used Units and Their Conversion

Surface

1 km^2 = 100 ha = 247.1 acre
1 ha = 2.471 acre

Values for geographical areas are usually expressed in km^2, while ha is used for agronomic or productive areas

1 m^2 = 1.196 sq yd

Mass

1 kg = 0.001 t = 2.20462 lb

Salinity (Electrical Conductivity—EC)

1 dS/m = 1 mS/cm = 1000 μS/cm = 1 mmho/cm
1 dS/m ≈ 10 meq/l

Concentration

1 mg/l = 1 ppm
1 $mmol_c$/l = 1 meq/l
1 dag/kg = 1%
1 $cmol_c$/kg = 1 meq/100 g

Contents

Editors and Contributors

About the Editors

Edith Taleisnik graduated as a biologist from the National University of Córdoba, in Argentina. She has an M.Sc. from Stanford University, USA, and a Ph.D. from the Ben Gurion University of the Negev, in Israel. She is currently a Researcher at the National Research Council of Argentina (CONICET), affiliated to the National Institute for Agricultural Technology (INTA), a professor at the Catholic University of Córdoba, Argentina and a member of the Argentine Academy of Agronomy and Veterinary (ANAV). She has been a grantee and, later, an active collaborator and member of the Board of Trustees of the International Foundation for Science (IFS).

Her research focuses on mechanisms of plant response and tolerance to salt and alkali conditions, mainly in forage grasses. She has taught undergraduate Plant Physiology courses for Agronomy students and national and international graduate courses on plant salt tolerance mechanisms. Along with other colleagues, she participated in the creation of the Argentine National Salinity Network (RAS, Red Argentina de Salinidad) and presided the institution from 2005 to 2016.

Raúl S. Lavado graduated from the College of Agronomy, University of Buenos Aires, Argentina (1968). He has had training/postdoctoral experiences in Spain (1971), Canada (1979) and USA (1980). From 1993 to 2010 he was Full Professor, and from 2010, he is Distinguished Professor at the College of Agronomy, University of Buenos Aires. He is a CONICET (National Research Council of Argentina) researcher and has been the Director of the Institute of Agricultural and Environmental Biosciences—INBA (2008–2014).

His research focuses mainly in soil, particularly soil salinization, soil fertility and soil contamination and has published more than 250 research papers, several books and chapters, technological publications and extension materials. He has been advisor of numerous Master and Ph.D. theses. He has been Associated Editor of the Journal of Soil and Water Conservation (USA) and other journals, and founder of the Argentinean journal Ciencia del Suelo. He is an honorary member of the

Argentine Society of Soil Science and received the Santelises Award from the Soil Science Society of Latin America, for the book "Soils of Argentina". He was the president of the Red Argentina de Salinidad (RAS) from 2016 to 2019.

Contributors

Liliana I. Allegretti Facultad de Ciencias Agrarias, Universidad Nacional de Cuyo, Mendoza, Argentina;
Instituto Argentino de Investigaciones de las Zonas Áridas (IADIZA)—CONICET Mendoza, Mendoza, Argentina

Carina Rosa Alvarez Soil Fertility and Fertilizer, School of Agronomy, University of Buenos Aires, Buenos Aires, Argentina

Juan A. Alvarez Instituto Argentino de Nivología, Glaciología y Ciencias Ambientales—CONICET Mendoza, Mendoza, Argentina;
Facultad de Ciencias Agrarias, Universidad Nacional de Cuyo, Mendoza, Argentina

Adalberto Benavides-Mendoza Departamento de Horticultura, Universidad Autónoma Agraria Antonio Narro, Saltillo, Mexico

Dolores A. Bustos Instituto de Fisiología y Recursos Genéticos Vegetales (IFRGV), Centro de Investigaciones Agropecuarias (CIAP), Instituto Nacional de Tecnología Agropecuaria (INTA), Córdoba, Argentina

José M. Cisneros Facultad de Agronomía y Veterinaria, Universidad Nacional de Río Cuarto, Río Cuarto, Argentina

Mariano A. Cony Instituto Argentino de Investigaciones de las Zonas Áridas (IADIZA)—CONICET Mendoza, Mendoza, Argentina

Christiano Rebouças Cosme Center for Agrarian Sciences, Department of Agronomic and Forest Sciences, Federal Rural University of the Semi-Arid (UFERSA), Mossoró, Brazil

Raimundo Nonato Távora Costa Universidade Federal do Ceará, Fortaleza, Ceará, Brazil

Luciana Couso Facultad de Agronomía, Cátedra de Genética, Universidad de Buenos Aires, Buenos Aires, Argentina

Jucirema Ferreira da Silva Center for Agrarian Sciences, Department of Agronomic and Forest Sciences, Federal Rural University of the Semi-Arid (UFERSA), Mossoró, Brazil

Marcia Regina Farias da Silva University of the Rio Grande do Norte State, Mossoró, Brazil

Cláudio Ricardo da Silva Federal University of Uberlândia, Uberlândia, Minas Gerais, Brazil

Nildo da Silva Dias Center for Agrarian Sciences, Department of Agronomic and Forest Sciences, Federal Rural University of the Semi-Arid (UFERSA), Mossoró, Brazil

Jorge Freire da Silva Ferreira United States Salinity Laboratory (USDA-ARS), Riverside, CA, USA

Francisco Vanies da Silva Sá Center for Agrarian Sciences, Department of Agronomic and Forest Sciences, Federal Rural University of the Semi-Arid (UFERSA), Mossoró, Brazil

Ivanice da Silva Santos Laboratory of Plant Production Technologies, Universidade Estadual da Paraíba, Catolé do Rocha, Paraíba, Brazil

Sofía Dágata Facultad de Ciencias Agrarias, Universidad Nacional de Cuyo, Mendoza, Argentina

Francisco Damiano Instituto de Clima y Agua, Instituto Nacional de Tecnología Agropecuaria (INTA), Hurlingham, Buenos Aires, Argentina

Claudivan Feitosa de Lacerda Universidade Federal do Ceará, Fortaleza, Ceará, Brazil

Geovani Soares de Lima Universidade Federal de Campina Grande, Pombal, Paraíba, Brazil

José Francismar de Medeiros Universidade Federal Rural do Semi-Árido, Mossoró, Rio Grande do Norte, Brazil

André Moreira de Oliveira Center for Agrarian Sciences, Department of Agronomic and Forest Sciences, Federal Rural University of the Semi-Arid (UFERSA), Mossoró, Brazil

Carla Natanieli de Oliveira Batista Center for Agrarian Sciences, Department of Agronomic and Forest Sciences, Federal Rural University of the Semi-Arid (UFERSA), Mossoró, Brazil

Edivan Rodrigues de Souza Agronomy Department, Federal Rural University of Pernambuco (UFRPE), Recife, Pernambuco, Brazil

Geocleber Gomes de Sousa Universidade da Integração Internacional da Lusofonia Afro-Brasileira, Redenção, Ceará, Brazil

Osvaldo Nogueira de Sousa Neto Multidisciplinary Center of Angicos, UFERSA, Angicos, Brazil

Lorena del R. Marinoni Programa de Documentación, Conservación y Valoración de La Flora Nativa (PRODOCOVA), Facultad de Ciencias Agrarias, Universidad Nacional del Litoral (FCA-UNL), Esperanza, Santa Fe, Argentina

Carla E. Di Bella IFEVA-CONICET, Facultad de Agronomía, Universidad de Buenos Aires, Buenos Aires, Argentina;
Cátedra de Forrajicultura, Departamento de Producción Animal, Facultad de Agronomía, Universidad de Buenos Aires, Buenos Aires, Argentina

Yésica Diaz Grupo de Estudios Ambientales—IMASL, Universidad Nacional de San Luis & CONICET, San Luis, Argentina

Cleyton dos Santos Fernandes Center for Agrarian Sciences, Department of Agronomic and Forest Sciences, Federal Rural University of the Semi-Arid (UFERSA), Mossoró, Brazil

María Emilia Fernández Instituto Argentino de Investigaciones de las Zonas Áridas (IADIZA)—CONICET Mendoza, Mendoza, Argentina

Miguel Ferreira Neto Center for Agrarian Sciences, Department of Agronomic and Forest Sciences, Federal Rural University of the Semi-Arid (UFERSA), Mossoró, Brazil

Fernando José Freire Agronomy Department, Federal Rural University of Pernambuco (UFRPE), Recife, Pernambuco, Brazil

Maria Betânia Galvão dos Santos Freire Agronomy Department, Federal Rural University of Pernambuco (UFRPE), Recife, Pernambuco, Brazil

Sheila A. C. Furquim Environmental Sciences Department, Universidade Federal de São Paulo (UNIFESP), Diadema-SP, Brazil

Nadia R. Gamboa Departamento Académico de Ciencias, sección Química; Grupo GRIDES, Pontificia Universidad Católica del Perú, Lima, Perú

Hans Raj Gheyi Federal University of Recôncavo da Bahia (UFRB), Cruz das Almas, Bahia, Brazil

Sergio Ghio Facultad de Agronomía, Cátedra de Genética, Universidad de Buenos Aires, Buenos Aires, Argentina

Julio Giavedoni Facultad de Ciencias Agrarias, Universidad Nacional del Litoral, Esperanza, Santa Fe, Argentina

Raúl Giménez Grupo de Estudios Ambientales—IMASL, Universidad Nacional de San Luis & CONICET, San Luis, Argentina;
Departamento de Geología, Facultad de Ciencias Físico Matemáticas y Naturales, Universidad Nacional de San Luis, San Luis, Argentina

Andrea Giordano Facultad de Agronomía, Cátedra de Genética, Universidad de Buenos Aires, Buenos Aires, Argentina;
Centre for AgriBioscience, AgriBio, La Trobe University, Bundoora, VIC, Australia

Silvina Greco Facultad de Ciencias Agrarias, Universidad Nacional de Cuyo, Mendoza, Argentina

Agustín A. Grimoldi IFEVA-CONICET, Facultad de Agronomía, Universidad de Buenos Aires, Buenos Aires, Argentina;
Cátedra de Forrajicultura, Departamento de Producción Animal, Facultad de Agronomía, Universidad de Buenos Aires, Buenos Aires, Argentina

Bárbara Guida-Johnson Instituto Argentino de Nivología, Glaciología y Ciencias Ambientales—CONICET Mendoza, Mendoza, Argentina;
Facultad de Ciencias Exactas y Naturales, Universidad Nacional de Cuyo, Mendoza, Argentina

Adriana Hernández-Rodríguez Facultad de Ciencia Agrotecnológicas, Universidad Autónoma de Chihuahua, Chihuahua, Chihuahua, México

Perla A. Imbellone Instituto de Geomorfología y Suelos, Universidad Nacional de La Plata, La Plata, Argentina

Silvia Imhoff ICiAgro Litoral, Facultad de Ciencias Agrarias, Universidad Nacional del Litoral & CONICET, Esperanza, Argentina

Dushmantha H. Jayawickreme Department of Earth Science, Southern Connecticut State University, New Haven, CT, USA

Esteban G. Jobbágy Grupo de Estudios Ambientales—IMASL, Universidad Nacional de San Luis & CONICET, San Luis, Argentina;
SARAS—South American Institute for Resilience and Sustainability Studies, Maldonado, Uruguay

Nerina B. Lana Instituto Argentino de Nivología, Glaciología y Ciencias Ambientales—CONICET Mendoza, Mendoza, Argentina;
Facultad de Ciencias Exactas y Naturales, Universidad Nacional de Cuyo, Mendoza, Argentina

Raúl S. Lavado Facultad de Agronomía, Universidad de Buenos Aires and INBA—CONICET/UBA, Buenos Aires, Argentina

Edna Lúcia Rocha Linhares Multidisciplinary Center of Caraúbas, UFERSA, Caraúbas, Brazil

Matheus Cardim Ferreira Lima Department of Agroforest Ecosystems, Polytechnical University of Valencia, Valencia, Spain;
Research and Extension Unit (AGDR), Food and Agriculture Organization of the United Nations (FAO), Rome, Italy

Dario Fernando Luna Instituto de Fisiología y Recursos Genéticos Vegetales (IFRGV), Centro de Investigaciones Agropecuarias (CIAP), Instituto Nacional de Tecnología Agropecuaria (INTA), Córdoba, Argentina

Cristiane E. C. Macedo Laboratory of Plant Biotechnology Studies, Universidade Federal do Rio Grande do Norte, Natal, Rio Grande do Norte, Brazil

Josemir Moura Maia Laboratory of Plant Production Technologies, Universidade Estadual da Paraíba, Catolé do Rocha, Paraíba, Brazil

Adolfo B. Marchese Facultad de Ciencias e Ingeniería, Pontificia Universidad Católica del Perú, Lima, Perú

Victoria Marchesini Grupo de Estudios Ambientales—IMASL, Universidad Nacional de San Luis & CONICET, San Luis, Argentina

Laura E. Martínez E.E.A. INTA Mendoza, Mendoza, Argentina

Pablo A. Meglioli Instituto Argentino de Nivología, Glaciología y Ciencias Ambientales—CONICET Mendoza, Mendoza, Argentina;
Facultad de Ciencias Agrarias, Universidad Nacional de Cuyo, Mendoza, Argentina

Yuri Lima Melo Laboratory of Ecophysiology of Cultivated Plants, Universidade Estadual da Paraíba, Campina Grande, Paraíba, Brazil

Maria Alejandra Moreno-Pizani Faculdade Pecege, Piracicaba, Brazil

Eduardo Musacchio Facultad de Agronomía, Cátedra de Genética, Universidad de Buenos Aires, Buenos Aires, Argentina

Amira Susana Nieva Instituto Tecnológico de Chascomús (INTECH), Chascomús, Argentina;
Max Planck Institute of Molecular Plant Physiology, Potsdam, Germany

Flavia Alonso Nogara Facultad de Agronomía, Cátedra de Genética, Universidad de Buenos Aires, Buenos Aires, Argentina

Marcelo D. Nosetto Grupo de Estudios Ambientales—IMASL, Universidad Nacional de San Luis & CONICET, San Luis, Argentina;
Cátedra de Climatología Agrícola, Facultad de Ciencias Agropecuarias (UNER), Entre Ríos, Argentina

Dámaris Ojeda-Barrios Facultad de Ciencia Agrotecnológicas, Universidad Autónoma de Chihuahua, Chihuahua, Chihuahua, México

Federico G. Olmedo E.E.A. INTA Mendoza, Mendoza, Argentina

Laura Raquel Orozco-Meléndez Facultad de Ciencia Agrotecnológicas, Universidad Autónoma de Chihuahua, Chihuahua, Chihuahua, México

José Luis Panigatti (Deceased), Instituto Nacional de Tecnología Agropecuaria (INTA), Buenos Aires, Argentina

Carlos B. Passera Facultad de Ciencias Agrarias, Universidad Nacional de Cuyo, Mendoza, Argentina

José F. Pensiero Programa de Documentación, Conservación y Valoración de la Flora Nativa (PRODOCOVA), Facultad de Ciencias Agrarias, Universidad Nacional del Litoral (FCA-UNL), Esperanza, Santa Fe, Argentina

Luiz Guilherme Medeiros Pessoa Academic Unit of Serra Talhada (UAST), Federal Rural University of Pernambuco (UFRPE), Serra Talhada, Pernambuco, Brazil

Ildefonso Pla Sentís Department of Environment and Soil Science, University of Lleida, Lleida, Spain

Jeane Cruz Portela Center for Agrarian Sciences, Department of Agronomic and Forest Sciences, Federal Rural University of the Semi-Arid (UFERSA), Mossoró, Brazil

Elena Ramos Facultad de Agronomía, Cátedra de Genética, Universidad de Buenos Aires, Buenos Aires, Argentina

M. Florencia Ricard INCITAP-CONICET, Santa Rosa, La Pampa, Argentina;
Facultad de Ciencias Exactas y Naturales, UNLPam, Santa Rosa, La Pampa, Argentina

Geraldina A. Richard Programa de Documentación, Conservación y Valoración de la Flora Nativa (PRODOCOVA), Facultad de Ciencias Agrarias, Universidad Nacional del Litoral (FCA-UNL), Esperanza, Santa Fe, Argentina

Helena Rimski Korsakov Soil Fertility and Fertilizer, School of Agronomy, University of Buenos Aires, Buenos Aires, Argentina

Pablo Peralta Roa Facultad de Agronomía, Cátedra de Genética, Universidad de Buenos Aires, Buenos Aires, Argentina

Andrés Alberto Rodríguez Consejo Nacional de Investigaciones Científicas y Técnicas (CONICET), Buenos Aires, Argentina;
Laboratorio de Estrés Abiótico y Biótico en Plantas, Unidad de Biotecnología 1, Instituto Tecnológico de Chascomús (INTECH), Chascomús, Argentina

Adriana M. Rodríguez Cátedra de Forrajicultura, Departamento de Producción Animal, Facultad de Agronomía, Universidad de Buenos Aires, Buenos Aires, Argentina

Mónica B. Rodríguez Facultad de Agronomía, Universidad de Buenos Aires, Buenos Aires, Argentina

Oscar Adolfo Ruiz Instituto Tecnológico de Chascomús (INTECH), Chascomús, Argentina

Pablo Rush Facultad de Agronomía, Cátedra de Genética, Universidad de Buenos Aires, Buenos Aires, Argentina

Esteban Sanchez CONACYT-Centro de Investigación en Alimentación Y Desarrollo A.C. Coordinación Delicias, Fraccionamiento Vencedores del Desierto, Delicias, Chihuahua, México

Santiago E. Sari E.E.A. INTA Mendoza, Mendoza, Argentina

Carmen E. Sartor Facultad de Ciencias Agrarias, Universidad Nacional de Cuyo, Mendoza, Argentina

Gustavo E. Schrauf Facultad de Agronomía, Cátedra de Genética, Universidad de Buenos Aires, Buenos Aires, Argentina

Matías F. Schrauf Facultad de Agronomía, Departamento de Métodos Cuantitativos, Universidad de Buenos Aires, Buenos Aires, Argentina

Joaquim A. G. Silveira Laboratory of Plant Metabolism, Universidade Federal do Ceará, Fortaleza, Ceará, Brazil

Ana Claudia Medeiros Souza Center for Agrarian Sciences, Department of Agronomic and Forest Sciences, Federal Rural University of the Semi-Arid (UFERSA), Mossoró, Brazil

Germán Spangenberg Centre for AgriBioscience, AgriBio, La Trobe University, Bundoora, VIC, Australia

Miguel A. Taboada Instituto de Suelos, Instituto Nacional de Tecnología Agropecuaria (INTA), CONICET, Buenos Aires, Argentina

Edith Taleisnik Consejo Nacional de Investigaciones Científicas y Técnicas (CONICET), Buenos Aires, Argentina;
Instituto de Fisiología y Recursos Genéticos Vegetales (IFRGV), Centro de Investigaciones Agropecuarias (CIAP), Instituto Nacional de Tecnología Agropecuaria (INTA), Córdoba, Argentina;
Facultad de Ciencias Agropecuarias, Universidad Católica de Córdoba, Córdoba, Argentina

Carlos H. Tavares Corrêa Departamento Académico de Humanidades, sección Geografía y Medio Ambiente; Grupo GRIDES, Pontificia Universidad Católica del Perú, Lima, Perú

Pablo Tomas Facultad de Ciencias Agrarias, Universidad Nacional del Litoral, Esperanza, Santa Fe, Argentina

Martín Torres Duggan Tecnoagro, Buenos Aires, Argentina

Rosana C. Vallone E.E.A. INTA Mendoza, Mendoza, Argentina;
Soil Science Department, National University of Cuyo (UNCuyo), Mendoza, Argentina

Cecilia Vega Riveros Instituto Argentino de Nivología, Glaciología y Ciencias Ambientales—CONICET Mendoza, Mendoza, Argentina

Thiago T. Vidoca Geography Department, Universidade de São Paulo (USP), São Paulo-SP, Brazil

Ernesto F. Viglizzo INCITAP-CONICET, Santa Rosa, La Pampa, Argentina;
Facultad de Ciencias Empresariales, Universidad Austral, Buenos Aires, Argentina

Pablo E. Villagra Instituto Argentino de Nivología, Glaciología y Ciencias Ambientales—CONICET Mendoza, Mendoza, Argentina;
Facultad de Ciencias Agrarias, Universidad Nacional de Cuyo, Mendoza, Argentina

Lisandro Voda Facultad de Agronomía, Cátedra de Genética, Universidad de Buenos Aires, Buenos Aires, Argentina

Juan M. Zabala Programa de Documentación, Conservación y Valoración de la Flora Nativa (PRODOCOVA), Facultad de Ciencias Agrarias, Universidad Nacional del Litoral (FCA-UNL), Esperanza, Santa Fe, Argentina

The Saline Environments in Latin America. Overview and Social Approach

Overview of Salt-Affected Areas in Latin America: Physical, Social and Economic Perspectives

Ildefonso Pla Sentís

Abstract In Latin América (LA), as well as in other parts of the world, salt-affected soils, both saline and sodic, are found under dryland and irrigated conditions, with negative consequences for the environment, for crop productivity and for animal and human health. Additionally, some tropical coastal and river delta areas have developed saline acid soils. Most of the salt-affected areas have extended under natural conditions. However, the development of affected areas as a result of human-induced processes, mainly associated with hydrological changes caused by irrigation and drainage practices, is increasing. This process negatively affects, sometimes irreversibly, the productive capacity of some of the best soils in many countries of LA, with important economic impacts and social consequences. Although recent estimates of the extension and distribution of human-induced salt-affected soils in LA are not available, there are clear indications that both problems, salinity and sodicity, under dryland and irrigated conditions, have been and are presently increasing in many LA countries. A country-by-country overview of soil and water salinity and sodicity is presented in this chapter, focusing mainly on irrigation and drainage problems.

Keywords Latin America · Salinity · Sodicity · Salt-affected soils · Irrigation · Drainage

1 Introduction

In general, all soils with problems directly or indirectly derived from the amount and kind of salts in solution are referred to as "salt-affected soils" (Pla Sentís 1983). Salt-affected soils, both saline and sodic, may develop both under dryland and irrigated conditions (Pla Sentís 2014, 2015), negatively affecting physical and chemical soil properties, crop production and animal and human health.

I. Pla Sentís (✉)
Department of Environment and Soil Science, University of Lleida, Lleida, Spain
e-mail: ildefonso.pla@udl.cat

E. Taleisnik and R. S. Lavado (eds.), *Saline and Alkaline Soils in Latin America*, https://doi.org/10.1007/978-3-030-52592-7_1

Salt-affected soils may develop through natural processes (primary salinization) or can be induced by human intervention (secondary salinization). The processes of secondary salinization become accelerated when the soil water regime is drastically changed with the introduction of irrigation with drainage restrictions (Fig. 1).

The problems of secondary soil salinity are most widespread in arid and semi-arid regions but salt-affected soils also occur extensively in sub-humid and humid climates, and in the coastal regions, where shallow saline and sodic groundwaters cause large-scale salinization and sodification. The most serious salinity problems are located in the irrigated arid and semiarid regions, yet it is in these regions that irrigation is essential to increase agricultural production to satisfy food requirements. Soil salinity is also a serious problem in areas where groundwater of high salt content is used for irrigation. Differences in the amount and kind of salts accumulated in the soil solution result in "salt-affected soils" of varied chemical, physical and physico-chemical properties, with different management strategies for their prevention, use and reclamation.

Saline soils are those where the salt content and osmotic potential of the soil solution prevent the crop from absorbing a large proportion of the soil water and do

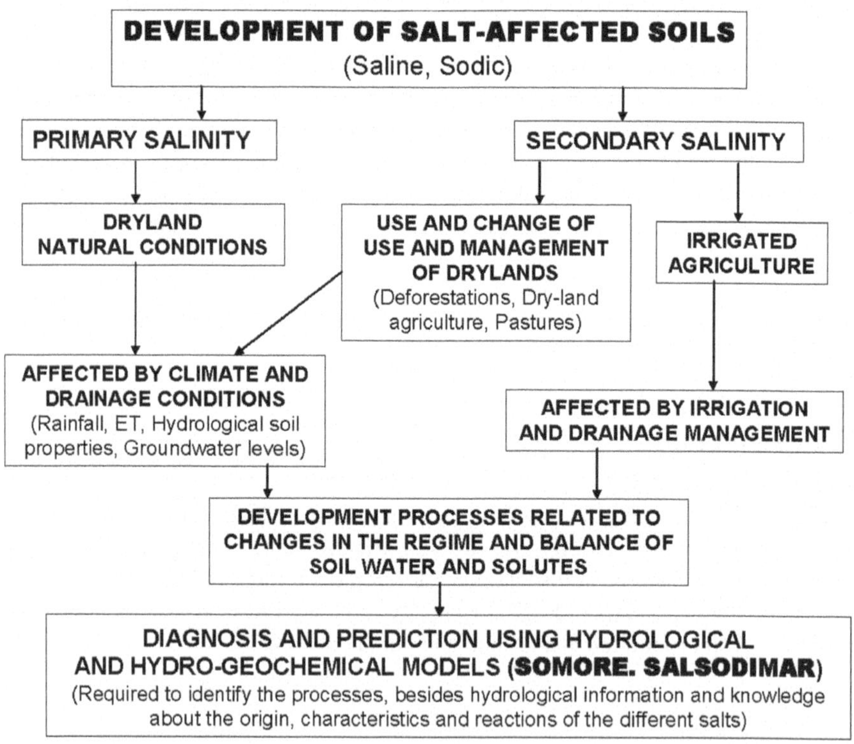

Fig. 1 Common factors and processes in the development of salt-affected soils under dryland and irrigated conditions. (SOMORE: Pla Sentís 2002, 2006) (SALSODIMAR: Pla Sentís 1988, 1997)

not show any direct effect on the soil physical properties. The main consequence is the partial or complete reduction in plant growth due to physiological water deficits.

Sodic soils are dominated by Na (and by Mg in some cases) on their cation exchange sites. Sodicity produces changes on the soil´s physical properties, both by dispersion and plugging of soil pores by the moving clay particles and by soil pore blockage by swelling clays. When surface soil disperses, the clay and silt particles clog surface pores, resulting in soil sealing, reduced infiltration and surface waterlogging. This affects land use and plant growth by decreasing the permeability of water and air through the derived soil waterlogging and impeding root penetration. Traditionally, the "sodic soils" have been called "alkali soils," although these include only the sodic soils with presence and accumulation of Na bicarbonates and carbonates and pH higher than 8.5–9.0. There are other soils with properties of sodic soils with lower pH and lower relative levels of Na than the so-called alkali soils (Pla Sentís 1968, 2015).

The expansion of irrigated agriculture is necessary for the sustainable production of the food required by a growing world population. Such development is limited by the increasing scarcity and low quality of available water resources and by the competitive use of those resources for other purposes. There are also increasing problems of contamination of surface and groundwaters. Taking also into consideration the high investments required for the development of irrigated agriculture, the degradation of irrigated lands through soil salinization and sodification become significant problems from the economic, social and environmental points of view (Qadir et al. 2014).

In many countries, irrigation has become a very important component of food production, sometimes the most important. The irrigated area in the world has increased from 50 million ha in 1900 to 100 million ha in 1950 and to 350 million ha in recent estimates (FAO-ITPS-GSP 2015). However, the yearly loss of productivity, mainly due to salinization of irrigated lands, amounts to 1.5 million ha, and salinity problems, of various degrees, presently affect almost 25–50% (depending on the evaluations) of all the irrigated land (Abrol et al. 1988). Although salinity-affected areas are much smaller than those affected by other degradation processes like erosion, the social, economic and environmental effects are of the same magnitude, as a consequence of the high value and productivity of irrigated lands and their coincidence with areas of large urban and industrial development. Precise information and recent estimates of the world extension and distribution of salt-affected soils do not exist. One of the most recent reports (FAO-ITPS-GSP 2015), mainly based on outdated information about the distribution of salt-affected soils in drylands in different continents, indicates that more than 10^7 km^2 of the world surface are covered by salt-affected soils, 40% with saline soils and 60% with sodic soils. The estimations of areas affected by secondary salinization are even less accurate and range from 0.6 to 1 million km^2. To this, approximately 0.3 million km^2 of human-induced dryland salinity and sodicity should be added (Marshall 1995). Some of these estimations are often the result of simple expert judgment, or of data which have been obtained and interpreted by different methods (Oldeman et al. 1991).

2 Saline and Sodic Soils in Latin America

2.1 General Analysis

Latin America is a vast and heterogeneous region. It comprises countries with diverse availability of natural resources and economies. Estimations of the extension and distribution of salt-affected soils in LA are not very updated or precise and partially based only on expert judgment. Oldeman et al. (1991) estimated that about 7×10^5 km^2 are affected by salinity and 6×10^5 km^2 by sodicity, giving a total extension of 1.3×10^6 km^2 (i.e., 130 million ha). The total irrigated area in LA amounts to 25–30 million ha. While some of the irrigated areas in large to medium production units are used for industrial and commercial crops like sugarcane, rice, cotton, etc., irrigated areas in smallholders production units presently produce 30–70% of the locally consumed food, depending on the region or country. It is estimated that 25–50% of that area is affected by human-induced secondary salinization and sodification. To this, it is necessary to add 4–5 million ha of human-induced dryland salt-affected (mostly sodic) soils.

Most of the secondary salt-affected soils in LA developed under irrigation are partially a consequence of low irrigation water quality, but mainly due to non-efficient water management and poor drainage conditions. Dryland salinity and sodicity usually develop in large plains and valleys, poorly drained, with shallow saline or sodic groundwaters mainly promoted by land use and land cover changes and by overgrazing (Lavado et al. 1990; Pla Sentís 2014; Taboada et al. 2017). Usually, human-induced salt-affected soils developed under arid to semiarid climates are saline, and those developed under sub-humid to humid climates are sodic (Fig. 2). Besides, in some coastal and swampy areas and deltas in tropical LA (Venezuela, Colombia, Brazil), saline acid soils have developed (Fig. 2). They have developed after drainage of fluvial-marine sediments deposited in continuously flooded saline environments (Pons 1973; Pla Sentís and Florentino 1985).

Research about salt-affected soils flourished in the 1960–1980 period, when salinity research (basic and applied) and evaluation were active at different levels in most countries. One result of this activity was the occurrence of important regional conferences, like the "VIII Reunión Latinoamericana de Fitotecnia" (Bogotá, Colombia), which held a special session dedicated to salt-affected soils in LA (Pla Sentís 1971), and the "International Workshop on Salt-Affected Soils in Latin America" held in Maracay (Venezuela) (Pla Sentís and Florentino 1985), where advances in research and evaluation of salt-affected soils in LA up to that date were reviewed. More recently, in a few LA countries, research has been activated again. Seven salinity conferences have been held since 2005 in Argentina and three in Brazil. In 2019, the First Latin American Salinity Symposium was held in Fortaleza, Brazil. Several books on regional salinity issues have been published in Spanish and in Portuguese (Taleisnik et al. 2008; Gheyi et al. 2016, Taleisnik and Lavado 2017). In Northern Brazil, a National Institute of Salinity Science and Technology (INCTSal) was created.

IRRIGATION WATER OR GROUNDWATER	DRAINAGE	CLIMATE	RESULTING** SOIL SOLUTION	POTENTIAL PROBLEM
EC: >2 dS/m (Cl >S>B) (Na≥CA) EC: 1-2 dS/m (S≥Cl>B) (CA>Na)	VARIABLE K: 1-50 mm/h GD: > 1.5 m	Arid-DSArid	EC: > 8 dS/m (Cl>>S>>B) (Na>CA) (Cl≥S>>B) (Na≥CA) pH: < 8.5 PS: CAC, CaS	SALINITY
EC: < 1 dS/m (B≥S>Cl) (CA>Na)	VERY RESTRICTED K: < 1 mm/h GD: < 0.5 m	Arid-DSArid	EC: > 4 dS/m (S>Cl>B) (Na>CA) pH: < 8.5 PS: CAC, CaS	SALINITY
		SArid-SHumid	EC: < 4 dS/m (S≥B>Cl) *→ (B>S>Cl) (Na>>CA) pH: > 7.5 PS: CAC	SODICITY
EC: < 1 dS/m (B>S>Cl, B>CA) (Na≥CA)	RESTRICTED K: < 5 mm/h GD: < 1.0 m	SArid-SHumid	EC: > 2 dS/m (B>S>Cl) (Na>>CA) pH: > 8.5 PS: CAC	SODICITY

EC: Electrical Conductivity; B: Bicarbonates; S: Sulphates; Cl: Chlorides; CA: Ca + Mg; K: Saturated Infiltration Rate; GD: Groundwater Depth or presence of soil layers restricting internal drainage; DSArid: Dry Semiarid; HSArid: Humid semiarid; SHumid: Subhumid; ** In saturation extract; * → Change in composition under anaerobic conditions ($2Na^+ + SO_4^= + 2C + 2H_2O \rightarrow S^= + 2NaHCO_3$); PS: Precipitated Salts (C: Carbonates; S: Sulphates)

DEVELOPMENT OF SALINE-ACID SOILS (Pla and Florentino 1985)

Salts (Cl^-, $SO_4^=$) + Fe^{++} + 4C + H_2O > (anaerobic conditions; sulfate-reducing bacteria) >>> **FeS_2 + $4CO_2$ + H_2O + Cl^-**; Improved drainage and aeration (natural or human induced) > Oxidation: FeS_2 + $4O_2$ + H_2O >>> H_2SO_4 + Fe (OH) SO_4 ; Fe(OH)SO_4 + H2O >>> Fe (OH)$_3$ + H_2SO_4 + Salts (Cl^-, $SO_4^=$): **(SULPHATE ACID SOIL) pH: 1-2 (Non stable)**

Reaction wih clay minerals (slow process): H_2SO_4 ($2H^+$ + $SO_4^=$) + Clay (X Al_2O_3-SiO_2) + H_2O → Clay-H^+ (unstable) + Al(OH)SO_4 + Hydrated Si + Salts (Cl^-, $SO_4^=$) → More stable clay minerals(with interstratifications of Hydrated Fe and Al) + Salts (Cl^-, $SO_4^=$) (Semiarid climate and soils with very low saturated hydraulic conductivity) → **(pH: 4-5; EC_{SA}>4 dS/m): SALINE ACID SOIL**

Fig. 2 Most common conditions leading to the development of different kinds of saline, sodic and saline acid soils, under irrigated and dryland conditions, in Latin America. Adapted from Pla Sentís (2015)

A country-by-country overview of soil and water salinity and sodicity problems in LA follows. Some countries are not included in this overview, because we have not been able to find references showing important processes of soil salinization in them.

2.2 *Salinity in LA Countries*

2.2.1 North America

México

México is the LA country with the largest area under irrigation, due to the prevailing climate conditions and population pressure. Irrigation is essential for Mexican economy, because it produces about 40% of the food, with irrigated crops yields 2–3 times higher than those of rainfed agriculture. The more commonly irrigated crops are wheat, sorghum, tomato, corn, safflower, cotton, potatoes and various fruits and vegetables. Currently, México has 6.5–7 million ha under irrigation, 3 million of them in 85 large irrigation systems and the rest in about 39,000 smallholders dispersed irrigation units. About 35% of the irrigation water comes from wells and aquifers, many over-exploited and with increasing salinity (one specific case is analyzed in the chapter by Ojeda-Barrios et al. in this book: "Causes, effects and management of salinity problems in pecan production in North Mexico"). In 85% of the area, furrow and flooding irrigation systems are used, in non very well-leveled lands, with very low water use efficiency (Palacios-Vélez and Pedraza-Oropeza 2015). This situation, along with non-lined water conduction ditches, poor drainage conditions and poor maintenance of the open drains, generally leads to rising groundwater levels. Underground drainage pipes have been installed in only about 80,000 ha of irrigated lands. Dumping urban sewage waters, with small or no treatment, in surface streams and irrigation ditches, has become more frequent in the last decades (Siebe and Cifuentes 1995; Pla Sentís 1997; Gallegos et al. 1999).

At the end of the 1980s, the large irrigation districts, previously under government control, became controlled by operator associations. Since then, some of the irrigation control and drainage infrastructures have not been well maintained, and the previous monitoring and control of groundwater levels and of soil salinity were mostly discontinued. The technical assistance and training to users, on irrigation and drainage management, decreased (Peña 1980). The final consequence of all the previous factors was a progressive soil salinization (and to a smaller extent, sodification), with reductions in crop productivity (Anaya and Noyola 1984). Although there are no up-to-date general evaluations of the areas with salt-affected soils in México, approximate estimations indicate that 10–20% (0.6–1.2 million ha) of the total irrigated area are affected by some degree of salinity or sodicity. In some irrigation systems of N México, with more arid climate and more saline irrigation water, the affected area may increase to 30–40% of the irrigated land (Mata-Fernandez et al. 2014). Besides the secondary salt-affected soils in irrigated fields, there are about 1 million ha of natural salt-affected lands, mainly in closed basins, arid zones and coastal areas.

2.2.2 Central America and the Caribbean

There are no reported important soil salinity problems affecting agricultural production in the Central American countries, probably due to the climatic conditions prevailing in the region. One exception is a report about salinity problems in the W Pacific Region of Nicaragua, in poorly drained soils with sugarcane and rice crops (Ortega et al. 1986).

Cuba

In Cuba, although evaluations of the extent of soils affected by natural salinity are lacking, probably they occupy more than 10,000 km^2, mainly in areas close to the seacoast. Most of the current salinity problems are secondary, human-induced and related to the development of irrigated agriculture in the past 50 years, mainly in Eastern Cuba, with semiarid climate (Alvarez et al. 2008). It is estimated that about 1 million ha (50%) of a total irrigated area of 2 million ha are affected by various levels of salinity and sodicity (Riverol et al. 2001). They are related to deficiencies in irrigation and drainage management and with the development of shallow saline groundwaters (González 2000). Many irrigated lands are in lowlands a few meters above sea level, in alluvial soils over very saline marine sediments, responsible for the high salinity of the groundwater, which is frequently very close to the soil surface, due to low efficiency of water use for irrigation, mainly by furrow and flooding systems. Irrigation is mainly with surface waters. These are usually of good quality except in the lower part of the valleys, where they get in contact with the saline groundwater in the unlined water conduction ditches (Pla Sentís 1985a).

Irrigated crops are mainly pastures, sugarcane, rice, fruits and vegetables, with production levels highly affected by the progressive increase in soil salinization (Table 1, Fig. 3). As the salinization problems increase, more irrigated areas are used for pastures tolerant to high salinity, affecting the production of other food crops (Martínez et al. 2017). Specially affected by salinization is the rice production, occupying more than 200,000 ha (Herrera et al. 2009). In general, the increasing salinization of irrigated lands in Cuba has very important negative social economic consequences, affecting the production and supply of food products for local and country population, and the important investments made in irrigation infrastructures (Ortega et al. 1986).

Presently, most of the evaluations of the salinity processes and problems are made through remote sensing (Lau Quan et al. 2005). The measures to control the salinization processes are mainly directed to improvement in drainage infrastructures, to the addition of mineral and organic amendments and to the selection and use of crops and varieties tolerant to salinity (Mesa 2003; Lamz and González 2013).

Table 1 Secondary saline soil, irrigation waters (IW) and groundwater (GW) ion composition in Eastern Cuba. Semiarid climate (author's data)

Soil/Water	pH	EC_{SE}	Ca^{++}	Mg^{++}	Na^{+}	K^{+}	HCO_3^{-}	Cl^{-}	$SO_4^{=}$	SAR
		dS/m	meq/l							$(mmol/l)^{1/2}$
Soil										
0–30 cm	8.9	17.0	30.0	39.0	115.0	–	7.0	112.0	62.0	19.0
IW (River)	7.5	0.7	2.0	1.0	4.5	–	4.0	2.7	1.0	4.0
IW (Well)	7.2	0.9	3.2	3.9	4.4	0.1	2.5	6.2	0.3	2.3
GW (90 cm)	8.4	12.0	19.0	26.0	90.0	–	7.0	91.0	37.0	19.0

EC_{SE} electrical conductivity, *SAR* sodium adsorption ratio. In soil, variables were measured in a saturation extract

Fig. 3 Secondary saline soil in Eastern Cuba. *Photos* I Pla

Puerto Rico

In Puerto Rico, the distribution of salinity and sodicity in the SW of the island was recently re-evaluated (Oliveras et al. 2018). A pioneer work on soil salinity in LA (Bonnet 1960), following USDA Handbook 60 (Richards 1954) published a few years earlier, had identified and evaluated the problems of soil salinity and sodicity associated with irrigated sugarcane in SW Puerto Rico. Most of the identified problems were remediated with the improvement of the irrigation and drainage infrastructure and with the use of amendments, resulting in increased sugarcane production. No other salinity problems have been reported in Puerto Rico.

Dominican Republic

About 17% of the agricultural area of the Dominican Republic is equipped for irrigation. Most of the irrigated areas are in valleys with semiarid to sub-humid climates, mainly in the Eastern part of the country. Surface water, usually of good quality, is the main source of irrigation, and only 10–15% of the area is irrigated with water from wells, with higher salt content (Table 2). Besides some large private irrigation schemes planted with sugarcane, other systems, covering about 50% of the whole irrigated area, are planted with crops such as plantains, corn, rice, banana, papaya and vegetables, in smaller farms.

Most of these irrigated systems are managed by more than 50,000 users, generally organized in water operator organizations, receiving government subsidies to cover operation and maintenance costs. Furrow and flooding irrigation practices are dominant, in not very well-leveled fields.

Although detailed studies and evaluations of salinity problems are not common, it has been estimated that about 80,000 ha, mainly in the Eastern part of the country, are affected by some degree of salinity and sodicity (Fig. 4). They are mainly attributed

Table 2 Secondary saline soils and irrigation waters (IW) ion composition in Western Dominican Republic. Semiarid climate (author´s data)

Soil/Water	pH	EC_{SE}	Ca^{++}	Mg^{++}	Na^{+}	K^{+}	HCO_3^{-}	Cl^{-}	$SO_4^{=}$	SAR
		dS/m	meq/l							$(mmol/l)^{1/2}$
Soil										
0.20 cm	7.9	7.0	30.0	20.0	36.0	24.0	3.0	32.0	51.0	7.0
20–40 cm	8.1	3.5	11.0	8.0	21.0	0.8	5.0	10.0	27.0	7.0
IW (River)	7.1	0.6	2.5	1.5	1.8	–	3.2	0.9	1.7	1.3
Soil										
0–20 cm	8.1	6.7	33.0	16.0	14.0	3.0	3.0	39.0	25.0	3.0
20–40 cm	8.3	3.4	20.0	7.5	6.5	0.9	2.0	18.0	15.0	2.0
IW (Well)	7.3	1.5	7.0	5.0	3.0	0.1	5.0	8.4	2.2	1.2

EC_{SE} electrical conductivity, *SAR* sodium adsorption ratio. In soil, variables were measured in a saturation extract

LEACHING TRIAL

Open drains 1,5 m depth, 100 m appart

Leaching Water: EC = 0.84 dS/m, 720 mm in 7 days

	EC_{SE} (dS/m)	
Depth (cm)	**Before leaching**	**After leaching**
0-30	14.2	1.5
30-60	13.8	5.9
60-90	13.4	8.4

(EC: Electrical Conductivity; SE: Saturation Extract)

Fig. 4 Secondary saline soil reclamation in Western Dominican Republic. *Photos* I Pla please notice the extra vertical line in the left lower panel. is it possible to remove it?

to poor control and low efficiency of water use, accompanied by poor maintenance of the drainage infrastructure. In sugarcane plantations of Eastern Dominican Republic, chloride toxicity effects have been identified on sugarcane with sprinkler irrigation using water from wells. Recently, reclamation of salinized lands by leaching has been successfully tested in sugarcane fields of the same plantation, after land leveling and improvement of an open drainage system, taking advantage of the high soil and subsoil permeability (Fig. 4) (Pla Sentís and Gómez 2019). Although the results are very promising, the application of the leaching practice in large extensions is limited due to the restricted availability of water.

In some sugarcane plantations of Eastern Dominican Republic, chloride toxicity effects appear on sprinkler-irrigated sugarcane, using well waters with more than 8-10 meq/liter Cl (Fig. 5). The problem has been solved using those waters with drip irrigation systems.

Fig. 5 Chloride toxicity in sugarcane using sprinkler irrigation with waters rich in Cl^- in a sugarcane plantation in Western Dominican Republic. *Photos* I Pla

2.2.3 South America

Colombia

In Colombia, it is estimated that about 9×10^4 km^2 are affected by natural salinity and sodicity, mainly in coastal areas and in the semiarid zones of the Caribbean region, in Andean Valleys and in highlands. Other evaluations suggest the area is actually 14×10^4 km^2 (IDEAM 2002). Saline acid soils have also been identified, developed in fluvial-marine sediments in the NE coastal areas of the Caribbean Sea and of the Pacific Ocean (Combatt Caballero et al. 2008).

The irrigated agriculture area is estimated to be 500,000–800,000 ha, with sugarcane and rice as the predominant crops, followed by fruits and vegetables. Most of the irrigation water used in large irrigation systems is of surface origin, generally of good quality, sometimes complemented with water extracted from deep wells, with higher salinity. Small-scale irrigation units, usually with fruit crops, vegetables and flowers, are mostly irrigated with well water and in some cases with highly contaminated (with residual waters) surface waters.

Irrigated lands, with human-induced secondary salt-affected soils, have been identified in zones with various crops (sugarcane, rice, cotton, bananas, etc.), but there are no general precise evaluations of the degree and extension of affected areas. The methodology of evaluation, including measurements and interpretation of results, differs in many cases (IDEAM 2002).

Most of the detailed evaluations are from Central Western Colombia, in areas with sugarcane as the main crop, where complementary irrigation is used under a sub-humid climate. There, about 80,000 from a cultivated total of 300,000 ha, are affected by salts (40–60% sodicity), mainly in areas with sugarcane as the predominant crop (González 2001) (Table 3, Fig. 6). Some areas with magnesic soils have been identified, having properties similar to the sodic soils (Madero et al. 2004). Poor management and very low efficiency of irrigation, in fields predominantly with furrow irrigation systems, with poor drainage conditions, have been identified as the main causing factors (Fig. 6). Drainage infrastructures, including open drainage ditches and in some cases underground "tile drainage", are generally not well maintained, making them non-functional and ineffective. In some areas with sodic soils, amendments have been used, including sub-products of the industrial sugar and ethanol production (vinasse), with promising results (Fig. 6).

Other human-induced salinity problems have been also identified in irrigated lands in the arid and semiarid regions, mainly in the Central and Eastern Caribbean regions. In those areas, predominant salt-affected soils are saline, with the exception of some areas with banana crops irrigated with surface waters where sodic soils have been identified. In any case, and especially in the Eastern Caribbean region, most of the salinity problems may be related with the high salinity of most of the scarcely available irrigation water, both of surface and well origin. Some of those affected soils have been identified as saline–sodic and sodic soils following strictly the original indices and criteria of the USDA Handbook 60 (Richards 1954), but they do not behave as sodic soils (very high pH, poor structure, low saturated hydraulic conductivity), because the prevailing salts in soil solution are sodium chlorides, with precipitated calcium and magnesium carbonates and calcium sulfates in the soil profile.

Venezuela

In Venezuela, close to 100,000 km^2, mainly located in the semiarid and arid coastal areas in the Caribbean NW and NE regions, and in the Central Western region, are covered by natural salt-affected soils, mostly saline. Additionally, areas with old and recently developed saline acid soils are found in coastal areas of the Maracaibo Lake (W of Venezuela), NW and NE Caribbean coastal regions and in the Orinoco Delta (Pla Sentís and Florentino 1985; Pla Sentís 1985b).

Early studies related with soil salinity in Venezuela date from the 1950s, identifying some salt-affected areas in irrigated fields with sugarcane in the Central Western region, and some evaluations and field trials of reclamation were made, using the criteria of the USDA Handbook 60 (Richards 1954). Since the 1960s, evaluations and diagnosis of actual and potential problems of salinity and specially sodicity were started in new and projected irrigated areas. Those studies developed and published a first version of a simulation model for the diagnosis of salt-affected soils and a

Table 3 Secondary sodic soil, irrigation waters (IW) and groundwater (GW) ion composition in Central Western Colombia. Sub-humid climate (author´s data)

Soil/Water	pH	EC_{SE}	Ca^{++}	Mg^{++}	Na^{+}	K^{+}	HCO_3^{-}	Cl^{-}	$SO_4^{=}$	SAR
		dS/m	meq/l							$(mmol/l)^{1/2}$
Soil										
0–20 cm	8.8	4.1	2.8	2.3	35.0	0.2	7.2	10.0	27.0	21.0
20–40 cm	8.9	3.4	1.6	1.8	31.0	0.3	4.6	8.0	22.0	2.0
IW (River)	7.3	0.5	2.4	1.1	2.4	0.1	3.9	0.2	1.8	2.0
GW (80 cm)	8.1	3.6	3.0	3.0	31.0	2.0	11.0	6.0	18.0	18.0

EC_{SE} electrical conductivity, *SAR* sodium adsorption ratio. In soil, variables were measured in a saturation extract

Fig. 6 Irrigation and development of a sodic soil in Cauca Valley sugarcane plantation and use of vinasse as amendment. *Photos* I Pla

map with the first approximation of the location and extension of areas with actual and potential salinity and sodicity problems related with irrigation in Venezuela (Pla Sentís 1983, 1985b, 1986). It included a complete characterization and detailed analysis of selected salt-affected soils (saline, sodic and saline acid) and irrigation waters in each of the affected zones (Pla Sentís 1985b). Aside from some isolated studies in specific irrigated zones, there is currently no updated estimation of the area under irrigation and about the area affected by salinity problems in Venezuela. Based on past evaluations (1970–1990s), when only part (250,000–300,000 ha) of the projected irrigated area (1.5 million ha) had been developed, it is estimated that

Fig. 7 Development of a sodic soil in a rice plantation in the Western Plains (Venezuela). *Photos* I Pla please notice the right panels are shorter than the left ones, and displaced to the left.

25–30% of the irrigated lands are affected to some degree by salinity and sodicity. There are evidences that the present situation is not far from those estimates.

The largest irrigated areas are in the Central and Western Plains, and in the North Central valleys, with dry sub-humid climates (dry season of six months/year), using mainly surface waters of low salinity. The main irrigated crops are pastures, rice and sugarcane. Under the frequent very poor drainage conditions and shallow groundwater levels, most of the developed salt-affected soils are sodic (Pla Sentís 1986, 2015) (Fig. 7, Table 4).

The rest of the irrigated areas in Venezuela are in regions with arid and semiarid climates, devoted mostly to fruit and vegetable crops in small to medium production units. Waters from deep wells are generally used, occasionally, with a proportion of surface water, of higher salinity than the water used in the more humid areas (Table 5).

Under these conditions, and especially when irrigation waters are more saline, salt-affected soils usually develop in a few years (Figs. 8 and 9). In these cases, when water resources are usually more limiting than available lands, a common practice is to abandon salinized areas, when crop productivity is critically affected, and to move to non-salinized neighboring lands that are then irrigated with the same available water. Under the prevailing arid to semiarid climatic conditions, with low average rainfall, some years concentrated in a few strong storms, if soils have adequate

Table 4 Secondary sodic soil, irrigation water (IW) and groundwater (GW) ion composition in a rice plantation in the Western Plains of Venezuela (author´s data)

Soil/Water	pH	EC_{SE}	Ca^{++}	Mg^{+++}	Na^{+}	K^{+}	HCO_3^{-}	Cl^{-}	$SO_4^{=}$	SAR
		dS/m	meq/l							$(mmol/l)^{1/2}$
Soil										
0–15 cm	9.7	4.5	2.2	1.2	45.0	2.0	18.0	20.0	12.0	37.0
15–30 cm	9.0	4.0	1.5	0.5	41.0	1.0	25.0	10.0	10.0	12.0
30–60 cm	8.5	3.5	1.2	0.4	34.0	0.5	20.0	7.0	9.0	38.0
IW (Well)	7.8	0.6	2.3	1.2	2.7	0.2	4.3	0.6	1.2	2.0
GW (100 cm)	8.4	1.8	1.0	2.0	16.0	1.5	9.6	4.6	4.0	13.0

EC_{SE} electrical conductivity, *SAR* sodium adsorption ratio. In soil, variables were measured in a saturation extract

Table 5 Saline acid soil and irrigation water (IW) ion composition in North Western Venezuela. Semiarid climate (author´s data)

Soil/Water	pH	EC_{SE}	Ca^{++}	Mg^{++}	Na^{+}	K^{+}	HCO_3	Cl^{-}	$SO_4^{=}$	SAR
		dS/m	meq/l							$(mmol/l)^{1/2}$
Soil										
0–30 cm	4.1	6.0	14.0	18.0	30.0	2.0	–	46.0	16.0	8.0
30–60 cm	4.4	9.3	28.0	30.0	46.0	1.0	–	62.0	31.0	9.0
60–90 cm	5.3	12.4	32.0	40.0	60.0	0.5	–	87.0	39.0	10.0
IW (River)	7.2	0.5	2.7	2.8	3.0	0.4	2.4	4.6	0.7	2.0

EC_{SE} electrical conductivity, *SAR* sodium adsorption ratio. In soil, variables were measured in a saturation extract

Fig. 8 Secondary "saline" soil in Central Western Venezuela

permeability, partial desalinization of the abandoned lands may happen, allowing them to be used again for cropping a few years later.

Brazil

In Brazil, it is estimated that at least 9 million ha are affected by natural salinity and sodicity problems. They are mainly in the NE region, with tropical semiarid climate, characterized by a very distinct and variable wet and dry seasons (Medeiros et al. 2019). The irrigation development in NE Brazil dates from the 1970–1980 period and was stimulated by strong government investments in irrigation infrastructures. The main objective of introducing irrigation in NE Brazil was to reduce extreme social economic problems in that region (one specific case is analyzed in the chapter by Dias et al. in this book: "Environmental, agricultural, and socio-economic impacts of salinization to family-based irrigated agriculture in the Brazilian semiarid"), derived from periodical droughts affecting negatively agricultural food production, required by a rapidly growing and very impoverished population (Heinze 2002). There was a very fast increase, sometimes uncontrolled, of the irrigated area, reaching in a few years an extension of more than 500,000 ha. It is estimated that further, mostly uncontrolled developments, have increased that area to almost 1 million ha (Denardin et al. 2018). The main irrigated crops are sugarcane, corn, sorghum, forages, beans, cotton and fruits (banana, melon, mango, coconut), partially for export. Surface waters, which were initially the main source of irrigation water, soon became inadequate, leading to an increased, frequently uncontrolled, exploitation of groundwater from shallow aquifers, with higher concentration of dissolved salts, mainly chlorides. Close to the seacoast, intrusions of sea water in the groundwater are frequent (Eschemback et al. 2014). As a result of the decreasing water resources and the increasing salinity of

Fig. 9 Development of an "acid sulfate soil" in the Delta of Orinoco and of a "saline acid soil" in NW Venezuela. *Photos* I Pla

the irrigation waters, there are evidences of increasing soil salinization processes, affecting at least 25–30% of the irrigated area, with important negative effects on the crop productivity, and on derived social economic consequences (Ferreira et al. 2012).

Current evaluation and general diagnosis at field level, of these serious salinization processes, are reduced to some partial studies (de Oliveira et al. 2006; Pedrotti et al. 2015; Coelho Castro and Marcos dos Santos 2020). There have been attempts to detect and evaluate the salt-affected areas in Brazil using a combination of remote sensing and geochemical-based measurements (Bouaziz et al. 2011). More efforts have been dedicated to the selection and use of crops, varieties and cultivars resistant to drought and salinity, as a management alternative for a satisfactory production in salinized areas, irrigated with low-quality waters, and to reduce the requirements for irrigation (Tabosa et al. 2007; Lacerda et al. 2011; Bezerra de Lima et al. 2017), and for reclamation of salinized areas (Alves dos Santos 2016).

In addition to the salinization processes in the semiarid region of NE Brazil, it is worth mentioning another development of saline acid soils, since 1900. In Rio Grande do Sul (SE Brazil), rice was irrigated with saline waters, in soils developed in fluvial-marine sediments under sub-humid climate (Denardin et al. 2018).

Ecuador

Ecuador has over 3 million ha dedicated to agricultural activities with crops like rice, corn, cassava, soybeans, palm oil, sugarcane, plantain, banana, cocoa, coffee and other fruit and vegetable crops, some with complementary irrigation. However, no important soil salinity and sodicity problems have been reported. Some publications comment that in irrigated areas, soil salinity caused by the use of water with medium content of salts in solution and excessive use of fertilizers is the main causes for crop yield reductions, but no other information or data are available. Natural saline soils have been identified in part of the Ecuador highlands (Jaramillo et al. 2014).

Peru

In Peru, besides natural salt-affected soils in some valleys in the highlands, most of the saline soils are found in the arid Western narrow coastal strip, between the Andes Mountains and the Pacific Ocean (De la Torre 2004). Most of the agricultural production under irrigation, which provides more than 50% of the food in Peru, is concentrated in about 750,000 ha in that region. Large production units, mainly for sugarcane and rice production, are not very common, and more than 500,000 people make a living from crop production (mainly fruits and vegetables) in smallholder plots of 1.5–2.5 ha. Surface water of relatively low salinity, sometimes stored in large dams, is the main source of irrigation.

The studies and evaluations of soil salinity problems in Peru, date back to the 1970s, when a fast development of salt-affected soils was observed, with negative effects on crop production. Problems of salinization were identified in about 300,000 ha of irrigated lands, 150,000 ha with high salinity levels. Poor irrigation management and deficient drainage, usually with rising groundwater levels, were considered the main causes of soil salinization (one specific case is analyzed in the chapter by Gamboa et al. in this book: "Salinization in Peruvian north coast soils: case study in San Pedro de Lloc"). Those problems were more common in the lower part of the valleys close to the seacoast, where, in some cases, intrusion of seawater affected groundwater salinity. This condition was more evident in irrigated rice areas.

In 1974, with the assistance of technical advisors from the Netherlands, a National Plan of Drainage and Land Reclamation was formulated. There were projects to

Fig. 10 Human-induced saline soil in NW Peru. *Photos* I Pla

improve the irrigation and drainage infrastructures, train Peruvian engineers and to give technical assistance to farmers and water users organizations (Alva et al. 1976). After a few years of activity, some drainage systems were improved, in some cases with tile drains, but those projects were discontinued. Since then, updated information about the salinity evolution in irrigated lands of the whole region has become very scarce.

There are isolated evidences that salinity problems have been increasing, affecting more than 50% of the irrigated lands, partially due to deficient maintenance of the irrigation and drainage infrastructures (Fig. 10). This is leading, in some cases, to the abandonment of nonproductive salinized irrigated lands, with migration of rural families to the main cities.

Bolivia

Salt-affected soils in Bolivia have been found in most of the regions with arid–semiarid climate. In the relatively flat highlands (3600–4000 m asl), with an average rainfall of 300–400 mm, concentrated (70%) in four months, there are salt-affected soils, including salt flats in some closed basins previously flooded with salty shallow lagoons, that have dried up. The high levels of salinity in those areas prevent or reduce the growth of pastures (Salm and Gehler 1987).

Some of those lands, with salts mainly concentrated in surface crusts, are used by poor farmers for very low productivity survival agriculture (Fig. 11). They collect and concentrate ("rain harvesting") the scarce rainfall falling in an area 5–10 times larger than the collecting area, in holes and trenches where crops are planted. The main crops are quinoa (*Chenopodium quinoa*), used for human food, and kauchi (*Suaeda foliosa*), a forage, both highly resistant to salinity (Jacobsen et al. 2000; Adolf et al. 2012) (Fig. 12). When the salinity in those points concentrating rainfall waters decreases, and in other areas with lower salinity, other crops like barley, potatoes, fava beans and alfalfa are included in the rotation (Geerts et al. 2008).

Fig. 11 Saline soil in the highlands of Bolivia. *Photos* I Pla

Fig. 12 Dryland cropping of kauchi (left) and quinoa (right) in the saline soils of the highlands in Bolivia. *Photos* I Pla

The water resources available for irrigation in the highlands of Bolivia are very scarce. It was estimated that there were close to 6,000 small irrigation units, covering about 320,000 ha, occupied and managed by 290,000 families. In the central highlands, some of the older irrigation systems use water from the river Desaguadero, coming from the Titicaca lake with very variable salt content (Ledezma and Ruiz 1995; Orsag et al. 1998) (Table 6).

In some cases, the introduction of irrigation decreased salinity in previously highly salinized soils, allowing production of crops with lower resistance to salinity than quinoa and kauchi, like barley, alfalfa, potatoes and fava beans. In many other irrigation units, water from usually shallow individual artesian wells is used, also providing water for human consumption (Gandarillas and Montaño 2013). Deeper wells are excavated for larger irrigation units managed by farmer associations. The more commonly used irrigation practices are furrow and flooding, but in not well-leveled lands. In most cases, due to low irrigation efficiency, poor drainage conditions and outlets for drainage waters and deficient supply of irrigation water of poor quality,

Table 6 Secondary saline soil and irrigation water (IW) in the river Desaguadero flood plain (*), and primary saline soil in the highlands (**) ionic composition. Bolivia, semiarid climate. (author´s data)

Soil/Water	pH	EC_{SE}	Ca^{++}	Mg^{++}	Na^{+}	K^{+}	HCO_3	Cl^{-}	$SO_4^{=}$	SAR
		dS/m	meq/l							$(mmol/l)^{1/2}$
**Soil*										
0–30 cm	7.3	15.0	25.0	6.0	120.0	2.0	–	–	–	31.0
*IW (River)	–	1.5	2.9	2.3	9.8	1.1	–	–	–	6.0
***Soil*										
0–15 cm	7.5	17.0	10.0	4.0	161.0	1.0	–	–	–	61.0
15–30 cm	8.0	29.0	11.0	5.0	256.0	1.5	–	–	–	90.0

EC_{SE} electrical conductivity, *SAR* sodium adsorption ratio. In soil, variables were measured in a saturation extract

many of the irrigated areas become salinized after some years, making irrigated agriculture in the highlands of Bolivia highly unsustainable (Hervé et al. 2002). Decrease in food production and migration of the impoverished farmers to the cities are some of the consequences.

Chile

In Chile, the areas with lands affected by actual and potential salinity problems, both natural and human induced, may develop in the semiarid and arid regions (40 million ha) from N of the capital Santiago up to the border with Peru. In those regions, there are irrigation systems in valleys of rivers descending from the Andes Mountains to the Pacific Ocean, in an area of about 27,000 km^2. Available surface irrigation waters are usually of poor quality, with medium to high content of soluble salts (Table 7). Some have high content of toxic elements like chlorides and boron, and sometimes they are contaminated with industrial and mining residues. The main irrigated crops are vineyards, olives, and a high variety of fruits and vegetables, partially for export (Sierra et al. 2001; Torres and Acevedo 2008).

Although there are only very few reported detailed evaluations of the type, levels and extension of soil salinity developed in those irrigated valleys, there are evidences of increasing problems affecting the economic productivity of some of the fruit crops not very tolerant to salinity or toxicity of chloride and boron (Torres and Acevedo 2008) The soil salinization processes are mainly attributed to the low quality of irrigation waters, and also to non-adequate irrigation and drainage management, leading to shallow saline groundwater levels in the lower part of the valleys.

Paraguay

Salt-affected soils in Paraguay are found in the Eastern Chaco region, under natural conditions, but especially due to human-induced changes in land cover and use in the last decades. It is mainly a dryland salinity problem, not related to irrigation, due to the presence of an extended area with shallow saline groundwater. The water reaches the soil surface in some places, salinizing soils and surface waters, and flooding some areas. In the last decades, due to extended deforestation, where natural forests with deep roots have been substituted by crops and pastures with shallow root systems and lower evapo-transpiration rates, the saline groundwater has come closer to the soil surface. Under those conditions, saline water reaches the soil surface by capillarity, causing the salinization of the surface soil (Fig. 13). In cropped areas, as the soil salinization increases, the original crops are being substituted by more tolerant crops and pastures, until salinity levels will only allow the growth of halophytes (Glatzle et al. 2020).

There is no updated detailed information about the evolution and extension of the affected areas, but there are references that they are increasing, and becoming a serious problem of irreversible land desertification in almost one million ha.

Table 7 Secondary saline soil and irrigation waters (IW) ionic composition in NW Chile. Arid climate. (author´s data)

Soil/water	pH	EC_{SE}	Ca^{++}	Mg^{++}	Na^{+}	K^{+}	HCO_3	Cl^{-}	$SO_4^{=}$	SAR
		dS/m	meq/l							$(mmol/l)^{1/2}$
Soil										
0–30 cm	8.2	10.0	35.0	18.0	60.0	1.0	4.0	66.0	52.0	12.0
IW (River)	8.1	0.6	5.0	1.0	1.0	–	2.5	0.5	4.0	0.6
(Well)	8.0	1.2	7.0	2.0	4.0	–	4.0	2.0	7.0	1.9

EC_{SE} electrical conductivity, *SAR* sodium adsorption ratio. In soil, variables were measured in a saturation extract

Fig. 13 Two consecutive levels of dryland "salinity" in the Chaco region of Western Paraguay. *Photos* I Pla

Argentina

Argentina is the Latin American country with the largest extension of natural salt-affected lands, both saline and sodic, which according to some reports reaches about 85×10^4 km^2 (Lavado 1984). They are located mainly in the semiarid regions in the NW, W and S of the country, but also in some sub-humid and humid areas with shallow groundwaters in the Central Eastern parts of Argentina (Casas and Pittaluga 1990). It is estimated that the area under irrigation has reached an extension of more than 2 million ha (5–6% of the total cropped area) of Argentina distributed in about 150,000 production units, 66% located in the arid and semiarid regions and the rest as complementary irrigation in more humid climates. Surface irrigation systems (furrow and flooding) are used in about 70% of the irrigated area, sprinkler irrigation in 20% and drip irrigation in 10% of the area. Sixty-five percent of the irrigated area uses water of surface origin, while the other 35% is irrigated with groundwater (Prieto et al. 2015).

Most evaluations report that about 630,000 ha (about 1/3) of the irrigated area is affected by some degree of salinization (60%) and sodification (40%) (Prieto et al. 2015). Most of the saline soils have developed in zones with less than 700 mm/year rainfall, while most of the sodic soils are found in zones with over 700 mm/year rainfall (Lavado 2007). Sodic soils have usually developed in sub-humid areas of the Pampean Region, where complementary irrigation is used for crops like soybeans, corn, wheat, sugar cane, fruits and vegetable crops, using shallow groundwaters rich in sodium bicarbonate (Andriulo et al. 1998; Génova 2011; Torres Duggan et al. 2012) (see also the chapter by Alvarez et al., in this book: "Effects of supplementary irrigation on soils and crops in humid and sub-humid areas in the Pampas region of Argentina"). Complementary irrigation demands for crops like soybeans, corn and wheat have increased, in part due to climate changes, but mainly due to the shallow root development, derived from other soil degradation processes like soil compaction, related to the prevailing agricultural management systems for such crops, in the last decades (Pla Sentís 2015). Soil salinization in irrigated areas of arid and semiarid regions is mainly attributed to low efficiency in the water use and management, to poor irrigation water quality and to deficient drainage conditions, usually associated with the development and occurrence of shallow saline groundwater levels (Sánchez et al. 2015).

In addition to the salinized area under irrigation, there are about 50,000 km^2 of dryland salt-affected (mostly sodic) soils, in the so-called Pampa Deprimida (Flooding Pampas) region (Taboada et al. 2017). It is estimated that in over 5,000 km^2 of those lands, the problems have been aggravated in the last decades, with increasing flooding and groundwater levels coming closer to the soil surface, partially due to changes in climate, but mainly to indirect human-induced factors. These include overgrazing by an increased cattle population and rising of the groundwater levels (Taboada et al. 2017) (see also the chapter by Imbellone et al. in this book:" Genesis, properties and management of salt-affected soils in the Flooding Pampas, Argentina"). The latter are partially derived from extensive (millions of ha) changes in land use and cover in the higher neighboring areas, where pastures have been replaced by annual crops like soybeans, with lower evapo-transpiration rates (Nosetto et al. 2012; Pla Sentís 2015) (see also the chapter by Jobbágy et al. in this book:" Salt accumulation and redistribution in the dry plains of Southern South America: Lessons from land use changes"). Field experiments with the objective of reducing and reclaiming these dryland salinization and sodification processes while improving grazing conditions for cattle have been carried out. The approach is to establish high water use, salt-tolerant and deep-rooted perennial pastures, like *Agropirum elongatum*, *Lotus tenuis*, *Melilotusalbus*, *Melilotus officinalis* and *Chloris gayana* (Casas and Pittaluga 1984). The expected beneficial effects of approach are reduction in capillary flow from the water table to the soil surface, increased infiltration and improvement of soil chemical and physical properties (Otondo et al. 2015). Other options such as draining water tables or pumping groundwater have limitations due to cost, disposal of drained water and soil hydrological conditions.

Problems of dryland salinity, resulting from saline groundwater getting closer to the soil surface, have also developed in large areas on the Chaco region in Argentina, with semiarid climate, in deforested lands for crops and pastures.

In a country where the economy is highly dependent on the agriculture and beef production and exports, the decrease in cattle and crop productivity and losses of very rich agricultural lands through salinization and sodification processes has and may have increasingly negative social economic consequences.

3 Conclusions

In Latin America, the area with natural salt-affected soils is smaller than in most of the other continents, and it is concentrated in some countries with larger arid and semiarid regions, such as México and Argentina. Areas under irrigation are also smaller in LA, but the relative levels of human-induced secondary soil salinization and sodification are similar to the irrigated lands in the rest of the world. Although total available water resources in LA are more abundant than in the other continents, they are very irregularly distributed, usually with lower availability in regions and countries where their requirements for irrigation are higher. Up to now, the increased agricultural production required to cover the food requirements of a rapidly increasing LA population and to provide the agricultural products for international markets, which is one of the main sources of income for several LA countries, has been mostly based on the increase of rainfed agricultural areas. The expected conditions of climate change, with increasingly erratic rainfall, in many LA regions, makes rainfed agricultural production very unreliable, especially when soils and lands, under the more common present management systems and practices, suffer progressive physical, chemical and biological degradation. Additionally, in some countries, the availability of land for future economic, productive and sustainable rainfed agricultural developments is becoming increasingly scarce. Under those conditions, the present and future developments of agricultural irrigated systems, potentially more productive, and not so dependent on changing and erratic rainfall, and as the only alternative for agricultural production in regions with arid and dry semiarid climates, have been considered in the last decades as a valid alternative in many LA countries. The main limitations for their development are and have been the required investments in irrigation and drainage structures, the availability of water resources (surface and subsurface) of good quality for irrigation, and the actual and potential salinity and sodicity problems.

The main development of irrigation in several countries of LA occurred in the 1960–1980 period and was mainly based initially on investments in large irrigation schemes. Usually, more attention was given to the engineering aspects of irrigation infrastructures (dams, distribution canals, etc.) than to installation of effective drainage systems and conditioning of irrigation fields (leveling, irrigation ditches, etc.). It went along with local research on soil and water management and technical assistance to new producers. This situation led to a rapid development, after a few years of irrigation, to problems of drainage, waterlogging and derived salinity.

We may conclude, mostly based on personal experience and information from local experts, that the development of large and expensive irrigation schemes has decreased in the last decades, mostly due to lack of funds. Instead, most of the new irrigation developments have been in small production units, using local surface and groundwater resources, in a frequently uncontrolled and disordered manner. In some extreme cases, due to competition for alternative uses of scarcely available good quality water resources, non-treated residual waters of urban and industrial origin are used for irrigation. In addition to these small irrigation units, mainly dedicated to production of food crops for self-consumption and for local markets, larger irrigation units have developed. These are managed by individual producers or farmer associations, mainly dedicated to produce crops for local and international markets. Often, such developments have not been sustainable for economic and marketing reasons, but mainly due to competing use of the scarcely available irrigation water, depletion of groundwater resources and productivity decrease due to growing salinity problems.

In general, it may be concluded that, currently and specially in the near future, with differences among countries and regions in LA, an increasing development of irrigated agriculture will be required, sometimes complementary to rainfall, to guarantee a sustainable and regular production of food to meet the requirements from a rapidly increasing population. It will also contribute to improve the social and economic conditions of poor, smallholders and to the sustainable production of some agricultural products for local and international markets. Prevention of salinity and sodicity problems, through appropriate irrigation and management for each particular combination of climate, soil, water and social economic conditions, will be essential for the economic and social sustainability of such agricultural production. This will require increased investments in local research and evaluations of the present and potential soil salinity and sodicity problems, judicious management of irrigation and drainage, increased training for those skills and technical assistance to producers. Failure to do so would lead to increased abandonment of the salinized nonproductive lands and migration of the impoverished farmers to large cities in LA countries, with all the derived negative economical and social consequences. Up-to-date estimates of the characteristics, extension and local regional distribution of saline soils in LA are required in order to better understand the extent of the problem and to develop soil use and management policies. Such estimates are essential, given the continuing decline of soil resources for food production.

Acknowledgements The author wishes to acknowledge the contributions to this chapter from Raúl S. Lavado (Argentina), P. I. Cairo (Chile), E. Madero (Colombia) and J. C. Loaiza (Colombia).

References

Abrol IP, Yadav JSP, Massoud FI (1988) Salt-affected soils and their management. FAO Soils Bulletin 39, Food and Agriculture Organization of the United Nations, Rome

Adolf VI, Shabala S, Andersen MN (2012) Varietal differences of quinoa's tolerance to saline conditions. Plant Soil 357:117–129

Alva CA, van Alphen G, Torre A, Manrique L (1976) Problemas de drenaje y salinidad en la costa peruana. Bulletin 16. ILRI. Wageningen

Alves Dos Santos M (2016) Adaptabilidade e potencial fitorremediador de espécies vegetais em solo salino. Dr. Thesis. UFRPE. Recife, Pernambuco. Brazil

Álvarez A, Baños R, Otero L (2008) Salinidad y uso de aguas salinas para la irrigación de cultivos y forrajes en Cuba. C Tecnol Ganadera 2(1):1–12

Anaya M, Noyola H (1984) Causas del ensalitramiento de los suelos y medidas para su prevención y control en México. In: Pla I, Florentino A (eds) Proceeding of International workshop on salt-affected soils of Latin América. UCV, Venezuela. pp 249–262

Andriulo A, Galetto ML, Ferreyra C, Cordone G, Sasal C, Abrego F, Galina J, Rimatori F (1998) Efecto de 11 años de riego complementario sobre un Argiudol típico pampeano. C Suelo 16:125–127

Bezerra de Lima Y, Vanies da Silva F, Ferreira M, Pereira de Paiva E, Gheyi H (2017) Accumulation of salts in the soil and growth of cowpea under salinity and phosphorus fertilization. Rev Ciência Agronô 48(5):765–773

Bonnet JA (1960) Edafología de Suelos Salinos y Sódicos. Estación Experimental Agrícola, UPR Río Piedras, Puerto Rico

Bouaziz M, Matschullat J, Gloaguen R (2011) Improved remote sensing detection of soil salinity from a semi-arid climate in Northeast Brazil. CR GEOSCI343 (11–12):795–803
Casas R, Pittaluga A (1984).Efecto de una pastura de Agropiro bajo clausura en la recuperación de suelos salinizados en el Partido de Carlos Tejedor. INTA; ACINTACNIA; 1 N° 12. Castelar Argentina
Casas R, Pittaluga A (1990) Anegamiento y salinización de los suelos en el noroeste de la provincia de Buenos Aires. In: Manejo de tierras Anegadizas. PROSA, FECIC. Buenos Aires
Coelho Castro F, Marcos dos Santos A (2020) Salinity of the soil and the risk of desertification in the Semiarid Region. Mercator, Fortaleza, v.19. https://doi.org/10.4215/rm2020.e19002.
Combatt Caballero E, Martínez Lara Z, Palenci M (2008) Generation of acidity by pyrit oxidation in inland acid sulphate soils of warm climate. Temas Agrarios 13(1):32–39
De la Torre A (2004) La degradación de tierras por salinidad en la región desértica de la costa Peruana. Ministerio de Agricultura, Instituto Nacional de Recursos Naturales, Intendencia de Recursos Hídricos. Lima
De Oliveira Aguiar Netto A, Machado R Cabral de Vasconcelos Barreto M (2006) Diagnóstico do processo de salino-sodificação no perímetro irrigado https://doi.org/10.15809/irriga.2006v11n4 p448-459
Denardin LG, Carmona F, Tiecher T, Posselty A, Alves LA, Weber EJ, Anghinoni I (2018) Salt-affected soils of the coastal plains in Rio Grande do Sul, Brazil. Geoderma Reg 14 https://doi.org/10.1016/j.geodrs.2018.e00186
Eschemback V, Bernert MR, Suchoronczek A, Jadoski SO, Lima A (2014) Características da Salinidade dos solos em cultivos agrícolas no Brasil Brazilian. J Appl Technol Agric Sci 7(3):115–124
FAO-ITPS-GSP (2015) Status of the world's soil resources. FAO-ITPS-GSP Main Report, Food and Agriculture Organization of the United Nations, Rome, Italy, pp 125–127
Ferreira J, Pardini R, Metzger JP, Fonseca CR, Pompeu PS, Sparovek G, Louzada J (2012) Towards environmentally sustainable agriculture in Brazil: challenges and opportunities for applied ecological research. J Applied Ecol 49:535–541
Gallegos E, Warren A, Robles E, Campoy E, Calderón A, Sainz MG, Bonilla P, Escolero O (1999) The effects of wastewater irrigation on groundwater quality in Mexico. Wat Sci Tech 40(2):45–52
Gandarillas H, Montaño G (2013) Riego Campesino Autogestionario. Presentation. https://www.waterweekla.com
Geerts S, Raes D, Garcia M, Condori O, Mamani J, Miranda R, Cusicanqui J, Taboada C, Vacher J (2008) Could deficit irrigation be a sustainable practice for quinoa (*Chenopodium quinoa* Willd.) in the southern bolivian altiplano. Agr Water Manage 95(8):909–917
Génova L (2011) Calidad del agua subterránea para riego complementario en la Pampa Húmeda argentina. Rev Fac Agronomía, La Plata 110(2):63–81
Gheyi HR, da Silva Dias N, de Lacerda C (eds) (2016) Manejo da salinidade na agricultura: estudos básicos e aplicados. 2nd edn. INCTSal Fortaleza
Glatzle A, Reimer L, Núñez J, Smeenk C, Musalem K, Laino R (2020) Groundwater dynamics, land cover and salinization in the dry Chaco in Paraguay. Ecohydrol Hydrobiol. 20 (2) 175-182. https://doi.org/10.1016/j.ecohyd.2019.10.003
González LM (2000) Extent, cause and management of salt affected soils in Cuba. FAO Newsletter on Sustainable Productive Use of Salt Affected Habitats 4: 8-11
González A (2001) Efecto de la aplicación de aguas bicarbonatadas sobre las propiedades físicas y químicas de algunos suelos del Valle del Cauca y Tolima. Tesis Doctoral. Universidad Nacional de Colombia , Palmira, p 138
Heinze BA (2002) Importância da agricultura irrigada para o desenvolvimento da região nordeste do Brasil, Monografia (MBA em Gestão Sustentável da Agricultura Irrigada). Ecobusiness School/FGV, Brasília, pp 48–53
Herrera J, Pujol R, Alarcón R, Espinosa E (2009) Calidad de las aguas de drenaje de procedentes de campos arroceros sobre suelos salinos en la Cuenca del Cauto. Rev C Técnicas Agropec 18(3):14–18

Hervé D, Ledezma R, Orsag V (eds) (2002) Limitantes y manejo de los suelos salinos y/a sodicos en el Altiplano Boliviano. IRD-CONDESAN. La Paz, Bolivia, p 266
IDEAM (2002) Zonificación de los Procesos de Salinización de los Suelos de Colombia. Subdirección de Geomorfología y Suelos, Bogotá, Colombia, p 43
Jacobsen SE, Quispe H, Mujica A (2000) Quinoa: an alternative crop for saline soils in the Andes. Scientists and Farmer—Partners in Research for the 21st Century. IPC program report
Jaramillo V, Arahana V, Torres ML (2014) Determination of the level of tolerance to salinity in in vitro conditions of the plants of tomate de árbol (*Solanum betaceum*) from different localities of the ecuatorian highlands. Av Cienc Ing Quito 6(1):B44–B50
Lacerda C, Sousa G, Silva F, Guimaraes F, Silva G, Cavalcante L (2011) Soil salinization and maize and cowpea yield in the crop rotation system using saline waters. Eng Agríc Jaboticabal 31(4):663–675
Lavado R (1984) Los suelos afectados por sales en la República Argentina. In: Pla I, Florentino A (eds) Proceeding of international workshop on salt-affected soils of Latin America. UCV, Venezuela, pp 263–268
Lavado R (2007) Visión sintética de la distribución y magnitud de los suelos afectados por salinidad en la Argentina. In: Taleisnik E, Grumberg K, Santa María G (eds) La salinización de suelos en la Argentina, Editorial EDUCC, Córdoba, Argentina pp 1–6
Lavado RS, Rubio G, Alconada M (1990) Changing soil salinity and alkalinity due to grazing. In: Transactions 14th International Congress of Soil Science. Kyoto, Japan. VII: 216–221
Lau-Quan A, Gaera-Llano E, Ruiz ME (2005) Estimation of soil salinity using a multispectral photo and GIS Telemap. Rev C Técnicas Agropec 14(1):47–54
Lamz A, González MA (2013) Salinity as a problem in agriculture: plant breeding an immediate solution. Cultivos Tropicales 34(4):31–42
Ledezma R, Ruiz P (1995) Opciones de riego en el Altiplano Central Boliviano; rio Desaguadero y agua subterrânea. In: Ribstein P, Francou B (eds) Aguas Glaciares y Cambios Cimâticos en los Andes Tropicales, ORSTOM-UMSA-SENAMHI-CONAPHI, pp 251–252.
Madero E, Malagón D, García A (2004) Una mirada al origen y las propiedades de los suelos magnésicos en el Valle del Cauca-Colombia. Acta Agron 53(3):1–21
Martínez F, García C, Gómez LA, Aguilar Y, Martínez-Viera R, Castellanos N, Riverol M (2017) Manejo sostenible de suelos en la agricultura cubana. Agroecología 12:25–38
Marshall AM (1995) Integrated soil management for sustainable use of salt-affected soils and network activities. In: Proceedings of the international workshop on integrated soil management for sustainable use of salt-affected soils. Bureau of Soils and Water Management, Manila, pp 55–75
Mata-Fernández I, Rodríguez-Gamiño ML, López-Blanco J, Vela-Correa G (2014) Dinámica de la salinidad en los suelos. Rev Digital del Dep El Hombre y su Ambiente 1(5):26–35
Medeiros LG, Galvao dos Santos MB, de Araùjo JC, Ribeiros P, Alves MF, Freire FJ (2019) Characterization and Classification of Halomorphic Soils in the Semiarid Region of Northeastern Brazil. J Agric Sci 11(4):405–418
Mesa D (2003) Obtención de plantas resistentes a la salinidad para los suelos salinos cubanos. Rev Cubana C Agrícola 37(3):217–230
Nosetto MD, Jobbagy EG, Brizuela AB, Jackson RB (2012) The hydrologic consequences of land cover change in central Argentina. Agric Ecosyst Environ 154:2–11
Oldeman LR, Hakkeling RTA, Sombroek WG (1991) World map of the status of human-induced soil degradation. An explanatory note. Second revised edition. In: International soil reference and information center (ISRIC), Wageningen
Oliveras M, Pérez L, Sotomayor D (2018) Geostatistical analysis for mapping soil salinity in the Lajas Valley agricultural reserve, South Western Puerto Rico. J Agric- Univ Puerto Rico 101(1):1–15
Orsag V, Gaita A, Ledezma R (1998). Limitaciones y potencialidades de los sectores de Chilahuala y el Choco (altiplano boliviano) para el riego con aguas del río Desaguadero. La Paz, Bolivia. Fac Agronomía. Universidad Mayor de San Andrés. Rev Jiltañani 3(3):42–56

Ortega SF, Peña F, Castillo N (1986) La salinidad de los suelos de Cuba. Aspectos económicos globales. Ciencias de la Agricultura. CU 27:137–144

Otondo J, Jacobo EJ, Taboada MA (2015) Mejora de propiedades físicas por el uso de especies megatérmicas en un suelo sódico templado. C Suelo 33(1):119–130

Palacios-Vélez OL, Pedraza-Oropeza F (2015) Drainage and salinity problems in the Mexican irrigation districts: an overview 1962–2013. Tecnología y Ciencias del Agua, VI (6):113–123. Instituto Mexicano de Tecnología del Agua. Morelos, México

Peña I (1980) Salinidad de los suelos agrícolas, su origen, clasificación prevención y rehabilitación. Boletín técnico No. 10. SARH. México, D. F

Pedrotti A, Chagas RM, Ramos VC, Prata APN; Lucas AAT, Santos PB (2015) Causas e consequências do processo de salinização dos solos. Rev Eletrônica Gest Educ Tecnol Ambient 19(2):1308–1324

Pla Sentís I (1968) Evaluation of the quality of irrigation waters with high bicarbonate content in relation to the drainage conditions. Trans 9th Int Congr Soil Sci Soc Adelaide. Australia 1:357–370

Pla Sentís I (1971) Calidad de aguas de riego como fuente de salinización y sodificación. In: Actas VIII Reunión Latinoamericana de Fitotecnia. Bogotá, pp 213–242

Pla Sentís I (1983) Sistema integrado agua-cultivo-suelo-manejo para evaluar la calidad de agua de riego. In: Isotopes and radiation techniques en soil physics and irrigation studies. IAEA. Viena, pp 191–206

Pla Sentís I (1985a) Informe de la Consulta en Edafología de Suelos Salinos. Programa de lucha contra la salinidad de los suelos en las provincias de Guantánamo y Holguin. Proyecto TCP/CUB/4403 (A) de Cooperación Técnica de la FAO. FAO. Roma.

Pla Sentís I (1985b) Origen, distribución y diagnóstico de suelos afectados por sales en Venezuela. Rev Fac Agron Maracay XIV (1–2):125–150

Pla Sentís I (1986) Diagnostic criteria for soil and water salinity in Venezuela. Agrok Talajtan 35(3–4):431–440

Pla Sentís I (1988) Riego y desarrollo de suelos afectados por sales en condiciones tropicales. Soil Technol 1(1):13–35

Pla Sentís I (1997) Evaluación de los procesos de salinización de suelos bajo riego. Edafología. Rev de la SECS Granada. Edición especial, España. pp 241–267

Pla Sentís I (2002) Hydrological approach to soil and water conservation. In: Rubio JL, Morgan RPC, Asins S, Andreu V (eds) Man and soil at the third millennium. Geoforma, Logroño (Spain) I, pp 65–87

Pla Sentís I (2006) Hydrological approach for assessing desertification processes in the Mediterranean Region. In: Kepner WG, Rubio JL, Mouat DA, Pedrazzini F (eds) Desertification in the mediterranean region: a security issue. Springer, The Netherlands, pp 579–600

Pla Sentís I (2014) Nuevas experiencias en la evaluación y diagnóstico de procesos de salinización y sodificación de suelos en América Latina. Suelos Ecuatoriales 44(2):125–137

Pla Sentís I (2015) Advances in the prognosis of soil sodicity under dryland and irrigated conditions. Intern Soil Water Conserv Res 2(4):50–63

Pla Sentís I, Florentino A (1985) Diagnóstico de problemas de salinidad en suelos salino-ácidos de las llanuras costeras al Norte de los Estados Falcón y Anzoátegui Rev Fac Agron Maracay XIV(1–2):117–123

Pla Sentís I, Gómez K (2019) Informe consulta sobre Riego y Salinización de Suelos en CAC. Barahona, República Dominicana

Pons LJ (1973) Outline of the genesis, characteristics, classification and improvement of acid sulphate soils. In: Proceding of International Symposium, on Acid Sulphate soils Wageningen, vol 1. The Netherlands, pp 3–27

Prieto D, Sánchez RM, Martínez RS (2015) Las áreas de riego y la degradación de los suelos. In: Casas R, Albarracín GF (eds) El Deterioro del Suelo y del Ambiente en la Argentina, vol I, pp 319–349

Qadir M, Quillerou E, Nangia V, Murtaza G, Singh M, Thomas RJ, Drechsel P, Noble AD (2014) Economics of salt-induced land degradation and restoration. Nat Res Forum 38(4):282–295
Richards LA (ed) (1954) Diagnosis and improvement of saline and alkali soils. USDA, Agriculture Handbook N°60, 160 p
Riverol M, Castellanos N, Peña F, Fuentes A (2001) Programa Nacional de Mejoramiento y Conservación de Suelos (PNMCS). Instituto de Suelos. Agrinfor. Minag, La Habana, Cuba
Salm H, Gehler E (1987) La salinización del suelo en el Altiplano Central de Bolivia y su influencia sobre la cobertura vegetal. Ecología en Bolivia La Paz 10:37–48
Sánchez RM, Dunel L, Scherger M (2015) Evaluación de las áreas bajo riego afectadas por salinidad y/o sodicidad en Argentina. Programa Nacional Agua. FAO-INTA-MAGP
Siebe C, Cifuentes E (1995) Environmental impact of wastewater irrigation in central Mexico: an overview. Int J Env Health Res 5:161–173
Sierra C, Césped R, Osorio A (2001) Caracterización de la salinidad de los suelos y aguas del valle del río Copiapó. Gobierno Regional de Atacama e Instituto de Investigaciones Agropecuarias (Chile). Centro Regional de Investigación Intihuasi (La Serena), Oficina Técnica Copiapó. Boletín INIA Nº 70
Taboada M, Damiano F, Lavado RS (2017) Suelos afectados por sales e inundaciones en la Pampa Deprimida y el Oeste Bonaerense. In: Taleisnik E, Lavado RS (eds) Ambientes salinos y alcalinos de la Argentina. Recursos y aprovechamiento productivo. Orientación Gráfica Editora, Universidad Católica de Córdoba, Buenos Aires, pp 56–87
Tabosa JN, Colkaço W, Dos Reis O, Simpliciol JB, De Carvalho H, Dias F (2007) Sorghum genotypes evaluation under salinity levels and gamma ray doses. Rev Bras Milho e Sorgo 6(3):339–350
Taleisnik E, Grunberg K, Santa Maria G (eds) (2008) La salinización de suelos en la Argentina: su impacto en la producción agropecuaria. Editorial Universidad Católica de Córdoba, Córdoba
Taleisnik E, Lavado RS (2017) Ambientes salinos y alcalinos en la Argentina. Recursos y aprovechamiento productivo. Orientación Gráfica Editora and Universidad Católica de Córdoba, Buenos Aires
Torres A, Acevedo E (2008) The salinity problem of the water and soil resources in Lluta and Azapa valleys in northern Chile. IDESIA Chil 26(3):31–44
Torres Duggan M, Álvarez CR, Taboada MA, Celesti T, Vignarolli F, D'ambrosio D (2012) Riego complementario en un Argiudol típico de la Pampa Ondulada argentina bajo siembra directa: efectos sobre algunas propiedades químicas y físicas. C Suelo 30:201:207

Environmental, Agricultural, and Socioeconomic Impacts of Salinization to Family-Based Irrigated Agriculture in the Brazilian Semiarid Region

Nildo da Silva Dias, Jucirema Ferreira da Silva, Maria Alejandra Moreno-Pizani, Matheus Cardim Ferreira Lima, Jorge Freire da Silva Ferreira, Edna Lúcia Rocha Linhares, Osvaldo Nogueira de Sousa Neto, Jeane Cruz Portela, Marcia Regina Farias da Silva, Miguel Ferreira Neto, and Cleyton dos Santos Fernandes

Abstract Soil salinity is one of the major abiotic factors causing a serious threat to global food security, mainly in arid and semiarid regions. Salinity brings socioeconomic impacts associated with low crop productivity and devaluation of agricultural lands. In this chapter, we approach agricultural, environmental, and socioeconomic

N. da Silva Dias (✉) · J. F. da Silva · J. C. Portela · M. Ferreira Neto · C. dos Santos Fernandes
Center for Agrarian Sciences, Department of Agronomic and Forest Sciences, Federal Rural University of the Semi-Arid (UFERSA), Mossoró, Brazil
e-mail: nildo@ufersa.edu.br

J. F. da Silva
e-mail: jucirema.ferreira@gmail.com

J. C. Portela
e-mail: jeaneportela@ufersa.edu.br

M. Ferreira Neto
e-mail: miguel@ufersa.edu.br

C. dos Santos Fernandes
e-mail: cleyton1959@hotmail.com

M. A. Moreno-Pizani
Faculdade Pecege, Piracicaba, Brazil
e-mail: morenom76@gmail.com

M. C. F. Lima
Department of Agroforest Ecosystems, Polytechnical University of Valencia, Valencia, Spain
e-mail: matcarfe@posgrado.upv.es

Research and Extension Unit (AGDR), Food and Agriculture Organization of the United Nations, (FAO), Rome, Italy

J. F. da Silva Ferreira
United States Salinity Laboratory (USDA-ARS), Riverside, CA, USA
e-mail: Jorge.ferreira@usda.gov

E. Taleisnik and R. S. Lavado (eds.), *Saline and Alkaline Soils in Latin America*, https://doi.org/10.1007/978-3-030-52592-7_2

impacts of soil salinization. We also report two case studies from irrigated areas of the Brazilian semiarid, where previously cultivated lands were abandoned due to increased soil salinity. A survey of the published literature showed that soil salinity became a global problem that is accelerated by human activities such as deforestation and lack of irrigation management. We conclude that socioeconomic impacts of soil salinity in agricultural lands translate directly as either loss or reduction of crop yield, profit margins, unemployment, and/or reduction of land commercial value in the long run due to soil infertility. Only governmental and private institutions have the financial capability to intervene and help small farmers to counteract this bleak scenario, including water and soil management improvement, to alter the future and improve quality of life of small farmers in the Brazilian semiarid region.

Keyword Peasant · Food security · Agricultural revenues · Droughts · Rural services

1 Salinization: A Threat to Agriculture and Food Security

Climate change affects mainly the arid and semiarid regions of the planet, making it difficult to impossible for farmers to achieve food security and limiting agricultural production in areas where rain is scarce or lacking. Food security only exists when all people have physical and economic access to nutritious and safe food resources at all times (World Food Summit 1996).

Thus, developmental policies for food security become essential to all who live in environments with low rainfall and vulnerable to climate changes. These policies may include social technologies of coexistence with the semiarid and, when appropriate, the development of efficient irrigation methods using alternative water sources such as treated domestic sewage effluents, drainage water, dams, underground dams, and water from tubular wells. Generally, these water sources contain high concentrations of salts that can salinize irrigated areas and hamper agricultural production. This situation can be aggravated by edaphoclimatic conditions and improper management practices applied to the agricultural areas (Medeiros et al. 2017; Brito et al. 2017).

In extreme cases, soil salinity can reduce the biodiversity of a region drastically and salts deposited on the soil surface may be transported by wind or by surface flow and salinize nearby water bodies and soils. Consequently, local vegetation growth

E. L. R. Linhares
Multidisciplinary Center of Caraúbas, UFERSA, Caraúbas, Brazil
e-mail: ednarocha@ufersa.edu.br

O. N. de Sousa Neto
Multidisciplinary Center of Angicos, UFERSA, Angicos, Brazil
e-mail: osvaldo.neto@ufersa.edu.br

M. R. F. da Silva
University of the Rio Grande do Norte State, Mossoró, Brazil
e-mail: mreginafarias@hotmail.com

will be hindered by the effects of salt on both soil and plants, thus affecting livestock and wild life due to the lack of natural pasture and arboreal shelter (Amorim and Porto 2001).

Socioeconomic impacts due to the salinization of irrigated areas are more pronounced in family-based agricultural areas. Lack of technical knowledge on saline water management by most farmers living in afflicted areas has led to increased soil salinization in a short term. In addition, family farmers have only one irrigated plot to perform agriculture and trade. This plot, once salinized beyond the threshold of most agricultural and horticultural crops, makes agriculture unfeasible for them and, because of the high cost of recovery of salinized soils, leads to abandonment of properties (Gheyi et al. 2016). On the other hand, agribusiness enterprises have the financial capability to abandon their salinized lands and move to newly purchased rural properties to maintain their profit margins on the commercialization of high-value agricultural products.

Salinity and periodic droughts are the most important issues for local farmers living in arid and semiarid areas of the world. Soil salinity is a serious global problem that threatens agricultural productivity with at least 20% of all irrigated areas being salt-affected, but with some estimates being as high as 50% (Pitman and Läuchli 2002). Estimates predict that 50% of all arable land will be impacted by salinity by 2050 (Butcher et al. 2016). This process is accelerated by human activities such as deforestation, use of saline water to irrigate crops, and subsequent inadequate irrigation management.

In this chapter, we discuss aspects related to environmental, socioeconomic, and agricultural impacts of soil salinization that threaten global food security, especially in arid and semiarid areas. Additionally, we report two case studies conducted in irrigated perimeters in Brazil where agricultural lands were abandoned due to salinization.

2 Impacts of Salinization

Several irrigated areas in Brazil and worldwide are affected by salts naturally originated in the soil or added by irrigation water, especially in arid and semiarid environments. Salinity impacts small and large agricultural communities through agricultural, environmental, and socioeconomic aspects.

2.1 Impacts on Agriculture

Soil salinization reduces plant growth, agricultural productivity (Fig. 1a), and fruit size, number, and quality (Machado and Serralheiro 2017; Ferreira et al. 2019). Water and soil Salinity limit the choice of crops that can be farmed, as the majority of our staple agricultural and horticultural crops are susceptible to salinity and may suffer a

Fig. 1 Impacts of soil salinization: **a** Burn on cherry tomatoes caused by salinity and the consequent loss of fruit yield (Mossoró, Brazil). **b** Weakly aggregated soils caused by excess sodium in soil profile of irrigation district Baixo-Açu (DIBA), Ipanguaçu, Brazil. *Source* The authors

significant drop in yield even when irrigated with waters of low to moderate salinity. Excess sodium can destroy soil structure (Fig. 1b), causing infiltration problems, erosion, and nutrient losses.

Salinization leads to reduction of water quality for domestic use and irrigation, damage to rural property infrastructure such as buildings, roads, fences, underground pipes, and agricultural equipment due to corrosion caused by salts. Poor animal health and loss of the native flora and fauna lead to reduced land value (Tenison 2009).

Thus, salinity effects on plant production, soil, farming equipment and livestock impacts rural farmers at different rates, depending on their economic level and their access (or lack of it) to rural development strategies, which are usually not available to low-income small farmers.

2.2 *Environmental Impacts*

The environmental impacts of soil and water salinity include the reduction of native vegetation (riparian forest and native forest) and biodiversity loss, decreased population of birds, and other wildlife due to lack of food and shelter, increased soil erosion, losses of aquatic fauna due to the absence of natural habitat, reduction of landscape aesthetic value that compromises recreational and tourist spaces (ecological parks and sanctuaries) and increased proliferation of colonizing species with undesirable changes in plant and landscape populations (Fig. 2).

Fig. 2 Predominance of salt-tolerant species such (*Blutaparon portulacoides* L.) that naturally colonizes salinized lands cultivated with banana trees (Burn on leaves indicates the effect of toxicity due to high salinity of the soil) irrigation district Baixo-Açu (DIBA), Ipanguaçu, Brazil. *Source* The authors

2.3 Social Impacts and the Local Economy

The impacts of increased soil salinity on the community social structure include the reduction of the aesthetic value of the landscape (Fig. 3), decreased agricultural revenue due to low crop productivity, and low commercial property value due to the

Fig. 3 Salinity induced by poor drainage and inadequate irrigation management. Irrigation district Baixo-Açu (DIBA), Ipanguaçu, Brazil. *Source* The authors

degradation of soil, environment, and lack of agricultural jobs that leads to population impoverishment. In extreme cases, this scenario results in rural exodus.

Tenison (2009) reported that soil salinization in family-based agricultural operations increases social pressure in the consolidation of rural property with a significant reduction in the availability of services in small rural towns. This problem results in increased costs of social assistance and public-income transfer policies in trying to reduce poverty and social inequalities. For this reason, many salinized lands are abandoned each year, triggering a process of mass migration to urban areas and, consequently, aggravating social problems such as unemployment, illiteracy, violence, and crime in affected areas. The reduction in crop yield caused by increased soil salinity and water scarcity increases poverty and reduces *per capita* income. The Brazilian Northeast region accommodates 22 million inhabitants, of which 9 million live in rural areas. The area has the lowest income rates in Brazil representing a GDP (gross domestic product) *per capita* of less than US$ 2,000/year or $5.50/day and the official Brazilian minimum wage (salary) was R$1,448.00/month (US$638.62/month or US$8,302.06/year) in 2015 (Aleixo et al. 2016). Some proposed measures to mitigate the salinity impact in family farms are the expansion and reform of rural advisory services, focusing on individual capacity building and leading to the development of multistakeholder strategies to promote adequate management practices of available resources, access to the market and, therefore, the sustainable use of the lands.

One must keep in mind that access to drinking water alone will not solve the socioeconomic situation of rural communities in the Brazilian semiarid (much like the situation of semiarid areas in other developing countries). When inequalities in water access conditions were assessed in the community of Cristais (Ceará state, Brazil), 31% of the households reported to earn less than official minimum wage per month, 37% of the household heads were illiterate, and 46% were retired or worked as pensioners and agricultural workers (Aleixo et al. 2016). Thus, besides drinking water, access to education, health services, and creation of jobs in afflicted rural communities is a must. In this sense, maybe the improvement and establishment of integrated agricultural systems using brine reject from water desalinization to generate animal protein (tilapia and small ruminant farming) for household consumption and commercialization may be part of the solution (Moura et al. 2016).

In the Brazilian semiarid, salinization (natural or anthropogenic) brings about social problems due to the yield losses of vegetables, grasses, cotton, and orchards. When these problems occur in irrigated perimeters, their effects contribute to the disorganization of the region's agro-food sector that affect the local economy and produce negative social effects on rural communities. Of course, this cannot be blamed only on natural causes, as it was evidenced during the severe drought of 2012–2015, which affected not only the Brazilian Northeast, but also the Southeast, including the states of São Paulo and Rio de Janeiro and even southern California. The lack of planning, proper infrastructure, accelerated population growth, and improper management of funds in Brazil (not uncommon in developing countries) led the Brazilian Southeast to a severe water scarcity similar to what would be seen in the Northeast (Slater 2019).

In Northeastern Thailand, Naing et al. (2013) evaluated the impacts of soil salinity on the agricultural production system and described socioeconomic conditions in areas affected by salts. Their results indicated that the reduction of productivity in crops due to salinization effects led to social tension, unemployment, and income reduction in all the locations studied. Additionally, the authors reported that large-scale soil salinization also affected smallholders, reducing their capability to produce staple crops and intensifying poverty. The authors concluded that although people are the main agents of secondary salinization, they are also their victims. Other studies involving lowland cropping systems in Northeast Thailand (Kabaki et al. 2003) indicated that low fertility of sandy soils in the region, along with frequent droughts, degradation and salinity are the main causes of low yield and, consequently, food safety problems for farmers.

In the United States, a study done by the NRCS-USDA (2012) indicated that soil salinity has become a major concern for counties of the upper James River watershed. The Natural Resources Conservation Service (NRCS), a branch of the United States Department of Agriculture (USDA) evaluated the potential economic impact of salinity in the upper James River, soils, and cropping patterns of the three top counties in the watershed (Brown, Spink, and Beadle Counties). The analysis of results indicated that saline impact in soils of the tri-county area was estimated to occur in 280 acres and lowered the average productivity of the four major crops raised in this area (corn, soybeans, spring wheat, and winter wheat) in 30%. The economic impact of this productivity decrease in the tri-country area was estimated at $26.2 million.

3 Salinity Affecting Irrigation Projects: Local Case Studies

3.1 Irrigated District of Baixo-Açu, Rio Grande Do Norte, Brazil

The first case study involves the project Irrigation District Baixo-Açu (DIBA), the body that brings together water users of the irrigated perimeter Low-Açu, founded on March 5, 1997. DIBA has the purpose of managing, operating, and maintaining the irrigation and drainage infrastructure commonly used in that perimeter. The major crops produced in this perimeter by family-managed farms are bananas, coconut, papaya, and zucchini.

A study conducted by Barreto (2019) shows that, after 23 years after the foundation of DIBA, from 20–30% of the lots were salinized due to the absence of planning to avoid the accumulation of salts in the soils, which will be discussed below (Fig. 4).

Figure 5 shows a schematic design with some land parcels managed by DIBA indicating that the main cause of soil accumulation of this irrigated perimeter, characterized by high evapotranspiration and superficial water table, is the absence of a drainage system to leach soil salts beyond the root zone. In Fig. 5a, the yellow line

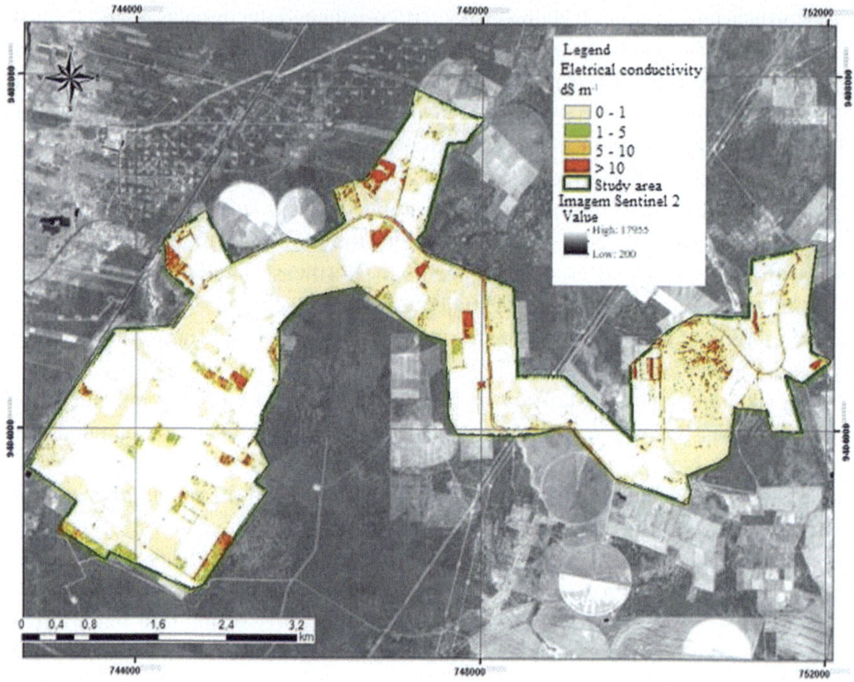

Fig. 4 A map of the Baixo-Açu project irrigation district soils with a high-risk potential for salinization (EC > 4 dS m^{-1}). *Source* Barreto (2019)

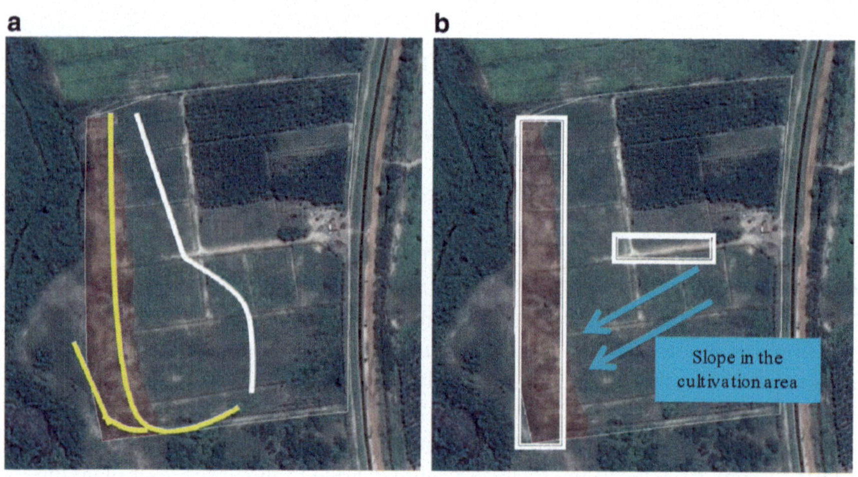

Fig. 5 Salinization of the parcels of the irrigated district of Baixo Assú (DIBA). **a**–Yellow lines show where the drainage system should be installed and white line show the progression of salinity. **b**–Blue lines show direction of the slope. *Source* Modified from Google Maps

indicates where the drainage system should be installed, but the area became a salt accumulation basin, being clearly identified by the presence of a white crust (salt) on the surface and of blackened spots (sodic humates). The white line shows the advance of salinity in the low areas.

Figure 5b shows the slope in the cultivation area. The soil of this area is classified as Cambissolo (EMBRAPA 2013), also known as inceptisol by soil survey staff (2014), which is naturally deficient in micro-drainage and, possibly, with a high degree of degradation (PST above 15%).

The absence of the macro- drainage of the area results in the accumulation of salts in the crop root zone, even using low electrical conductivity water (CE < 0.6 dSm^{-1}), because semi-prevented drainage raises the ground level of the lower area, accumulating salts on the soil surface by capillary action.

3.2 Apodi Valley, Rio Grande Do Norte, Brazil

The rural producers of DIBA have a different profile in relation to peasants who practice irrigated agriculture in the "Sertão" of Apodi. In the irrigated perimeters, land ownership is varied, and farmers acquire land through purchase, exchanges, and leases, while Apodi farmers practice family farming, almost always in irrigated lands inherited from their parents. Although the succession system is carried out by sharing, the family nucleus remains preserved. In both cases, the advance of land salinization is evident and has severe social impacts in the agricultural production system, threatening food security of rural families.

In this second case study, the difficulties are more evident, since the farmers have a small financial capital to acquire new lands. In DIBA, producers work as employers. So, farmers often abandon their parcels because of the financial inability to pay accumulated debts. The low profitability is a consequence of the high cost of electricity and low productivity, resulting in the end of agricultural activities and in land abandonment.

In the Apodi valley, rice is the main crop but, according to farmers, the cultivated areas are decreasing each year due to the low productivity (Fig. 6).

In the Apodi valley, impoverished farmers must cope with soil infertility due to salinization because land ownership does not change easily and will probably pass from parents to their children through inheritance. The land use also changes and parcels where red rice was once cultivated become pasture for cattle during the rainy season.

In addition to the problem of salinization, drought also reduced the cultivated areas for rice production between the years of 2013 and 2017. It is worth noting that from 2012 to 2017, the Brazilian northeast went through the worst drought in 50 years. However, the rice yield per hectare reached approximately 7 tons in 2018 (Globo 2019), which is considered a good yield for the region. This high yield is attributed to the rainfall above average for the period.

Fig. 6 Salinized soils–**a** and rice culture with characteristic symptoms of toxicity effect–**b**. Vazia da Salina, Apodi, Brazil. *Source* The authors

In both cases reported above, declining in agricultural production is leading to land abandonment, reducing rural population, and increasing unemployment. This problem impacts the local economy so that rural workers who do not have land ownership, but who develop agricultural activities, live an occupational duality: they work in agriculture part-time supplementing their income with non-agricultural activities. This reality often creates bigger problems when farmers reach the retirement age.

4 Final Considerations

Salinization of irrigated areas is one of the important issues for local farmers living in the arid and semiarid zones of the world. Soil salinity has become a serious problem around the world and this process is accelerated by human activities such as deforestation and poorly managed irrigation.

The socioeconomic impacts of soil salinity in an area of agricultural production are directly influenced by a reduced crop yield driven by salinity effects on plants and soil. These factors eventually lead to the loss of commercial value of agricultural land.

Irrigated perimeters face enormous problems of salt accumulation and therefore the degradation of water resources. Despite the lack of technical assistance, farmers continue their efforts to manage periodic droughts and to cultivate the land to increase food security for their families.

Salinity problems are more severe in irrigated areas with arid and semiarid conditions because insufficient precipitation to leach salts away from root zone, associated with increased evapotranspiration, severely limits crop growth and productivity. This scenario is seen especially when there is no management associated with irrigation and no agricultural practices that mitigate the negative effects of salt accumulation in soils leading to poor crop performance or death.

Governments, institutions, and the private sector should not only provide technical assistance for farmers, specially emphasizing water and soil management, but also should consider the proper investment in infrastructures for water storage and distribution that will be crucial to face unpredictable droughts in the dawn of global warming. The advisory services should focus their efforts in promoting the adoption of salt- and drought-tolerant crops, and the development of social technologies that could increase food production and ensure food and nutrition security throughout these regions.

References

Aleixo B, Rezende S, Pena JL, Zapata G, Heller L (2016) Human right in perspective: inequalities in access to water in a rural community of the Brazilian Northeast. Ambien Soc 19:63–84

Amorim MCC, Porto ER (2001) Avaliação da qualidade bacteriológica das águas de cisternas: Estudo de caso no município de Petrolina-PE. In: 3º Simpósio Brasileiro de Captação e Manejo de Agua da Chuva no Semiarido

Barreto AC (2019) Modelagem da salinidade do solo com a utilização de técnicas de sensoriamento remoto. Thesis, Universidade Federal Rural do Semi-Árido, Mossoró

Brito CFB, Santos MR, Fonseca VA, Arantes AM (2017) Physiological characteristics and yield of 'Pérola' pineapple in the semi-arid region. Rev Bras Eng Agríc Ambien 21:834–839

Butcher K, Wick AF, DeSutter T, Chatterjee A, Harmon J (2016) Soil salinity: a threat to global food security. Agron J 108:2189–2200

EMBRAPA (2013) Sistema brasileiro de classificação de solos. Centro Nacional de Pesquisa de Solos, Rio de Janeiro

Ferreira JFS, Liu X, Suarez DL (2019) Fruit yield and survival of five commercial strawberry cultivars under field cultivation and salinity stress. Sci Hort 243:401–410

Gheyi HR, Dias NS, Lacerda CFL, Gomes Filho E (2016) Manejo da salinidade na agricultura: estudos básicos e aplicados. INCTSal, Fortaleza

Globo (2019) Produção de arroz do município de Apodi. https://g1.globo.com/rn/rio-grande-do-norte/noticia/safra-melhora-e-produtores-colhem-20-mil-quilos-de-arroz-vermelho-por-dia-em-apodi-no-rn.ghtml. Acessed 20 Dec 2019

Kabaki N, Tamura H, Yoshihashi T, Miura K, Tabuchi R, Fujumori S, Morita H, Wungkahart T, Watanavitawas P, Uraipong B, Arromratana U (2003) Development of a sustainable lowland cropping system in Northeast Thailand. Jpn Agric Res Q 37:37–44

Machado RMA, Serralheiro RP (2017) Soil salinity: effect on vegetable crop growth. management practices to prevent and mitigate soil salinization. Hortic 3:1–13

Medeiros JF, Cordão Terceiro Neto CP, Dias NS, Gheyi HR, da Silva MV, Loiola AT (2017) Salinidade e pH de um Argissolo irrigado com água salina sob estratégias de manejo. Rev Bras Agric Irrig 11:1407–1419

Moura ESR, Cosme CR, Dias NS, Portela JC, Souza ACM (2016) Yield and forage quality of saltbush irrigated with reject brine from desalination plant by reverse osmosis. Rev Caatinga 29:1–10

Naing A, Iwai CB, Saenjan P (2013) Food security and socio-economic impacts of soil salinization in northeast Thailand. Int J Environ Rural Dev 4:76–81

NRCS-USDA (2012) Economic impact of saline soils XE "Saline soils" in upper James River. NRCS-USDA, Washington

Pitman MG, Läuchli A (2002) Global impact of salinity and agricultural ecosystems. In: Läuchli A, Lüttge U (eds) Salinity: environment—plants—molecules. Springer, Dordrecht
Slater D (2019) Water scarcity in Brazil: a case study. Marine Corps University, Quantico Station
Soil Survey Staff (2014) Keys to soil taxonomy. U. S. Department of Agriculture—Natural Resources Conservation Service, Washington
Tenison K (2009) Salinity training manual—NSW Department of Primary Industries, Orange
World Food Summit (1996) Rome declaration on world food security. World Food Summit, Rome

Soils, Water, Agriculture and Management in Arid and Semi-Arid Regions

Salt Accumulation and Redistribution in the Dry Plains of Southern South America: Lessons from Land Use Changes

Esteban G. Jobbágy, Raúl Giménez, Victoria Marchesini, Yésica Diaz, Dushmantha H. Jayawickreme, and Marcelo D. Nosetto

Abstract We synthesize research on the magnitude and dynamisms of salt stocks in the Chaco-Pampas. While current soil maps characterize one-fourth of the region's soils as saline; integrated soil-hydrology-vegetation studies show salinity to be more widespread and dynamic due to the presence of shallow and stagnant groundwater systems. Two salt retention mechanisms that are turned "off" or "on" by land use changes are proposed: "arrested drainage" and "landscape stagnation." In the Chaco and Espinal, deforestation and cultivation have relaxed arrested drainage, raising the groundwater and bringing a diluted salt pool to the surface that damages crops

E. G. Jobbágy (✉) · R. Giménez · V. Marchesini · Y. Diaz · M. D. Nosetto
Grupo de Estudios Ambientales—IMASL, Universidad Nacional de San Luis & CONICET, San Luis, Argentina
e-mail: jobbagy@gmail.com; jobbagy@unsl.edu.ar

R. Giménez
e-mail: gimenezgea@gmail.com

V. Marchesini
e-mail: victoriamarchesini@gmail.com

Y. Diaz
e-mail: diazyesik@gmail.com

M. D. Nosetto
e-mail: marcelonosetto@gmail.com

E. G. Jobbágy
SARAS—South American Institute for Resilience and Sustainability Studies, Maldonado, Uruguay

R. Giménez
Departamento de Geología, Facultad de Ciencias, Físico Matemáticas y Naturales, Universidad Nacional de San Luis, 5700, San Luis, Argentina

D. H. Jayawickreme
Department of Earth Science, Southern Connecticut State University, New Haven, CT, USA
e-mail: jayawickred1@southernct.edu

M. D. Nosetto
Cátedra de Climatología Agrícola, Facultad de Ciencias Agropecuarias (UNER), Entre Ríos, Argentina

E. Taleisnik and R. S. Lavado (eds.), *Saline and Alkaline Soils in Latin America*, https://doi.org/10.1007/978-3-030-52592-7_3

and soils. Over the shallower and fresher water tables of the Pampas, tree plantations trigger local arrested drainage and in the landscape stagnation context work as virtual lowlands, consuming groundwater and increasing salt pools. While land use changes highlight the dominant role of vegetation dictating the dynamics of salt pools, key fundamental questions about long-term salt dynamics remain open. Rapid and widespread land use changes offered a unique laboratory to understand saline environments showing that they are more dynamic and sensitive to our interventions, than previously thought.

Keywords Salinization · Groundwater · Land degradation · Afforestation · Deforestation · Dryland salinity · Landscape ecohydrology

1 Introduction

The South American continent hosts one of the most extensive and flat sedimentary plains on the planet (Jobbágy et al. 2008). Known as the Chaco-Pampas plain, it is framed by the positive relief of the Cordillera de los Andes and the Sierras Pampeanas ranges to the west and by the Brazilian shield to the east and north. Its landscapes, which have very low regional slopes, have received a massive contribution of Andean materials, including loess and volcanic ash, and have been shaped during the late quaternary by the combination of wind and alluvial forces (Sayago et al. 2001). The main grain production and exportation foci of the continent have developed on the fertile soils of this plain. While being almost completely unirrigated, these cultivation hotspots are still expanding over natural vegetation which is composed of xerophytic forests (Chaco and Espinal) and grasslands (Pampas) (Baldi and Paruelo 2008; Houspanossian et al. 2016; Volante et al. 2016), accompanied by riparian and lacustrine wetlands (Fig. 1).

Saline land systems including their soils, continental water bodies and ecosystems are a natural component of the global diversity of environments (Bui 2013), yet their expansion or even just their natural presence are considered a serious problem under the pressure to increase agricultural productivity while sustaining other ecosystem services (Setia et al. 2013; Butcher et al. 2016). The presence of saline environments has been documented both in the Pampas and in the Chaco and Espinal, mainly based on soil profile descriptions, plant communities and surface water bodies (Lavado and Taboada 1988; Lewis 1991; Batista and Leon 1992; Cisneros et al. 1999; Quirós and Drago 1999; Perelman et al. 2001; Feldman et al. 2008). Being a very flat and extensive plain with a subhumid to semi-arid climate, conditions for salt accumulation at the surface and at depth can be expected in most of its territory (Schofield and Kirby 2003), particularly when considering recent findings on the strong degree of groundwater-ecosystem coupling in flat landscapes (Fan et al. 2017). The first global characterization of phreatic groundwater systems, based on numerical modelling and a collection of thousands of observations of water table depths (Fan et al. 2013), shows that large sedimentary plains, including the Chaco-Pampas, host water tables that

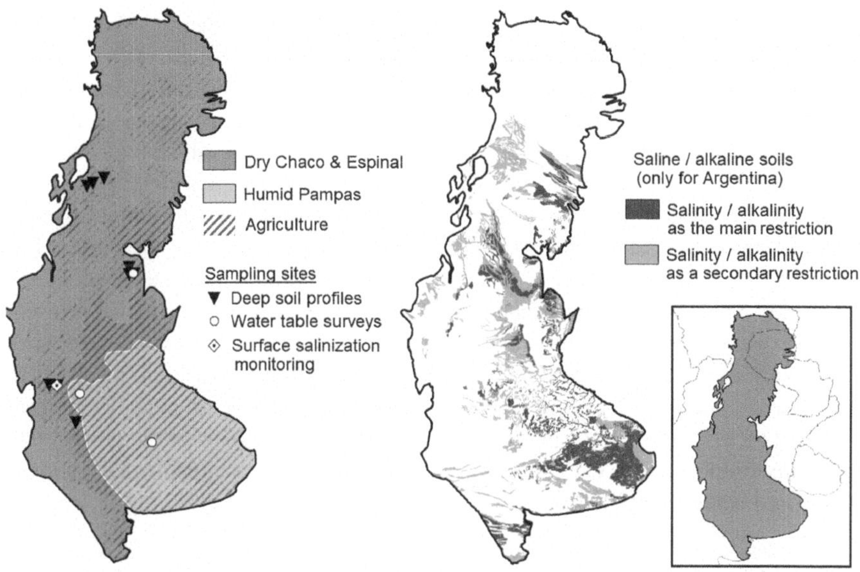

Fig. 1 Map of the Chaco-Pampas plains. The left map shows the original distribution of Chaco and Espinal dry forests and Pampas grassland with the current agricultural areas (shaded) and the location of three types of sampling sites: (i) deep soil profiles (El Morro, Fraga, Parera, Salta, Bandera), (ii) water table surveys (Bandera, Vicuña Mackenna, Hortensia), and (iii) surface salinization monitoring (El Morro). The right map shows salinity/alkalinity-affected soils in the area according to the 1:500,000 soil charts of the National Institute of Agricultural Technology of Argentina (GeoINTA 2020). Ecoregions limits in the left map were defined sensu Olson et al. (2001) and downloaded from the World Wildlife website (https://www.worldwildlife.org), while the area of agricultural crops was obtained from the Copernicus Global Land Service product "Dynamic Land Cover map at 100 m resolution, CGLS-LC100" (Buchhorn et al. 2019). Soil maps for the right map were downloaded from the GeoINTA website (https://www.geointa.inta.gob.ar)

tend to be very close to the surface, thanks to the slow groundwater horizontal fluxes (i.e., stagnation), creating widespread opportunities for direct contact between terrestrial ecosystems and groundwater. The fact that these conditions, which are mainly dictated by lithology and geomorphology, are combined with negative climatic water balances in which precipitation is exceeded by potential evapotranspiration, makes the Chaco-Pampas plain prone to accumulate salts both in its deep sediments and in the surface of its low landscape sectors.

The rapid and intense land use changes occurred in the Chaco-Pampas plain in the last two decades have shown how vegetation, through its influence on water dynamics, can exert a strong control over salt accumulation and distribution processes. For this reason, land use changes in the plain can unleash salinization processes that are difficult to predict if the stagnant nature of the hydrological cycle of these regions and its high sensitivity to vegetation is not recognized. These salinization processes are framed in the concept of "dryland salting," coined in Australia (Peck 1978); since they do not derive from irrigation, which is practically null in the region, but result instead from functional and structural changes of vegetation (Hillel 2000).

In this chapter, we focus on the semi-arid to sub-humid fraction of the Chaco-Pampas plain, seeking to (i) highlight that there is a general accumulation of salts in the soil profile of the plains where they are occupied by woody ecosystems, (ii) conceptualize the main mechanisms of salt retention in sedimentary plains, and (iii) synthesize the available research about the effects of intense vegetation changes on salt accumulation and redistribution in soils and groundwater of the region. For these purposes, we use evidence generated at the plot and landscape level over a decade and a half by the authors and other colleagues. This evidence includes direct observations of deep soil profiles, phreatic groundwater and surface water bodies; as well as geoelectric measurements and satellite information for different locations in the Chaco, Espinal and La Pampa. Also, we (iv) review the scientific challenges that remain open in relation to the previous points and the articulation of the perspective on salinity proposed here with the search for guidelines for sustainable management of land and water in dry plains.

2 Visible and Hidden Salts

The fraction of the territory that has been classified as saline or saline/alkaline in the Chaco-Pampas plain ranges between 12 and 27% according to available soil surveys from Argentina (GeoINTA 2020, Fig. 1). These boundary values consider only the mapping units in which saline/alkaline soils are the primary component (low estimate) and those where they are the primary or the secondary component (high estimate). While a coarse figure of one-fourth of the land being currently classified as salty to some degree can be applied to the whole region, there are areas within this plain where saline soils are much more abundant, particularly in association with surface water bodies such as lagoons, many of which are landlocked, and rivers (Fig. 1). Existing descriptions suggest that the presence of abundant salts is limited to local or intrazonal situations and is fed by the rest of the landscape. While this appears valid for the Pampas, a region corresponding to the wettest and temperate fraction of the plain which was originally covered by grasslands, it is less clear for the rest of the plain that includes the Chaco and Espinal forests, where the few available deep soil surveys (> 6 m of depth, Table 1) suggest a much more widespread or zonal presence of salts that were not recognized by traditional surveys limited to describing profiles up to the top of the C horizon (typically < 1.5 m).

Available deep profile observations suggest that the predominant situation in the matrix of the dry forest landscapes of the Chaco and Espinal is that of a high accumulation of salts in the unsaturated zone (Santoni et al. 2010; Kim 2011; Amdan et al. 2013; Jayawickreme et al. 2011; Contreras et al. 2013; Gimenez et al. 2016). The compilation of six studies covering the provinces of La Pampa, San Luis, Santiago del Estero and Salta, gathering a total of 28 observation sites, shows that the profiles under native vegetation display a strong increase in the chloride content below the

Table 1 Available characterizations of deep salt storage in native ecosystems of the Chaco and Espinal plains. Each case corresponds to an individual study that includes several sampling sites. All profiles are at least 6 m deep and have been analyzed for extractable chloride. Chloride stocks are calculated down to 6 m of depth. The dominant vertical pattern of salt distribution is described by showing the depth of the top (start) and maximum concentration (peak) of the salt bulge as well as the shape of its deep decline function (tail)

Ecoregion	Province	Sites	Cl-stock	Vertical distribution			References
		n	Kg/m^2	Start	Peak	Tail	
Espinal	La Pampa	3	0.2	1 m	2 m	Sharp decline	Kim (2011)
Espinal	San Luis	5	0.67	2 m	2–4 m	No decline	Jayawickreme et al. (2011)
Espinal	San Luis	3	0.68	2 m	5 m	Slight decline	Contreras et al. (2013)
Espinal	San Luis	5	2.6	1 m	2–4 m	Slight decline	Santoni et al. (2010)
Chaco	Salta	6	0.77	2 m	2–4 m	Slight decline	Amdan et al. (2013)
Chaco	San Luis	3	5.3	1 m	1–4 m	Slight decline	Marchesini et al. (2013)
Chaco	Santiago del Estero	3	12.4	1 m	2 m	Sharp decline	Giménez et al. (2016)

second meter, and in some cases even below the first meter (Table 1). The maximum values are reached between the second and the fourth meter in general and then slight decreases are observed in the case of the sites of San Luis and Salta, while more pronounced declines are found in the rest of the sites. The most remarkable aspect of these measurements is that they document a total storage of chloride down to six meters of depth of more than 0.5 kg/m^2 in the case of San Luis and more than 5 kg/m^2 in Salta and Santiago del Estero. In La Pampa, the lower levels of accumulation coincide with sites that have recently undergone a transition from grassland to forest or savanna (i.e., woody encroachment). These measurements provide a low boundary of the actual content of total soluble salts in these environments. Assuming that chloride is the only anion and that it is accompanied by an even combination of sodium and magnesium, the values presented here should be doubled to reach a figure of total salt content. However, at the sites where sulfate concentration has been measured, it has been very high, exceeding that of chlorides by a factor of 1.5–9 (Jayawickreme et al. 2011), indicating that the natural stock of total salts in these ecosystems is likely much higher. Similarly, the comparison of chloride concentrations and the values of electrical conductivity in 1:2 soil:water extracts for the profiles in Santiago del Estero and Salta indicates that less than half of the electrical conductivity can be explained by the chloride anion, with the rest being necessarily explained by other anions, presumably sulfate, as suggested by the neutral pH levels (Gimenez 2016; Amdan et al. 2013).

In San Luis, a strong textural gradient of sediments across the study locations has revealed that salt stocks decrease with increasing sand content (Santoni et al. 2010). The overwhelming role of texture became evident in a much drier but extremely sandy environment (sand dunes) corresponding to the Monte shrubland in Mendoza, where deep soil chloride accumulation bulges were not found (Jobbágy et al. 2011). These observations suggest that soil texture, through its effect on deep drainage, would exert a dominant control over that of the climate by dictating salt accumulation patterns in dry plains.

Although there are still just a few deep soil observation points to perform a reliable extrapolation for the entire region, the studies collected here represented in all cases the dominant soils and ecosystems of the focus areas, which, according to the available soil maps have been characterized as non-saline. From these new studies, the notion that the Chaco and the Espinal host a very large but invisible salt stock has emerged. Only if soil sampling is deepened below the second meter or, as we will see in the following sections, these salts move to the surface after a change in water circulation patterns, they will become visible to those in charge of managing production and conservation in these ecosystems.

3 Two Key Mechanisms

A fundamental aspect of soil studies that help explain the presence and regional distribution of saline environments is focused on soil genesis, including the identification of the primary sources of soluble materials and the geological processes associated with them (Downes 1954; Mahjoory 1979; Sidhu et al. 1991; Furquim et al. 2010). Here, we focus on a complementary and possibly more practical aspect, which is the understanding of the salt retention mechanisms both at the level of soil profiles and whole landscapes. In both cases, the keys are dictated by the circulation of water. The first mechanism explains the widespread accumulation of salts in the dry forest profiles shown in Table 1 and involves the total restriction of deep drainage or "arrested drainage." The natural woody vegetation of dry plain environments seems to meet this condition in many regions of the world (Edmunds and Gaye 1994; Scanlon et al. 2006; Ibrahim et al. 2014) and would achieve it thanks to their ability to use rainfall inputs exhaustively either in near real time or with seasonal or interannual deferrals. Even throughout periods in which precipitation exceeds evapotranspiration, these ecosystems prevent any drainage by taking advantage of occasional water surpluses in subsequent dry periods due to the large storage capacity that results from the combination of deep sedimentary soils and deep root systems. This mechanism would restrain any water escaping below the root front and, as shown with global root distribution syntheses, plants could be very flexible deepening roots in response to exceptionally large wetting pulses (Fan et al. 2017). Therefore, the soils of these ecosystems host a deep stagnant vadose zone where solutes can accumulate.

For how long can this mechanism operates and maintain salt accumulation in dry plain ecosystems? Studies in North America and the observations reported here for San Luis indicate that this condition can last for millennia. In the case of North America, the characterization of very deep profiles (>30 m) using chloride and radionuclide tracers indicates that throughout the Holocene the semiarid ecosystems of the sedimentary plain of the central-southern USA maintained an "arrested drainage" condition (Scanlon et al. 2005; Walvoord et al. 2002). In San Luis, current measurements of atmospheric chloride deposition and thermoluminescence dating of sediments were used to estimate the fraction of the total chloride deposition received throughout the soil profile lifespan that was still in place. Retention levels declined from 95 to 4% as the sand content of the sediments increased, suggesting an almost complete retention in finer materials (Santoni et al. 2010).

A study of the temporal dynamics of deep soil moisture and salinity in dry forests of San Luis (Marchesini et al. 2013) showed that the stock of salts contributes to preserve a partially humid zone in the profiles thanks to the high osmotic potential that prevents plant absorption of salty soil water. Yet, drainage pulses triggered by the strong disturbance of roller chopping (partial deforestation technique) in this area allowed a more exhaustive use of moisture in that soil zone thanks to the slight downward displacement of salts. Despite the intensity of these disturbances, the recovery of natural vegetation seemed to maintain the "arrested drainage" condition, helping to explain the ability of many of these ecosystems to preserve their deep salt load received throughout the Holocene within the unsaturated zone (Walvoord et al. 2002).

The second mechanism explaining salt retention in dry sedimentary plains operates at a larger spatial scale and is referred herein as "landscape stagnation." Extremely flat sedimentary landscapes, especially those shaped by wind action, severely limit lateral water transport both above and belowground (Fan et al. 2013), unless very humid conditions that sustain long lasting and flowing floods prevail (Jolly et al. 1998). Hence, sedimentary plains can still retain salts even if the arrested drainage condition is relaxed and salts are able to move downward, thanks to the mechanisms of landscape stagnation. The topographic and lithological conditions of the plains that lead to very slow lateral flows favor widespread shallow groundwater levels (Fan et al. 2013) and the development of short hydrological pathways between locally high (recharge) and low (discharge) areas. These local flows prevail over longer regional ones, favoring the redistribution of salts rather than their net removal from the landscape (Tóth 1963; Jobbágy et al. 2017).

Although surface flows are less important for removing salts, they can play a relevant role in the local positioning of salt accumulating zones. For example, lower landscape positions tend to accumulate salts when evaporative water losses from groundwater exceed direct rainfall inputs, yet they can avoid this accumulation if they receive "extra" water inputs via run-on from their higher surroundings. Geoelectric profiles along a toposequence of the dry Pampas occupied by agricultural crops (Fig. 2a, b) show the intervention of these two forces by revealing how salts tend to accumulate in the intermediate zone of the toposequence and not in the lowest portion during a relatively humid growing season in which a period of positive water

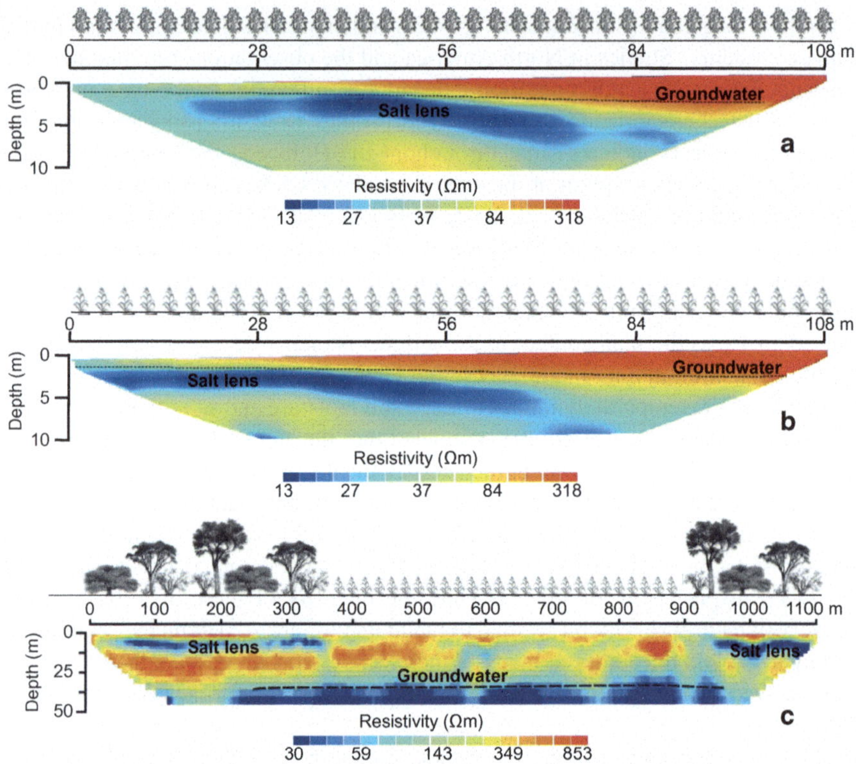

Fig. 2 Electrical-resistivity imaging along sedimentary landscapes of the Pampas and Espinal. **a–b** A toposequence under full cultivation in Vicuña Mackenna, Córdoba (see Nosetto et al. 2013 for details), showing a relatively wet period (a, February 2009 under a developed soybean canopy) and a relatively dry period (b, November 2009 under recently sown maize). Both profiles show fresh deep groundwater with a saturated salty lens on top that has its highest salinity (lowest resistivity) where water tables are most accessible to crops. Yet in the first sampling (a, wet) but not in the second (b, dry), the lowest zone has fresher groundwater, likely as a result of run-on dilution. **c** Long transect covering a full cultivated plot (70 years after clearing) and its neighboring forest plots in Fraga, San Luis (see Jayawickreme et al. 2011). Dry Espinal forest plots have an unsaturated salt lens that matches soil core observations and seems to decline below ~10 m of depth. The cultivated plot has lost the unsaturated salty lens and generated an apparent lift of the (still deep) groundwater level. Resistivity imaging was performed with a multielectrode resistivity unit (Syscal Junior, Iris Instruments). Standard electrode-pairing methods (Wenner, Dipole–Dipole; or Schlumberger) were used to capture the optimal two-dimensional distribution of ground conductivity. In Vicuña Mackenna and Fraga, electrodes were installed in a straight line with 1.5 versus 4-m separation distances between them, allowing an imaging depth of 12–15 versus 40 m. Field data were inverted with RES2DINV software. Within the saturated zone (below the groundwater level), resistivity can be interpreted as the inverse of water salinity. In the unsaturated zone, salt content interacts with moisture content making interpretation more complex

balance generated run-off/on redistribution and focused recharge in the lowest area (Nosetto et al. 2013). During a drier growing season in which negative water balance conditions prevailed (i.e., groundwater discharge exceeding recharge), the same place showed increased groundwater salinity toward the lower landscape position. These observations illustrate how landscape stagnation operates by favoring the local redistribution of salts in detriment of its regional evacuation, opening the possibility for other factors beyond topography, that can affect recharge and discharge fluxes, to redefine groundwater level gradients and salt transport in time and space.

The way in which local landscape controls salt distribution in the saturated zone is revealed by the relationships between the variability of surface and groundwater levels and salinity in cultivated landscapes of the Pampas and Chaco where arrested drainage is not operating (Fig. 3). In the most humid site (sub-humid Pampas), wetlands show opposite salinity conditions compared to groundwater (Fig. 3c). The high lagoon in this landscape (small depression within a relatively high area) is fresher

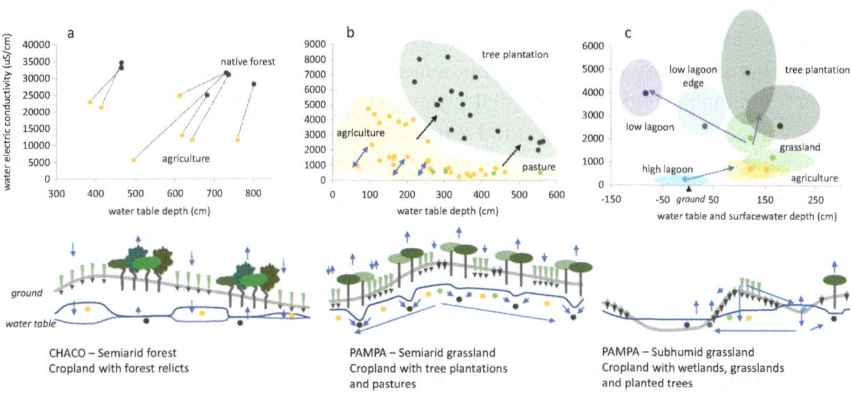

Fig. 3 Characterization of phreatic groundwater depth and salinity gradients in three landscapes of the Chaco-Pampas plain in which cultivation prevails. **a** Bandera, Santiago del Estero (see Gimenez et al. 2016 for details), a semiarid Chaco location on an alluvial landscape that hosts one of the earliest agricultural hotspots of the region that become massively deforested over the past 30 years. Groundwater was sampled in paired position along the regional W-E topographic gradient including plots with native forest relicts and adjacent agricultural plots. The lines connect each pair of forest and agriculture plots. **b** Vicuña Mackenna, Córdoba (see Nosetto et al. 2009, 2013 for details), a semiarid Pampas location on an eolian dune landscape dominated by crops and hosting several planted eucalyptus strips. Groundwater was sampled in multiple plots including fully cultivated toposequences and toposequences with paired plots of agriculture and tree plantation strips. Black arrows link groups of neighboring plots with agriculture and tree plantations. Blue arrows represent temporal changes in the same groundwater wells associated with recharge (fallow) and discharge (growing season) cycles. **c** Hortensia, Buenos Aires, a subhumid Pampas locations on an eolian landscape with abundant blowout lagoons. Repeated sampling at fixed locations (9 positions, 8 dates in two years) is summarized by mean values (markers) and maximum–minimum ranges (ellipses) for (i) wetlands including a high elevation lagoon and a low elevation lagoon and its surrounding natural vegetation, (ii) croplands and adjacent (iii) successional grasslands, and (iv) eucalyptus and elm tree plots. For each landscape, a simple scheme illustrating the prevailing landforms and the dominant water fluxes are shown

than its surrounding water tables, while the low lagoon and its surrounding wetlands are saltier. The balance between recharge and discharge is positive in the high wetland thanks to the dominance of run-on over groundwater seepage, whereas the opposite is true for the low lagoon. As a result, salts tend to flow from the high lagoon to the water table and from there to the low lagoon in response to local topographic effects combined with the overall landscape stagnation. In the semiarid Pampas site, the agricultural matrix of the landscape shows greater salinity fluctuations in low positions where the water tables are shallower and recharge and discharge cycles more intense (Fig. 3b, Nosetto et al. 2009). On the semiarid Chaco site, the arrested drainage condition of native dry forests has been interrupted (as explained in the next section) and the effects of landscape stagnation begin to manifest as an inverse relationships between groundwater salinity and depth (Fig. 3a).

In landscapes of the plain with higher regional slopes and groundwater transport capacity (i.e., high transmissivity resulting from high sand content in thick saturated layers) and/or natural networks of channeling of water table surpluses, the "landscape stagnation," and therefore, the accumulation of salts will be reduced. This is observed in medanos systems of Western Pampas, where despite the strongly negative water balance, there is a very limited development of saline environments in the low areas and an appreciable regional flow that evacuates them (Bogino and Jobbágy 2011; Echegoyen et al. 2018). In these sand dune systems, both the arrested drainage and the landscape stagnation cease to be a limitation (Jobbagy et al. 2011; Echegoyen et al. 2018), and therefore, salinity does not emerge as a dominant problem. In more sloped dry landscapes of Australia, run-off pulses can also be an important vector of salt transport and release from basins (Callow et al. 2019).

4 The Imprint of Vegetation Change on Salinity

Land use and vegetation changes can reshape vertical and lateral water fluxes to such an extent that the salt retention mechanisms of arrested drainage and landscape stagnation described above can become "turned on" or "off." In this section, we illustrate these transformations with several field examples that introduce profile, plot, and landscape level observations. We reorganize the existing evidence derived from multiple ecohydrological studies that explore the complex interplay between water, salt and vegetation dynamics focusing on the following specific aspects of salt transport: downward transport in the unsaturated zone (i.e., leaching), water table level shifts and vertical transport within the saturated zone, lateral groundwater gradients and horizontal transport within the saturated zone, upward transport in the saturated–unsaturated contact zone and (sub)surface accumulation, and geomorphological alterations of landscape water/salt transport.

The pervasive salt retention of different natural vegetation formations in the dry sedimentary plains of the Chaco and Espinal ends with the onset of annual crop cultivation. Arrested drainage is no longer sustained under agriculture, as multiple paired site comparisons of chloride profiles have revealed (Marchesini et al. 2017).

After two decades of agricultural use, sites in the Espinal and Chaco loose most of their chloride stock and chronosequences of cultivation reveal a progressive downward displacement of the salt bulge (Marchesini et al. 2017), with a more retarded transport of sulfate compared to chloride, likely related to the lower solubility of sulfate-bearing gypsum salts (Jayawickreme et al. 2011). Geoelectrical profiling provides complementary evidence about this salt leaching process bringing a less quantitative characterization of the vertical salt distribution shifts but a more robust horizontal integration of the process that single soil cores can only capture at a point (Jayawickreme et al. 2014, Fig. 2c). The most plausible causes of this hydrological shift caused by cultivation are the reduction of annual evapotranspiration, its increased seasonality brought by the alternation of long fallow periods with very active growing periods, and the dramatic reduction of plant rooting depths that accompany the replacement of perennial native trees, shrubs and grasses with annual crops (Schenk and Jackson 2002; Fan et al. 2017). While evapotranspiration cuts favor the development of drainage pulses during wet periods, shallower roots prevent their recapture by plants in dry periods (Ridley et al. 1997, 2001). As a result, drainage starts and salts are released from their vadose deposit toward greater depths.

Once salt stocks move downward, they may remain out of reach to surface ecosystems as they go through the vadose zone. Yet, once they arrive at the saturated zone, two contrasting outcomes may emerge. Salts can eventually leave the landscape through regional groundwater transport or they may remain in the area and "bounce up" toward the surface. The second outcome will take place if lateral water outputs are not sufficient to alleviate growing vertical inputs (i.e., a manifestation of landscape stagnation).

In the El Morro location, a survey of static levels in phreatic water wells showed an average rise of 0.15 m/year in one of the oldest cultivated hotspots of the Espinal-Pampas ecotone, with "jumps" of up to 10 m in 35 years (Contreras et al. 2013). While it has been tempting to relate these water table raises to the higher than average rainfall experienced in parts of that time interval, local observations demonstrated that arrested drainage was never interrupted under forest relicts while leached vadose zones were found across the fence in their neighboring cultivated plots (Santoni et al. 2010; Contreras et al. 2013). Circumstantial evidence of long-term water table level raises linked to massive land use changes have also been presented for the Chaco (Gimenez et al. 2016) and the Pampas (Jobbágy et al. 2019) plains by sporadic records and century-old infrastructure damaged by rising groundwater for the first time on record.

While rising water tables in stagnant landscapes do not generate a net removal of salts, they can redistribute them and change their dilution/precipitation level. Vadose salts that were not interacting with plants or groundwater and prevailed in a precipitated form become dissolved and get moved closer to the surface, and in that new state, they may deteriorate land and water resources. Where phreatic groundwater was initially salty, as in most of the Chaco and Espinal, recharge and vadose salt mobilization tends to result in fresher groundwater because the balance between vadose water and salt inputs yields lower concentrations than those of the hosting water tables (Gimenez et al. 2016). This appears to be the case in the Bandera landscape in the Chaco plains, where cultivation has resulted in fresher water tables (Fig. 3a). It is important to remark that having these waters closer to the surface represents a problem rather than an opportunity in this dry region, given that the crops of this region cannot tolerate the salinity levels that, even after such significant change, remain above 10,000 μS/cm in most plots (Fig. 3a). In the currently highly cultivated state of the Bandera landscape, stable isotopes in shallow groundwater have shown almost identical signatures under forest relicts, likely reflecting the old stagnant systems that were widespread before deforestation. Agricultural plots, instead, show diverse isotopic signatures likely associated with their variable individual histories of recharge pulses (Gimenez et al. 2016). As water table levels have risen in this landscape, a heterogenous patchwork of "new" fresher water layer has formed over the homogeneous "old" saltier groundwater table that can only be accessed from the surface under the forest relicts where local recharge has not taken place yet (Fig. 3).

As the water tables approach the surface, they can become a source of water for plant transpiration and direct soil or pond evaporation. As a result, the local salt pool can concentrate close to the surface and get supplemented by the load of remote salts brought laterally by the short distance fluxes that operate between neighboring recharge and discharge areas and/or from deeper and saltier groundwater levels (Jobbágy and Jackson 2007; Mujica et al. 2019). Under the semiarid climate of the El Morro location, the sustained water table level rises described above (Contreras et al. 2013) have led to a rapid surface salinization process that can be easily detected through remote sensing (Fig. 4). As water tables approach the surface of the lowest portions of this relatively sloped fraction of the plains, cultivation ceases and a spontaneous succession toward a novel stage of salty wetlands proceeds. At the core of these wetlands expanding salty surfaces become evident during dry periods. The area covered by these systems has more than tripled over the last ten years (Fig. 4), indicating that in spite of the relatively dry conditions of that decade (10% lower than the 100-year average precipitation), water tables are still rising and mobilizing salts to the surface.

Without becoming visible at the surface, salts can still accumulate in soils and groundwater in response to land use changes in the stagnant landscapes of the Chaco-Pampas plains. This situation is exemplified by studies in the semiarid (Fig. 3) and sub-humid Pampas (Fig. 3) in which tree islands within agricultural or grassland matrices worked as hydrological sinks in which the recharge from adjacent areas has been locally discharged by transpiration. As a result of this hydrological effect, salts accumulated in a similar manner to what is naturally observed in the wetlands located at the lowest landscape positions (Fig. 3), but at greater depth. The strong effect that vegetation can have switching the water balance from positive to negative is illustrated by several studies in the flooding Pampas, where tree plantation have halted drainage (i.e., onset of arrested drainage mechanism) for decades even through the rainiest period on records (Jobbágy and Jackson 2004; Engel et al. 2005) and in spite of the relatively wet climate of that part of the plains. As long as these "islands" of trees within the herbaceous matrix of crops and pastures have water tables at an accessible depth, they trigger a strong local groundwater and deep soil salinization process that is invisible at the surface but obvious in the capillary zone of the profiles

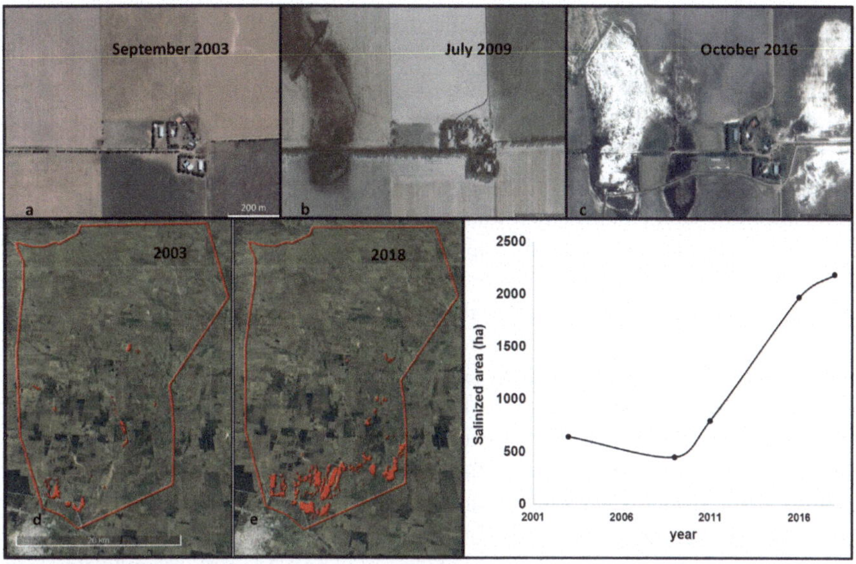

Fig. 4 Surface salinization process in the Espinal—Pampas ecotone in San Luis. **a–c** Development of dryland salinity patches in a farm close to Villa Mercedes. The area, which was fully cultivated in 2003, experienced a salinization and secondary succession process that covered approximately one half of its extension. **d–e** Expansion of salty and wet environments in the lower part of the "El Morro" watershed (2003–2016). The salinized area increased from 600 ha in 2003 to 2100 ha in 2018. Salinized areas were mapped through visual interpretation of Google Earth imagery

and that seems to reach a limit only when the biological salt tolerance of the planted trees is reached (Nosetto et. al 2008). Under a 50 year old tree plantation in the flooding Pampas a double assessment based on groundwater discharge estimates and soil + plant elemental storage shifts showed an accumulation of 3–4 Kg/m^2 of sodium (presumably at least two times more total salts) in the 0–6 m depth range (Jobbágy and Jackson 2007), highlighting how fast the onset of arrested drainage combined with lateral and/or deeper groundwater subsidies can develop salt stocks.

The relationship between water table depth and salinity described in the sandy landscapes of the Mackenna locations (Fig. 3b) illustrate how increased groundwater consumption by tree plantations raise salinity, and does so more intensely as water tables become closer to the surface. Native trees in the Chaco and Espinal are also able to consume salty groundwater down to at least 8 m of depth, as evidenced by strong diurnal water table level fluctuations registered at several locations (Jobbágy et al. 2011; Gimenez et al. 2016). Forest relicts in the agricultural landscapes are likely triggering salt concentration beyond the slow effect of arrested drainage as the native plants start consuming groundwater in response to its closer proximity to the surface. Depending on their coverage and distribution, these forest relicts may prevent further water table level rises, as suggested by models and observations in other dryland salinity contexts (Salama et al. 1993; George et al. 2012). Yet, if water table levels come too close to the surface, tree die-off can be expected in the

Chaco and Espinal plains, more as a result of anoxia rather than salinity (Bogino and Jobbágy 2011).

Besides the vegetation shifts discussed above, direct natural and (planned or unplanned) human influences on landscape stagnation can affect the salinization processes of the plains. The spontaneous coalescence of surface water bodies during wet periods in the subhumid portions of the plains (Aragón et al. 2011) or the establishment of artificial channels and drainages networks (Duncan et al. 2008; García et al. 2018a, b) can reduce landscape stagnation leading to salt removal through surface or subsurface transport, respectively. A very peculiar spontaneous geomorphological transformation illustrates the potential of these mechanisms in one of the study areas introduced here (Contreras et al. 2013). In the western fringe of the plains, the El Morro region hosts loessic landscapes with higher slopes than those displayed by the other sites presented here. After more than half a century of cultivation, this region not only unfolded the dryland salting process presented before (Fig. 4), but has also shown a very fast development of deep landscape incisions resulting from groundwater sapping and stimulated by the generalized water table level raises (Contreras et al. 2013; Gallardo et al. 2017). Twenty years after their formation, this network of incisions keeps growing and draining groundwater with its salt load. A comparison of base streamflow salinity measurements over an eight-year period (2008–2016) reveals a gradual decline of water salinity, likely responding to a rapid and widespread leaching and lateral removal of salts from the cultivated fraction of the landscape. This process seems to be operating an order of magnitude faster than previously anticipated (Contreras et al. 2013). Landscape stagnation in this eolian landscape has been abruptly unlocked by the spontaneous development of a drainage network showing that the salt accumulation and release process at this spatial scale respond to unplanned geomorphological changes triggered by human activities in a similar manner to what long dated artificial drainage infrastructure has attempted in irrigated plains (Hillel 2000; Qadir and Oster 2004).

5 Knowns, Unknowns and Practical Challenges

Fifteen years of ecohydrological studies under the most intense vegetation changes displayed by the sedimentary plains of Southern South America have provided general insight about the magnitude, distribution and dynamism of salt pools in soils and shallow groundwater (Fig. 5). Native woody ecosystems in these plains including the Chaco and Espinal regions appear to have massive and widespread salt stocks in the unsaturated zone that remained overlooked by traditional soil surveys that were too shallow to detect them. These salt stocks can be as old as the sediments where they sit and remain there thanks to what we coin as arrested drainage, which is the full restriction of downward water and salt flows in soil profiles occupied by natural woody vegetation. The relatively deep water tables of the Chaco and Espinal have very high salinity indicating that the saturated zone makes an important additional contribution to the total salt pool of that part of the plains (Fig. 5). Under

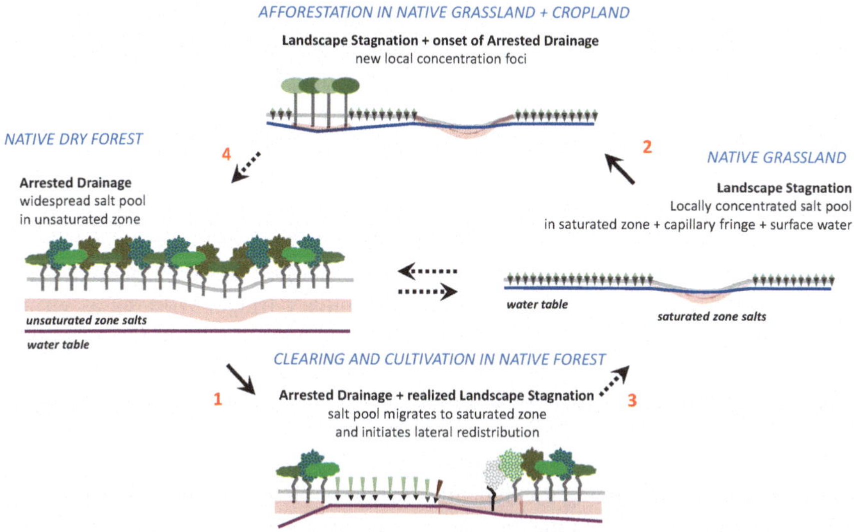

Fig. 5 Conceptual scheme of alternative ecohydrological stages and transitions for the Chaco-Pampas plains based on observations (trajectories 1 and 2) and speculations (trajectories 3 and 4). The representation of native grasslands and their transition to landscapes of grasslands and croplands with isolated "islands" of planted trees is based on studies performed in the Pampas. The representation of native forests and their transition to cultivated landscapes with forest relicts is based on studies performed in the Chaco and Espinal. Observations are used to speculate how the same given sedimentary plain subject to reciprocal switches between forests and grasslands would redistribute its salt pool as the dominant salt retention mechanisms shift from arrested drainage to landscape stagnation. Whether the Chaco/Espinal and the Pampas would follow, respectively, trajectories 3 and 4 are uncertain and depend on unknowns such as the magnitude of the total active salt pool in the Pampas, the timing of salt leaching and lateral redistribution in the Chaco/Espinal and the degree of landscape stagnation that this last region may display once water tables reach the surface in most of its extension. The pink belts illustrate lateral and vertical salt distribution without intending to provide a comparison of total stocks among regions

a more positive water balance dictated by a rainier and cooler climate but also by the less exhaustive water use capacity of the native grasslands that it has hosted, the Pampas show a more diluted and localized salt stocks after a history of vertical and lateral transport. Salts still remain in the Pampas landscapes however, thanks to a second retention mechanism that we call landscape stagnation. Unless effective regional subsurface water evacuation takes place (as it does in some sloped sandy areas of the plains), the sedimentary landscapes of the Pampas tend to retain their salt load, mainly within the saturated zone and in surface water bodies, particularly in low landscape areas that work as net hydrological sinks in which evapotranspiration exceeds water inputs from the surface (Fig. 5).

The widespread expansion of annual crops over woody ecosystems in the Chaco and Espinal and the more localized but intense transformation of the grasslands and croplands of the Pampas planted with trees provide useful lessons about the dynamisms of salt pools. In the Chaco and Espinal region, forest clearing and cultivation has relaxed the arrested drainage condition and mobilized salts toward the water tables, making these areas realize now their landscape stagnation condition with groundwater getting closer to the surface and bringing closer to the surface a diluted (yet still risky for crops) salt pool. In specific regions, this process has already caused surface salinization degrading agricultural land. Over the shallower and less salty water tables of the Pampas, tree plantations within the grassland and cropland matrix triggered locally an arrested drainage condition that halts salt leaching. In addition, they start a net groundwater consumption flux that rapidly inflates the salt pool at the water table and its overlaying capillary fringe. Working as virtual lowlands, these hydrological sinks accumulate salts in the groundwater pool up to the biological limit set by the specific salinity tolerance of the planted trees, with uptake (and salinization) being slowed down when this threshold is achieved (Nosetto et al. 2008).

While the land use changes described above highlight the dominant role that vegetation can have dictating the dynamics of salt pools in sedimentary plains, key fundamental questions about the long-term development and transformation of salty environments remain open (Fig. 5). Would a widespread conversion of the herbaceous ecosystems of the Pampas into planted and encroached woody systems create a new condition more similar to the Chaco and Espinal with deeper water tables and saltier unsaturated zones? If possible, would that happen in decades, centuries or millennia? Would the massive deforestation and cultivation of the Chaco and Espinal eventually lead to a fully redistributed salt pool that leaves most of the landscape but a few low hydrological sinks salt-free, like in the subhumid Pampas? If possible, would that happen in decades, centuries or millennia? (Fig. 5). Also central to these issues is the quantification of the global salt stocks (besides their current degree of dilution and mobility) in the landscapes of the plains, which require a careful integration across the vertical (unsaturated–saturated continuum) and horizontal (upland-lowland) dimensions. Are the Pampas less salty than the Chaco and Espinal or they just have a more diluted and redistributed salt pool? Are salt pools strongly stratified in the saturated zone? Is this stratification disturbed or preserved during intense recharge or discharge episodes? Only a deeper research program that integrates the study of soils, surface and groundwater systems across managed and natural vegetation as well as spontaneous and intentional geomorphological transformations will be able to provide light for those unknowns. Data to improve the horizontal and vertical representation of the plains and to capture most of its land use transitions is still needed, but more so are parsimonious modeling approaches that can represent salt pools in more hydrologically explicit context (García et al. 2018a, b; Garcia et al. 2019; Mujica et al. 2019).

Far from falling into a theory versus practice choice, these research challenges gains when they sit at the interface of management and science. Land use changes in the Chaco-Pampas plain have provided an invaluable landscape experiment while bringing at the same time wicked environmental degradation problems. Theory-guided restoration initiatives such as forest strip conservation in the Chaco and Espinal or biodrainage with tree plantation in the Pampas particularly focused on degraded and degrading zones could provide a new generation of experiments. This type of research in the Chaco-Pampas plain will complement what has been learned about salinization in a similar context of management-science interaction in other dry plains of the world (Allan 2008; Jobbágy et al. 2008; Allan and Curtis 2003; Marchesini et al. 2017).

References

Allan C (2008) Can adaptive management help us embrace the Murray-Darling Basin's wicked problems? In: Pahl-Wostl C, Kabat P, Möltgen J (eds) Adaptive and integrated water management: coping with complexity and uncertainty. Springer, Berlin, pp 61–73

Allan C, Curtis A (2003) Regional scale adaptive management: lessons from the north east salinity strategy (NESS). Australas J Environ Manag 10:76–84. https://doi.org/10.1080/14486563.2003.10648576

Amdan ML, Aragón R, Jobbágy EG, Volante JN, Paruelo JM (2013) Onset of deep drainage and salt mobilization following forest clearing and cultivation in the Chaco plains (Argentina). Water Resour Res 49:6601–6612. https://doi.org/10.1002/wrcr.20516

Aragón R, Jobbágy E, Viglizzo E (2011) Surface and groundwater dynamics in the sedimentary plains of the Western Pampas (Argentina). Ecohydrology 4(3):433–447

Baldi G, Paruelo JM (2008) Land-use and land cover dynamics in South American Temperate grasslands. Ecol Soc 13(2):6

Batista W, León R (1992) Asociación entre comunidades vegetales y algunas propiedades del suelo en el centro de la Depresión del Salado. Ecología Austral 2:47–55

Bogino SM, Jobbágy EG (2011) Climate and groundwater effects on the establishment, growth and death of *Prosopis caldenia* trees in the Pampas (Argentina). For Ecol Manage 262(9):1766–1774

Buchhorn M, Smets B, Bertels L, Lesiv M, Tsendbazar N-E, Herold M, Fritz S (2019) Copernicus global land service: land cover 100m: epoch 2015: Globe (Version V2.0.2). Tim Jacobs. https://doi.org/10.5281/zenodo.3243508

Bui EN (2013) Soil salinity: a neglected factor in plant ecology and biogeography. J Arid Environ 92:14–25. https://doi.org/10.1016/j.jaridenv.2012.12.014

Butcher K, Wick AF, DeSutter T, Chatterjee A, Harmon J (2016) Soil salinity: a threat to global food security. Agron J 108(6):2189–2200. https://doi.org/10.2134/agronj2016.06.0368

Callow JN, Hipsey MR, Vogwill RIJ (2019) Surface water as a cause of land degradation from dryland salinity. Hydrol Earth Syst Sci Dis 1–32. https://doi.org/10.5194/hess-2019-405

Cisneros JM, Cantero JJ, Cantero A (1999) Vegetation, soil hydrophysical properties, and grazing relationships in saline-sodic soils of Central Argentina. Can J Soil Sci 79:399–409. https://doi.org/10.4141/S98-055

Contreras S, Santoni CS, Jobbágy EG (2013) Abrupt watercourse formation in a semiarid sedimentary landscape of central Argentina: the roles of forest clearing, rainfall variability and seismic activity. Ecohydrology 6(5):794–805

Downes RG (1954) Cyclic salt as a dominant factor in the genesis of soils in south-eastern Australia. Aust J Agric Res 5:448–464. https://doi.org/10.1071/AR9540448

Duncan RA, Bethune MG, Thayalakumaran T, Christen EW, McMahon TA (2008) Management of salt mobilisation in the irrigated landscape—a review of selected irrigation regions. J Hydrol 351(1):238–252

Echegoyen C, Lecomte K, Campodonico V, Yaciuk P, Jobbágy E, Heider G, Pasquini (2018) Uso de Radón-222 para determinar el flujo de agua en una laguna freática de la llanura medanosa de San Luis. In: ALHSUD (ed) XIV Congreso Latinoamericano de Hidrogeología Salta, Argentina

Edmunds WM, Gaye CB (1994) Estimating the spatial variability of groundwater recharge in the Sahel using chloride. J Hydrol 156(1):47–59

Engel V, Jobbágy EG, Stieglitz M, Williams M, Jackson RB (2005) Hydrological consequences of Eucalyptus afforestation in the Argentine Pampas. Water Resour Res 41. https://doi.org/10.1029/2004WR003761

Fan Y, Li H, Miguez-Macho G (2013) Global patterns of groundwater table depth. Science 339:940–943. https://doi.org/10.1126/science.1229881

Fan Y, Miguez-Macho G, Jobbágy EG, Jackson RB, Otero-Casal C (2017) Hydrologic regulation of plant rooting depth. Proc Natl Acad Sci USA 114:10572–10577. https://doi.org/10.1073/pnas.1712381114

Feldman SR, Bisaro V, Biani NB, Prado DE (2008) Soil salinity determines the relative abundance of C3/C4 species in Argentinean grasslands. Glob Ecol Biogeogr 17:708–714. https://doi.org/10.1111/j.1466-8238.2008.00403.x

Furquim SAC, Graham RC, Barbiero L, Queiroz Neto JP, Vidal-Torrado P (2010) Soil mineral genesis and distribution in a saline lake landscape of the Pantanal Wetland, Brazil. Geoderma 154:518–528

Gallardo AH, Giaccardi A, Aguilera D, Battistino YN, Martinez A (2017) Gully erosion and streams development in the río quinto basin, midwest Argentina. Revista de la Sociedad Geológica de España 30:21–30

García PE, Badano ND, Menéndez AN, Bert F, García G, Podestá G, Rovere S, Verdin A, Rajagopalan B, Arora P (2018a) Influencia de los cambios en el uso del suelo y la precipitación sobre la dinámica hídrica de una cuenca de llanura extensa. Caso de estudio: Cuenca del Río Salado, Buenos Aires, Argentina. Ribagua 5:92–106. https://doi.org/10.1080/23863781.2018.1495990

García PE, Menéndez AN, Podestá G, Bert F, Arora P, Jobbágy E (2018b) Land use as possible strategy for managing water table depth in flat basins with shallow groundwater. Int J River Basin Manag 16:79–92. https://doi.org/10.1080/15715124.2017.1378223

García GA, García PE, Rovere SL, Bert FE, Schmidt F, Menéndez ÁN, Nosetto MD, Verdin A, Rajagopalan B, Arora P, Podestá GP (2019) A linked modelling framework to explore interactions among climate, soil water, and land use decisions in the Argentine Pampas. Environ Modell Softw 111:459–471. https://doi.org/10.1016/j.envsoft.2018.10.013

GeoINTA (2020) INTA Argentina, Buenos Aires. https://www.geointa.inta.gob.ar/

George SJ, Harper RJ, Hobbs RJ, Tibbett M (2012) A sustainable agricultural landscape for Australia: a review of interlacing carbon sequestration, biodiversity and salinity management in agroforestry systems. Agric Ecosys Environ 163:28–36. https://doi.org/10.1016/j.agee.2012.06.022

Giménez R (2016) Impacto hidrológico de distintas estrategias agrícolas en el Chaco semiárido. University of Buenos Aires, Doctoral Dissertation, Faculty of Agronomy, University of Buenos Aires

Giménez R, Mercau J, Nosetto M, Páez R, Jobbágy E (2016) The ecohydrological imprint of deforestation in the semiarid Chaco: insights from the last forest remnants of a highly cultivated landscape. Hydrol Process 30:2603–2616. https://doi.org/10.1002/hyp.10901

Hillel D (2000) Salinity management for sustainable irrigation. Salinity Manage Sustain Irrig. https://doi.org/10.1596/0-8213-4773-x

Houspanossian J, Giménez R, Baldi G, Nosetto M (2016) Is aridity restricting deforestation and land uses in the South American Dry Chaco? J Land Use Sci 11:369–383. https://doi.org/10.1080/1747423X.2015.1136707

Ibrahim M, Favreau G, Scanlon BR, Seidel JL, Le Coz M, Demarty J, Cappelaere B (2014) Augmentation sur le long terme de la recharge diffuse des aquifères suite à l'expension des cultures pluviales dans le Sahel, Afrique de l'Ouest. Hydrogeol J 22:1293–1305. https://doi.org/10.1007/s10040-014-1143-z

Jayawickreme DH, Santoni CS, Kim JH, Jobbágy EG, Jackson RB (2011) Changes in hydrology and salinity accompanying a century of agricultural conversion in Argentina. Ecol Applic 21:2367–2379. https://doi.org/10.1890/10-2086.1

Jayawickreme DH, Jobbágy EG, Jackson RB (2014) Geophysical subsurface imaging for ecological applications. New Phytol 201(4):1170–1175. https://doi.org/10.1111/nph.12619
Jobbágy E, Jackson R (2004) Groundwater use and salinization with grassland afforestation. Glob Chang Biol 10 (1299–1312)
Jobbágy EG, Jackson RB (2007) Groundwater and soil chemical changes under phreatophytic tree plantations. J Geophys Res: Biogeosci 112. https://doi.org/10.1029/2006JG000246
Jobbágy EG, Nosetto MD, Santoni CS, Baldi G (2008) El desafío ecohidrológico de las transiciones entre sistemas leñosos y herbáceos en la llanura chaco-pampeana. Ecología Austral 18:305–322
Jobbágy EG, Nosetto MD, Villagra PE, Jackson RB (2011) Water subsidies from mountains to deserts: their role in sustaining groundwater-fed oases in a sandy landscape. Ecol Applic 21:678–694. https://doi.org/10.1890/09-1427.1
Jobbágy EG, Tóth T, Nosetto MD, Earman S (2017) On the fundamental causes of high environmental alkalinity (pH $\geq$ 9): an assessment of its drivers and global distribution. Land Degrad Dev 28:1973–1981. https://doi.org/10.1002/ldr.2718
Jobbágy E, Nosetto MD, Gimenez R, Mercau GL (2019) El (dis)servicio de la regulación hídrica en la llanura chaco-pampeana. In: Paruelo J, Laterra P (eds) El Lugar de la naturaleza en la toma de decisiones vol B5. CICCUS. pp 212–221
Jolly ID, Narayan KA, Armstrong D, Walker GR (1998) The impact of flooding on modelling salt transport processes to streams. Environ Modell Softw 13:87–104. https://doi.org/10.1016/S1364-8152(98)00003-6
Kim JH (2011) Trading carbon and water through vegetation shifts. Duke University, Durham
Lavado RS, Taboada MA (1988) Water, salt and sodium dynamics in a Natraquoll in Argentina. CATENA 15:577–594. https://doi.org/10.1016/0341-8162(88)90008-2
Lewis JP (1991) Three levels of floristical variation in the forests of Chaco, Argentina. J Veg Sci 2:125–130. https://doi.org/10.2307/3235905
Mahjoory RA (1979) The nature and genesis of some salt-affected soils in Iran. Soil Sci Soc Amer J 44:1019–1024
Marchesini VA, Fernández RJ, Jobbágy EG (2013) Salt leaching leads to drier soils in disturbed semiarid woodlands of central Argentina. Oecologia 171:1003–1012. https://doi.org/10.1007/s00442-012-2457-y
Marchesini VA, Giménez R, Nosetto MD, Jobbágy EG (2017) Ecohydrological transformation in the Dry Chaco and the risk of dryland salinity: following Australia's footsteps? Ecohydrology 10:e1822. https://doi.org/10.1002/eco.1822
Mujica CR, Milione GM, Bea SA, Jobbágy EG (2019) Modeling soil chemical changes induced by afforestation of natural grasslands in large plain ecosystems. Ecologia Austral 29:433–445
Nosetto MD, Jobbágy EG, Tóth T, Jackson RB (2008) Regional patterns and controls of ecosystem salinization with grassland afforestation along a rainfall gradient. Glob Biogeochem Cycles 22. https://doi.org/10.1029/2007GB003000
Nosetto MD, Jobbágy EG, Jackson RB, Sznaider GA (2009) Reciprocal influence of crops and shallow ground water in sandy landscapes of the Inland Pampas. Field Crops Res 113:138–148. https://doi.org/10.1016/j.fcr.2009.04.016
Nosetto MD, Acosta AM, Jayawickreme DH, Ballesteros SI, Jackson RB, Jobbágy EG (2013) Land-use and topography shape soil and groundwater salinity in central argentina. Agric Water Manage 129:120–129. https://doi.org/10.1016/j.agwat.2013.07.017
Olson DM, Dinerstein E, Wikramanayake ED, Burgess ND, Powell GVN, Underwood EC, D'amico JA, Itoua I, Strand HE, Morrison JC, Loucks CJ, Allnutt TF, Ricketts TH, Kura Y, Lamoreux JF, Wettengel WW, Hedao P, Kassem KR (2001) Terrestrial ecoregions of the world: a new map of life on earth: a new global map of terrestrial ecoregions provides an innovative tool for conserving biodiversity. BioScience 51(11):933–938. 10.1641/0006-3568(2001)051[0933:teotwa]2.0.co;2
Peck AJ (1978) Salinization of non-irrigated soils and associated streams: a review. Aust J Soil Res 16:157–168. https://doi.org/10.1071/SR9780157
Perelman SB, Leon RJC, Oesterheld M (2001) Cross-scale vegetation patterns of flooding pampa grasslands. J Ecol 89:562–577. https://doi.org/10.1046/j.0022-0477.2001.00579.x

Qadir M, Oster JD (2004) Crop and irrigation management strategies for saline-sodic soils and waters aimed at environmentally sustainable agriculture. Sci Total Environt 323(1):1–19. https://doi.org/10.1016/j.scitotenv.2003.10.012

Quiros R, Drago E (1999) The environmental state of Argentinean lakes: an overview. Lakes Reservoirs: Res Manage 4:55–64. https://doi.org/10.1046/j.1440-1770.1999.00076.x

Ridley AM, White RE, Simpson RJ, Callinan L (1997) Water use and drainage under phalaris, cocksfoot, and annual ryegrass pastures. Austr J Agric Res 48:1011–1023

Ridley AM, Christy B, Dunin FX, Haines PJ, Wilson KF, Ellington A (2001) Lucerne in crop rotations on the Riverine Plains 1. The soil water balance. Austr J Agric Res 52:263–277. https://doi.org/10.1071/AR99165

Salama RB, Laslett D, Farrington P (1993) Predictive modelling of management options for the control of dryland salinity in a first-order catchment in the wheatbelt of Western Australia. J Hydrol 145:19–40. https://doi.org/10.1016/0022-1694(93)90218-X

Santoni CS, Jobbágy EG, Contreras S (2010) Vadose zone transport in dry forests of central Argentina: Role of land use. Water Resour Res 46:2009WR008784. https://doi.org/10.1029/2009WR008784

Sayago JM, Collantes MM, Karlson A, Sanabria J (2001) Genesis and distribution of the Late Pleistocene and Holocene loess of Argentina: a regional approximation. Quat Int 76–77:247–257. https://doi.org/10.1016/S1040-6182(00)00107-5

Scanlon BR, Reedy RC, Stonestrom DA, Prudic DE, Dennehy KF (2005) Impact of land use and land cover change on groundwater recharge and quality in the southwestern US. Glob Chang Biol 11:1577–1593. https://doi.org/10.1111/j.1365-2486.2005.01026.x

Scanlon BR, Keese KE, Flint AL, Flint LE, Gaye CB, Edmunds WM, Simmers I (2006) Global synthesis of groundwater recharge in semiarid and arid regions. Hydrol Process 20(15):3335–3370

Schenk HJ, Jackson RB (2002) Rooting depths, lateral root spreads and below-ground/above-ground allometries of plants in water-limited ecosystems. J Ecol 90:480–494. https://doi.org/10.1046/j.1365-2745.2002.00682.x

Schofield RV, Kirkby MJ (2003) Application of salinization indicators and initial development of potential global soil salinization scenario under climatic change. Glob Biogeochemi Cycles 17:n/a-n/a. https://doi.org/10.1029/2002gb001935

Setia R, Gottschalk P, Smith P, Marschner P, Baldock J, Setia D, Smith J (2013) Soil salinity decreases global soil organic carbon stocks. Sci Total Environ 465:267–272. https://doi.org/10.1016/j.scitotenv.2012.08.028

Sidhu PS, Sharma PK, Bajwa MS (1991) Characteristics, distribution and genesis of salt-affected soils in Punjab. J Indian Society of Remote Sensing 19:269–276. https://doi.org/10.1007/BF03023974

Tóth J (1963) A theoretical analysis of groundwater flow in small drainage basins. J Geophys Res 68:4795–4812. https://doi.org/10.1029/jz068i016p04795

Volante JN, Mosciaro MJ, Gavier-Pizarro GI, Paruelo JM (2016) Agricultural expansion in the Semiarid Chaco: poorly selective contagious advance. Land Use Policy 55:154–165. https://doi.org/10.1016/j.landusepol.2016.03.025

Walvoord MA, Plummer MA, Phillips FM, Wolfsberg AV (2002) Deep arid system hydrodynamics 1. Equilibrium states and response times in thick desert vadose zones. Water Resour Res 38:44–41. https://doi.org/10.1029/2001wr000824

Strategies for the Use of Brackish Water for Crop Production in Northeastern Brazil

Claudivan Feitosa de Lacerda, Hans Raj Gheyi, José Francismar de Medeiros, Raimundo Nonato Távora Costa, Geocleber Gomes de Sousa, and Geovani Soares de Lima

Abstract Agriculture in the Brazilian Northeastern region, particularly in semi-arid climates, depends on scarce water resources. This has promoted interest in the use of marginal quality waters for crop production. In Northeastern Brazil, the occurrence of brackish and saline water sources is common and their use in agriculture without adequate management can reduce crop productivity and negatively impact soil properties. Long-term strategies that ensure the socioeconomic and environmental sustainability of agricultural systems are required. We present an overview of water sources in Northeastern Brazil and several management strategies evaluated in this region, including: salt tolerant crops, soil and water management, appropriate cropping systems, application of mineral and organic amendments, and plant x microorganism interactions. Two case studies on the application of on farm-strategies are presented, highlighting the technical feasibility of using brackish water in irrigated agriculture. These results also underscore the need to use management strategies, which allow the use of these water sources with less impact on crops and soils. We conclude that the use of marginal quality water is an important alternative

C. F. de Lacerda (✉) · R. N. T. Costa
Universidade Federal do Ceará, Fortaleza, Ceará, Brazil
e-mail: cfeitosa@ufc.br

R. N. T. Costa
e-mail: rntcosta@ufc.br

H. R. Gheyi
Universidade Federal de Recôncavo da Bahia, Cruz das Almas, Bahia, Brazil
e-mail: hgheyi@gmail.com

J. F. de Medeiros
Universidade Federal Rural do Semi-Árido, Mossoró, Rio Grande do Norte, Brazil
e-mail: jfmedeir@ufersa.edu.br

G. G. de Sousa
Universidade da Integração Internacional da Lusofonia Afro-Brasileira, Redenção, Ceará, Brazil
e-mail: sousagg@unilab.edu.br

G. S. de Lima
Universidade Federal de Campina Grande, Pombal, Paraíba, Brazil
e-mail: geovanisoareslima@gmail.com

E. Taleisnik and R. S. Lavado (eds.), *Saline and Alkaline Soils in Latin America*, https://doi.org/10.1007/978-3-030-52592-7_4

for agricultural production in Northeastern Brazil. The simultaneous application of several strategies may be the most convenient approach.

Keywords Brazilian semi-arid · Irrigation · Salinity · Water management · Crop yield

1 Introduction

Irrigated agriculture in Northeastern Brazil, especially in semi-arid areas, has contributed to strengthening other aspects of the economy through diversified cropping practices, stimulation of agroindustry and export of products. However, the high requirement of water for agriculture and the scarcity of good quality water to meet the growing demands of the population (consumption, industries, among others) has even made several agricultural enterprises unfeasible. This situation has promoted interest in the use of lower quality waters, including brackish waters and wastewaters, as well as in the multiple use of water sources in agricultural activities (Seckler et al. 1998; Ayers and Westcot 1999; Beltrán 1999; Rebouças et al. 2013; Andrade et al. 2018b).

Water sources can be classified as fresh (up to 0.05% salts), brackish (0.05–3.0% salts) and saline (above 3.0% salts). In Northeastern Brazil, the occurrence of groundwater and surface water with salinity problems is common (Medeiros 1992), most of them classified as brackish, which limits their use for irrigation and also for other purposes. On the other hand, in small and medium-sized dams, in shallow wells and in riverbeds and streams, there is considerable variation in discharge rates throughout the year, with the highest values of salinity at the end of the dry season. These variations may become more pronounced, including in medium-sized and large reservoirs, in periods of water crisis caused by long periods of drought, such as those observed between 2012 and 2016 (Marengo et al. 2017).

The use of lower quality water sources in agriculture, notably of brackish waters, can reduce crop productivity and impact soil properties (Beltrán 1999; Medeiros et al. 2009; Pessoa et al. 2019). Therefore, irrigation with such waters depends on long-term strategies that ensure the socioeconomic and environmental sustainability of agricultural systems (Beltrán 1999). These strategies must contribute to improvements in chemical, physical and biological conditions of soil, reductions in the concentration and entry of salts into the root environment, reduction in the impacts on the plant, and increments in water use efficiency (Glenn et al. 1998; Sharma and Rao 1998; Malash et al. 2005; Murtaza et al. 2006; Alves et al. 2011; Lacerda et al. 2011b; Oliveira et al. 2016; Dias et al. 2016; Santos et al. 2016; Sousa et al. 2018a, b; Soares et al. 2019; Lira et al. 2020).

2 Water Sources in Northeastern Brazil

Waters for irrigation are usually of superficial origin (rivers, lakes and dams) or ground origin (deep and shallow wells) (Almeida 2010), and variations in their ion composition depend on the source, path followed, season of the year, degree of weathering of the basin rocks and type of soil on which they flow. The main ions in water are cations sodium, calcium and magnesium, and anions chloride, sulfate and bicarbonate. Potassium, nitrate and carbonate ions are also present in considerable concentrations in some cases. In addition, it is also possible to find other elements, such as lithium, silicon, bromide, iodine, fluorine, copper, cobalt, nickel, zinc, chromium, lead, manganese, molybdenum, as well as phosphates, and organic matter particles, associated with the pollution of water bodies. Boron is also found in waters, mainly originating from volcanic rocks, and it is considered harmful to the growth of most plants even at concentrations below 0.5 mg L^{-1} in irrigation water.

The composition of the waters is directly related to the nature of the local substrate, specifically to the nature of the rock and type of soil, and to its type of exploitation, so that groundwaters are notably more concentrated than surface waters (Leprun 1983). The waters of the Brazilian semi-arid region have low sulfate concentrations, except some groundwaters in sites where gypsiferous rocks occur (Medeiros et al. 2012). The waters with high salinity are chloride-sodic (Medeiros 1992; Silva Júnior et al. 1999; Neves et al. 2017).

Under the conditions of the region, water salinity varies most often in the following order: dams < rivers < shallow wells, with the following ion composition, $Na^{+} > Ca^{2+} > Mg^{2+} > K^{+}$ and $Cl > HCO_3^{-} > SO_4^{2-}$. Waters from dams and wells mostly present Ca^{2+}/Mg^{2+} ratio > 1, whereas for shallow wells and rivers that have electrical conductivity above 0.7 dS m^{1}, this ratio is < 1.0 (Leprun 1983). Table 1 presents the cation and anion composition of some waters in Northeastern Brazil.

3 Strategies for the Use of Saline Water in Agriculture

Strategies to address the problem of salinity in waters intended for irrigation must take into account water quality, soil conditions and plant salt tolerance. In general, the management strategies used for cultivation under salinity conditions can be divided into two groups, non-specific and specific. Non-specific strategies are used under various cultivation conditions and can increase land productivity under both saline and non-saline conditions. They include: application of organic matter, application of liquid biofertilizers (cattle and crab), use of chemical amendments (fertilizers and correctives), hydroponics, mycorrhization, foliar application of organic and inorganic substances, crop rotation, increased plant density, intercropping systems, soil coverage, among others. On the other hand, specific strategies are those that are directly related to the problem of salinity. These strategies include: use of tolerant and moderately tolerant crops, cultivation of halophytes, mixture of waters of different

Table 1 Cation and anion composition of water sources in the semi-arid region of Northeast Brazil

Source—Site	pH	EC	Cations ($mmol_c\ L^{-1}$)				Anions ($mmol_c\ L^{-1}$)			
		$dS\ m^{-1}$	Ca^{2+}	Mg^{2+}	Na^+	K^+	Cl^-	HCO_3^-	CO_3^-	SO_4^{2-}
Dam—Pau dos Ferros, RN	8.2	0.5	1.6	1.3	2.0	0.2	2.4	2.4	0.0	–
Assu dam—São Rafael, RN	8.1	0.3	1.0	0.8	1.0	0.1	1.6	1.6	0.0	–
São Gonçalo dam—Sousa, PB	7.1	0.2	1.0	0.8	0.6	0.2	0.6	1.8	0.0	–
São Francisco river—Petrolina, PE	–	0.1	0.4	0.2	0.2	0.0	0.4	0.6	0.0	0.2
Açu river—Ipanguaçu, RN	7.8	0.3	0.8	0.6	1.0	0.1	0.9	1.7	0.0	0.1
Pequeno river—Pombal, PB	7.7	0.3	0.9	0.5	1.2	0.1	1.0	1.8	0.0	0.1
Shallow well—Angicos, RN	7.6	4.8	9.0	12.1	28.0	0.3	43.7	3.1	0.3	1.1
Shallow well—Condado, PB	8.1	0.5	1.9	1.0	3.1	0.1	3.7	2.2	0.1	0.1
Jandaíra aquifer—Mossoró, RN	7.2	3.11	13.5	4.10	8.10	0.1	17.8	6.2	0.0	–
Jandaíra aquifer—Baraúnas, RN	6.7	1.5	8.5	3.6	4.7	0.1	6.1	9.6	0.0	1.1
Jandaíra aquifer – Baraúnas, RN	6.7	1.1	7.2	2.6	2.3	0.1	4.1	6.3	0.0	–
Mossoró river—Mossoró, RN	8.1	0.82	3.10	2.75	3.15	0.1	5.6	2.8	0.4	0.5
Açu sandstone aquifer—Mossoró, RN	7.7	0.60	2.8	1.1	2.3	0.4	2.2	3.8	0.0	–
Serra Grande aquifer—Parnaíba, PI	7.3	1.9	16.0	1.5	1.4	0.5	2.0	2.8	–	15.6
Irrigation canal—Pentecoste, CE	7.5	0.5	1.5	1.6	1.7	0.2	2.9	2.3	–	–

(continued)

Table 1 (continued)

Source—Site	pH	EC	Cations ($mmol_c\ L^{-1}$)				Anions ($mmol_c\ L^{-1}$)			
		$dS\ m^{-1}$	Ca^{2+}	Mg^{2+}	Na^{+}	K^{+}	Cl^{-}	HCO_3^{-}	CO_3^{-}	SO_4^{2-}
Amazonas well—Iguatu, CE	7.0	2.3	3.3	3.6	16.0	0.1	14.0	4.2	–	1.6
Desalination reject—Pentecoste, CE	7.1	8.0	6.0	5.0	50.0	0.7	90.0	1.8	–	–

Sources Medeiros J.F., Lacerda C.F. (personal communication)

salinities, cyclic use of water, use of saline waters at the phonological stages in which the crop has higher tolerance, bio-drainage, establishment of specific conditions for germination (pre-treatment of seeds, use of higher seed rate in sowing, high matric potential), among others.

3.1 Cultivation of Salt-Tolerant Crops in Northeastern Brazil

Plant species can be grouped into halophytes and glycophytes in relation to their responses to salinity. Halophytes are native to saline environments and obviously tolerate high concentrations of salts in the root environment. Glycophytes or non-halophytes include most cultivated species, whose growth is inhibited even at low levels of salinity (Greenway and Munns 1980). However, such distinction is not absolute, since glycophyte species range from tolerant to very sensitive (Maas and Hoffman 1977; Ayers and Westcot 1999; Vieira et al. 2005; Melo et al. 2006; Pereira et al. 2017; Oliveira et al. 2018a).

Some glycophytes that are tolerant or moderately tolerant to salinity have been studied in Northeastern Brazil, with promising results, such as: coconut, cotton, sorghum, melon, passion fruit, cowpea, sunflower, West Indian cherry, and some forage species (Dantas et al. 2002; Ferreira Neto el al. 2002; Terceiro Neto et al. 2013; Sousa et al. 2014a; Feitosa et al. 2016; Sá et al. 2019; Lira et al. 2020). Some tropical ornamental species irrigated with brackish waters also show promising results, according to initial studies carried out in the region (Neves et al. 2018; Oliveira et al 2018a). Some management strategies in crops of high socioeconomic importance and considered sensitive or moderately sensitive such as rice, maize and citrus, have also been attempted (Melo et al. 2006; Barbosa et al. 2012; Brito et al. 2014). For citrus species, a study carried out in Northeastern Brazil has sought to identify rootstocks tolerant to salinity, as well as to evaluate the tolerance of the graft–rootstock combinations in the different stages of plant development (Brito et al. 2014, 2017, 2018; Soares et al. 2015; Sousa et al. 2016).

Among the cultivated perennial species, coconut may be one of those with highest potential for cultivation under irrigation with brackish waters (Santos et al. 2020). Many studies with young and adult plants of this crop have demonstrated that water salinity of up to 5.0 dS m^{-1} does not cause significant effects on its growth and production (Ferreira Neto et al. 2002, 2007; Marinho et al. 2006; Silva et al. 2016; Lima et al. 2017; Santos et al. 2020). Such higher tolerance of coconut may be associated with its natural occurrence in coastal areas, where plants are constantly exposed to sea spray. Sandy soils and the total rainfall during the rainy season of these regions leach salts, a factor that should be considered when this crop is irrigated with brackish waters. In soils with high clay content, the continuous use of brackish waters in irrigation can cause economic and environmental damage in these perennial crops, especially when there is no drainage system installed in the area (Sousa et al. 2014a; Lira et al. 2018).

Under some circumstances, the cultivation of glycophytes is virtually unfeasible, particularly in areas where only highly saline water is available or where the water table is saline and shallow and soil permeability is low. In some of these cases, the use of cultural practices, such as deep plowing or subsoiling, which favor the infiltration of water in the profile and reduce salinity in the surface soil layer, can be recommended (Sousa et al. 2014a). In other cases, however, the best alternative is the cultivation of halophyte species. The potential use of these species for the production of energy, oil and fodder has been evaluated in some semi-arid regions of the world, both under high soil salinity conditions and under irrigation with highly saline water (Gomes et al. 2015; Bonilla et al. 2019).

The most studied halophyte is the saltbush, a common name given to plants of the genus *Atriplex*. This genus belongs to the *Chenopodiacea* family, which has more than 400 species distributed in several arid and semi-arid regions of the world, of which about 15% are of interest for animal production. *Atriplex nummularia* is one of the most important as forage (Porto et al. 2001; Melo et al. 2018). This species stands out for being able to produce and maintain an abundant biomass even in environments of high aridity and salinity, adapting very well to regions with rainfall around 100–250 mm $year^{-1}$. *Atriplex* plants can also prosper in saline areas or under irrigation with saline water, showing high forage yield with high protein content under these conditions. This species also behaves as a hyperaccumulator of Na^+, with potential use in the phytoextraction of this element from the soil (Leal et al. 2008, 2019). Qadir et al. (2007) report that phytoextraction is an efficient strategy to recover saline–sodic soils, with performance comparable to that of chemical correctives. Similarly, Souza et al. (2011) found a promising effect of *Atriplex nummularia* on a saline–sodic Entisol.

3.2 *Mixtures and Cyclic Use of Saline Waters*

A strategy widely used by Brazilian farmers is mixing waters of different qualities, which reduces the concentration of salts and increases the volume of water available for irrigation (Malash et al. 2005; Neves et al. 2015). The final quality of the water obtained with the mixture can be estimated by:

$$\mathrm{EC}_{wf} = \frac{(\mathrm{EC}_{w1}.V_{w1})}{(V_{w1+w2})} + \frac{(\mathrm{EC}_{w2}.V_{w2})}{(V_{w1+w2})}$$

where:

EC_{wf}	Final concentration of the mixture, dS m^{-1}
EC_{w1}	Electrical conductivity of the water 1, in dS m^{-1}
EC_{w2}	Electrical conductivity of the water 2, in dS m^{-1}
V_{w1}	Volume of water 1
V_{w2}	Volume of water 2

V_{w1+w2} Final volume of the mixture
V_{w1}/V_{w1+w2} Represents the proportion of water 1 (P_{w1})
V_{w2}/V_{w1+w2} Represents the proportion of water 2 (P_{w2}).

The equation can be rewritten as:

$$EC_{wf} = (EC_{w1}.P_{w1}) + (EC_{w2}.P_{w2}) \text{ or}$$
$$EC_{wf} = (EC_{w1}.[1 - P_{w2}] + (EC_{w2}.P_{w2}), \text{ where } P_{w1} + P_{w2} = 1.$$

Obviously the mixture of waters requires more than one available source of water. Alternatively, farmers may opt for cyclic use of water sources with different salt concentrations (Flowers et al. 2005; Terceiro Neto et al. 2013). This strategy reduces the impacts of salinity on soil and plant, besides increasing the efficiency in the use of good quality water. Barbosa et al (2012) found that the cyclic use of high- and low-salinity water in the irrigation of maize reduced soil salinity in the 0–30 cm layer and did not negatively influence crop yield. Similar results were obtained by Neves et al. (2015) with cowpea and by Terceiro Neto et al. (2013) with melon.

The use of different management strategies to guarantee the sustainable use of saline waters in irrigation seems to be something consensual among different researchers. However, economic viability also needs to be taken into account. A study conducted for three years in Pakistan sought to evaluate the beneficial effects of the use of gypsum, organic fertilizer and cyclic use of water with high and low salinity/sodicity in a crop rotation with wheat and cotton (Murtaza et al. 2006). It was found that irrigation with only saline/sodic water reduced crop yield and the treatments tested were efficient in reducing the impacts of water salinity on crop yield and on salt accumulation in the soil. However, the cyclic use of low- and high-salinity water had the best benefit/cost ratio, followed by the treatments using canal water (control), saline/sodic water + gypsum, saline/sodic water + organic fertilizer, and finally the treatment with only saline/sodic water.

3.3 *Use of Brackish Waters at the Stages of Higher Salt Tolerance*

An important aspect that should be considered when using water sources with different levels of salinity is that genotypes of the same species can respond differently to salinity effects at the various stages of their development. In addition, most cultivated species of economic importance are relatively sensitive to salinity in the seedling establishment stage and the early stage of growth (Sharma and Minhas 2005; Soares Filho et al. 2016). However, information about the most sensitive and tolerant stages for most crops is unknown, mainly under field conditions (Shannon and Grieve 1998; Neves et al. 2015). Moreover, it is unclear whether the effects observed are due to the sensitivity of the crop to salinity at a given stage or to the duration of the stage in which the plant was exposed to the saline substrate, or even whether this is due to the combination of these factors (Gheyi et al. 2005).

Results under greenhouse conditions show that sorghum, wheat and cowpea are more sensitive during the vegetative growth stage and early reproductive stage, less sensitive at the flowering stage and tolerant during grain filling (Shalhevet et al. 1995). Other studies have shown that the effects of salinity on the production of wheat, cotton and cowpea can be significantly reduced when irrigation with saline water begins after seedling establishment (Murtaza et al. 2006; Chauhan and Singh 2008; Neves et al. 2009a, 2015). Thus, it is possible to irrigate many annual crops with saline water during the less sensitive stages and use low-salinity water during the most sensitive stage.

Several studies have shown that salinity tolerance of melon, a crop of great economic importance for the Northeast region of Brazil, varies according to the development stage (Botía et al. 2005; Porto Filho et al. 2006), and the application of saline waters in the fruiting stage can improve fruit quality (Botía et al. 2005). The results demonstrate that the application of saline water during fruiting does not affect the marketable production of the crop, and improvements in fruit quality with increasing contents of total soluble solids are observed.

For cowpea, a crop of great socioeconomic relevance for the populations of the Brazilian semi-arid region, it was found that the use of saline water only at the stages of higher tolerance can limit the impacts of salinity on the environment and on plant development, increasing the water and nutrients use efficiencies (Lacerda et al. 2009). Plants that were continuously irrigated with saline water after germination showed reductions in biomass production and in water use efficiency. On the other hand, the application of saline water in periods of intense growth and at the flowering and fruiting stages did not affect vegetative growth and grain yield by plants compared to irrigation only with well water, keeping the water use efficiencies. These results demonstrate the importance of strategy of using water of different qualities, taking into account the tolerance of each stage of crop development (Murtaza et al. 2006).

3.4 *Use of Densely Planted Crops*

Although the reduction of leaf area in plants under salt stress conditions is an important mechanism for reducing water losses, it is not beneficial because the production of photoassimilates by whole plants depends on the interception of light energy, which depends on leaf area index. A study conducted by Assis Júnior et al. (2007) demonstrated that salinity reduces the vegetative growth more than the grain yield of cowpea, and each plant under this condition occupies an area smaller than that occupied by those irrigated with low-salinity water. As a consequence, plants under salt stress have mean photosynthesis rates higher than those of plants irrigated with low-salinity water. This is due to the shorter time of exposure and the shading of plants grown under salt stress (Lacerda et al. 2011a; Freitas et al. 2013) suggesting that, under this condition, cowpea can be cultivated using a smaller spacing (Gomes et al. 2011).

Maize and cowpea plants grown at different spacing and irrigated with low- and high-salinity water showed different photosynthetic rates as a function of salinity and leaf location on the plant (Gomes et al. 2011; Lacerda et al. 2011a). This is due to the fact that plants under salt stress have lower vegetative growth and lower leaf area index, which alters the distribution of radiation within the crop. In the apical leaves, photosynthesis is limited only by the salt stress caused by irrigation water salinity, since radiation did not vary between treatments (Lacerda et al. 2011a). On the other hand, the use of saline water in irrigation increased the amount of radiation intercepted by the basal leaves, increasing their photosynthetic rate and temperature (Lacerda et al. 2011b). Thus, the photosynthetically active radiation values varied between the leaves analyzed, as a function of the spacing and level of irrigation water salinity.

Accordingly, it would be possible to grow cowpea at a higher planting density under saline conditions, maintaining the leaf area index and the distribution of photosynthetically active radiation at adequate values for the photosynthetic process. Under these conditions, the reduction of planting spacing results in gains of yield and water use efficiency by cowpea plants, increasing the land use efficiency (Lacerda et al. 2011a). This is a very promising strategy for annual crops. It can be associated with crop rotation and other management strategies, besides being applicable under saline conditions (soil and water). However, it is necessary to know the responses under different levels of salinity; in the case of furrow irrigation, spacing may influence the accumulation of salts in the soil. Finally, its use requires an economic/financial analysis.

3.5 Crop Rotation and Intercropping

Crop rotation is beneficial for the improvement in physical, chemical and biological conditions of soil, control of weeds, as well as diseases and pests, replacement of organic remains, and protection of soil against the action of climatic agents. The management of organic matter through crop rotation, green manure and intercropping can promote better use of chemical fertilizers and enable reductions in costs of mineral nitrogen fertilization, since it increases soil biological activity.

The use of crop rotation may be another alternative for semi-arid regions, which face problems of salinity (Murtaza et al. 2006; Lacerda et al. 2011b; Sousa et al. 2014a; Feitosa et al. 2016). This strategy is very promising for annual crops, especially for soils with good drainage or in association with other strategies that favor leaching, and can contribute to increasing land use efficiency and soil conservation. In order to obtain better results with crop rotation using saline waters, the salt tolerant species should be cultivated during the dry season, especially when using waters of higher salinity.

With the use of crop rotation in semi-arid regions, it is possible to produce fodder or grain crops throughout the year, using saline water in the dry season and rainwater in the wet season, without significantly altering the environment (Murtaza et al. 2006; Bezerra et al. 2010a; Lacerda et al. 2011b; Neves et al. 2015). The data of Fig. 1 show that the practice of using saline water in irrigation during the dry season caused accumulation of sodium ions in the soil, and this effect was mostly reverted by the total rainfall (1000–1400 mm), in the coast of Ceará State, Brazil. However, under semi-arid climates, the total rainfall may be insufficient for leaching salts from the soil (Meireles et al. 2003). Nevertheless, a study conducted in Brazilian semi-arid showed a complete leaching of salt during the rainy season (Neves et al. 2015). But, when sodium-rich waters are used, chemical conditioners should be employed preventively in order to enhance the efficiency of the leaching of this ion and reduce the impacts on soil physical properties (Mitchell et al. 2000; Andrade et al. 2018b).

In India, the feasibility of using crop rotation was demonstrated with wheat, sorghum and millet, associated with the use of a subsurface drainage system (Sharma and Rao 1998). Studies conducted in Brazil, particularly in the State of Ceará, to evaluate cycles of crop rotation between cowpea and maize demonstrated that cultivation during the dry season using saline water in irrigation caused increments in the EC and exchangeable sodium percentage of soil; however, the values of EC measured after the cultivation of the rainy season decreased in all treatments (Lacerda et al. 2011b). The effect of soil leaching, due to the rainy season, was similar to that observed in another similar study with cotton and wheat, conducted in Pakistan (Murtaza et al. 2006).

Irrigation with saline water in the dry season also caused reductions in the yields of corn and cowpea; however, the accumulation of sodium and salts in the soil during

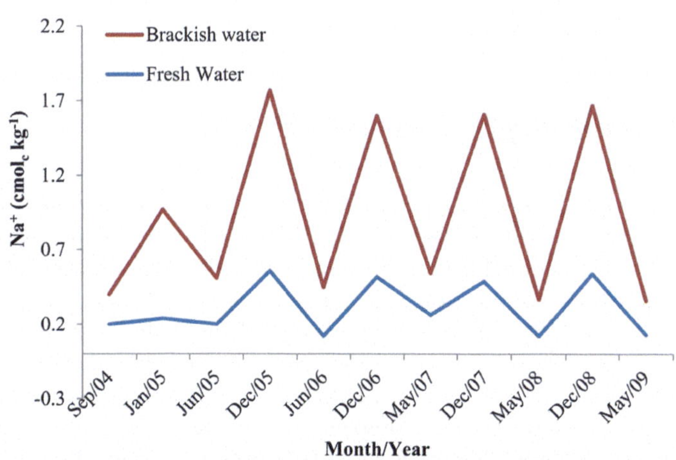

Fig. 1 Mean content of Na^+ in the 0–0.6 m soil layer from September 2004 to May 2009, in Fortaleza, Ceará, Brazil. During the dry seasons, crops were grown under furrow irrigation (sorghum and cowpea) with well water (0.8 dS m^{-1}) and saline water (ECw of 5.0 dS m^{-1}); during the rainy seasons, crops were grown under rainfed conditions. *Source* Lacerda CF (personal communication)

the cultivation in the dry season was not enough to cause significant effects on crop yield during the rainy season. According to Santos et al. (2005) and Assis Júnior et al. (2007), the surplus of irrigation water or rain leached excess salts from the soil profile, resulting in a reduced effect of salinity on the root environment, which favors the growth and development of the crop. Similar results have been observed in studies with rotation between cowpea and sunflower (Neves et al. 2015), sunflower and maize (Feitosa et al. 2016) and cotton and wheat (Murtaza et al. 2006). However, the results can be changed by environmental conditions, especially rains at the beginning of the cultivation in the rainy season (Feitosa et al. 2016).

Intercropping is the practice of growing two or more crops in the same area, during the same period. Due to this association, it is possible to verify changes in the microclimate and physiological responses of the plants involved, depending on the density, shading, leaf area, season, time of day and sampling point of the microclimatic elements. These changes in microclimate in intercropped systems may also alter the responses to some abiotic factors, including salinity. But, almost all of the research on plant responses to salinity is carried out under monocropping conditions.

In one of the few studies carried out in this area with maize/cowpea intercropping irrigated with brackish waters, Araújo (2015) demonstrated that the effects of salts on total biomass production and yield were more pronounced in plants under monocropping, notably in the cowpea plants. According to the author, the microclimatic condition of the intercropped system may have contributed to reducing the influence of salinity on yield, especially of cowpea, resulting in higher values of land use efficiency. It is important to highlight that studying plant tolerance in association (intercropping) represents a new approach in comparison to traditional research studies in the area (monocropping), highlighting the need for further studies to better understand plant responses.

3.6 Crops in Hydroponic Systems

The use of brackish water in hydroponic cultivation is an interesting strategy for small farmers of Northeastern Brazil (Soares et al. 2016), especially in the nutrient flow technique (NFT) system, which can be associated with other strategies mentioned in this chapter. The use of this technique minimizes the effects of applying brackish water directly to the soil, minimizing the negative effects on the environment. In this system, ions in nutrient solution are readily available, including that of potentially toxic elements, such as Na^+ and Cl^-. However, it is believed that the effects of salinity are less pronounced in a hydroponic environment, since there is no effect of the soil matric potential (Santos et al. 2010; Soares et al. 2016). In addition, part of the nutrients, such as Ca^{2+} and Mg^{2+}, are already contained in the water, which reduces fertilizer costs.

Experiments with hydroponic systems have been relatively promising, with results available for the several crops, particularly leafy vegetables such as lettuce

(Santos et al. 2010; Alves et al. 2011; Covas et al. 2017), coriander (Rebouças et al. 2013; Silva et al. 2018), arugula (Silva et al. 2012), melon (Dias et al. 2010), basil (Bione et al. 2014; Gondim Filho et al. 2018), pepper (Furtado et al., 2017; Santos et al. 2018) and ornamental sunflower (Maciel et al. 2014; Santos Junior et al. 2016). For most of these crops, a low-cost hydroponic module can be used, which reduces energy costs and can contribute to increase the small farmer's net income (Santos Junior et al. 2016).

3.7 Plant Response to Fertilization Under Saline Conditions

Conventional fertilization techniques based on the use of mineral fertilizers (Bezerra et al. 2019; Dias et al. 2019) and alternative forms (Sousa et al. 2014b; Mesquita et al. 2015; Andrade et al. 2018a; Oliveira et al. 2018b; Sousa et al. 2018a, b; Sá et al. 2019; Suddarth et al. 2019), have been frequently employed to increase salt tolerance of crops. Increased doses of fertilizers could result in higher tolerance to salinity, reducing the absorption of potentially toxic ions, such as Na^+ and Cl^-, and increasing the availability of essential nutrients such as K, Ca and N. However, regardless of the form of application, the ionic interactions that affect the availability, absorption and transport of nutrients are highly complex even in the absence of salinity and other stresses, which often justify the obtaining of conflicting results.

Salinity increases the level of complexity for the mineral nutrition of crops, affecting the activity of ions in solution and the processes of absorption, transport, assimilation and distribution. This complexity is explained by the differences in the concentration and ionic composition of the saline media to which plants are subjected, by the number of essential nutrients involved, and by the different responses in terms of salt tolerance and nutrient demand by plants. This generates a not always explained range of interactions. The different cultivation conditions, time of exposure to the stress and the type of tissue sampled, increase the difficulties to interpreting the results (Lacerda et al. 2006a).

The most likely results of the effects of salinity on mineral nutrition, besides reduction of growth, are changes in the plant quality, such as an increase in the degrees Brix of fruits, excessive accumulation of salts in forage plants, accumulation of nitrate and other nutrients, changes in fruit shape and size, and reduction in the concentrations of Ca^{2+} and K^+. On the other hand, the amount of fertilizers required by plants grown in saline soils may be lower than that required by plants grown in non-saline ones, due to the reduction in plant growth and yield (Neves et al. 2009b; Lacerda et al. 2016). This can lead to water table contamination, because a greater than normal amount of nutrients will be subjected to leaching (Lacerda et al. 2016, 2018).

According to most authors, salinity reduces plant growth due to osmotic, toxic and nutritional effects. However, the reduction in growth is initially affected by osmotic effects and subsequently by excessive accumulation of toxic ions (Munns 2002). Although it is obvious that the excess of certain ions, such as Na^+ and Cl^-,

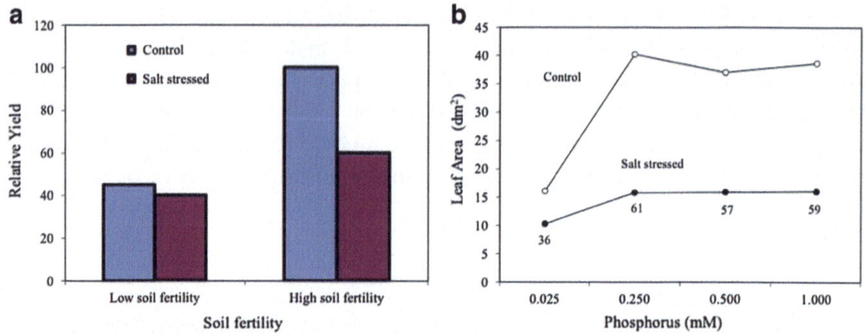

Fig. 2 **a** Hypothetical responses (Relative yield %) of plants as a function of salinity and soil fertility, **b** Leaf area of cowpea plants as a function of phosphorus concentration, in the absence (control) or presence of 72 mM NaCl. Values inside the figure indicate the percentage of reduction compared to the control treatment. Plants were grown in a greenhouse in pots containing 3 L of nutrient solution, in the absence (control) or presence of 72 mM NaCl (salt stressed). Plants were collected 20 days after the addition of the salts. Phosphorus was applied in the form of KH_2PO_4 and the concentration of K^+ was adjusted to 2.0 mM, using KCl. The data are means of four replicates

influences the acquisition of nutrients by the plant, it is not known whether changes in the contents of mineral nutrients contribute to the reduction in growth associated with salinity, or whether they are mere consequences of the reduction in growth.

Fertilization in plants irrigated with saline water increases plant growth, but this response depends on the salt level tested. In general, the degree of tolerance to salinity is not increased and positive responses are obvious only in soils poor in mineral nutrients (Grattan and Grieve 1998; Lacerda et al. 2003, 2016; Braz et al. 2019). Under low soil fertility conditions, plants are limited by nutritional stress, so the application of fertilizers favors their growth. However, the responses to fertilization usually tend to be higher in non-stressed plants (Fig. 2).

The high levels of nutrients in saline media and the maintenance of absorption could lead to nutrient accumulation in the tissues, resulting from an effect of concentration. This may lead to a lack of equilibrium between the acquisition and assimilation of a given nutrient, causing toxicity and intensifying the deleterious effects caused by salinity. It is possible, therefore, that the optimal level of some nutrients, such as phosphorus, in the absence of salts, may be toxic to some plants when grown in saline medium (Grattan and Grieve 1998; Lacerda et al. 2006a, 2006b; Sousa et al. 2010), that is, there may be an imbalance caused by excess, which can change the dose x response curve (Fig. 3). In addition, there may be a reduction in economic returns, since yield is reduced by the excess of salts, even when high doses of fertilizers are used (Feitosa et al. 2016).

Many studies demonstrate that plant responses to the application of fertilizers (mineral and organic) decrease with increased salinity, in line with the previous comment (Lacerda et al. 2006b, 2016). In other words, the use of mineral and organic fertilizers increases the availability of nutrients and the plant response will be higher

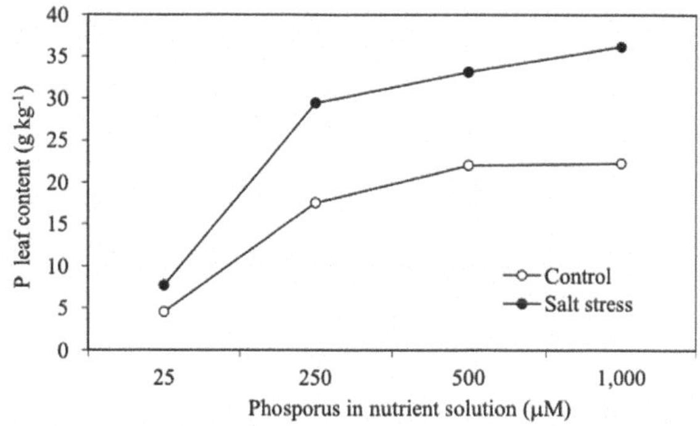

Fig. 3 Phosphorus contents in cowpea leaves as a function of P concentration, in the absence (control) or presence of 72 mM NaCl. For more information, see legend of Fig. 2b. *Source* Lacerda et al. (2006b)

the poorer the soil and will be lower at high levels of salinity (Fig. 4). Other roles attributed to organic fertilizers are yet to be scientifically proven.

3.8 *Plant and Microorganism Interactions*

Studies on the symbiotic association of plants with mycorrhizal or endophytic fungi show that inoculated plants have higher tolerance to various types of abiotic stresses, including salinity (Lúcio et al. 2013; Farias et al. 2019). Such higher tolerance is due to possible protection mechanisms provided by the fungi, including the greater absorption of nutrients, alteration in root morphology (higher number of adventitious roots) and the influence of arbuscular mycorrhizal fungi (AMF) on the EC of the soil (Giri et al. 2003; Bezerra et al. 2010b; Lúcio et al. 2013; Farias et al. 2019).

In a study with melon plants irrigated with saline water, it was found that the symbiotic association between AMF and melon plants promoted an increase in extraction of N, P and K, especially at the low and medium levels of salinity, and reduction in the absorption of potentially toxic ions (Na^+ and Cl^-) from irrigation water of 3 dS m^{-1} salinity (Lúcio et al. 2013). The association with AMF also promoted an increase

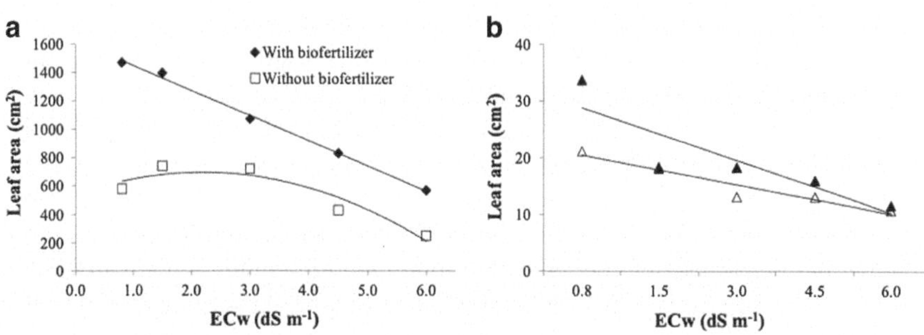

Fig. 4 **a** Leaf area of cowpea plants irrigated with different levels of salinity, in the absence and presence of bovine biofertilizer. **b** Leaf area of sunflower irrigated with saline water in soil with (▲) and without (Δ) bovine biofertilizer. *Sources* **a** Silva et al. (2011); **b** Gomes et al. (2015)

in vegetative growth and photosynthesis rate, but the beneficial effect of mycorrhiza decreased with the increase in salinity. Salinity reduced mycorrhizal colonization, especially at levels higher than 3 dS m^{-1}, which may justify, in part, the reduction in the efficiency of the association as salinity increased.

Despite the results that show that mycorrhizal and dark septate endophytic (DSE) fungi act as mitigating agents in plants under abiotic stress, examples of successful practical application in agriculture are lacking. This indicates the need for more research related to this topic.

3.9 Application of Several Compounds to Mitigate Saline Stress in Plants

Exogenous application of several substances has been used in the form of sprays, in nutrient solution or in the pre-treatment of seeds, in order to mitigate the effects of salinity on plants, such as: proline, glycinebetaine, ascorbic acid, kaolin-based films, hydrogen peroxide, salicylic acid, among others.

Proline and glycinebetaine are two solutes that accumulate in plants under salt stress and it is assumed that the accumulation of these substances in the cytoplasm can help in the osmotic balance of the cell, since the inorganic ions prevalent in saline environments tend to accumulate in vacuoles. They also protect enzymes and membranes. The rate of synthesis and accumulation of these substances varies widely among species. Ashraf and Foolad (2007) have tested the exogenous application of proline and glycinebetaine, as ways to reduce the effects of salt stress. These authors stressed that although many studies have shown success in the exogenous application of proline and glycinebetaine reducing the effects of abiotic stresses, they are not unanimous. For example, foliar spraying with proline at concentrations up to 20 mM did not reverse the effects of salinity on eggplant plants (Shahbaz et al. 2013). Lima et al. (2016) found that the foliar application of proline at concentrations up to 30 mM intensified the deleterious effects of 3.0 dS m^{-1} water salinity on the 'All Big' pepper production components. It appears that the positive effect is not as significant as the authors' claim, which may possibly compromise the practical application of these products by farmers, taking into account the cost–benefit ratio.

The use of hydrogen peroxide (H_2O_2) in seeds has resulted in the acceleration of the germination and initial establishment of seedlings (Silva et al. 2019c), which seems to give greater vigor to the seed when exposed to salt stress. Changes in the responses of the enzymes of the antioxidant protection system have been verified, contributing to increase the tolerance of seedlings to salt stress. On the other hand, spraying of maize plants with H_2O_2 induced acclimatization to salt stress, partially reversing the deleterious effects of salinity on growth (Gondim et al. 2011). Silva et al. (2019a) also observed that the pretreatment of soursop seeds with H_2O_2 at a concentration of 20 μM minimized the deleterious effects of irrigation water salinity on gas exchange and plant growth. The pre-treatment of plants with small amounts of H_2O_2 has been tested in a nutrient solution (Azevedo Neto et al. 2005; Carvalho et al. 2011), in the seed pre-treatment (Gondim et al. 2010; Silva et al. 2019b) and via leaf spray (Gondim et al. 2011, 2013; Andrade et al. 2019).

4 Sustainable Use of Saline Waters in Agriculture

4.1 Case Study 1: Melon Production Under Irrigation with Brackish Water in Mossoró, RN, Brazil

The Mossoró region is located at the Apodi Plateau covering an area about 5000 km^2, in the Rio Grande do Norte state and a small part of the Ceará state. In the Plateau area and in the valleys and tableland areas near rivers, there are large areas under irrigation and with potential for irrigation, with soils without problems of drainage. Groundwater and surface water resources have potential to irrigate more than 100,000 ha, without considering the water supply with the transfer from the São Francisco River.

Under the current conditions, due to the periods of prolonged droughts that occur cyclically, the perennial sources of water throughout and over the years are the groundwater from the Jandaíra limestone and Açu sandstone aquifers. These waters guarantee an area cultivated under irrigation in the region larger than 15,000 ha. Due to the climatic conditions, the region has specialized in melon cultivation, accounting for more than 75% of the national production and more than 95% of the melon exported by Brazil.

However, among the existing groundwater, that from the limestone aquifer has the highest potential for use, due to the larger exploitable volume of reserve and the low costs for obtaining it. The wells that exploit the Açu sandstone aquifer, except those on the edge of Plateau, are deeper, with higher costs of investment, maintenance and electricity consumption. In addition, the recharge potential of this aquifer is low, so it is constantly lowering, unlike the limestone aquifer.

In the lower areas of the region, which are closer to the coast, water availability has remained constant over the years, even in periods of prolonged droughts. However, the waters have higher levels of salinity (Alencar 2007; Vasconcelos et al. 2013), which has increased over the years with consecutive droughts. In this area, water salinity ranges from 2.5 to 6.0 dS m^{-1}, and this higher value was observed after a long period of drought between 2012 and 2017, associated with the greater exploitation of the waters of this aquifer. Figure 5 shows the fluctuation of water salinity in some wells of this region over the years.

With the reduction in the volume of water in the Jandaíra limestone aquifer after the last prolonged drought in the region, irrigated agriculture has been more concentrated where the waters from this aquifer have higher salt concentration. In the last years of drought, the salinity of these waters has increased significantly and affected the production of several crops, such as melon and watermelon. Consequently, medium and large producers began to invest in the construction of deep wells to explore the

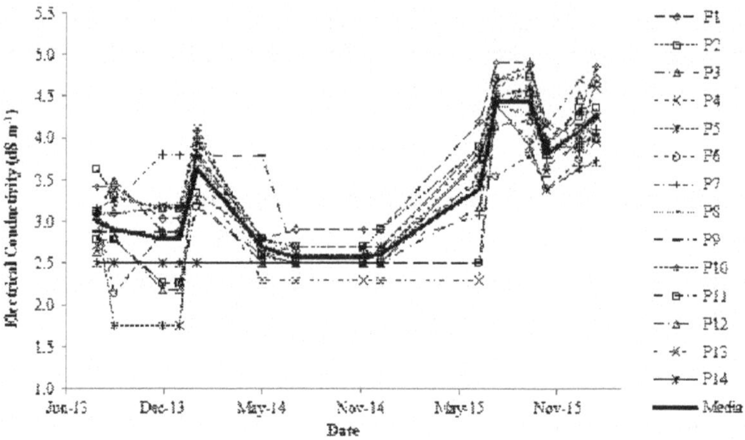

Fig. 5 Electrical conductivity of water from different wells (P1–P14) that exploit the Jandaíra limestone aquifer for melon production, in Mossoró, RN, between June/2013 and Feb/2016. *Source* Medeiros JF (personal communication)

waters from the Açu sandstone aquifer, which present low salinity and can be blended with brackish waters drawn from the limestone aquifer. However, due to the high cost of this water, it is necessary to optimize its use in order to have economic and environmental viability.

Studies carried out by Porto Filho et al. (2011) and Terceiro Neto et al. (2013) indicated that the use of water with higher salinity applied at the final stages of development reduces damage caused by salinity to melon. This management may be more important than mixing fresh water and brackish water in a certain proportion, and applying this mixture throughout the crop cycle. However, there was a need to further optimize the use of water from deeper wells, so tests were conducted in the region, with the objective of using the minimum amount of water from the sandstone aquifer during the crop cycle, without a significant loss in yield. Thus, over two years (2017 and 2018), experiments were conducted to study different strategies for the use of water from both aquifers, associated with strategies of irrigation management, aiming to reduce as much as possible the use of water from the sandstone aquifer.

Melons were cultivated for two successive years with combinations of salt concentrations (different mixtures) at various stages of the cycle: increase of salinity throughout the cycle at different intensities, use of the water of highest salinity from flowering with application of a leaching fraction using fresh water every week, use of blended water along the crop cycle, and the use of traditional farmer management. In the two years, irrigation was managed using different strategies: irrigation depths determined by the estimation of crop evapotranspiration, according to Allen et al. (2006), estimated from the soil moisture measurement performed daily, and using traditional farmer's management. Farm management commonly adopted for the crops, applied 309 mm water during the first crop cycle (74 days from sowing to harvest), 221 mm from the Açu sandstone well and 88 mm from the Jandaíra limestone well, that is, more than 70% of fresh water). When irrigation was adjusted according to the growth of the plant, 224 mm water were applied, using 145 to 57 mm of freshwater, representing 65–26% of the total irrigation.

The production in the first year for the water management practiced by the farm did not differ from other management performed, despite the substantial reduction in the total volume of water applied, mainly of freshwater. The mixture of waters applying only 26% of freshwater throughout the cycle, by climatological management or control based on soil moisture, represented 73 and 54% of the water applied by the farm management and did not differ from any other treatment studied. Marketable yields data obtained in the second year were on average 70% of those obtained in the first year, due to the high incidence of pests. Farm management applied 336 mm, a value well above those in the other management schemes. Treatments with management, however, resulted in melon yields equal to or higher than those obtained under the farm management.

This case study showed the feasibility of using brackish waters with electrical conductivity higher than 4.0 dS m^{-1} for melon production under semiarid Northeastern Brazil, employing the rational water management, even considering that the melon cultivar used, the most cultivated in Brazil, is one of the least tolerant to salt stress.

4.2 *Case Study 2: Reuse of Drainage Water in the Irrigated Perimeter of Curu-Pentecoste, Ceará, Brazil*

The Curu Pentecoste Irrigated Perimeter is located in the State of Ceará, Brazil, covering about 60 km from the North coast. The predominant soils are Entisols, with a medium to heavy texture. There are 175 farmers, with individual plots of 5.0 ha. The water for irrigation is supplied through the Federal Public dams General Sampaio and Pereira de Miranda, with water storage capacities of 322 and 360 hm^3, respectively. The water distribution in the plots is carried out by gravity through furrow irrigation, with water distribution by siphons in unlined tertiary channels. In this sense, the water conservation strategy within the Irrigated Perimeter using the reuse of excess water from furrow irrigation is an alternative to alleviate the shortage of surface water, especially during periods of drought (Macêdo et al. 2020).

4.2.1 Case Study with Papaya Crop

The study was conducted with the objective of evaluating the yield of papaya (*C. papaya* L.) as a function of irrigation levels, using surplus water from furrow irrigation (reuse) in a localized micro-sprinkler irrigation system. The soils of the area show a typically flat relief, of sandy loam texture. The water source was classified as C_3S_1, having a high risk of salinity (C_3) and slight risk of sodicity (S_1).

Four irrigation levels, corresponding to the replacements of 50, 100, 150 and 200% crop evapotranspiration (ETc), were tested. Water requirement was calculated based on the daily evaporation data measured in a Class A pan. Irrigation water productivity was obtained by the ratio between crop yield and the total volume of water applied along the crop cycle. The lowest mean yield of papaya was obtained in the treatment with 50% ETc, which received 1804.45 m^3 ha^{-1} of water, whereas the highest mean yield was obtained in the treatment with 150% ETc, which was associated with the application of 5413.37 m^3 ha^{-1} of water.

Irrigation water productivity data showed that in the treatment associated with the highest crop yield, it was possible to obtain 9.9 kg of papaya for every m^3 of water applied.

4.2.2 Case Study with Watermelon Crop

The study aimed to analyze the effect of water levels, potassium fertilization and the interaction between these two factors on the yield of watermelon (*Citrullus lanatus*), 'Crimson Sweet' variety, using a localized drip irrigation system with reuse of water from the surplus of furrow irrigation. The data of soil and water attributes were the same as those referenced in the experimentation unit with papaya. The irrigation treatments applied in the plots consisted of five water levels: 25, 75, 100, 150 and 200% of the maximum depth evapotranspired by the crop. The secondary treatments consisted of applying the following potassium doses: 0, 30, 60 and 120 kg ha^{-1} of K_2O, which were divided and applied at 15, 25 and 50 days after germination.

The maximum mean yield was 38.6 t ha^{-1}, obtained in the treatment corresponding to 150% of the evapotranspired depth and potassium dose of 60 kg ha^{-1}. These results demonstrate that both factors are limiting crop yield. Significant increases in watermelon production as a function of potassium fertilization have also been observed by Locascio and Hochmuth (2002). An increase in the mean values of water productivity with the increase in potassium doses, up to 60 kg ha^{-1}, and a reduction above this values, a result also observed by Monteiro et al. (2008) and Barros et al. (2002), evaluating the irrigation water productivity under various levels of fertilization.

The results demonstrate that the reuse of drainage water from the surplus of furrow irrigation in localized irrigation systems enabled increments in irrigation water productivity above 60% for papaya and watermelon crops, when compared to the mean values obtained in furrow irrigation. This can be extrapolated to other situations in the Brazilian semi-arid region, as a way to increase the water supply for irrigated agriculture. The level of water salinity was not a limiting factor for crop yield under this situation, possibly because it is water from surface flow, which typically results in lower accumulation of salts compared to soil drainage.

5 Conclusions

Agriculture is an activity that requires large volumes of water, particularly in semi-arid tropical regions. In this sense, the use of marginal quality water is an important alternative for agricultural production in Northeastern Brazil, as well as for industrial activities. The results presented in this chapter show the technical feasibility of using brackish water in irrigated agriculture. However, these same results demonstrate the need to use management strategies, which allow the use of these water sources with less impact on crops and soils. The simultaneous application of several strategies can be the best way to live with the problem, as demonstrated in the case studies carried out on farms. Certainly, the demonstration of environmental and economic advantages is an important way to increase the interest of farmers and governments in the use of brackish waters for crop production.

Acknowledgements Acknowledgments are due to the 'Conselho Nacional de Desenvolvimento Científico e Tecnológico (CNPq),' 'Instituto Nacional de Ciência e Tecnologia em Salinidade (INCTSal),' and 'Coordenação de Aperfeiçoamento de Pessoal de Nível Superior, Brazil, for the financial support for the research projects presented in this chapter.

References

Alencar RD (2007) Monitoramento da qualidade da água de poços no calcário Jandaíra e restrições na agricultura irrigada. Master Dissertation, Universidade Federal Rural do Semi-Árido

Allen RG, Pereira LS, Raes D, Smith J (2006) Evapotranspiración del cultivo: guías para la determinación de los requerimientos de agua de los cultivos. FAO, Rome

Almeida AO (2010) Qualidade da água de irrigação. Embrapa Mandioca e Fruticultura, Cruz das Almas

Alves MS, Soares TM, Silva LT, Joseane P, Fernandes JP, Oliveira MLA, Paz VPS (2011) Estratégias de uso de água salobra na produção de alface em hidroponia NFT. Rev Bras Eng Agríc Ambient 15:491–498. https://doi.org/10.1590/S1415-43662011000500009

Andrade EMG, Lima GS, Lima VLA, Gheyi HR, Silva AAR, Fernandes PD, Souza LP, Silva SS (2018a) Physiology and growth of cashew - Anão precoce (*Anacardium occidentale* L.) subjected to salt stress and organic fertilization. Aust J Crop Sci 12:1150–1158

Andrade JJ, Oliveira FJM, Pessoa LGM, Nascimento SAS, Souza ES, Barros Junior G, Miranda MFA, Oliveira AC, Freire MBGS (2018b) Effects of elemental sulfur associated with gypsum on soil salinity attenuation and sweet sorghum growth under saline water irrigation. Aust J Crop Sci 12:221–226. https://dx.doi.org/10.21475/ajcs.18.12.02.pne664

Andrade EMG, Lima GS, Lima VLA, Silva SS, Gheyi HR, Araújo AC, Gomes JP, Soares LAA (2019) Production and postharvest quality of yellow passion fruit cultivated with saline water and hydrogen peroxide. AIMS Agric Food 4:907–920. https://doi.org/10.3934/agrfood.2019.4.907

Araújo APB (2015) Influência da irrigação com água salina em sistema consorciado milho/feijão de corda. Ph.D. Thesis, Universidade Federal do Ceará

Ashraf M, Foolad MR (2007) Roles of glycine betaine and proline in improving plant abiotic stress resistance. Env Exp Bot 59:206–216. https://doi.org/10.1016/j.envexpbot.2005.12.006

Assis Júnior JO, Lacerda CF, Silva FB, Silva FLB, Bezerra MA, Gheyi HR (2007) Produtividade do feijão-de-corda e acúmulo de sais no solo em função da fração de lixiviação e da salinidade da água de irrigação. Eng Agríc 27:702–713. https://doi.org/10.1590/S0100-69162007000400013

Ayers RS, Westcot DW (1999) A qualidade da água na agricultura (Estudos FAO). UFPB, Campina Grande

Azevedo Neto AD, Prisco JT, Enéas-Filho J, Medeiros JR, Gomes-Filho E (2005) Hydrogen peroxide pre-treatment induces salt-stress acclimation in maize plants. J Plant Physiol 162:1114–1122. https://doi.org/10.1016/j.jplph.2005.01.007

Barbosa FS, Lacerda CF, Gheyi HR, Farias GC, Silva Júnior RJC, Lage YA, Hernandez FFF (2012) Productivity and íon content in maize irrigated with saline water in a continuous or alternating system. Cienc Rural 42:1731–1737. https://doi.org/10.1590/S0103-84782012001000003

Barros VS, Costa RNT, Aguiar JV (2002) Função de produção de cultura do melão para níveis de água e adubação nitrogenada no vale do Curu-CE. Irriga 7:98–105. https://doi.org/10.15809/irriga.2002v7n2p98-105

Beltrán JM (1999) Irrigation with saline water: benefits and environmental impact. Agric Water Manag 40:183–194. https://doi.org/10.1016/S0378-3774(98)00120-6

Bezerra AKP, Lacerda CF, Hernandez FFF, Silva FB, Gheyi HR (2010a) Rotação cultural feijão caupi/milho utilizando-se águas de salinidades diferentes. Cienc Rural 40:1075–1082. https://dx.doi.org/10.1590/S0103-84782010000500012

Bezerra MEJ, Lacerda CF, Sousa GG, Gomes VFF, Mendes Filho PF (2010b) Biomassa, atividade microbiana e FMA em rotação cultural milho/feijão-de-corda utilizando-se águas salinas. Rev Ciênc Agron 41:562–570. https://dx.doi.org/10.1590/S1806-66902010000400008

Bezerra MAF, Pereira WE, Bezerra FTC, Cavalcante LF, Medeiros SAS (2019) Nitrogen as mitigator of salt stress in yellow passion fruit seedlings. Semina: Ci Agrarias 40:611–622. https://dx.doi.org/10.5433/1679-0359.2019v40n2p611

Bione MAA, Paz VP, Silva F, Ribas RR, Soares TM (2014) Crescimento e produção de manjericão em sistema hidropônico NFT sob salinidade. Rev Bras Eng Agrí Ambient 18:1228–1234. https://doi.org/10.1590/1807-1929/agriambi.v18n12p1228-1234

Bonilla OH, Dias FYEC, Lucena EMP, Lacerda CF, Loiola MIB (2019) Comunidade halofítica herbáceo-arbustiva em Perímetro Irrigado do município de Pentecoste-CE. Rev Bras Geog Física 12:1934–1951. https://doi.org/10.26848/rbgf.v12.5.p1934-1951

Botía P, Navarro JM, Cerdá A, Martínez V (2005) Yield and fruit quality of two melon cultivars irrigated with saline water at different stages of development. European J Agron 23:243–253. https://doi.org/10.1016/j.eja.2004.11.003

Braz RS, Lacerda CF, Assis Júnior RN, Ferreira JFS, Oliveira AC, Ribeiro AA (2019) Growth and physiology of maize under water salinity and nitrogen fertilization in two soils. Rev Bras Eng Agríc Ambient 23:907–913. https://doi.org/10.1590/1807-1929/agriambi.v23n12p907-913

Brito MEB, Fernandes PD, Gheyi HR, Melo AS, Soares Filho WS, Santos RT (2014) Sensibilidade à salinidade de híbridos trifoliados e outros porta-enxertos de citros. Rev Caatinga 27:17–27

Brito MEB, Sá FVS, Silva LA, Soares Filho WS, Gheyi HR, Moreira RCL, Fernandes PD, Figueiredo LC (2017) Saline stress onto growth and physiology of trifoliate citrus hybrids during rootstock formation. Biosc J 33:1523–1534. https://doi.org/10.14393/BJ-v33n6a2017-37155

Brito MEB, Soares LAA, Soares Filho WS, Fernandes PD, Silva ECB, Sá FVS, Silva LA (2018) Emergence and morphophysiology of Sunki mandarin and other citrus genotypes seedlings under saline stress. Spanish J Agric Res 16:e0801–e0815. https://doi.org/10.5424/sjar/2018161-9400

Carvalho FEL, Lobo AKM, Bonifácio A, Martins MO, Lima Neto MC, Silveira JAG (2011) Aclimatação ao estresse salino em plantas de arroz induzida pelo pré-tratamento com H_2O_2. Rev Bras Eng Agríc Ambient 15:416–423. https://doi.org/10.1590/S1415-43662011000400014

Chauhan CPS, Singh RB (2008) Suplemental irrigation of wheat with saline water. Agric Water Manag 95:253–258. https://doi.org/10.1016/j.agwat.2007.10.007

Covas AMW, Freitas FTO, Viana PC, Rafael MRS, Azevedo Neto AD, Soares TM (2017) Teor de solutos inorgânicos em alface cultivada com água salobra em diferentes sistemas hidropônicos. Rev Bras Eng Agri Ambient 21:150–155. https://doi.org/10.1590/1807-1929/agriambi.v21n3p150-155

Dantas JP, Marinho FJL, Ferreira MMM, Amorim MSN, Andrade SIO, Sales AL (2002) Avaliação de genótipos de caupi sob salinidade. Rev Bras Eng Agríc Ambient 6:425–430. https://doi.org/10.1590/S1415-43662002000300008

Dias NS, Lira RB, Brito FR, Sousa Neto ON, Ferreira Neto M, Oliveira AM (2010) Produção de melão rendilhado em sistema hidropônico com rejeito da dessalinização de água em solução nutritiva. Rev Bras Eng Agrí Ambient 14:755–776. https://doi.org/10.1590/S1415-43662010000700011

Dias NS, Ferreira JFS, Liu X, Suarez D (2016) Jerusalem artichoke (*Helianthus tuberosus* L.) maintains high inulin, tuber yield, and antioxidant capacity under moderately-saline irrigation waters. Ind Crops Prod 94:1009–1024. https://doi.org/10.1016/j.indcrop.2016.09.029

Dias AS, Lima GS, Pinheiro FWA, Gheyi HR, Soares LAA, Nobre RG, Souza LP, Silva SS, Oliveira SG (2019) Gas exchanges and growth of sesame (*Sesamum indicum*, L.) cultivated under saline waters and nitrogen-potassium fertilizers. Aust J Crop Sci 13:1526–1532. https://doi.org/10.21475/ajcs.19.13.09.p1794

Farias GC, Nunes KG, Soares MA, Siqueira KA, Lima WC, Neves ALR, Lacerda CF, Gomes-Filho E (2019) Dark septate endophytic fungi mitigate the effects of salt stress on cowpea plants. Braz J Microbiol. https://doi.org/10.1007/s42770-019-00173-4

Feitosa HO, Lacerda CF, Carvalho CM, Costa RNT, Barbosa AB, Gheyi HR (2016) Productivity and economic analysis of sunflower/maize crop rotation under different levels of salinity and nitrogen. African J Agric Res 11:1999–2006. https://doi.org/10.5897/AJAR2016.1095

Ferreira Neto M, Gheyi HR, Holanda JS, Medeiros JF, Fernandes PD (2002) Qualidade do fruto verde de coqueiro em função da irrigação com água salina. Rev Bras Eng Agríc Ambient 6:69–75. https://doi.org/10.1590/S1415-43662002000100013

Ferreira Neto M, Gheyi HR, Fernandes PD, Holanda JS, Blanco FF (2007) Emissão foliar, relações iônicas e produção do coqueiro irrigado com água salina. Cienc Rural 37:1675–1681. https://doi.org/10.1590/S0103-84782007000600026

Flowers TJ, Ragab R, Malash N, Gawad GA, Cuartero J, Arslan A (2005) Sustainable strategies for irrigation in salt-prone Mediterranean: SALTMED. Agric Water Manag 78:3–14. https://doi.org/10.1016/j.agwat.2005.04.014

Freitas VS, Marques EC, Bezerra MA, Prisco JT, Gomes-Filho E (2013) Crescimento e acúmulo de íons em plantas de cajueiro anão precoce em diferentes tempos de exposição à salinidade. Semina: Ciências Agrárias 34:3341–3352. https://dx.doi.org/10.5433/1679-0359.2013v34n6Supl1p3341

Furtado GF, Cavalcante AR, Chaves LHG, Santos Júnior JA, Gheyi HR (2017) Growth and production of hydroponic pepper under salt stress and plant density. Am J Plant Sci 8:2255–2267. https://doi.org/10.4236/ajps.2017.89151

Gheyi HR, Correia KG, Fernandes PD (2005) Salinidade do solo e crescimento e desenvolvimento das plantas. In: Nogueira RJC, Araújo EL, Willadino LG, Cavalcante UMT (eds) Estresses ambientais: danos e benefícios em plantas. UFRPE, Recife, pp 138–148

Giri B, Kapoor R, Mukerji KG (2003) Influence of arbuscular mycorrhizal fungi and salinity on growth, biomass, and mineral nutrition of Acacia Acacia auriculiformis. Biol Fertil Soils 38:70–175. https://doi.org/10.1007/s00374-003-0636-z

Glenn E, Tanner R, Miyamoto S, Fitzsimmons K, Boyer J (1998) Water use, productivity and forage quality of the halophyte *Atriplex nummularia* grown on saline waste water in a desert environment. J Arid Environ 38:45–62. https://doi.org/10.1006/jare.1997.0320

Gomes KR, Amorim AV, Ferreira FJ, Andrade Filho FL, Lacerda CF, Gomes-Filho E (2011) Respostas de crescimento e fisiologia do milho submetido a estresse salino com diferentes espaçamentos de cultivo. Rev Bras Eng Agríc Ambient 15:365–370. https://doi.org/10.1590/S1415-43662011000400006

Gomes KR, Sousa GG, Lima FA, Viana TVA, Azevedo BM, Silva GL (2015) Irrigação com água salina na cultura do girassol (*Helianthus annuus* L.) em solo com biofertilizante bovino. Irriga 20:680–693. https://dx.doi.org/10.15809/irriga.2015v20n4

Gondim Filho H, Silva PCC, Silva MG, Pereira MM, Soares TM, Azevedo Neto AD, Vasconcelos RS, Gheyi HR (2018) Growth, production and essential oil content of basil genotypes in hydroponic conditions under salt stress. J Exp Agric Int 25:1–10. https://doi.org/10.9734/JEAI/2018/43023

Gondim FA, Gomes-Filho E, Lacerda CF, Prisco JT, Azevedo Neto AD, Marques EC (2010) Pretreatment with H_2O_2 in maize seeds: effects on germination and seedling acclimation to salt stress. Braz J Plant Physiol 22:103–112. https://doi.org/10.1590/S1677-04202010000200004

Gondim FA, Gomes-Filho E, Marques EC, Prisco JT (2011) Efeitos do H_2O_2 no crescimento e acúmulo de solutos em plantas de milho sob estresse salino. Rev Ciênc Agron 42:373–381. ttps://doi.org/10.1590/S1806–66902011000200016

Gondim FA, Miranda RS, Gomes-Filho E, Prisco JT (2013) Enhanced salt tolerance in maize plants induced by H_2O_2 leaf spraying is associated with improved gas exchange rather than with non-enzymatic antioxidant system. Theor Exp Plant Physiol 25:251–260. https://doi.org/10.1590/S2197-00252013000400003

Grattan SR, Grieve CM (1998) Salinity-mineral nutrient relations in horticultural crops. Sci Hort 78:127–157. https://doi.org/10.1016/S0304-4238(98)00192-7

Greenway H, Munns R (1980) Mechanisms of salt tolerance in nonhalophytes. Ann Rev Plant Physiol 31:149–190. https://doi.org/10.1146/annurev.pp.31.060180.001053

Lacerda CF (2005) Interação salinidade x nutrição mineral. In: Nogueira RJC, Araújo EL, Willadino LG, Cavalcante UMT (eds) Estresses ambientais: danos e benefícios em plantas. UFRPE, Recife, pp 127–137

Lacerda CF, Oliveira HPM, Oliveira TS, Gomes-Filho E (2003) Crescimento e acúmulo de íons em folhas de sorgo forrageiro submetido a soluções iso-osmóticas de sais (NaCl + KCl). Rev Ciênc Agron 34:7–12

Lacerda CF, Morais HMM, Prisco JT, Gomes-Filho E, Bezerra MA (2006b) Interação entre salinidade e fósforo em plantas de sorgo forrageiro. Rev Ciênc Agron 37:258–263

Lacerda CF, Neves ALR, Guimarães FVA, Silva FLB, Prisco JT, Gheyi HR (2009) Eficiência de utilização de água e nutrientes em plantas de feijão-de-corda irrigadas com água salina em diferentes estádios de desenvolvimento. Eng Agríc 29:221–230. https://dx.doi.org/10.1590/S0100-69162009000200005

Lacerda CF, Silva FB, Neves ALR, Silva FLB, Gheyi HR, Ness RLL, Gomes-Filho E (2011a) Influence of plant spacing and irrigation water quality on a cowpea-maize cropping system. Int Res J Agric Sc Soil Sci 1:163–167

Lacerda CF, Sousa GG, Silva FLB, Guimarães FVA, Silva GL, Cavalcante LF (2011b) Soil salinization and maize and cowpea yield in the crop rotation system using saline waters. Eng Agríc 31:663–675. https://dx.doi.org/10.1590/S0100-69162011000400005

Lacerda CF, Ferreira JFS, Liu X, Suarez DL (2016) Evapotranspiration as a criterion to estimate nitrogen requirement of maize under salt stress. J Agron Crop Sci 202:192–202. https://doi.org/10.1111/jac.12145

Lacerda CF, Ferreira JFS, Suarez DL, Freitas ED, Liu X, Ribeiro AA (2018) Evidence of nitrogen and potassium losses in soil columns cultivated with maize under salt stress. Rev Bras Eng Agríc Ambient 22:553–557. https://doi.org/10.1590/1807-1929/agriambi.v22n8p553-557

Leal IG, Accioly AMA, Nascimento CWA, Freire MBGS, Montenegro AAA, Ferreira FL (2008) Fitorremediação de solo salino-sódico por *Atriplex nummularia* e gesso de jazida. R Bras Ci Solo 32:1065–1072. https://doi.org/10.1590/S0100-06832008000300015

Leal LSG, Pessoa LGM, Oliveira JP, Santos NA, Silva LFS, Barros Júnior G, Freire MBGS, Souza ES (2019) Do applications of soil conditioner mixtures improve the salt extraction ability of *Atriplex nummularia* at early growth stage? Inter J Phytorem 21 https://doi.org/10.1080/15226514.2019.1678109

Leprun JC (1983) Primeira avaliação das águas superficiais do Nordeste: Relatório final do Convênio Manejo e Conservação do Solo do Nordeste Brasileiro. SUDENE/DRN, Recife, pp 91–141

Lima GS, Santos JB dos, Soares LAA, Gheyi HR, Nobre RG, Pereira RF (2016) Irrigação com águas salinas e aplicação de prolina foliar em cultivo de pimentão 'All Big'. Com Sci 7:513–522. https://doi.org/10.14295/CS.v7i4.1671

Lima BLC, Lacerda CF, Ferreira Neto M, Ferreira JFS, Bezerra AME, Marques EC (2017) Physiological and ionic changes in dwarf coconut seedlings irrigated with saline water. Rev Bras Eng Agríc Ambient 21:122–127. https://doi.org/10.1590/1807-1929/agriambi.v21n2p122-127

Lira RM, Silva EFF, Simoes Neto DE, Santos Junior JA, Lima BLC, Silva JS (2018) Growth and yield of sugarcane irrigated with brackish water and leaching fractions. Rev Bras Eng Agríc Ambient 22:170–175. https://doi.org/10.1590/1807-1929/agriambi.v22n3p170-175

Lira JB, Andrade AP, Magalhães ALR, Campos FS, Araújo GGL, Deon DS, Gois GC, Regitano Neto A, Cunha DS, Tabosa JN, Silva TGF, Nagahama HJ (2020) Production of pearl millet irrigated with different levels of brackish water and organic matter. Comm Soil Sci Plant Anal. https://doi.org/10.1080/00103624.2020.1729794

Locascio SJ, Hochmuth GJ (2002) Watermelon production as influenced by lime, gypsum, and potassium. Hort Sci 2:322–324. https://doi.org/10.21273/HORTSCI.37.2.322

Lúcio WS, Lacerda CF, Mendes Filho PF, Hernandez FFF, Neves ALR, Gomes-Filho E (2013) Crescimento e respostas fisiológicas do meloeiro inoculado com fungos micorrízicos arbusculares sob estresse salino. Semina: Ci Agrárias 34:1587–1602. https://doi.org/10.5433/1679-0359.2013v34n4p1587

Maas EV, Hoffman GJ (1977) Crop salt tolerance-current assessment. J Irrig Drain Div ASCE 103:115–134

Macêdo ABM, Costa RNT, Araújo DF, Nunes KG (2020) Water productivity with localized irrigation and groundwater and reuse water in the cultivation of plant species. Rev Bras Eng Agríc Ambient 24. https://dxdoi.org/10.1590/1807-1929/agriambi.v24n4pxxx-xxx

Maciel MP, Soares TM, Gheyi HR, Rezende EPL, Oliveira GXS (2012) Produção de girassol ornamental com uso de águas salobras em sistema hidropônico NFT. Rev Bras Eng Agrí Ambient 16:165–172. https://doi.org/10.1590/S1415-43662012000200006

Malash N, Flowers TJ, Ragab R (2005) Effect of irrigation systems and water management practices using saline and non-saline water on tomato production. Agric Water Manag 78:25–38. https://doi.org/10.1016/j.agwat.2005.04.016

Marengo JA, Torres RR, Alves LM (2017) Drought in Northeast Brazil—past, present, and future. Theor Appl Climatol 129:1189–1200. https://doi.org/10.1007/s00704-016-1840-8

Marinho FJL, Gheyi HR, Fernandes PD, Holanda JS, Ferreira Neto M (2006) Cultivo de coco 'Anão Verde' irrigado com águas salinas. Pesq Agropec Bras 41:1277–1284. https://doi.org/10.1590/S0100-204X2006000800010

Medeiros JF (1992) Qualidade da água de irrigação utilizada nas propriedades assistidas pelo "GAT" nos Estados do RN, PB, CE e avaliação da salinidade dos solos. Master Dissertation, Universidade Federal da Paraíba

Medeiros JF, Gheyi HR, Dias NS, Carmo GA (2009) Riscos e medidas mitigadoras para a salinidade em áreas irrigadas. In: Paz VPS, Oliveira AS, Pereira FAC, Gheyi HR (eds) Manejo e sustentabilidade da irrigação em regiões áridas e semi-áridas. UFRB, Cruz das Almas, pp 187–219

Medeiros JF, Gheyi HR, Nascimento IB (2012) Salinidade de solo e da água e seus efeitos na produção agrícola. In: Gheyi HR, Medeiros SS, Galvão CO (eds) Recursos hídricos em regiões semiáridas. INSA/UFRB, Cruz das Almas, pp 90–218

Meireles ACM, Andrade EM, Cruz MGM, Lemos Filho LCA (2003) Avaliação do impacto da fertirrigação em Cambissolos na Chapada do Apodi, Ceará. Rev Ciênc Agron 34:207–212

Melo PCS, Anunciação Filho CJ, Oliveira FJ, Bastos GQ, Tabosa JN, Santos VF, Melo MRCS (2006) Seleção de genótipos de arroz tolerantes à salinidade durante a fase vegetativa. Cienc Rural 36:58–64. https://doi.org/10.1590/S0103-84782006000100009

Melo HF, Souza ER, Almeida BG, Mulas M (2018) Water potential in soil and *Atriplex nummularia* (phytoremediator halophyte) under drought and salt stresses. Int J of Phytorem 20:249–255. https://doi.org/10.1080/15226514.2017.1374334

Mesquita FO, Cavalcante LF, Nunes JC, Luna Souto AG, Medeiros RF, Rodrigues RM (2015) Formação de mudas de nim com aplicação de biofertilizante bovino submetido à drenagem e estresse salino. Biosci J 31:47–54. https://doi.org/10.14393/BJ-v31n1a2015-18055

Mitchell JP, Shennan C, Singer MJ, Peters DW, Miller RO, Prichard T, Grattan SR, Rhoades JD, May DM, Munk DS (2000) Impacts of gypsum and winter cover crops on soil physical properties and crop productivity when irrigated with saline water. Agric Water Manag 45:55–71. https://doi.org/10.1016/S0378-3774(99)00070-0

Monteiro ROC, Costa RNT, Leão MCS, Aguiar JV (2008) Eficiência do uso da água e nitrogênio na produção de melão. Irriga 13:367–377. https://doi.org/10.15809/irriga.

Munns R (2002) Comparative physiology of salt and water stress. Plant Cell Environ 25:239–250. https://doi.org/10.1046/j.0016-8025.2001.00808.x

Murtaza G, Ghafoor A, Qadir M (2006) Irrigation and soil management strategies for using saline-sodic water in a cotton Cotton–wheat rotation. Agric Water Manag 81:98–114. https://doi.org/10.1016/j.agwat.2005.03.003

Neves ALR, Lacerda CF, Guimarães FVA, Hernandez FFF, Silva FB, Prisco JT, Gheyi HR (2009a) Trocas gasosas e teores de minerais no feijão-de-corda irrigado com água salina em diferentes estádios. Rev Bras Eng Agríc Ambient 13:873–881. https://dx.doi.org/10.1590/S1415-43662009000700009

Neves ALR, Lacerda CF, Guimarães FVA, Hernandez FFF, Silva FB, Prisco JT, Gheyi HR (2009b) Acumulação de biomassa e extração de nutrientes por plantas de feijão-de-corda irrigadas com água salina em diferentes estádios de desenvolvimento. Cienc Rural 39:758–765. https://dx.doi.org/10.1590/S0103-84782009005000014

Neves ALR, Lacerda CF, Sousa CHC, Silva FLB, Gheyi HR, Ferreira FJ, Andrade Filho FL (2015) Growth and yield of cowpea/sunflower crop rotation under different irrigation management strategies with saline water. Cienc Rural 45:814–820. https://doi.org/10.1590/0103-8478cr20131657

Neves ALR, Pereira MA, Lacerda CF, Gheyi HR (2017) Aspectos socioambientais e qualidade da água de dessalinizadores nas comunidades rurais de Pentecoste-CE. Rev Amb Água 12:124–135. https://doi.org/10.4136/ambi-agua.1722

Neves ALR, Lacerda CF, Oliveira AC, Sousa CHC, Oliveira FIF, Ribeiro MSS (2018) Quantitative and qualitative responses of *Catharanthus roseus* to salinity and biofertilizer. Rev Bras Eng Agric Ambient 22:22–26. https://doi.org/10.1590/1807-1929/agriambi.v22n1p22-26

Oliveira PGF, Moreira OC, Branco LMC, Costa RNT, Dias CN (2012) Eficiência de uso dos fatores de produção água e potássio na cultura da melancia irrigada com água de reuso. Rev Bras Eng Agríc Ambient 16:153–158. https://doi.org/10.1590/S1415-43662012000200004

Oliveira FA, Medeiros JF, Cunha RC, Souza MWL, Lima LA (2016) Uso de bioestimulante como agente amenizador do estresse salino na cultura do milho pipoca. Rev Ciênc Agron 47:307–315. https://doi.org/10.5935/1806-6690.20160036

Oliveira EV, Lacerda CF, Neves ALR, Gheyi HR, Oliveira DR, Oliveira FIF, Viana TVA (2018a) A new method to evaluate salt tolerance of ornamental plants. Theor Exp Plant Physiol 30:173-180. https://doi.org/10.1007/s40626-018-0112-7

Oliveira FM, Batista RO, Cavalcante LF, Costa FGB, Souto AGL, Coelho DCL, Silva KB, Oliveira Filho FX (2018b) Behavior of neem seedlings (*Azadirachta indica*) irrigated with saline water in the soil with biofertilizer and drainage. Aust J Crop Sci 12:1950–1956. https://dx.doi.org/10.21475/ajcs.18.12.12.p1324

Pereira FAL, Medeiros JF, Gheyi HR, Dias NS, Preston W, Vasconcelos CBL (2017) Tolerance of melon cultivars to irrigation water salinity. Rev Bras Eng Agríc Ambient 21:846–851. https://doi.org/10.1590/1807-1929/agriambi.v21n12p846-851

Pessoa LGM, Freire MBGS, Santos RL, Freire FJ, Santos PR, Miranda MFA (2019) Saline water irrigation in semiarid region: I - effects on soil chemical properties. Aust J Crop Sci 13:1169–1176. https://doi.org/10.21475/ajcs.19.13.07.p1686

Porto Filho FQ, Medeiros JF, Sousa Neto ER, Gheyi HR, Matos JA (2006) Viabilidade da irrigação do meloeiro com águas salinas em diferentes fases fenológicas. Cienc Rural 36:453–459. https://doi.org/10.1590/S0103-84782006000200015

Porto Filho FQ, Medeiros JF, Gheyi HR, Dias NS, Sousa PS, Dantas DC (2011) Evolução da salinidade e pH de solo sob cultivo de melão irrigado com água salina. Rev Bras Eng Agríc Ambient 15:1130–1137. https://doi.org/10.1590/S1415-43662011001100004

Porto ER, Amorim MCC, Silva Júnior LGA (2001) Uso do rejeito da dessalinização de água salobra para a irrigação da erva-sal (*Atriplex nummularia*). Rev Bras Eng Agríc Ambient 5:111–114. https://doi.org/10.1590/S1415-43662001000100020

Prisco JT, Bezerra MA (2006a) Morpho-physiological responses of cowpea leaves to salt stress. Braz J Plant Physiol 18:455–465. https://dx.doi.org/10.1590/S1677-04202006000400003

Qadir M, Oster JD, Schubert S, Noble AD, Sahrawat KL (2007) Phytoremediation of sodic and saline-sodic soils. Adv Agron 96:197–247. https://doi.org/10.1016/S0065-2113(07)96006-X

Rebouças JRL, Ferreira Neto M, Dias NS, Souza Neto ON, Diniz AA, Lira RB (2013) Cultivo hidropônico de coentro com uso de rejeito salino. Irriga 18:624–634. https://doi.org/10.15809/irriga.2013v18n4p624

Santos Júnior JA, Gheyi HR, Cavalcante AR, Dias NS, Medeiros S de S (2016) Produção e pós-colheita de flores de girassóis sob estresse salino em hidroponia de baixo custo. Eng Agrí 36:420–432. https://doi.org/10.1590/1809-4430-Eng.Agric.v36n3p420-432/2016

Santos DB, Ferreira PA, Oliveira FG, Garcia GO, Carneiro JES, Cano MAO (2005) Estratégias de manejo da irrigação do feijoeiro com água salina. Rev Bras Eng Agríc Ambient 9:16–20

Santos NA, Soares TM, Silva EFF, Silva DJR, Montenegro AAA (2010) Cultivo hidropônico de alface com água salobra subterrânea e rejeito da dessalinização em Ibimirim, PE. Rev Bras Eng Agrí Ambient 14:961–969. https://doi.org/10.1590/S1415-43662010000900008

Santos AN, Silva EFF, Silva GF, Barnabe JMC, Rolim MM, Dantas DC (2016) Yield of cherry tomatoes as a function of water salinity and irrigation frequency. Rev Bras Eng Agríc Ambient 20:107–112. https://doi.org/10.1590/1807-1929/agriambi.v20n2p107-112

Santos AC, Silva MG, Boechat CL, Chagas DS, Mendes WS (2018) Brackish water: an option for producing hydroponic *Capsicum annuum* in laminar flows of mineral nutrients. Rev Colomb Cienc Hortic. 12:147–155. https://dx.doi.org/10.17584/rcch.2018v12i1.7446

Santos MMS, Lacerda CF, Neves ALR, Sousa CHC, Ribeiro AA, Bezerra MA, Araújo ICS, Gheyi (2020) Ecophysiology of the tall coconut growing under different coastal areas of northeastern Brazil. Agric Water Manag 232. https://doi.org/10.1016/j.agwat.2020.106047

Sá FVS, Gheyi HR, Lima GS, Paiva EP, Silva LA, Moreira RCL, Fernandes PD, Dias AS (2019) Ecophysiology of West Indian cherry irrigated with saline water under phosphorus and nitrogen doses. Biosci J 35:211–221. https://doi.org/10.14393/BJ-v35n1a2019-41742

Seckler D, Amarasinghe U, Molden D, Silva R, Barker R (1998) World water demand and supply, 1990 to 2025: scenarios and issues. Research Report 019, International Irrigation Management Institute, Colombo, https://dx.doi.org/10.3910/2009.019

Shahbaz M, Mushtaq Z, Andaz F, Masood A (2013) Does proline application ameliorate adverse effects of salt stress on growth, ions and photosynthetic ability of eggplant (*Solanum Melongena* L.)? Sci Hort 164:507–511. https://doi.org/10.1016/j.scienta.2013.10.001

Shalhevet J, Morris GH, Scroeder BP (1995) Root and shoot growth response to salinity in maize and soybean. Agron J 87:512–516. https://doi.org/10.2134/agronj1995.00021962008700030019x

Shannon MC, Grieve CM (1998) Tolerance of vegetable crops to salinity. Sci Hort 78:5–38. https://doi.org/10.1016/S0304-4238(98)00189-7

Sharma BR, Minhas PS (2005) Strategies for managing saline/alkali waters for sustainable agricultural production in South Asia. Agric Water Manag 78:136–151. https://doi.org/10.1016/j.agwat.2005.04.019

Sharma DP, Rao KVGK (1998) Strategy for long term use of saline drainage water for irrigation in semi-arid regions. Soil Till Res 48:287–295. https://doi.org/10.1016/S0167-1987(98)00135-4

Silva Júnior LGA, Gheyi HR, Medeiros JF (1999) Composição química de águas do cristalino do nordeste brasileiro. Rev Bras Eng Agríc Ambient 3:11–17. https://doi.org/10.1590/1807-1929/agriambi.v3n1p11-17

Silva FLB, Lacerda CF, Sousa GG, Neves ALR, Silva GL, Sousa CHC (2011) Interação entre salinidade e biofertilizante bovino na cultura do feijão-caupi. Rev Bras Eng Agríc Ambient 15:383–389. https://doi.org/10.1590/S1415-43662011000400009

Silva AO, Soares TM, Silva EFF, Santos AN, Klar AE (2012) Consumo hídrico da rúcula em cultivo hidropônico NFT utilizando rejeitos de dessalinizador em Ibimirim-Pe. Irriga 17:114–125. https://doi.org/10.15809/irriga.2012v17n1p114

Silva ARA, Bezerra FML, Lacerda CF, Araújo MEB, Lima RMM, Sousa CHC (2016) Estabelecimento de plantas jovens de coqueiro "anão verde" em solos afetados por sais e sob deficiência hídrica. Rev Bras Frutic 38:1–12. https://doi.org/10.1590/0100-29452016206

Silva MG, Oliveira IS, Soares TM, Gheyi HR, Santana G, Pinho JS (2018) Growth, production and water consumption of coriander in hydroponic system using brackish waters. Rev Bras Eng Agri Ambient 22:547–552. https://doi.org/10.1590/1807-1929/agriambi.v22n8p547-552

Silva PCC, Azevedo Neto AD, Gheyi HR (2019) Mobilization of seed reserves pretreated with H $_2$ O $_2$ during germination and establishment of sunflower seedlings under salinity. J Plant Nutr 42:2388–2394. https://doi.org/10.1080/01904167.2019.1659349

Silva AAR, Lima GS, Azevedo CAV, Veloso LLS, Capitulino JD, Gheyi HR (2019a) Induction of tolerance to salt stress in soursop seedlings using hydrogen peroxide. Com Sci 10:484–490. https://doi.org/10.14295/cs.v10i4.3036

Silva AAR, Lima GS, Veloso LLS, Azevedo CAV, Gheyi HR, Fernandes PD, Silva LA (2019b) Hydrogen peroxide on acclimation of soursop seedlings under irrigation water salinity. Semina: Ciênc Agrár 40:1441–1454. https://doi.org/10.5433/1679-0359.2019v40n4p1441

Soares Filho WS, Gheyi HR, Brito MEB, Nobre RG, Fernandes PD, Miranda RS (2016) Melhoramento genético vegetal e seleção de cultivares tolerantes à salinidade. In: Gheyi HR, Dias NS, Lacerda CF, Gomes Filho E (eds) Manejo da salinidade na agricultura: estudos básicos e aplicados. 2ed. INCTSal, Fortaleza, pp 259–274

Soares LAA, Brito MEB, Fernandes PD, Lima GS, Soares Filho WS, Oliveira ES (2015) Crescimento de combinações copa - porta-enxerto de citros sob estresse hídrico em casa de vegetação. Rev Bras Eng Agríc Ambient 19:211–217. https://doi.org/10.1590/1807-1929/agriambi.v19n3p211-217

Soares TM, Duarte SN, Silva EFF, Paz VPS, Oliveira JLB (2016) Uso de águas salobras em sistemas hidropônicos de cultivo. In: Gheyi HR, Dias NS, Lacerda CF, Gomes Filho E (eds). Manejo da salinidade na agricultura: estudos básicos e aplicados. 2ed. INCTSal, Fortaleza, pp 373–393

Soares HRE, Santos Júnior JA, Silva EFF, Rolim MM, Silva GF (2019) Water and physiological relationships of lettuce cultivated in hydroponics with brackish waters. Rev Cienc Agron 50:216–222. https://doi.org/10.5935/1806-6690.20190025

Sousa GG, Lacerda CF, Cavalcante LF, Guimarães FVA, Bezerra MEJ, Silva GL (2010) Nutrição mineral e extração de nutrientes de planta de milho irrigada com água salina. Rev Bras Eng Agríc Ambient 14:1143–1151. https://doi.org/10.1590/S1415-43662010001100003

Sousa CHC, Lacerda CF, Silva FLB, Neves ALR, Costa RNT, Gheyi HR (2014a) Yield of cotton/cowpea and sunflower/cowpea crop rotation systems during the reclamation process of a saline-sodic soil. Eng Agríc 34:867–876. https://dx.doi.org/10.1590/S0100-69162014000500006

Sousa GG, Viana TVA, Lacerda CF, Azevedo BM, Silva GL, Costa FRB (2014b) Estresse salino em plantas de feijão-caupi em solo com fertilizantes orgânicos. Rev Agro@mbiente 8:359–367. https://doi.org/10.5327/Z 1982–8470201400031824

Sousa JRM, Gheyi HR, Brito MEB, Xavier DA, Furtado GF (2016) Impact of saline conditions and nitrogen fertilization on citrus production and gas exchanges. Rev Caatinga 29:415–424. https://doi.org/10.1590/1983-21252016v29n218rc

Sousa GG, Rodrigues VS, Sales JRS, Cavalcante F, Silva GL, Leite KN (2018a) Estresse salino e cobertura vegetal morta na cultura do milho. Rev Bras Agric Irrig 12:3078–3089. https://doi.org/10.7127/RBAI.V12N700889

Sousa GG, Rodrigues VS, Soares SC, Damasceno IN, Fiusa JN, Saraiva SEL (2018b) Irrigation with saline water in soybean (Glycine max (L.) Merr.) in a soil with bovine biofertilizer. Rev Bras Eng Agríc Ambient 22:604–609. https://doi.org/10.1590/1807-1929/agriambi.v22n9p604-609

Souza ER, Freire MBGS, Nascimento CWA, Montenegro AAA, Freire FJ, Melo HF (2011) Fitoextração de sais pela *Atriplex nummularia* lindl. sob estresse hídrico em solo salino sódico. Rev Bras Eng Agríc Ambient 15:477–483. https://doi.org/10.1590/S1415-43662011000500007

Suddarth SRP, Ferreira JFS, Cavalcante LF, Fraga VS, Anderson RG, Anderson JJH, Bezerra FTC, Medeiros SAS, Costa CRG, Dias NS (2019) Can humic substances improve soil fertility under salt stress and drought conditions? J Environ Qual 48:1605–1613. https://doi.org/10.2134/jeq2019.02.0071

Terceiro Neto CPC, Gheyi HR, Medeiros JF, Dias NS, Campos MS (2013) Produtividade e qualidade de melão sob manejo com água de salinidade crescente. Pesq Agropec Trop 43:354–362. https://doi.org/10.1590/S1983-40632013000400007

Vasconcelos NS, Dantas Neto J, Medeiros JF, Lima CJGS (2013) Qualidade das águas subterrâneas de área irrigada da comunidade de Pau Branco, em Mossoró (RN). Holos 1:47–63. https://doi.org/10.15628/holos.2013.1271

Vieira MR, Lacerda CF, Cândido MJD, Carvalho PL, Costa RNT, Tabosa JN (2005) Produtividade e qualidade da forragem de sorgo irrigado com água salina. Rev Bras Eng Agríc Ambient 9:42–46

Potential Agricultural Use of Reject Brine from Desalination Plants in Family Farming Areas

Nildo da Silva Dias, Cleyton dos Santos Fernandes, Osvaldo Nogueira de Sousa Neto, Cláudio Ricardo da Silva, Jorge Freire da Silva Ferreira, Francisco Vanies da Silva Sá, Christiano Rebouças Cosme, Ana Claudia Medeiros Souza, André Moreira de Oliveira, and Carla Natanieli de Oliveira Batista

Abstract After drought, salinity is the second most important hindrance to sustain agriculture in the semiarid. Subterranean waters extracted from wells are often high in salts and, during dry years, this dependency on saline ground water precludes water and food security for small farmers and their families. Water desalination offers a potential solution to this problem, but the process results in a reject brine

---

N. da Silva Dias (✉) · C. dos Santos Fernandes · F. V. da Silva Sá · C. R. Cosme · A. C. M. Souza · A. M. de Oliveira · C. N. de Oliveira Batista
Center for Agrarian Sciences, Department of Agronomic and Forest Sciences, Federal Rural University of the Semi-Arid (UFERSA), Mossoró, Brazil
e-mail: nildo@ufersa.edu.br

C. dos Santos Fernandes
e-mail: cleyton1959@hotmail.com

F. V. da Silva Sá
e-mail: vanies_agronomia@hotmail.com

C. R. Cosme
e-mail: christianoreboucas@ufersa.edu.br

A. C. M. Souza
e-mail: anaclaudia.gambiental@hotmail.com

A. M. de Oliveira
e-mail: andremoreira@ufersa.edu.br

C. N. de Oliveira Batista
e-mail: carlanataliely@hotmail.com

O. N. de Sousa Neto
Multidisciplinary Center of Angicos, UFERSA, Angicos, Brazil
e-mail: osvaldo.neto@ufersa.edu.br

C. R. da Silva
Federal University of Uberlândia, Uberlândia, Minas Gerais, Brazil
e-mail: claudio.ricardo@ufu.br

J. F. da Silva Ferreira
United States Salinity Laboratory (USDA-ARS), Riverside, CA, USA
e-mail: Jorge.ferreira@usda.gov

E. Taleisnik and R. S. Lavado (eds.), *Saline and Alkaline Soils in Latin America*, https://doi.org/10.1007/978-3-030-52592-7_5

that needs to be properly disposed of to prevent increasing soil salinity and environmental degradation. This chapter considers desalination of naturally saline well waters as a potential solution to water and food security when used in conjunction with an integrated production system involving reject brine for farm-raised fish and the use of fish pond water to grow organic salt-tolerant vegetables and forage crops for small ruminants. We present results on the recovery of desalination systems in different small communities in the Brazilian northeast and chemical analyses of the saline water input, of the desalinized water, of the resulting reject brine, and of soils that received the desalinized water. Our results indicate that the use of desalination reject brine in family agricultural production is technically, economically, and socio-environmentally feasible, especially when using integrated and sustainable production systems.

Keywords Water desalination · Water security · Reverse osmosis · Fish farming · Family farming

1 Introduction

Throughout the rural areas of the Brazilian semiarid region, the great challenge is to ensure that families have access to good-quality water both for domestic and agricultural use (Souza et al. 2015). One of the economically feasible solutions is the use of groundwater, although, in most cases, its higher salt level restricts its use for human consumption and irrigation (Hach 2002; Knapp and Baerenklau 2006; Panagopoulos et al. 2019).

Reverse osmosis desalination has been the most commonly used method to purify brackish groundwater, and in this context, the Brazilian government program known as "Água doce" (fresh water) sponsored approximately 2000 reverse osmosis desalination plants in local communities and rural land settlements of the Brazilian semiarid region. The use of this technology has benefited 2.5 million people, alleviating the scarcity of freshwater supplies, a chronic condition that afflicts the Brazilian semiarid (Soares et al. 2006).

The desalination of water has been practiced since ancient times but has not been widely adopted due to technological limitations, high capital costs, high energy consumption, and finally, very high unit cost when compared to conventional municipal water (Tsiourtis 2001). Advances in technology in recent years have greatly reduced capital and energy costs, so that desalination projects can be considered as a way for acquiring good-quality water (Zotalis et al. 2014). However, besides potable water, desalination produces a hypersaline effluent (hereafter, referred to as reject brine) that can salinize soils if not properly discarded. In coastal regions, the reject brine can be disposed into the sea, but in remote inland rural locations, this is not possible due to the distance from the sea. In Brazil, studies have shown that reject brine is improperly discharged into soil and water bodies, causing major environmental impacts such as soil erosion, salinization, and contamination of water bodies

(Antas et al. 2019; Oliveira et al. 2018; Mohamed et al. 2005). Thus, the major challenge of using reverse osmosis lies in the disposal or reuse of the reject brine while avoiding environmental damage (Oliveira et al. 2017).

Converting the reject brine from a waste to a resource through treatment and beneficial use may minimize both costs and environmental impacts (Cath et al. 2013). There are reports of successful experiments demonstrating the use of reject brine in productive activities such as shrimp production, tilapia hatchery farming, vegetable and fodder production, laundry, and vehicle washing (Neves et al. 2017). The possibility of reuse is of great importance, considering that the number of desalination plants installed in northeastern Brazil has generated a large volume of reject brine.

In Brazil, some recent studies have pointed out that the reject brine has a potential for various agricultural uses (Dias et al. 2010). However, it is noteworthy that when used for these purposes, it requires the adoption of appropriate management strategies, because it is a highly saline water source, and its misuse can lead to great damage to the environment, such as soil salinization and desertification.

2 Water Desalination in the Brazilian Semiarid Region: Benefits and Impacts

2.1 Water Security in Isolated Communities

Groundwater is a water security alternative for isolated communities in the Brazilian semiarid region through deep well drilling under public policies. However, due to the high salinity of groundwater commonly observed, reverse osmosis desalination is an effective treatment widely used to reduce water salinity. In the Brazilian semiarid region, approximately 2500 desalination plants have already been installed, directly benefiting over 100,000 people in 212 municipalities. Each desalinizing unit produces approximately 10,000 L of desalinized water per week, enough to meet the needs of approximately 30 families. However, the efficiency of this system is variable, and several factors must be considered in a desalination project, such as system resilience, which result from the preliminary system design (Monteiro et al. 2009).

Reject brine volume is a function of the desalination plant size and water recovery rate, expressed as the percentage of the volume of freshwater produced to the total volume of saline water input (Panagopoulos et al. 2019). This rate is dependent on several factors such as membrane surface scale formation, osmotic pressure, and the quality of the water input. The higher the recovery rate of a system, the larger the volume of freshwater, and the smaller the volume of reject brine produced. The average water recovery rate is estimated to be around 45% and 80% for seawater and brackish reverse osmosis plants, respectively (Panagopoulos et al. 2019). Antas et al. (2019) evaluated the recovery rate of reverse osmosis from the desalination system in seven rural communities in the state of Rio Grande do Norte, Brazil, during 2013

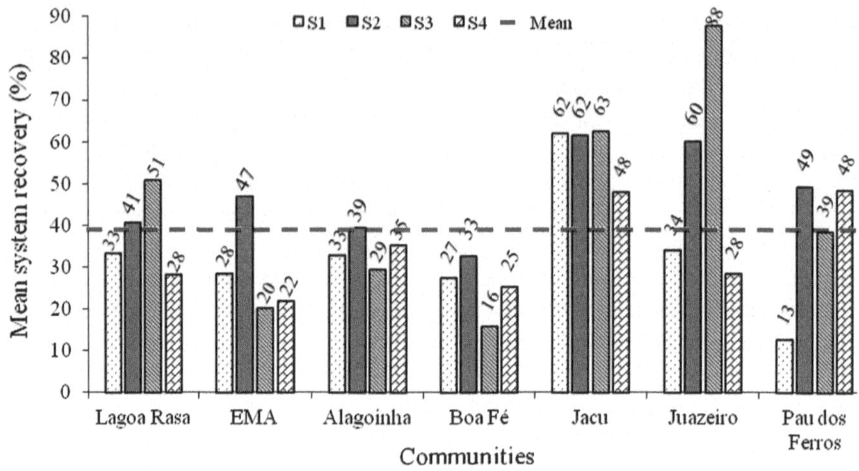

Fig. 1 Average water recovery rates of reverse osmosis desalination systems in seven rural communities of the state of Rio Grande do Norte, Brazil, during dry and rainy seasons in 2013 and 2014. The dotted red line shows the total overall mean across seasons. *Source* Antas et al. (2019). S1 = Season 1 (October/November-2013); S2 = Season 2 (February/March-2014); S3 = Season 3 (June/July-2014) and; S4 = Season 4 (October/November-2014)

and 2014. The authors found that in these, the values ranged from 13 to 88%, with an average of 39.3% (Fig. 1).

Another important aspect to be considered is the salt rejection of the membranes, that is, the ability of membranes to reject dissolved salts during water permeation, which indicates the effectiveness in removing salts and other chemical species (Antas et al. 2019). In general, the salt rejection rate ranges from 90 to 88.8% for most ions dissolved in the water (Hydranautics 2002). However, this ability is influenced by a wide variety of factors such as solute dimensions, retained component morphology, membrane pore size, chemical properties of the solution to be filtered, and hydrodynamic factors, which determine the drag stress and shear forces on the membrane surface (Schneider and Tsutiya 2001).

Antas et al. (2019) showed that 71% of the analyzed samples were within the acceptable range for salt rejection, that is, they had values above 90%, estimated by the electrical conductivity of the desalinated water (Table 1). This fact indicates that the salt rejection system in 29% of desalination plants did not have the required minimum efficiency. Problems of this type are usually related to lack of equipment maintenance. Yet, overall, reverse osmosis is a very efficient technology.

The chemical analysis of saline groundwater feed, reject brine, and drinking water (desalinized water) from the Santa Elza rural settlement is shown below (Table 2). Almost all salts were removed by desalination in the process of transforming feed water into fresh (potable) water.

Table 1 Salt rejection rate (%) in desalination plants from rural communities of the state of Rio Grande do Norte, Brazil

Locality	EC[a]	K^+	Na^+	Ca^{2+}	Mg^{2+}	Cl^-	CO_3^{2-}	HCO_3^-
	Rejection rate							
Lagoa Rasa	98.99	100.00	97.22	98.57	97.93	82.35	100.00	97.62
Ema	95.37	100.00	96.46	98.15	94.29	96.05	100.00	95.89
Alagoinha	84.40	89.47	74.91	95.00	77.78	87.10	100.00	86.00
Boa Fe	94.68	81.82	94.19	98.78	94.95	96.89	100.00	90.91
Jacu	88.58	93.10	91.76	98.25	66.90	95.69	100.00	91.89
Juazeiro	93.03	95.00	93.59	99.00	97.99	95.88	–	80.00
Pau dos Ferros	92.38	91.67	92.20	98.06	100.00	94.34	100.00	96.88

Source Antas et al. (2019)
[a]Electrical conductivity

2.2 *Environmental Impacts of Reject Brine from Desalination Plants*

Despite the limited public understanding of the environmental benefits and challenges of desalination, this technology can be a valuable regional development tool for the Brazilian semiarid region. However, it is necessary to consider the environmental risks associated to the brine effluent co-produced during the desalination due to higher saline concentration than the feed water of the system. Hence, brine reject disposal, when carried out improperly, has great potential for negative impacts on the environment, including salinization of soil and contamination of water bodies (Moura et al. 2016). Reject brine, in addition to its high salinity, may contain dangerous pretreatment chemicals, organic compounds, and heavy metal (Panagopoulos et al. 2019).

Some studies have been performed to analyze the environmental impacts caused by the improper disposal of brine from reverse osmosis water plants. Anders (2013) evaluated these impacts in ten locations in Mossoró, Brazil (rural communities and settlements), and found that, at the point of discharge, 84.6% of soil samples collected were saline or sodic (Fig. 2a), while at 0.8 m and 1.6 m away from the discharge point, 81% and 66.7% of soil samples were saline, respectively (Fig. 2b, c). The authors concluded that there are risks of desertification; however, due to the large variability of soil analysis between the sites studied, it is suggested that desertification risk assessment should be made individually.

Studies at two brackish water treatment plants in western Rio Grande do Norte, Brazil, were conducted by Oliveira et al. (2018) to evaluate the potential agricultural use of reject brine and to identify problems concerning the salinization of soils that receive reject brine generated in the desalination plants. Samples of reject brine and soils in the area where this waste is disposed of were collected for physicochemical characterization. The results indicated that both samples of reject brine have highly

Table 2 Chemical analysis of feed saline groundwater, reject brine, and freshwater from Santa Elza rural settlement, Mossoró, Brazil

Water source	pH	EC	K^+	Na^+	Ca^{2+}	Mg^{2+}	Cl^-	CO_3^{2-}	HCO_3^-	PO_4^-	SAR	Hardness	Cations	Anions
		dS m^{-1}	$mmol_c\ L^{-1}$							mg L^{-1}		mg L^{-1}	$mmol_c\ L^{-1}$	
Feed water	8.9	2.04	11.6	10.2	7.70	12.60	4.50	0.0	3.4	0.03	1.8	630	17.4	9.0
Reject brine	8.0	2.23	5.1	18.3	8.50	15.20	4.70	0.2	5.8	0.05	2.9	760	23.3	15.4
Freshwater	7.6	0.08	1.3	1.1	0.25	0.36	0.03	0.0	0.0	0.004	0.3	18	0.8	0.6

EC = Electrical conductivity, SAR = Sodium adsorption ratio

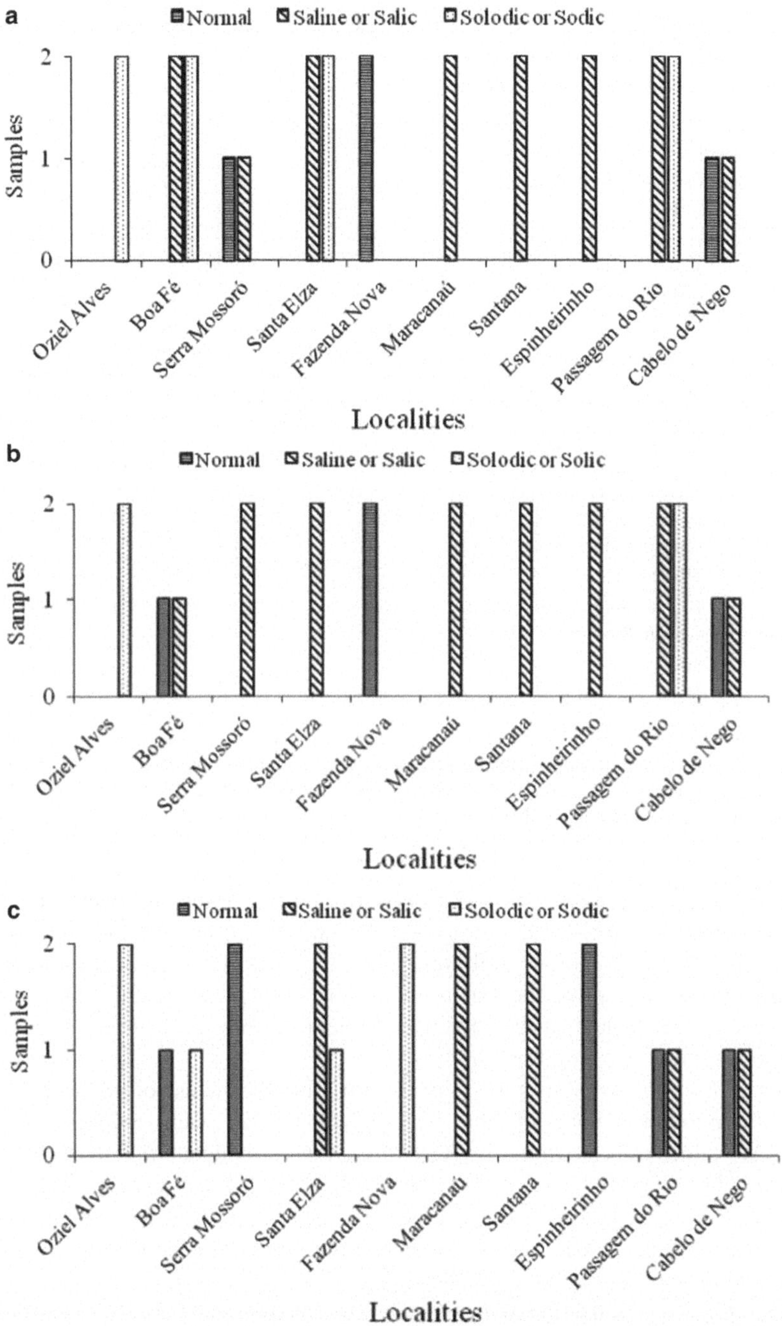

Fig. 2 Classification of soil samples for salinity and sodium saturation at the brine discharge point (**a**), at 0.8 m (**b**) and at 1.6 m (**c**) from the discharge point of reject brine in ten communities in the state of Rio Grande do Norte, Brazil. *Source* Anders (2013)

Table 3 Chemical analysis of the water from the reject brine from the communities Boa Fé and Lagoa Rasa in four periods of sampling

Locality	Seasonal climatic conditions	pH	EC	Na^+	Cl^-	*SAR	USSL[1]	S[2]	T[3]
			$dS\ m^{-1}$	$mmol_c\ L^{-1}$					
Lagoa Rasa	Dry season 2013	7.63	1.48	10.24	5.00	6.00	C3	S1	T3
	Beginning of rainy season 2014	8.00	1.80	13.01	7.40	6.90	C3	S1	T3
	End of rainy season 2014	7.70	1.15	8.93	6.20	5.20	C3	S1	T2
	Dry season 2014	7.60	1.31	19.25	6.00	10.93	C3	S2	T3
Boa Fé	Dry season 2013	7.49	8.41	33.70	92.00	6.50	C4	S1	T3
	Beginning of rainy season 2014	7.02	9.30	38.90	100.0	7.10	C4	S1	T3
	End of rainy season 2014	7.20	7.30	40.09	87.00	7.12	C4	S1	T3
	Dry season 2014	7.35	7.56	72.63	82.00	12.11	C4	S1	T3

*Sodium Adsorption Ratio. $SAR = Na^+/(Ca^{++} + Mg^{++})^{0.5}$

[1,2,3]Classification of waters for irrigation with respect to potential risks of salinity (C), problems of infiltration—sodicity (S), and toxicity by ions (T), respectively (Ayers and Westcot 1999). *Source* of Data from table Oliveira et al. (2018)

restricted use for irrigation, being classified as C_3 or C_4 (Ayers and Westcot 1999) (Table 3). Due to the risks of salinization by the reject brine and the consequent deleterious effects of salts on soil and plants, these can be used in the irrigation of agricultural crops provided that a set of soil-water-plant system management strategies are established, such as the use of tolerant species, subsurface irrigation, and application of leaching fractions.

Due to the low SAR values, there is no restriction for the use of the reject brine in both communities regarding the risks of reduction in water infiltration in the soil. SAR values indicate that this water offers low to moderate risk of infiltration problems. However, the samples have moderate (Lagoa Rasa) to severe (Boa Fé) restrictions of use regarding toxicity, especially the samples from Boa Fé due to the high concentrations of Cl^- and Na^+ ions and the electrical conductivity, which is unsuitable for most agricultural and horticultural crops (Table 3).

Regarding soil salinization in the reject brine disposal area (Table 4), it was found that both localities had saline or saline-sodic soils (0.12–6.75 $dS\ m^{-1}$), with greater accumulation of salts in the dry period. The high values of exchangeable sodium percentage (ESP) determined indicate the predominance of sodium adsorbed in the

Table 4 Characteristics of soils that receive reject brine from the desalination plants from two localities in different seasons of the year. Data are from discharge point distances (0, 1, and 2 m) and from two soil depths (0–20 and 20–40 cm)

Season	Discharge point (m)	Layer depth (m)	Locality							
			Boa Fé (Mossoró[2])				Lagoa Rasa (Apodi[2])			
			pH	EC_e ($dS\ m^{-1}$)	ESP (%)	Soil classification[1]	pH	EC_e ($dS\ m^{-1}$)	ESP (%)	Soil classification[1]
Dry 2013	0	0–0.20	7.04	3.75	6	Saline	8.54	7.10	49	Saline-Sodic
	0	0.20–0.40	7.19	3.45	13	Saline	8.50	2.56	19	Saline-Sodic
	1	0–0.20	5.33	3.62	7	Saline	8.90	13.90	70	Saline-Sodic
	1	0.20–0.40	5.14	4.04	8	Saline	8.60	13.20	56	Saline-Sodic
	2	0–0.20	4.50	0.13	0	Normal	8.53	5.86	33	Saline-Sodic
	2	0.20–0.40	4.39	0.12	0	Normal	8.51	6.25	39	Saline-Sodic
Beginning of rainy season 2014	0	0–0.20	7.60	6.25	12	Saline	8.60	2.57	26	Saline-Sodic
	0	0.20–0.40	7.76	5.71	18	Saline-Sodic	8.70	2.70	28	Saline-Sodic
	1	0–0.20	7.84	5.25	16	Saline-Sodic	9.00	11.56	61	Saline-Sodic
	1	0.20–0.40	7.85	5.71	15	Saline-Sodic	8.88	8.95	49	Saline-Sodic
	2	0–0.20	8.15	5.37	13	Saline	8.93	7.65	46	Saline-Sodic
	2	0.20–0.40	8.25	5.00	17	Saline-Sodic	8.70	7.77	39	Saline-Sodic
End of rainy season 2014	0	0–0.20	7.53	6.75	13	Saline	8.83	3.51	18	Saline-Sodic
	0	0.20–0.40	7.50	7.02	21	Saline-Sodic	8.72	3.61	25	Saline-Sodic
	1	0–0.20	7.70	3.62	10	Saline	8.86	5.93	34	Saline-Sodic
	1	0.20–0.40	7.75	3.57	14	Saline	8.67	7.15	32	Saline-Sodic
	2	0–0.20	6.21	0.63	1	Normal	9.11	7.42	49	Saline-Sodic
	2	0.20–0.40	6.17	0.71	2	Normal	8.65	7.08	37	Saline-Sodic

(continued)

Table 4 (continued)

Season	Discharge point (m)	Layer depth (m)	Locality							
			Boa Fé (Mossoró[2])				Lagoa Rasa (Apodi[2])			
			pH	EC_e (dS m^{-1})	ESP (%)	Soil classification[1]	pH	EC_e (dS m^{-1})	ESP (%)	Soil classification[1]
Dry in 2014	0	0–0.20	7.43	4.25	10	Saline	9.11	2.81	20	Saline-Sodic
	0	0.20–0.40	7.62	4.16	11	Saline	9.17	3.47	28	Saline-Sodic
	1	0–0.20	7.73	3.50	11	Saline	9.49	3.43	32	Saline-Sodic
	1	0.20–0.40	8.31	4.16	17	Saline-Sodic	9.00	3.81	31	Saline-Sodic
	2	0–0.20	8.05	5.87	16	Saline-Sodic	9.53	4.53	49	Saline-Sodic
	2	0.20–0.40	8.23	3.92	16	Saline-Sodic	8.81	6.52	40	Saline-Sodic

[1]Classification of salt-affected soils (Bohn et al. 1985)

[2]Mossoró and Apodi are located in the state of Rio Grande do Norte, northeastern Brazil. EC_e = Electrical conductivity of the soil saturated paste

0–40 cm soil layer. Another problem of these soils where reject brine has been disposed of is the high values of pH (above 8.50), which can hinder the availability of plants to absorb important minerals such as P, Mg, Fe and Zn, which are unavailable or have decreased availability in soils with pH above 7.5.

It is important to point out that, despite the restriction on the use of reject brine, its safe utilization depends on the practices and management strategies adopted, which include the selection of tolerant plants, mixture with fresh water, leaching fraction applied, etc.

2.3 Potential Agricultural Use of Reject Brine

Due to the large number of reverse osmosis water treatment plants installed in northeastern Brazil, studies are needed to enable the proper disposal of the waste generated (Oliveira et al. 2017). Alternatives of use of the reject brine are being studied, such as the cultivation of halophyte species like saltbush (*Atriplex nummularia* L.). For being native to arid regions, this species adapts well to the climatic conditions of northeastern Brazil, managing to produce a large amount of phytomass.

In developed countries, reject brine is usually discharged into the oceans, but other disposal options such as evaporation ponds, reduction of reject brine volume with the cultivation of aquatic plants, percolation ponds and irrigation of halophyte plants have been studied. In the USA, the reject brine is used in the irrigation of several crops, but according to Mickley (2004), this requires a lot of land available, and the reject brine is usually mixed with good-quality water to reduce the concentration of salts, being limited by the availability of good-quality water for dilution, climate, and soil absorption rates. In addition, it can be used in leisure areas such as lawns, parks, golf courses, open spaces, and green belts for soil conservation and environmental preservation.

Riley et al. (1997) considered the cultivation of halophyte plants as the best option to dispose of reverse osmosis reject brine. Other authors such as Dubon and Pinheiro (2001) also observed promising results when investigating the growth of tilapia (*Oreochromis*) in highly saline waters. In addition to fish farming, shrimp farming and/or multi-cropping (the practice of growing two or more crops on the same area during a single growing season) have also been employed to make use of the reverse osmosis reject brine.

Another option for the desalination reject brine is its use in the composition of the nutrient solution to grow vegetables in hydroponic systems, as demonstrated in the study conducted by Dias et al. (2010). Hydroponic crops are advantageous when brackish water is used because, due to the absence of matric potential, the effects of induced drought are avoided, increasing plant tolerance to salinity (Soares et al. 2006).

According to Mickley (2004), the choice of the best option to dispose of the desalination reject brine must meet, among other factors, the local availability (land, compatibility of receiving waters, and distance), regional availability (geology, state laws, geography, and climate), reject brine volume, costs involved, public opinion, and permissibility.

Antia (2015) evaluated the feasibility of desalination projects for exclusive use in the irrigation of crops and concluded that, due to the annual fluctuation of income in most agricultural units, due to climate variations and commodity prices, it is difficult to justify the financial investment to install a reverse osmosis desalination unit to produce a constant rate of high-quality desalinated water. Tchiadje (2007) studied strategies to reduce water salinity by desalination on production of crops, such as rice, cotton and pepper, and concluded that the yields increased as water salinity decreased and, for many crops, a relatively small reduction in soil or irrigation water salinity (for instance, 0.5–5 g L^{-1}) was sufficient to increase yields by 25–75%. Therefore, it can be inferred that an agricultural unit can substantially increase profitability by partially reducing the use of desalination reject brine.

3 Management of Reject Brine: Case Study in a Family Farming Unit

3.1 Experimental Location and Description

The project was conducted in the Santa Elza rural settlement (5°06′50.29″ S; 37°31′9.86″ W), located in the rural area of Mossoró, Rio Grande do Norte, Brazil (Fig. 3).

The action included integrated and sustainable subsystems with the purpose of giving an agricultural use to the brine coming from the desalination: initially, the saline well water (EC = 1.9 dS m^{-1} and pH = 7.6) was pumped to the treatment plant, benefiting the families with drinking water (EC = 0.12 dS m^{-1} and pH = 6.8); the reject brine (EC = 3.2 dS m^{-1} and pH = 7.4) from reverse osmosis was then directed to two fish hatcheries built for the breeding of tilapia; the effluents from fish farming (EC = 3.88 dSm^{-1} and pH = 7.71) enriched in organic matter were used as a source of water and nutrition for the cultivation of forage plants and organic vegetables. Finally, the forage produced was used in the feeding and fattening of goats and/or sheep, closing the sustainable production system (Fig. 4).

3.2 Results

3.2.1 Tilapia Farming

In a four-month cycle, 400 tilapias with an average weight of 630 g were collected from each hatchery, increasing family income and protein supply. The average feed conversion rate found was 1.5:1, which is considered high by Lovshin (1997), who

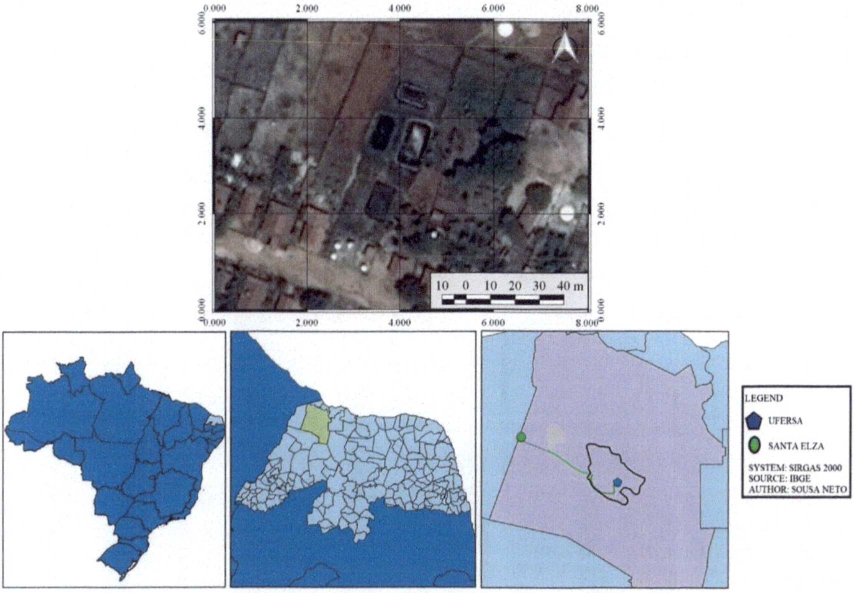

Fig. 3 Location of the Santa Elza settlement and the rural community Serra Mossoró, Mossoró, Brazil. *Source* Modified from Google Maps

states that it is essential to obtain a high-feed conversion rate, so that the intensive system of tilapia farming be economically viable. Within the limits of tolerance, tilapia grows and reproduces in brackish and salty waters, adapting to low levels of dissolved oxygen content, and coexists in a wide range of water acidity and alkalinity, tolerating high concentrations of toxic ammonia compared to most fish.

3.2.2 Production and Quality of Vegetables Fertigated with Fish Farming Effluent

The yields and nutritional quality of the main vegetables cultivated and irrigated with fish farming effluent were quantified (Table 5).

Although crop yield was reduced due to the salinity of the effluent water used in the fertigation of these vegetables, yield losses are economically acceptable considering the water restrictions in the region. Santos et al. (2012) concluded that there is a relative reduction of 10% in the production of arugula when irrigated with highly saline water (3.5–5.5 dS m^{-1}). However, the relative yield loss depends on crop tolerance to salinity. For example, lettuce and beet crops have threshold salinity of 1.3 and 4.0 dS m^{-1}, respectively, being salt-sensitive and salt-tolerant species, respectively. However, there are beet cultivars that tolerate water salinity above 4 dS m^{-1} (Lv et al. 2019) and other crops like that can tolerate salinity of irrigation water of 7 (Jerusalem artichoke) and 9 dS m^{-1} (spinach) with only 11% tuber yield or no

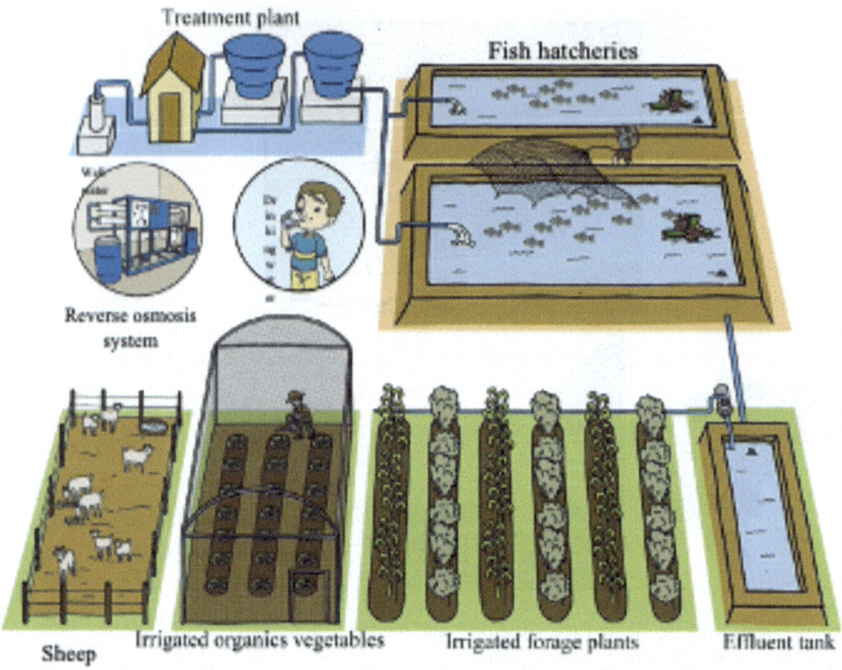

Fig. 4 Schematic representation of the integrated system used to desalinize well water, using the brine reject from desalination to raise fish. The mineral enriched water from fish production is then directed to an effluent tank and used to produce organic vegetables and salt-tolerant forage crops to feed small ruminants. The system is adjusted according to the salinity of the water input, of the local soil, and may integrate different crops according to the final salinity of the reject brine

Table 5 Average weekly yield of vegetables irrigated with fish farming effluent

Vegetables	Average weekly yield
Lettuce (*Lactuca sativa*)	50 units
Coriander (*Coriandrum sativum*)	45 bunches
Chives (*Allium schoenoprasum*)	26 bunches
Cabbage (*Brassica oleracea*)	40 bunches
Arugula (*Eruca sativa*)	30 bunches
Tomato (*Solanum lycopersicum*)	10.0 kg
Bell pepper (*Capsicum annuum*)	12.0 kg
Carrot (*Daucus carota*)	7.0 kg
Beet (*Beta vulgaris* subsp. *esculenta*)	6.0 kg

Table 6 Nutritional values of vegetables fertigated with fish farming effluent

Vegetable	Protein (g 100 g^{-1})	Vitamin C (mg 100 g^{-1})	Dietary fiber (g 100 g^{-1})
Lettuce	0.79	19.13	2.18
Coriander	17.20	10.12	35.50
Chives	3.60	55.12	2.60
Cabbage	3.20	95.26	3.61
Arugula	1.60	38.70	1.50
Tomato	0.83	28.10	1.16
Bell pepper	0.80	190.20	2.03
Carrot	1.41	5.83	4.23
Beet	1.36	4.70	1.98

reduction in shoot biomass or nutritional value, respectively (Dias et al. 2016; Ferreira et al. 2018). Research work recently presented at the INOVAGRI 2019 showed that spinach cultivars Raccoon and Gazelle can be cultivated with water of salinity as high as 13 dS m^{-1} in sandy medium without any significant loss in shoot biomass (J. Ferreira, personal communication).

The nutrients contained in the effluent from fish farming, due to the excretion of fish and the feed supplied in the nurseries, stimulate the vegetative growth of plants because of the improvement in soil fertility, especially the incorporation of organic matter (Andrade Filho et al. 2013). In addition, irrigation with effluent attenuates the deleterious effects of salt stress on plants.

The nutritional quality of vegetables grown under irrigation with effluent is similar to those found in the literature for organic or conventional vegetables (Table 6) (UNICAMP 2011). However, it is worth pointing out that the chemical composition of foods of plant origin depends not only on an isolated production factor, but on a set of factors and their interactions, such as fertilization, type of soil, occurrence of pests and diseases, climate, harvest, and genetic characteristics of the plant.

3.2.3 Growth, Yield, and Quality of Forage Plants Fertigated with Fish Farming Effluent

Table 7 shows the results obtained for the growth, yield, and forage quality variables of elephant grass, sorghum, and heirloom corn fertigated with fish farming effluent. In general, the dry matter yield found for the three forage species and grain yield for sorghum and heirloom corn were satisfactory and similar to those reported in the literature for these species.

Regarding the quality of the three forage plants, the average values recorded for crude protein (CP) were above the minimum necessary for ruminal fermentation to occur (Table 7), as described by Minson (1984), who established a minimum CP concentration of 7% for the process to occur satisfactorily.

Table 7 Average values obtained for growth, yield, and forage quality variables of elephant grass, sorghum, and heirloom corn fertigated with fish farming effluent from desalination reject brine

Elephant grass (*Pennisetum purpureum*)					
PH (cm)	TDM (ton ha^{-1})	CP (%)	Ca	Na	Cl
98.5	8.8	9.8	5.3	1.5	17.3
Sorghum (*Sorghum bicolor*)					
TDM* (ton ha^{-1})		CP (%)	GY (ton ha^{-1})	TSS (Brix)	JV (ton ha^{-1})
18.8		12.4	5.5	16.1	9.6
Heirloom corn (*Zea mays*)					
TDM* (ton ha^{-1})		CP (%)	GY (ton ha^{-1})	TSS (°Brix)	NGE (unit)
23.89		9.8	10.6	8.4	380.0

*Leaves + stem + head
PH = Plant height; TDM = Total dry matter; CP = Crude protein; Ca = Calcium; Na = Sodium; Cl = Chlorine; GY = Grain yield; TSS = Total soluble solids; JV = Juice volume in the stem; NGE = Number of grains per ear

The yield and quality of elephant grass, sorghum, and heirloom corn found are consistent with the average values reported in the literature (Minson 1984; Albuquerque et al. 2012; Vale and Azevedo 2013). This indicates that, despite the high salinity of the effluent, the yield and forage quality of the species were not reduced. Both yield and forage quality are related to the tolerance of the species to salinity and especially to the nutritional supply, mainly of organic matter, provided by the effluent. Although grasses are, in general, more tolerant than dicots to salinity, legume crops are higher in crude protein (CP) than grasses and may tolerate high salt concentration in irrigation water. For instance, alfalfa was reported to have from 20–30% CP and maintains its mineral composition and antioxidant capacity when irrigated with waters of salinity up to 24 dS m^{-1} (Ferreira et al. 2015). Another study with 15 alfalfa populations reported that biomass decreased to 50% of control when irrigation water had an ECiw of 18.3 dS m^{-1} with greater reduction at 24 dS m^{-1} (Cornacchione and Suarez 2017). Alternatively, forage trees with vigorous growth and tolerance to salinity (e.g., *Leucaena leucochepala*) may be evaluated under irrigation with brine deject.

4 Final Considerations

Desalination of saline and brackish waters by reverse osmosis benefits rural communities in the Brazilian semiarid region, but there is an environmental concern about the disposal of reject brine, due to its potential negative impacts on the environment if not properly managed.

Studies indicate that there is technical, economic, and socio-environmental feasibility for the use of desalination reject brine in family agricultural production, especially when using integrated and sustainable production systems. The use of reject brine for agricultural purposes can be profitable in rural communities and settlements while contributing to environmental conservation of soil and water resources. However, one must consider the salinity of the water input for desalinizers and of the local soils where the reject brine will be applied to grow agricultural and horticultural crops.

References

Albuquerque CJB, Tardin FD, Parrella RAC, Guimarães A, de Oliveira, sus Silva KM (2012) Sorgo sacarino em diferentes arranjos de plantas e localidades de Minas Gerais, Brasil. Rev Bras Milho Sorgo 11:69–85

Anders CR (2013) Caracterização química da água de dessalinizadores e dos solos sob a influência do rejeito salino em Mossoró – RN. Dissertation, Universidade Federal Rural do Semi-Árido

Andrade Filho J, Sousa Neto ON, Dias NS et al (2013) Atributos químicos de solo fertirrigado com água residuária no semiárido brasileiro. Irriga 18:661–674

Antas FPS, Dias NS, Gurgel GCS, Oliveira N, dos Santos FC, Oliveira AM, Filhoc JC, Sousa Neto ON, Freitas JM, Andraded LM (2019) Analysis of recovery by desalination systems in the west of Rio Grande do Norte, Brazil. Desalin Water Treat 138:230–236

Antia DDJ (2015) Desalination of water using ZVI (Fe0). Water 7:3671–3831

Ayers RS, Westcot DW (1999) A qualidade da água na agricultura. UFPB, Campina Grande

Bohn HL, Mcnell BL, O'connor GA (1985) Soil chemistry. Wiley, New York

Cath TY, Elimelech M, McCutcheon JR, McGinnis RL, Achilli A, Anastasio D, Brady AR, Childress AE, Farr IV, Hancock NT, Lampi J (2013) Standard methodology for evaluating membrane performance in osmotically driven membrane processes. Desalination 312:31–38

Cornacchione MV, Suarez DL (2017) Evaluation of alfalfa (*Medicago sativa* L.) populations' response to salinity stress. Crop Sci 57:137–150. https://doi.org/10.2135/cropsci2016.05.0371

Dias NS, Ferreira JFS, Liu X, Suarez DL (2016) Jerusalem artichoke (*Helianthus tuberosus,* L.) maintains high inulin, tuber yield, and antioxidant capacity under moderately-saline irrigation waters. Ind Crops Prod 94:1009–1024. https://doi.org/10.1016/j.indcrop.2016.09.029

Dias NS, Lira RB, Brito RF, Sousa Neto OND, Ferreira Neto M, Oliveira AMD (2010) Produção de melão rendilhado em sistema hidropônico com rejeito da dessalinização de água em solução nutritiva. Rev Bras Eng Agríc Amb 14:755–761

Dubon JAM, Pinheiro JCV (2001) Aproveitamento de Águas residuais provenientes de dessalinizadores instalados no Estado do Ceará. In: Encuentro de las Aguas, Santiago, Chile

Ferreira JFS, Sandhu D, Liu X, Halvorson JJ (2018) Spinach (*Spinacea oleracea* L.) response to salinity: nutritional value, physiological parameters, antioxidant capacity, and gene expression. Agriculture 8:1–17. https://doi.org/10.3390/agriculture8100163

Ferreira JFS, Cornacchione MV, Liu X, Suarez DL (2015) Nutrient composition, forage parameters, and antioxidant capacity of alfalfa (*Medicago sativa,* L.) in response to saline irrigation water. Agriculture 5:577–597

Hach C (2002) Water analysis handbook. Loveland, Colorado

Hydranautics. Technical Service Bulletin (2002) Índice de Qualidade da Água. https://www.cetesb.sp.gov.br/Agua/rios/indice_iap_iqa.asp. Accessed 12 Dec 2016

Knapp KC, Baerenklau KA (2006) Ground water quantity and quality management: agricultural production and aquifer salinization over long time scales. J Agr Resour Econ 31:616–641

Lovshin LL (1997) Worldwide tilapia culture. In: Workshop International de Aquicultura, São Paulo

Lv X, Chen S, Wang Y (2019) Advances in understanding the physiological and molecular responses of sugar beet to salt stress. Front Plant Sci 10. https://doi.org/10.3389/fpls.2019.01431
Mickley MC (2004) Membrane concentrate disposal: practices and regulation. U.S. Department of the Interior, Denver
Minson DJ (1984) Effects of chemical and physical composition of herbage eaten upon intake. In: Hacker JB (ed) Nutritional limits to animal production from pasture. CAB, Farnham Royal, pp 167–182
Mohamed AMO, Maraqaa M, Al Handhalyb J (2005) Impact of land disposal of reject brine from desalination plants on soil and groundwater. Desalination 182:411–433
Monteiro GS, Silva JN, Lôbo HLL (2009) Simulação e análise de sistema de dessalinização via osmose inversa: Considerações para análise da qualidade da água. Encicl Biosf 5:1–12
Moura ESR, Cosme CR, Dias NS, Portela JC, Souza ACM (2016) Yield and forage quality of saltbush irrigated with reject brine from desalination plant by reverse osmosis. Rev Caatinga 29:1–10
Neves ALR, Pereira MA, Lacerda CF, Gheyi HR (2017) Aspectos socioambientais e qualidade da água de dessalinizadores nas comunidades rurais de Pentecoste-CE. Rev Amb Água 12:124–135
Oliveira AM, Dias NS, Freitas JJR, Martins DF, Rabelo LN (2017) Avaliação físico-química das águas do processo de dessalinização de poços salobros e salinos em comunidades rurais do oeste potiguar. Águas Subter 31:58–73
Oliveira AM, Dias NS, Gurgel GCS (2018) Impactos físico-químicos da disposição do rejeito da dessalinização das águas de poços salinos em Neossolo e Chernossolo do Oeste Potiguar. Irriga 23:413–425
Panagopoulos A, Haralambous KJ, Loizidou M (2019) Desalination brine disposal methods and treatment technologies—a review. Sci Total Environ 693:1–23
Riley JJ, Fitzsimmons KM, Glenn EP (1997) Halophyte irrigation: an overlooked strategy for management of membrane fraction concentrate. Desalination 110:197–211
Santos RSS, Dias NS, Duarte SN, Lima CJGS (2012) Uso de águas salobras na produção de rúcula cultivada em substrato de fibra de coco. Rev Caatinga 25:113–118
Schneider RT, Tsutiya MT (2001) Membranas filtrantes para o tratamento de água, esgoto e águas de reuso. ABAS, São Paulo
Soares TM, Silva IJO, Duarte SN, Silva EFF (2006) Destinação de águas residuárias provenientes do processo de dessalinização por osmose reversa. Rev Bras Eng Agríc Amb 10:730–737
Souza FI, Dias NS, Sousa Neto ON, Portela JC, Medeiros Souza AC, Nascimento Porto VC, Silva CR, de Sousa Junior FS (2015) Agricultural potential of reject brine from water desalination. Afr J Agric Res 10:4713–4717
Tchiadje NFT (2007) Strategies to reduce the impact of salt and crops (rice cotton and chilli) production: a case study of the tsunami-affected area of India. Desalination 206:524–530
Tsiourtis NX (2001) Desalination and the environment. Desalination 141:223–236
UNICAMP—Universidade Estadual de Campinas (2011) Tabela Brasileira de composição de alimentos. NEPA/UNICAMP, Campinas. www.unicamp.br/nepa/taco. Accessed 4 Sept 2015
Vale MB, Azevedo PV (2013) Avaliação da produtividade e qualidade do capim elefante e do sorgo irrigados com água do lençol freático e do rejeito do dessalinizador. Holos 3:181–195
Zotalis K, Dialynas EG, Mamassis N, Angelakis AN (2014) Desalination technologies: Hellenic Experience. Water 6:1134–1150

Salt Affected Soils in the Brazilian Semiarid and Phytoremediation as a Reclamation Alternative

Maria Betânia Galvão Santos Freire, Fernando José Freire, Luiz Guilherme Medeiros Pessoa, Edivan Rodrigues de Souza, and Hans Raj Gheyi

Abstract Under arid and semiarid climates, the natural process of soil formation can cause the accumulation of salts that limit plant growth and development. The Northeast region of Brazil has an extensive area under semiarid climate with drought most of the time. Additionally, the inadequate management of irrigation has promoted the accumulation of salts in soils, degrading them. Some of such areas are becoming unproductive and are abandoned. Reclamation techniques that involve drainage, use of chemical and organic conditioners, are expensive and difficult to implement. Phytoremediation of salt-affected soils is a low cost alternative. Plants adapted to the environment, which tolerate high levels of salts, can grow and produce biomass, should be studied. It is also important that the plants used are able to absorb the salts, extracting them from the soils. However, phytoremediation results are not observed in the short term. But, over time, phytoremediation promotes the return of vegetation to degraded soils, as well as the associated microbiota and protects the soil surface. This chapter reports trials in Brazil, evaluating some plant species for their ability to survive and improve soil quality.

Keywords Saline soils · Sodic soils · Reclamation techniques · Halophytes · *Atriplex nummularia*

M. B. G. S. Freire (✉) · F. J. Freire · E. R. de Souza
Agronomy Department, Federal Rural University of Pernambuco (UFRPE), Recife, Pernambuco, Brazil
e-mail: maria.freire@ufrpe.br

F. J. Freire
e-mail: fernando.freire@ufrpe.br

E. R. de Souza
e-mail: edivan.rodrigues@ufrpe.br

L. G. M. Pessoa
Academic Unit of Serra Talhada (UAST), Federal Rural University of Pernambuco (UFRPE), Serra Talhada, Pernambuco, Brazil
e-mail: luiz.pessoa@ufrpe.br

H. R. Gheyi
Federal University of Recôncavo da Bahia (UFRB), Cruz das Almas, Bahia, Brazil
e-mail: hgheyi@gmail.com

E. Taleisnik and R. S. Lavado (eds.), *Saline and Alkaline Soils in Latin America*, https://doi.org/10.1007/978-3-030-52592-7_6

1 Salt Affected Soils in Brazil

1.1 Overview

Salt concentration in soils depends on a number of factors such as parent rock, clay content and type, presence of organic matter, soil management, fertilization, irrigation aspects, relief and climate. If rainfall exceeds evapotranspiration and there are no drainage problems, salts are naturally leached from the soil profile. In the Northeast region of Brazil there is a large area under semiarid climate. Under such conditions, some soils are more susceptible to salinization, depending on the physical, chemical and mineralogical characteristics of the soil as well as the management adopted.

Very weathered soils such as Oxisols and Ultisols do not have salt accumulation problems because water can easily percolate and leach salts. Nevertheless, Entisols, Alfisols, Aridisols and Inceptisols, commonly found soils in Brazilian semiarid region are susceptible to salinization (Fig. 1). In general, these soils do not have a

Fig. 1 Saline-sodic soils in Pernambuco State (Brazil): Aerial view and soil profile of Fluvisol (**a**) and Cambisol (**b**). Photos: Luiz Guilherme Medeiros Pessoa, Pessoa (2012)

Fig. 2 Salt crust on the soil surface in Cachoeira II Irrigated Perimeter, Serra Talhada (Pernambuco, Brazil). Photo: Maria Betânia Galvão dos Santos Freire

good structure, have high density and low porosity. Sometimes they have subsurface deposition or salt crusts on the surface (Fig. 2).

Large areas have been degraded by salinization in Northeastern Brazil. This region is characterized by shallow soils, low quality irrigation waters, lack of drainage and often shallow groundwater (Pessoa et al. 2016). Salt crusts retard water infiltration and promote runoff, contributing to soil degradation. Salinity assessment depends on soil surveys that require time and financial resources, soil sampling and analysis. Alternative techniques using geoprocessing and remote sensing tools are being used successfully to replace traditional surveys in various parts of the world, including in Brazil. Pessoa et al. (2016) studied spectral properties of saline soils in four watersheds in semiarid region of Pernambuco State and characterized six groups of soils (Table 1). The chemical analysis revealed different levels of degradation among the groups, ranging from non-degraded to strongly degraded conditions, as defined by salinity and sodicity degrees. They found mean soil pH values ranging from 6.5 to 10.2, electrical conductivities of saturation extract (ECse) from 0.8 to 66.1 dS m^{-1}, exchangeable sodium percentage (ESP) from 2.8 to 31.8%, and sodium adsorption ratio (SAR) from 2.6 to 585.1 $(mmol_c\ L^{-1})^{0.5}$. Most of the variables evaluated had high variability, both exchangeable cations and those present in the saturation extract (Table 1).

This high variability in short distances is another characteristic of these soils. Especially in irrigated perimeters, productive areas are found in the vicinity of saline-sodic soils. Accumulation of salts can occur on the surface or subsurface, depending on the movement of water in the soil profile. In rainy periods, water translocate salts to deeper layers. During drought, the capillary rise of saline water promotes the deposition of salts as superficial crusts on the soil (Fig. 2).

The reclamation of salt affected soils has not been common in Brazil. In general, degraded areas are left fallow and cultivation proceeds to other areas, generating an environmental degradation liability. The return of vegetation in these areas would initiate a slow reclamation process. On the other hand, salinity assessment depends on soil surveys that require time and financial resources, soil sampling and analysis.

Table 1 Chemical properties of six groups of soils, established through cluster analysis (based on salinity and sodicity variables—pH, electrical conductivity, sodium adsorption ratio, and exchangeable sodium percentage) in four watersheds in Pernambuco State, Brazil (Pessoa 2012)

Property	Soil group											
	1		2		3		4		5		6	
	Mean	SD[1]	Mean	SD	Mean	SD	Mean	SD	Mean	SD	Mean	SD
Exchangeable cations												
pH	7.2	0.56	8.3	0.90	10.2	0.64	8.2	0.43	6.5	0.29	7.2	0.20
Ca^{2+} ($cmol_c\ kg^{-1}$)	14.0	12.80	6.1	4.70	8.9	3.22	5.6	3.62	3.2	2.98	6.4	6.72
Mg^{2+} ($cmol_c\ kg^{-1}$)	3.3	2.85	2.7	2.34	1.8	0.84	1.7	1.35	1.7	1.59	1.8	2.0
Na^{+} ($cmol_c\ kg^{-1}$)	1.5	1.79	4.1	3.12	2.0	0.93	0.3	0.52	0.2	0.17	0.4	0.39
K^{+} ($cmol_c\ kg^{-1}$)	0.5	0.33	0.3	0.21	0.5	0.29	0.6	0.75	0.4	0.23	0.5	0.68
CEC[2] ($cmol_c\ kg^{-1}$)	19.9	16.28	16.1	12.5	13.3	2.86	11.3	5.73	10.0	8.53	9.7	8.7
S[3] ($cmol_c\ kg^{-1}$)	19.3	16.4	13.2	9.99	13.2	2.83	8.4	4.81	5.5	4.31	8.9	8.63
BS[4] (%)	90.4	18.65	83.5	18.14	99.4	0.13	73.9	24.09	60.4	21.1	87.4	18.8
ESP[5] (%)	5.8	3.23	31.8	10.44	16.4	10.57	3.2	2.68	2.8	1.69	6.3	5.71
TOC[6] ($dag\ kg^{-1}$)	1.4	7.9	0.8	4.1	0.7	1.5	1.4	6.5	1.6	11.7	1.1	6.1
Saturation extract												
ECse[7] ($dS\ m^{-1}$)	64.5	21.33	28.4	23.14	66.1	43.88	3.7	7.1	0.8	1.0	2.1	4.1
Ca^{2+} ($mmol_c\ L^{-1}$)	205.6	189.53	58.5	71.72	14.3	9.08	20.4	53.0	1.9	2.48	6.6	13.3
Mg^{2+} ($mmol_c\ L^{-1}$)	265.3	161.96	106.3	146.37	5.8	6.06	26.6	54.7	3.8	3.64	7.5	10.6
Na^{+} ($mmol_c\ L^{-1}$)	937.4	755.54	779.8	671.09	1856.1	1099.78	108.2	306.9	5.0	8.35	8.0	20.58
K^{+} ($mmol_c\ L^{-1}$)	12.4	16.45	1.9	1.47	143.6	90.2	1.6	2.9	0.5	0.35	0.8	1.26

(continued)

Table 1 (continued)

Property	Soil group											
	1		2		3		4		5		6	
	Mean	SD[1]	Mean	SD	Mean	SD	Mean	SD	Mean	SD	Mean	SD
SAR[8] $(mmol_c\ L^{-1})^{0,5}$	90.6	79.90	97.4	102.92	585.1	122.3	10.3	27.2	3.7	6.57	2.6	4.56

[1]Standard deviation
[2]Cation exchange capacity by index cation method
[3]Sum of bases
[4]Base saturation
[5]Exchangeable sodium percentage
[6]Total organic carbon
[7]Electrical conductivity of saturation extract
[8]Sodium adsorption ratio
Source Pessoa et al. (2012)

Alternative techniques using geoprocessing and remote sensing tools are being used successfully to replace traditional surveys in various parts of the world, including in Brazil.

1.2 Saline and Sodic Soils Monitoring and Mapping

It is widely known that salinity decreases soil quality and its identification and monitoring are essential for the management and sustainability of soils and waters. Remote sensing techniques and geographic information system (GIS) have been widely used to evaluate and to characterize salt affected areas in large scale in several countries around the world (Evans and Caccetta 2000; Farifteh et al. 2006). These tools are powerful for mapping soil salinization due to their fast update, broad coverage and abundant spectral information (Gorji et al. 2017; Triki Fourati et al. 2017). However, studies involving monitoring of soil salinization by these techniques in Brazil are still scarce and appropriate methodologies are needed to accurately assess concentrations and types of salts present in these soils.

Spectral characteristics related to salts can be measured by hyperspectral sensors only if they do not match with strong bands of atmospheric absorption. Pessoa et al. (2016) evaluated the effect of surface salinity (0–5 cm depth) of agricultural areas on spectral responses of soils by multivariate statistics. Hyperspectral data were taken from soil samples of 78 sampling sites in Pernambuco State (Brazil), under different salinity levels. Despite the large difference between saline sites (Table 1), no significant difference in spectral responses was observed. The authors also concluded that exchangeable sodium percentage had high correlation with spectral reflectance.

Using multispectral (TM/Landsat-5 e OLI/Landsat-8) and hyperspectral sensors (Hyperion/EQ-1), Moreira et al. (2015) investigated irrigation-induced soil salinization in a study area used for rice cultivation, in Ceará State (Brazil).

Salinity indices of narrow band from Hyperion/EQ-1 sensor produced a root mean square error (RMSE) lower to estimate electrical conductivity, better discrimination between saline and non-saline soils using spectral measures, as well as higher support vector machine (SVM) classification accuracy (Moreira et al. 2015). Exposed soils detected by the sensor OLI/Landsat-8 had saline areas with smaller variations of NDVI in time (1984–2011) due to inhibition in vegetation growth. The authors concluded that soil salinity detection was much more associated with soil brightness than with absorption bands because most salinity indices express brightness to some extent, and linearly correlated with the electrical conductivity values of saline soils.

The risk/susceptibility to soil salinization in Santa Maria da Boa Vista (Pernambuco State, Brazil) was evaluated by Castro (2018) (Fig. 3). She applied a methodology that uses physical, natural and environmental attribute data through geoprocessing tools. Data were collected in the field for later validation of the laboratory mapping by Kappa statistical method. The soils with very high, high and medium risk to salinization occupied 21.84, 15.72 and 59.71% of the total area, respectively

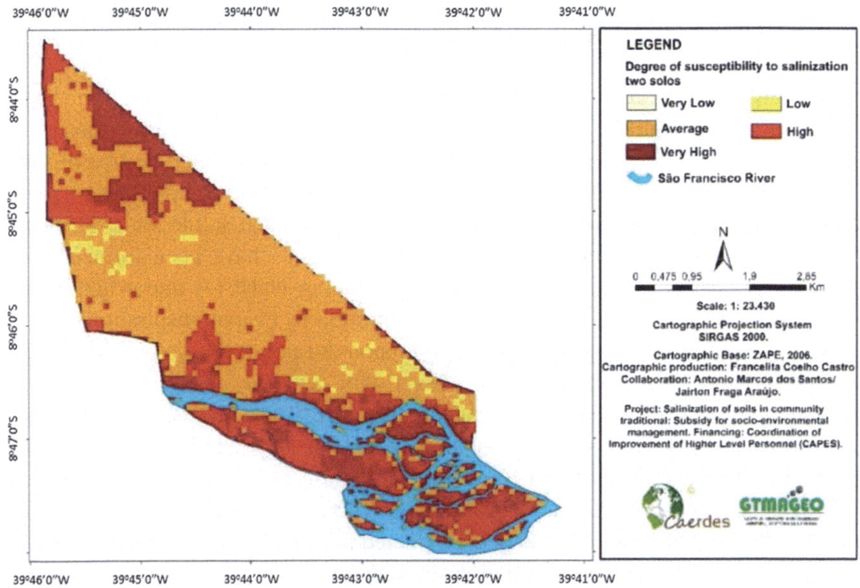

Fig. 3 Map of susceptibility to salinization in Cupira Quilombola Community—Santa Maria da Boa Vista (Pernambuco State, Brazil). *Source* Castro (2018)

(Castro 2018). According to the author, the combination of soil susceptibility to salinization, intensive water use in agricultural production and natural attributes favored the high proportions of degraded soils in these community lands.

One of the main remote sensing techniques used to monitor environmental problems such as soil salinization is the linear spectral unmixing (LSU) method according to Bouaziz et al. (2011). They applied this technique to identify and quantify soil salinization in Brazilian semiarid using wide-range MODIS data. For that, 18 spectral indices were used combined with data from chemical analysis of 112 soil samples from Brazilian semiarid to reveal the salt content in the soil surface layer (0–15 cm). In this study, it was possible to verify a moderate correlation between electrical conductivity and spectral indices, but this correlation was improved by applying the LSU method. They suggested that the LSU method plays an important role in finding more accurate information regarding soil salinity, even about low spatial resolution sensors such as MODIS. They also performed multiple linear regressions between spectral indices and electrical conductivity data of the soil analysis and obtained a moderate multiple determination coefficient ($R^2 = 0.40$). Thus, the authors proposed that the combination of spectral indices and soil analysis is an efficient way to create large-scale soil salinity prediction maps.

Because soil salinization advances in the Brazilian semiarid region, efforts still need to be made regarding mapping and monitoring of soil salinity by remote sensing and GIS techniques.

1.3 Soil and Water Quality in the Brazilian Semiarid Region

Most soluble salts are geographically distributed by water movement over the surface of the land or through the soil mantle from their original areas to lower positions in the landscape, where they naturally contaminated soils that could be free of salts (Metternicht and Zinck 2008). The genesis of sodic soils is more complex. Some sodic soils can be formed by the natural leaching of sodium-rich saline soils, resulting in a high exchangeable sodium percentage. Others may have been formed from the chemical weathering of igneous rocks, producing $NaHCO_3$ under non-saline conditions and forming Na_2CO_3 due to concentration by evapotranspiration.

The climate contributes when there is not enough rainfall for water to percolate through the soil; in topographic depressions with drainage problems that hinder leaching of salts. Additionally, under inadequate management, the use of saline water in irrigation increases the accumulation of salts in soils and their degradation. In the irrigated perimeters of Northeast Brazil, accumulation of salts in soil by irrigation water is common and there is no consideration with regard to salt balance. This has contributed to salinization of soils in the irrigated perimeters in the semiarid region.

Lopes et al. (2014) identified variations in water quality of the Orós reservoir in the semiarid region of Ceará (Brazil) and observed that the main factors influencing water quality were the natural weathering of geological soil components, suspended solids through surface runoff from surrounding agricultural areas and anthropogenic effects in the basin. Part of the reservoir waters were at risk of promoting the accumulation of sodium salts in soils.

Additionally, water quality can vary in time and space, which is very common in the semiarid region of Brazil. Fernandes et al. (2009) monitored waters of rivers, wells and dams in four seasons of the year to assess the quality of water used for irrigation of an irrigated perimeter in Pernambuco, Brazil. They observed the predominance of sodium and chloride in the waters, greater salinity in well water that induced a greater risk of soil salinization in all periods of evaluation. Carbonate contents in water were low, not representing a restriction for their use in irrigation.

However, water degradation goes beyond salinity, is highly influenced by urbanization and land use. To evaluate water quality in surface reservoirs in Upper Jaguaribe Basin, in the Southwest of Ceará State (Brazilian Semiarid), Chaves et al. (2019) collected water samples in three large reservoirs, in dry and rainy periods. They concluded that anthropization had a direct influence on the water quality and the total degradation of those surface waters due to land use in the watersheds is a matter of time. They found EC values ranging from 0.1 to 0.3 dS m^{-1}, in general lower in the rainy season, but the waters were within the salinity limits to be used in irrigation (0.7–3.0 dS m^{-1}). On the other hand, the pH ranged between 5.3 and 9.0. The authors associated these results with the rocks and geological formation of the soils, and to factors related to anthropogenic influence around the reservoirs.

In irrigated perimeters located inside the semiarid region of Brazil, soils have been degraded on account of inadequate irrigation. Water is often saline and the management of irrigation is not followed by drainage. The exposed soils are fragile and degradation expands around huge areas, where desertification is becoming a problem. The reclamation of these areas is complicated and expensive by the techniques extensively used in other countries. Phytoremediation is a way to protect the soil at low cost, in addition to maintaining environmental balance.

2 Phytoremediation as an Alternative Reclamation Technique

The reclamation of salt affected soils has not been common in Brazil. In general, degraded areas are left fallow and cultivation proceeds to other areas, generating an environmental degradation liability. The return of vegetation in these areas would initiate a reclamation process. Phytoremediation is an appropriate, low-cost option, often used when soils or wastewater are contaminated by heavy metals, but it can also be used in sodic or saline-sodic soils (Souza et al. 2014; Miranda et al. 2018). In such case, the phytoextraction technique involves plant species which absorb and accumulate salts in their shoots, allowing the contaminant to be removed from the soil via biomass harvesting (Souza et al. 2014; Melo et al. 2018). Qadir et al. (2007) highlight that phytoextraction is a more effective strategy for the reclamation of saline-sodic soils than the use of chemical amendments.

Hence, phytoremediation of salt-affected soils has several benefits, including lower costs than chemicals, secondary uses of the cultivated species, increased soil aggregate stability, biomass production and macropores development, which optimize the hydraulic properties of soil and enhance root growth. Phytoremediation also allows for soil reclamation at greater depths than other techniques and increases nutrient availability after the completion of remediation process (Souza et al. 2014).

Qadir et al. (2007) proposed five mechanisms shown in Fig. 4 and described below.

2.1 Mechanisms Contributing to Phytoremediation

Several mechanisms are responsible for soil phytoremediation.

(a) Biological activity of plant roots and microorganisms promotes an increase in the partial pressure of CO_2 in the root zone. In calcareous soils it enhances the dissolution of calcite and the release of exchangeable calcium and calcium bicarbonate, and carbonate in the soil solution. In non-calcareous soils, the increase in the partial pressure of CO_2 results in the release of exchangeable hydrogen and reduction in soil pH. Exchangeable calcium in the first situation, and exchangeable hydrogen in the second, act to replace the exchangeable sodium in the soil exchange complex, minimizing the dispersion promoted by this element and its effects on the physical properties of the soil.

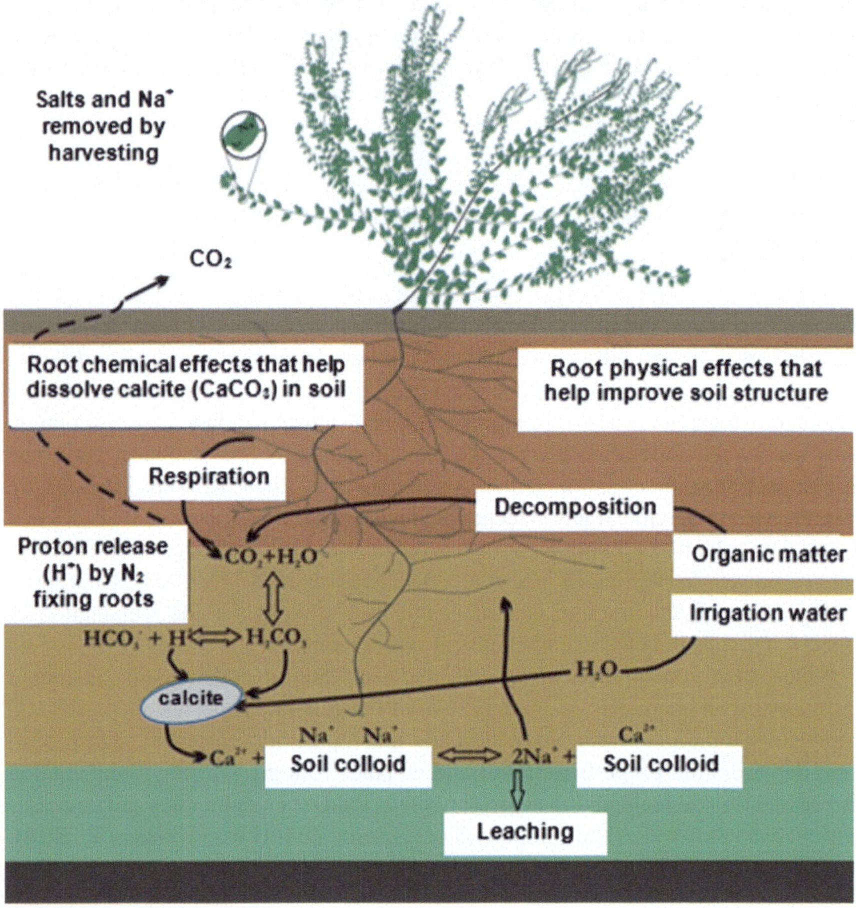

Fig. 4 Factors and reactions occurring during phytoremediation of salt affected soils: increase in CO_2 partial pressure of the root zone, release of protons (H^+), physical effects of the roots, increase in the content of organic matter and removal of salts by harvesting the aerial part of the plants. *Source* Modified from Qadir et al. (2007)

(b) Proton release by the roots in the process of cation absorption contributes to the reduction of pH in the vicinity of the roots. The exchangeable hydrogen extrusion process contributes to reducing pH and to replace the exchangeable sodium in the soil exchange complex, releasing it into the soil solution and facilitating its leaching and absorption by plants.

(c) The physical effect of the roots, related to the formation of macropores and maintenance of the soil structure, enabling the infiltration and penetration of water in the soil profile. In addition to the previous mechanisms, with sodium being released from the exchange complex and gas exchanges reaching deeper layers of the soil, plant cultivation improves the soil in depth, reaching layers that are difficult to be reached by other soil reclamation techniques of saline and sodic soils.

(d) The increase in organic matter content with the cultivation of plants also acts in the aggregation of soil particles, allowing greater stability of aggregates and promoting increase in infiltration rate.
(e) The extraction of salts by plants and the production of aerial biomass, which can be used in animal feed.

The beneficial effects of plants in a saline-sodic soil in semiarid region of Pernambuco was observed in a field experiment with *Atriplex nummularia* L. The plants were cultivated on a Cambisol for 42 months after transplanting, when the roots were distributed up to 70 cm depth and spread laterally (Fig. 5). In this root system, the processes cited in Fig. 4 were occurring, improving soil quality and providing organic matter to the soil microbiota (Fig. 6).

However, the time required for the reclamation of a saline soil is related to many aspects that depend on the soil, the phytoextraction plant and the salinization process. Reclamation will take longer on more saline and sodic soils; reclamation will be faster with plants with high potential for phytoextracting salts; the more continuous the supply of salts to the area, the longer it takes to recover the soil.

Knowledge of the processes of salt accumulation in soils is important for phytoremediation to be effective. If the soil is saline and the source of salt accumulation is of primary origin, the phytoremediation process must be continuous, i.e. phytoremediation species should remain in the degraded area and it is important to reconcile the management of these species with agricultural crops, preferably in intercropping. When salinity and sodicity come from anthropic causes, if the accumulation process is curtailed after phytoremediation, it is recommended to start agricultural activities with salt-tolerant crops, gradually to be replaced by more sensitive ones. However, it must be ensured that the management used after the reclamation of the saline area does not result in new exposure to the problem.

Fig. 5 Soil profile beside an *Atriplex nummularia* L. plant (42 months old), highlighting distribution of root system, experimental area in Serra Talhada (Pernambuco State, Brazil). Photo: Maria Betânia Galvão dos Santos Freire

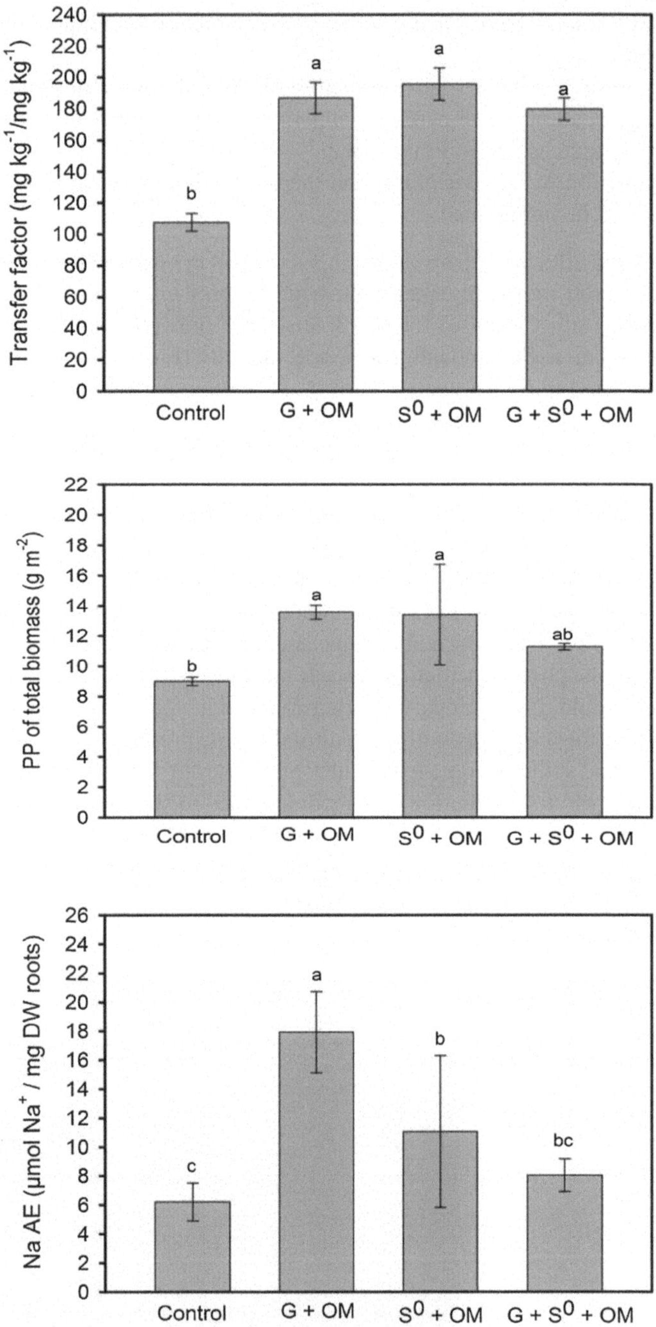

Fig. 6 Transfer factor (**a**), phytoremediation potential (**b**) and sodium absorption efficiency (Na AE) (**c**) of *Atriplexnummularia* L. plants at 60 days after transplanting as a function of mixture of amendments tested. Means followed by the same letters do not differ from each other by Tukey test at $p < 0.05$ probability level. *Source* Leal et al. (2019)

According to Mishra and Sangwan (2016), bioremediation technology must involve successive cultivations of hyper-accumulating plants until the salinity of the soil reaches acceptable levels for other species of greater economic interest. Then, agricultural crops more tolerant to salts, such as cotton, sorghum, sugar beet, fodder grasses and cowpea can be introduced in association with the selected hyper-accumulators, enabling the agricultural use of the soils. These authors suggest that phytoremediation plants should be harvested and, if possible, used as animal feed or production of biosalts, wood, energy, or even incinerated for further industrial use of extracted salts.

2.2 *Suitable Plants for Phytoremediation. Trials in Brazil*

Several plant species can be used in phytoremediation (Hasanuzzaman et al. 2014; Mishra and Sangwan 2016), among them herbaceous plants, shrubs and trees, as long as they are tolerant, have good salt absorption capacity and high biomass production. However, when the objective is the specific removal of chloride and sodium salts from the soil, halophytes are the better choice, due to their ability to survive in soils with very high levels of sodium and chloride. In a review of the potential use of halophytes to recover saline soils, Hasanuzzaman et al. (2014) cite many halophyte species, with illustrations, limits of salt tolerance and their possible uses in human food.

The selection of species to be used in phytoremediation should be based not only on the ability to tolerate high saline concentrations in the soil, but also on the ability to provide a useful product for agricultural activity. Another prerequisite for plants to be used in phytoremediation in salt affected soils is their ability to tolerate oxygen-deficient environments. Floods often occur in such soils due to the limited natural ability of water to infiltrate during rainy seasons, keeping the soil saturated. Thus, plant genotypes which have both these conditions should be preferred for use in phytoremediation of saline and sodic soils.

The species of the genus *Atriplex*, belonging to Chenopodiaceae family, are among the most commonly used plants for phytoremediation of salt affected soils. The Chenopodiaceae family occupies the first place among the top ten plant families in terms of the number of halophytes and xerohalophytes. This family includes species with different types of photosynthetic metabolism (C3 and C4). They often inhabit arid, semiarid, and saline areas, thus experiencing either osmotic stress or a combination of osmotic and ionic stress (Rakhmankulova et al. 2019).

Halophytes have many productive applications: rehabilitating degraded lands, preventing desertification, providing firewood and timber, creating shade and shelter, and producing industrial crops and animal fodder (Shiri et al. 2015). Under drought and salinity conditions, xerohalophytes and halophytes developed numerous physiological and biochemical adaptation strategies, including the accumulation of osmolytes, control of water balance, and maintenance of ionic homeostasis (Rakhmankulova et al. 2019).

Halophytes can regulate the concentration of salts in some ways, such as by extruding salts through glands and trichomes, as observed in the genus *Atriplex* (Souza et al. 2012; Kiani-Pouya et al. 2017). The extraction of salts by *Atriplex nummularia* L. is significant, research results conducted under protected and field environments in Brazil are presented in Table 2. Variability in plant productivity per hectare is due to the evaluation period, plant density and pruning.

Melo et al. (2018) evaluated water relations of *Atriplex nummularia* L. under two soil moisture levels (equivalent to 70 and 37% of field capacity). The plants were irrigated with solutions of NaCl and a mixture of NaCl, KCl, $MgCl_2$ and $CaCl_2$ reproducing six electrical conductivity levels (EC): 0, 5, 10, 20, 30, and 40 dS m^{-1}. The proportion among different ions was based on the composition of irrigation waters used by farmers in the semiarid region of Pernambuco State. The type of ion in the irrigation water did not influence the soil potential. *Atriplex nummularia* survived even at EC close to 40 dS m^{-1} and water potential was approximately −8 MPa. Water potentials determined for different moisture levels, EC levels and salt types may be relevant for the management of this species in semiarid regions and for reclaiming salt-affected soils.

In a field experiment in soil under an intense salinization process in Northeastern Brazil, Souza et al. (2014) studied the management of *Atriplex nummularia* L. plants,

Table 2 Dry matter production, time of evaluation and extraction of salts by *Atriplex nummularia* L., in field condition and in greenhouse

Purpose	Time (days)	Condition	Dry matter (kg ha^{-1})	Extraction (kg ha^{-1})	Reference
Phytoremediation[1]	135	Greenhouse	7283.0	966.0 (Na, K, Cl)	Souza et al. (2011)
Phytoremediation[2]	130	Greenhouse	3000.0	300.0 (Na)	Leal et al. (2008)
Phytoremediation[3]	544	Field	12540.0	728.0 (Na, Cl)	Silva et al. (2016b)
Phytoremediation[4]	544	Field	7330.0	431.2 (Na, Cl)	Silva et al. (2016b)
Phytoremediation[5]	480	Field	8610.0	654.8 (Na, Cl)	Souza et al. (2014)
Phytoremediation[6]	480	Field	8040.0	505.2 (Na, Cl)	Souza et al. (2014)
IDW[7]	375	Field	13819.8	2222.0 (ash)	Porto et al. (2006)
IDW[8]	380	Field	9436.0	1053.0 (ash)	Porto et al. (2001)

[1]Estimation based on production in 12 kg pot, plant kept at moisture equivalent to 75% of field capacity (stem + leaf)
[2]Estimation based on production in 8 kg pot, plant kept at moisture equivalent to 75% of field capacity (stem + leaf)
[3]Plants in 1.0 × 1.0 m spacing; pruning at 6 and 12 months, plant harvested 18 months after transplanting seedlings
[4]Plants in 2.0 × 2.0 m spacing; pruning at 6 and 12 months, plant harvested 18 months after transplanting seedlings
[5]Plants in 2.5 × 2.5 m spacing, pruning at 6 and 12 months, plant harvested 16 months after transplanting seedlings
[6]Plants in 2.5 × 2.5 m spacing, no pruning, plant harvested 16 months after transplanting seedlings
[7]Irrigation with desalination waste (IDW); volume applied per plant 300 L $week^{-1}$ (stem + leaf)
[8]Irrigation with desalination waste (IDW); volume applied per plant 75 L $week^{-1}$ (stem + leaf)

particularly in terms of plant growth, periodic pruning, salt extraction, and soil attributes before and after cultivation. The growth rate was uniform until 6 months after transplanting (MAT) for both the pruned and unpruned plants. At 6 MAT, the average height of the pruned plants was 130.75 cm. At this stage, all aerial material (leaves and stems) above 80 cm, was pruned. The mean height of the plants was 105.25 cm four months after the second pruning, with 45.25 cm of this height being post-pruning growth. Considering this growth, it can be inferred that in a longer period, the pruned plants would have reached the height of the unpruned plants.

These results highlight the capacity of *Atriplex nummularia* L. to regrow. The plants also grow laterally, closing spaces between plants and protecting the soil between the lines. It is noteworthy that the regrowth material is less lignified and more tender and palatable as forage for animals, especially for goats. *Atriplex* plants can grow to over 2.0 m in height in one year (Porto et al. 2001, 2006) and vary in height from 2–3 m after five years (Glenn et al. 2009).

The ideal spacing for the cultivation of *Atriplex* is unclear. Maximization of *Atriplex* biomass yields can be obtained by reducing spacing between plants. However, such spacing reduction should prevent plant overlap. Until 12 MAT, the 2.5 × 2.5 m spacing (1600 plants ha^{-1}) could be reduced to 2.0 × 2.0 m (2500 plants ha^{-1}). Such procedure would increase the crop population by 900 plants ha^{-1} and, hence, would increase biomass yield and salt extraction by plants. It is worthwhile to mention that if lateral branches were pruned, crop spacing could be further reduced to 1.5 × 1.5 m spacing (4444 plants ha^{-1}). It has been verified that plant overlapping would occur after 16 months of cultivation, when the mean crown diameter of unpruned plants would reach 258 cm.

Investigating *Atriplexnummularia* L. field-growth in Northeastern Brazil, Silva et al. (2016b) tested it in two spacing (1 × 1 and 2 × 2), pruning every six months). They recommended *Atriplexnummularia* L. at 1 × 1 m spacing for salt phytoextraction per area, plants extracted more sodium and chloride from the soil than calcium, magnesium and potassium. To improve soil physical conditions they suggested the management of this plant under 2 × 2 m spacing, however the experimental time was not enough to improve soil physical properties. For this purpose, the plant must be maintained on soil for periods longer than 18 months, probably for many years.

Evaluating the effect of mixtures of amendments on the initial growth and salt extraction efficiency of *Atriplexnummularia* L., Leal et al. (2019) applied gypsum+organic matter, elemental sulphur+organic matter, and gypsum+elemental sulphur+organic matter. They did not observe growth improvement with those amendments, but Na and Cl contents in the dry matter of *Atriplex* plants increased, improving their ability to extract salt. They also found that *Atriplex* plants biomass production was leaf > stem > root, accumulating more sodium and chloride in leaves. A use for this biomass could possibly be in animal feeding. Finally, they evaluated transfer factor (TF), phytoremediation potential of the total biomass (PP) and sodium absorption efficiency (Na AE) of *Atriplexnummularia* plants 60 days after transplanting (Fig. 6). All the amendments improved the potential for using *Atriplex* plants in phytoremediation, and the authors highlighted the gypsum+organic matter treatment. The application of some amendments can improve the capacity to extract salt from the soil and it is important for soil rehabilitation.

Previously, Leal et al. (2008) had tested gypsum for improving phytoremediation by *Atriplex* plants, with 96% increase in root dry matter in response to gypsum application. Sodium accumulation in shoots was higher with gypsum, which promoted an increment in the sodium transfer factor from the soil to the plant.

The farmers in Brazil are not familiar with *Atriplex* plants and prefer other tree species. Many of the common species are not tolerant to salts and cannot grow on degraded soils but there are a few recent studies being conducted on this issue. Leite (2018) tested four species, in a degraded area of Brazilian semiarid region, the Cachoeira II Irrigated Perimeter (Serra Talhada, Pernambuco). The plants were: *Atriplexnummularia* L., *Leucaena leucocephala* (Lam.) de Wit, *Azadirachta indica* A. Juss. and *Mimosa caesalpiniifolia* Benth, either in monocrop, or in intercropping treatments: (1) *A. nummularia* and *L. leucocephala*; (2) *A. nummularia* and *A. indica*; (3) *A. nummularia* and *M. caesalpiniifolia*. After 30 months of cultivation, biomass production was evaluated (Table 3). The only native species of Caatinga

Table 3 Production of dry biomass (stem, leaf, flower, fruit and total) of *Atriplexnummularia* L., *Leucaena leucocephala* (Lam.) de Wit, *Azadirachta indica* A. Juss. and *Mimosa caesalpiniifolia* Benth plants under monocrop and intercropped cultivation, at 30 months

Plant	Plant part				
	Stem	Leaf	Flower	Fruit	Total
	kg plant^{-1}		g plant^{-1}		kg plant^{-1}
A. nummularia[1]	4.15ab	0.95a	–	–	5.10ab
A. nummularia[2]	4.28ab	1.27a	–	–	5.55ab
A. nummularia[3]	6.67ab	1.43a	–	–	8.10ab
A. nummularia[4]	5.88ab	1.58a	–	–	7.46ab
L. leucocephala[5]	1.64ab	0.52a	5.93b	736.95a	2.90ab
A. indica[6]	7.39a	3.22a	32.74a	135.41b	10.78a
M. caesalpiniifolia[7]	0.52b	0.18a	–	–	0.73b
L. leucocephala[8]	1.94ab	0.29a	6.45b	501.29a	2.74ab
A. indica[9]	3.57ab	1.52a	34.55a	124.65b	5.22ab
M. caesalpiniifolia[10]	1.58ab	0.88a	33.20a	–	2.46ab

[1]*Atriplexnummularia* plant in monocrop cultivation; [2]*Atriplex nummularia* plant in intercropped cultivation (*A. nummularia* x *L. leucocephala*); [3]*Atriplex nummularia* plant in intercropped cultivation (*A. nummularia* x *A. indica*); [4]*Atriplex nummularia* plant in intercropped cultivation (*A. nummularia* x *M. caesalpiniifolia*); [5]*Leucaena leucocephala* plant in intercropped cultivation (*A. nummularia* x *L. leucocephala*); [6]*Azadirachta indica* plant in intercropped cultivation (*A. nummularia* x *A. indica*); [7]*Mimosa caesalpiniifolia* plant in intercropped cultivation (*A. nummularia* x *M. caesalpiniifolia*); [8]*Leucaena leucocephala* plant in monocrop cultivation; [9]*Azadirachta indica* plant in monocrop cultivation; [10]*Mimosa caesalpiniifolia* plant in monocrop cultivation. Means followed by the same letter in the column do not differ statistically by Tukey test at $p < 0.1$ probability level. *Source* Leite (2018)

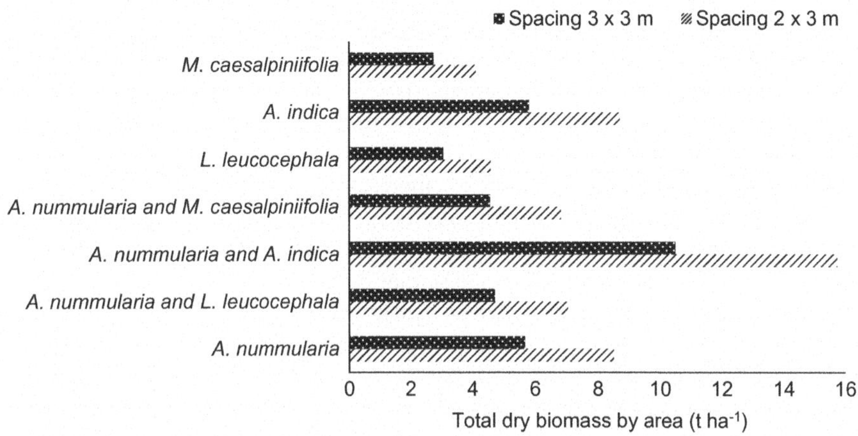

Fig. 7 Total shoot dry biomass production per hectare (t ha^{-1}) estimated at two spacing cultivation (3 × 3 m and 2 × 3 m) using single plants [*Atriplexnummularia* L., *Leucaena leucocephala* (Lam.) de Wit, *Azadirachta indica* A. Juss. and *Mimosa caesalpiniifolia* Benth)] and intercropping treatments of: *A. nummularia and M. caesalpiniifolia, A. nummularia* and *A. indica,* and *A. nummularia* and *L. leucocephala. Source* Leite (2018)

Biome (*Mimosa caesalpiniifolia*) had lower biomass production especially in intercrop with *A. nummularia.* Although it is widely found in Brazilian Semiarid, it cannot grow properly in that salt affected soil and, possibly, it is not a good species for phytoremediation. The exotic species had better growth and *Azadirachta indica* remarkably benefited from intercropping with *A. nummularia* in biomass production (Table 3).

It was also possible to estimate biomass production by area at 3 × 3 and 2 × 3 m spacing. Intercropping *A. nummularia* and *A. indica* would produce almost 16 t ha^{-1} of aerial plant biomass at 2 × 3 m spacing and more than 10 t ha^{-1} at 3 × 3 m (Fig. 7). The root system was not considered although it is the most active part in phytoremediation, because it was not possible to remove the roots from the soil without considerable losses. Leite (2018) also studied soil properties and microorganisms associated with these plants, and reported that *Atriplexnummularia* in an intercropping production system, was effective to improve soil quality, enabling the cultivation of species less tolerant to salts (Leite et al. 2020).

In another study bioprospecting bacteria associated to *Atriplexnummularia* L. in two field experiments located in two localities of Pernambuco State, Silva et al. (2016a) found 107 salt-tolerant isolates. The authors tested the isolates for phosphate solubilization rate, biological nitrogen fixation, indole acetic acid production, exopolysaccharides and quorum sensing molecules, and tolerance to salinity. They suggested some bacteria that are promising to contribute for rehabilitation of salt affected soils by *Atriplexnummularia.* Santos et al. (2011) found that pruning stimulated microbial activity close to the root system.

In relation to soil physical attributes, phytoremediation also plays a fundamental role in improving soil infiltration, contributing to the leaching of salts. Miranda et al. (2018) compared the response to organic (bovine and sheep manure) and chemical conditioners (gypsum and polymer) with *Atriplexnummularia* cultivation on the physical properties of a saline-sodic soil at Irrigated Perimeter of Custódia (State of Pernambuco). Although the polymer was more efficient in improving the physical properties of the studied soil, the cultivation of *Atriplex* was also effective and the authors highlight its importance as a more sustainable alternative that can be used as complementary fodder in animal feed in semiarid regions.

For phytoremediation purposes, *Atriplex* plants can be grown without the addition of fertilizers. However, the use of products that provide greater growth and absorption of salts can be beneficial. Cunha et al. (2017) evaluated dry matter production and sodium extraction capacity of *Atriplex* plants, in response to nitrogen doses, with and without phosphorus application (Fig. 8). *Atriplexnummularia* was responsive to N fertilization potentiating the extraction of Na, and they concluded that it can be a viable technique in the management of salt affected soils through phytoremediation. Bioconcentration factor (BCF) was calculated as the ratio between Na concentrations in leaves and soil exchangeable Na, in g kg^{-1}, and makes it possible to understand the phytoaccumulation potential of the plant in relation to a contaminant. There was a linear increase in BCF with the increase in N doses, but the application of P decreased BCF (Fig. 8). For this reason, the authors considered the application of N to be a positive factor in phytoremediation, but the addition of P was not indicated.

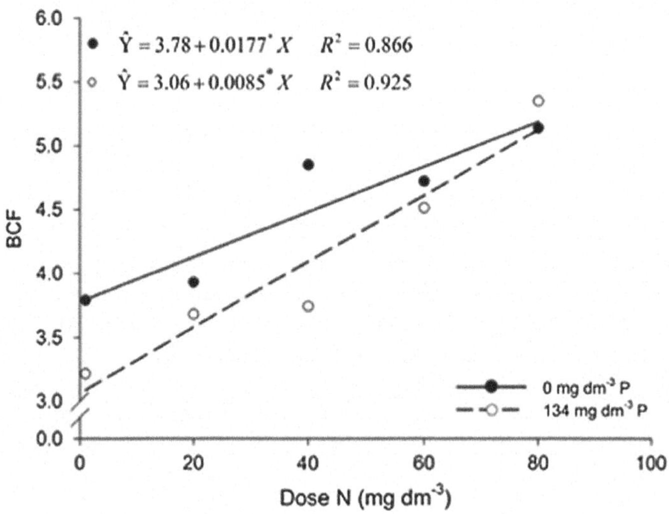

Fig. 8 Bioconcentration factor (BCF) according to the applied nitrogen doses, with and without phosphate fertilization. Significant with $P < 0.05$ by t-test. *Source* Cunha et al. (2017)

3 Future Perspectives

Phytoremediation research is growing in Brazil but many questions still remain unanswered. Plants reaction depends on the type of soil, as there is a great variability of soils, it is necessary to expand these studies. Native species in saline areas of the semiarid region and their associated microorganisms are candidates for phytoremediation.

On the other hand, species providing products that can be commercially exploited are also interesting, even if they are not native. The phytoremediation potential of established plants with tolerance to salt and water stress should be assessed.

The management of plants for phytoremediation must be improved, considering aspects such as spacing between plants, crop rotation, consortium among species, use of irrigation, fertilization and other cultural treatments. Such studies will contribute to promote phytoremediation as an environmentally friendly option for reclamation of saline or contaminated soils at low cost.

References

Bouaziz M, Matschullat J, Gloaguen R (2011) Improved remote sensing detection of soil salinity from a semi-arid climate in Northeast Brazil. C R Géosci 343:795–803. https://doi.org/10.1016/j.crte.2011.09.003

Castro FC (2018) Salinização de solos em comunidade tradicional – subsídio à gestão socioambiental. Universidade do Estado da Bahia, Programa de Pós-Graduação em Ecologia Humana e Gestão Socioambiental, Juazeiro. Dissertação (Mestrado), 84 p

Chaves LCG, Lopes FB, Maia ARS, Meireles ACM, Andrade EM (2019) Water quality and anthropogenic impact in the watersheds of service reservoirs in the Brazilian semi-arid region Rev Ciên Agron 50(2):223–233. https://doi.org/10.5935/1806-6690.20190026

Cunha JC, Freire MBGS, Ruiz HA, Silva JDG (2017) Phytoextraction potential of *Atriplex nummularia* plants under nitrogen and phosphate fertilization. Commun Soil Sci Plan 48(1):20–36. https://doi.org/10.1080/00103624.2016.1253716

Evans FH, Caccetta PL (2000) Broad-scale spatial prediction of areas at risk from dryland salinity. Cartograhy 29(2):33–40. https://doi.org/10.1080/00690805.2000.9714106

Farifteh J, Farshad A, George RJ (2006) Assessing salt affected soils using remote sensing, solute modelling, and geophysics. Geoderma 130(3–4):191–206. https://doi.org/10.1016/j.geoderma.2005.02.003

Fernandes JG, Freire MBGS, Cunha JC, Galvíncio JD, Corrêa MM, Santos PR (2009) Qualidade físico-química das águas utilizadas no Perímetro Irrigado Cachoeira II, Serra Talhada, Pernambuco. Rev Brás Ciênc Agrar 4:27–34

Glenn EP, Mckeon C, Gerhart V, Nagler PL, Jordan F, Artiola J (2009) Deficit irrigation of a landscape halophyte for reuse of saline waste water in a desert city. Landscape Urban Plan 89:57–64. https://doi.org/10.1016/j.landurbplan.2008.10.008

Gorji T, Sertel E, Tanik A (2017) Monitoring soil salinity via remote sensing technology under data scarce conditions: a case study from Turkey. Ecol Indic 74:384–391. https://doi.org/10.1016/j.ecolind.2016.11.043

Hasanuzzaman M, Nahar K, Alam MM, Bhowmik PC, Hossain MA, Rahman MM, Prasad MNV, Ozturk M, Fujita M (2014) Potential use of halophytes to remediate saline soils. Bio Med Res Intern 589341:1–12. https://doi.org/10.1155/2014/589341

Kiani-Pouya A, Roessner U, Jayasinghe NS, Lutz A, Rupasingue T, Bazihizina N, Bohm J, Alharbi S, Hedrich R, Shabala S (2017) Epidermal bladder cells confer salinity stress tolerance in the halophyte quinoa and *Atriplex* species. Plant Cell Environ 40:1900–1915. https://doi.org/10.1111/pce.12995

Leal IG, Accioly AMA, Nascimento CWA, Freire MBGS, Montenegro AAA, Ferreira FL (2008) Fitorremediação de solo salino sódico por *Atriplex nummularia* e gesso de jazida. Rev Bras Ciênc Solo 32:1065–1072. https://doi.org/10.1590/S0100-06832008000300015

Leal LSG, Pessoa LGM, Oliveira JP, Santos NA, Silva LFS, Barros Júnior G, Freire MBGS, Souza ES (2019) Do applications of soil conditioner mixtures improve the salt extraction ability of *Atriplex nummularia* at early growth stage? Int J Phytoremediat 21:1–8. https://doi.org/10.1080/15226514.2019.1678109

Leite MCBS (2018) Atributos de solo sob revegetação no semiárido brasileiro e extração de elementos pelas plantas. Universidade Federal Rural de Pernambuco, Programa de Pós-Graduação em Ciência do Solo, Recife. Tese (Doutorado), 123 p

Leite MCBS, Freire MBGS, Queiroz JVJ, Maia LC, Duda GP, Medeiros EV (2020) Mycorrhizal *Atriplex nummularia* promote revegetation and shifts in microbial properties in saline Brazilian soil. Appl Soil Ecol 153:1–10. https://doi.org/10.1016/j.apsoil.2020.103574

Lopes FB, Andrade EM, Meireles ACM, Becker H, Batista AA (2014) Assessment of the water quality in a large reservoir in semiarid region of Brazil. Rev Bras Eng Agr Ambient 18:437–445. https://doi.org/10.1590/S1415-43662014000400012

Melo HF, Souza ER, Almeida BG, Mulas M (2018) Water potential in soil and *Atriplex nummularia* (phytoremediator halophyte) under drought and salt stresses. Int J Phytoremediat 20:249–255. https://doi.org/10.1080/15226514.2017.1374334

Metternicht G, Zinck JA (2008) Spectral behavior of salt types In: Metternicht G, Zinck JA (eds) Remote sensing of soil salinization: impact on land management. CRC Press, Boca Raton, pp 21–37

Miranda MF, Freire MBGS, Almeida BG, Freire AG, Freire FJ, Pessoa LG (2018) Improvement of degraded physical attributes of a saline-sodic soil as influenced by phytoremediation and soil conditioners. Arch Agron Soil Sci 64:1–15. https://doi.org/10.1080/03650340.2017.1419195

Mishra NK, Sangwan A (2016) Phytoremediation of salt-affected soils: a review of processes, applicability, and the impact on soil health in Hisar, Haryana. IJESIRD 2(7):418–423

Moreira LCJ, Teixeira AD, Galvão LS (2015) Potential of multispectral and hyperspectral data to detect saline-exposed soils in Brazil. GISci Remote Sens 52(4):416–436. https://doi.org/10.1080/15481603.2015.1040227

Pessoa LGM (2012) Analysis of salt affected soils in semiarid landscapes of Pernambuco, Brazil. Recife, Federal Rural University of Pernambuco, Graduation Program in Agronomy—Soil Science. Ph.D. Thesis, 124 p

Pessoa LGM, Freire MBGS, Wilcox BP, Green CHM, Araújo RJT, Araújo Filho JC (2016) Spectral reflectance characteristics of soils in northeastern Brazil as influenced by salinity levels. Environ Monit Assess 188(616):1–11. https://doi.org/10.1007/s10661-016-5631-6

Porto ER, Amorim MCC, Dutra MT, Paulino RV, Brito LTL, Matos ANB (2006) Rendimento da *Atriplex nummularia* irrigada com efluentes da criação de tilápia em rejeito da dessalinização de água. Rev Bras Eng Agríc Ambient 10(1):97–103. https://doi.org/10.1590/S1415-43662006000100015

Porto ER, Amorim MCC, Silva Júnior LGA (2001) Uso do rejeito da dessalinização de água salobra para irrigação da ervasal (*Atriplex nummularia*). Rev Bras Eng Agríc Ambient 5(1):111–114. https://doi.org/10.1590/S1415-43662001000100020

Qadir M, Oster JD, Hubert S, Noble AD, Sahrawat KL (2007) Phytoremediation of sodic and saline-sodic soils. Adv Agron 96:197–247. https://doi.org/10.1016/S0065-2113(07)96006-X

Rakhmankulova ZF, Shuyskaya EV, Yu Voronin P, Yu Usmanov I (2019) Comparative study on resistance of C_3 and C_4 xerohalophytes of the genus *Atriplex* to water deficit and salinity. Russ J Plant Physiol 66:250–258

Santos KCF, Silva MSL, Silva LE, Miranda MA, Freire MBGS (2011) Atividade biológica em solo salino-sódico saturado por água sob cultivo de *Atriplex nummularia*. Rev Ciênc Agron 42(3):619–627. https://doi.org/10.1590/S1806-66902011000300007

Shiri M, Rabhi M, Amrani E, Abdelly C (2015) Cross-tolerance to abiotic stresses in halophytes: application for phytoremediation of organic pollutants. Acta Physiol Plant 37(10):209. https://link.springer.com/article/10.1007/s11738-015-1954-0

Silva FG, Santos IB, Sousa, AJ, Farias ARB, Diniz WPS, Sobral JK, Freire MBGS (2016a) Bioprospecting and plant growth-promoting bactéria tolerant to salinity associated with *Atriplex nummularia* L. in saline soils. Afr J Microbiol Res 10(31):1203–1214. https://doi.org/10.5897/AJMR2016.8202

Silva YJAB, Silva YJAB, Freire MBGS, Lopes EAP, Santos MA (2016b) *Atriplex nummularia* Lindl. as alternative for improving salt-affected soils conditions in semiarid environments: a field experiment. Chil J Agr Res 76:343–348. https://doi.org/10.4067/S0718-58392016000300012

Souza ER, Freire MBG, Melo DVM, Montenegro AAA (2014) Management of *Atriplex nummularia* Lindl. in a salt affected soil in a semiarid region of Brazil. Intern J Phytorem 16:73–85. https://doi.org/10.1080/15226514.2012.759529

Souza ER, Freire MBGS, Cunha KPV, Nascimento CWA, Ruiz HA, Lins CMT (2012) Biomass, anatomical changes and osmotic potential in *Atriplex nummularia* Lindl. cultivated in sodic saline soil under water stress. Environ Exp Bot 82:20–27. https://doi.org/10.1016/j.envexpbot.2012.03.007

Souza ER, Freire MBGS, Nascimento CWA, Montenegro AAA, Freire FJ, Melo HF (2011) Fitoextração de sais pela A*triplex nummularia* Lindl. sob estresse hídrico em solo salino-sódico. Rev Bras Eng Agríc Ambient 5(15):477–483. https://doi.org/10.1590/S1415-43662011000500007.

Triki Fourati H, Bouaziz M, Benzina M, Bouaziz S (2017) Detection of terrain indices related to soil salinity and mapping salt-affected soils using remote sensing and geostatistical techniques. Environ Monit Assess 189(4):177. https://doi.org/10.1007/s10661-017-5877-7

Salinization in Peruvian North Coast Soils: Case Study in San Pedro de Lloc

Nadia R. Gamboa, Adolfo B. Marchese, and Carlos H. Tavares Corrêa

Abstract Soils in San Pedro de Lloc have been used for agriculture since pre-Columbian times. Rice crops are still currently irrigated by flooding using water from the Jequetepeque River. Therefore, soil degradation by salinization is a permanent threat to soil fertility and production. This chapter presents the results of studies carried out in the 1960s, 70s, and 80s, as well as recent works on surface water, underground water, and soil quality monitoring. Physicochemical parameters, such as electrical conductivity, pH, total dissolved solids, hardness, chemical oxygen demand, as well as nitrate, sulfate, phosphate, and chloride concentrations were measured in both surface water and underground water. Organic matter, texture class, and SAR were additionally measured in the soil samples. Electrical conductivity was spatially represented for easy comparison. The isopleth curves of the water table and a map showing the expansion of the agricultural area are included. It is concluded that the increase in the volume of water used for irrigation after the construction of the Gallito Ciego dam has raised the water table and soil salinization has been stimulated. This has led to the abandonment of cultivation fields and groundwater wells between 2003 and 2019 due to intense salinization.

Keywords Shallow water table · Soil degradation · Rice · Fruit trees · Water quality

N. R. Gamboa (✉)
Departamento Académico de Ciencias, sección Química; Grupo GRIDES, Pontificia Universidad Católica del Perú, Lima, Perú
e-mail: ngamboa@pucp.pe

A. B. Marchese
Facultad de Ciencias e Ingeniería, Pontificia Universidad Católica del Perú, Lima, Perú

C. H. Tavares Corrêa
Departamento Académico de Humanidades, sección Geografía y Medio Ambiente; Grupo GRIDES, Pontificia Universidad Católica del Perú, Lima, Perú

E. Taleisnik and R. S. Lavado (eds.), *Saline and Alkaline Soils in Latin America*, https://doi.org/10.1007/978-3-030-52592-7_7

1 Introduction

Soil salinity is a problem that affects approximately 40% of the total agricultural area of the Peruvian coast (ONERN 1988). This region is highly productive, dynamic, and commercially successful (World Bank 2017). This is due to the presence of broad plains with fertile soils, water availability, and nearby markets. However, the arid conditions of the coast force agriculture to develop only by irrigation (Alva et al. 1976).

Agriculture is oriented to both the export and supply of the internal market through non-traditional products, such as grapes and asparagus, or traditional products, such as sugar and rice. Rice is one of the main traditional crops on the northern coast (which corresponds to 33.5% of the national production). It is developed from flood irrigation, it consumes between 15,000 and 18,000 m^3/ha and generates coastal soil degradation by salinization. Salzgitter (1969), Lostao (1971), ONERN (1973), Alva et al. (1976), ONERN (1988), Fox (2013), and Marchese (2015) reported the state of soil salinization on the Peruvian coast and, especially, in the Jequetepeque Valley and San Pedro de Lloc district.

2 Soil Degradation in the Peruvian Northern Coast

The Peruvian coast, known as "*costa*", is an important natural region that occupies a narrow strip of land 15–100 km wide and 3080 km long between the Andes Mountains Range and the Pacific Ocean. In this region, the dominant landforms are a flat to undulating plains interrupted by river valleys and coastal massifs (Wilson 1984; Pulgar Vidal 1987). The climate in the coast is arid, semi-warm, with a deficiency of rain in all seasons and with high relative humidity (SENAMHI 1988). Aridity in the Peruvian Coast is due to the high stability of the atmosphere because of the influence of the cold surface ocean water near the coast, related to the Humboldt Current, and the intense coastal upwelling (Schulz et al. 2011). Sixty-two river basins flow through the western slope of the Andes Mountains and finally arrive at the Pacific Ocean. From a continuous or intermittent supply of freshwater, rivers that reach the desert coast give rise to valleys with fertile soils where most of the country's productive agriculture develops. During ENSO (El Niño-Southern Oscillation) events, unusual warming of surface waters may cause heavy rainfall on the northern Peruvian coast (Trenberth 1997) that changes the aridity to a wetter condition and promotes flooding and recharge of the coastal aquifers.

The hard accessibility in a complex geography produces a very unequal occupation of the territory. Because of this, the costa region concentrates 57% of the population (INEI 2017), and although it has only 23% of the entire country's agricultural land, it produces 46% of the national agricultural gross domestic product (GDP) (MINAGRI 2015). The costa region offers favorable conditions for agriculture due to climatic conditions, flat terrain, relative water availability, and proximity to the consumption and export centers (World Bank 2017). The coastal valleys provide water for irrigation systems that have allowed a variety of crops such as asparagus, peppers, onions, sugarcane, yellow corn, rice, and fruit trees to be cultivated in the desert. In the costa region, 87% of the crops are under irrigation and correspond to a dynamic, highly productive, and commercially successful agricultural system (World Bank 2017).

Despite the success of agriculture, the excessive use of irrigation water combined with extreme aridity is an important cause of drainage and salinization problems in the costa region. Concerns about drainage and salinization problems in Peru were attended during the 1960s and 1970s, mainly by institutions, such as the National Office of Natural Resources Assessment (ONERN) and the Drainage and Land Reclamation Center (Cendret). ONERN (1973) reviewed the knowledge about soil salinization until the 1970s. As part of a soil resources evaluation program in Peru during the 1960s, ONERN initiated the execution of a project to delimitate and quantify the areas affected by salinization problems in the Peruvian Coast. The US Salinity Laboratory (USSL-USDA 1954) classification of saline and sodium soils was used to evaluate 39 valleys on the Peruvian coast. The results indicated that 37% of the agricultural soils analyzed (146,361 ha) corresponded to soils affected by incipient up to evident salinity (ONERN 1973). On the other hand, Cendret elaborated a systematic study on drainage problems and salinization of 42 valleys at the Peruvian coast between 1972 and 1974 (Lostao 1971; Alva et al. 1976). Based on the technical reports, they concluded that 34% of croplands presented strong to moderate problems of drainage and salinity.

3 Soil Salinization in San Pedro de Lloc

The study area for this contribution is the San Pedro de Lloc Hydraulic Sub-sector belonging to the Jequetepeque River regulated basin (Fig. 1). San Pedro de Lloc is the capital city of the province of Pacasmayo. This hydraulic sub-sector is northwest of the district and covers 6055.92 ha (ca. 60.56 km^2) of cultivation areas. The study area is located in an inter-basin area south of the Jequetepeque River.

Andean Foothills, an alluvial plain, low hills, sand sheets, dune fields, and the current beach dominate the landscape of the study area. The Andean Foothill is composed of volcanic rocks of the Llama Group and prevails east and northeast; this landform unit is present west of the Andes Mountains and reaches the highest elevation of the study area (over 200 m above sea level). To the southwest of the Andean Foothills, the alluvial plain (named pampa) arises, being the most extensive landform unit. It consists of a wide plain composed of poorly consolidated conglomerates deposited in the form of alluvial fans. Flows of materials that come from streams running over the western slope of the Andes Mountains form these conglomerates (Wilson 1984). The alluvial plain has gentle slopes that extend for 15–20 km toward the Pacific Ocean. Near the coast, to the southwest of the study area, isolated low-altitude hills formed by intrusive igneous rocks are present. The aeolian deposits cover southern and southeastern parts of coastal plains and, as it approaches the

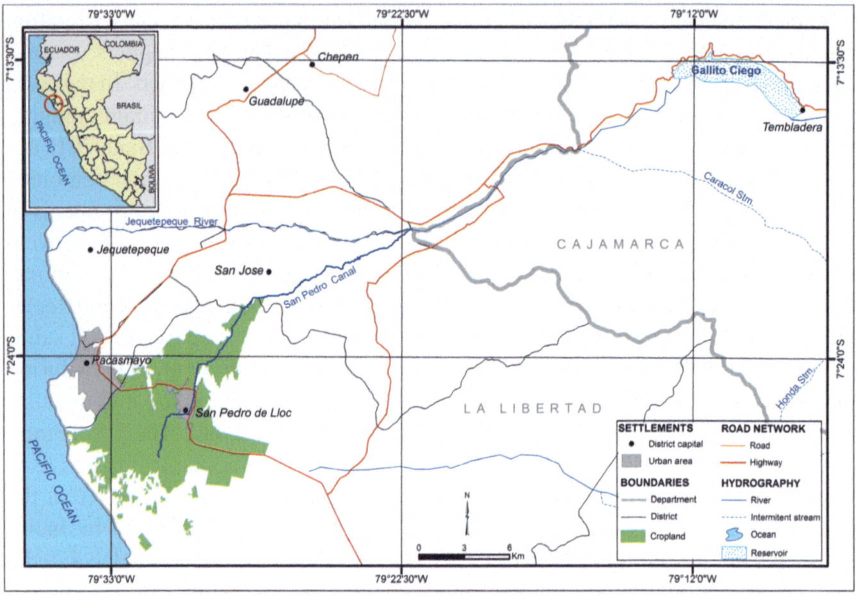

Fig. 1 Study area location (Tavares Corrêa, for this publication)

Andean Foothills, they accumulate as dune fields where the succession of barchan dunes, barchanoids, and complex dunes is noted. Sand sheets and dunes accumulate on the windward slope of the hills as ramps and exceed the summits of the hills advancing to the leeward slope in the form of shadow dunes. Finally, the coastal strip consists of beaches of fine sand and pebbles (Tavares Corrêa 2004).

San Pedro de Lloc and agriculture have been associated for centuries. Cupisnique (BC 700), Salinar (BC 300), Moche (between AD 200 and 800), Chimu (between AD 1100 and 1450), and Inca (between AD 1470 and 1532) societies occupied the Jequetepeque valley successively. There is geological and archaeological evidence about the widespread practice of floodplain cultivation dating to the Late Moche through Chimu periods, particularly along with seasonal or low-discharge drainage systems, as well as reuse, rebuilding and relocation of major canals in the area (Dillehay et al. 2004). The canals for irrigation, defensive overflow weirs, and aqueducts built by Chimus were abandoned after the Spanish Conquest. Subsequently, the Spaniards established the indigenous reductions, and a fishermen village in San Pedro, in the domains of native chief Lloco, was renamed as San Pedro de Lloc, as it is known today (Burga 1976; Dillehay et al. 2004).

There are historical records that indicate the oasis condition of this area and the predominance of algarrobo (*Prosopis juliflora*) and espino (*Acacia tortuosa*). Corn, cotton, beans, and fruit trees grew and coexisted with algarrobo and espino during the seventeenth century. In the eighteenth century, the native algarrobo forests diminished, and rice replaced wheat. Between 1905 and 1923, rice became the almost

Fig. 2 Rice paddies in San Pedro de Lloc (Photographs by Nadia Gamboa 2007)

exclusive crop, and since 1935, this area became the rice pantry of the country (Burga 1976) (Fig. 2).

Rice (*Oryza sativa* L.) grows up to 2500 m above sea level, needs minimum temperatures from 10 to 13 °C for germination, with an optimum between 30 and 35 °C. The optimum temperature for stem, leaf, and root growth is 23 °C. Above it, plants grow faster, but are susceptible to diseases. Above 50 °C, blossoming does not occur. Low night temperatures favor the ripening of the grains (MINAGRI and SENAMHI 2018).

The soil varies from sandy to clayey, with fine and medium textures that are usually characteristic of flooded plains. Fine-textured soils make tillage difficult (Fig. 3), but in contrast, they are more fertile because of their higher clay content, organic matter, and nutrients. Therefore, soil texture is a crucial parameter in irrigation and fertilizer management. The water flow for irrigation varies between 12,000 and 18,000 L/hm^2, and the water consumption in gravity irrigation is 13,000 m^3/ha (MINAGRI and SENAMHI 2018).

Fig. 3 Tillage in San Pedro de Lloc (Photograph by Nadia Gamboa 2011)

Fig. 4 Gallito Ciego Reservoir (Photographs by Nadia Gamboa 2011)

Fig. 5 San Pedro Catchment (left), irrigation canal (middle), and drainage in Santa Elena Beach (right) (Photographs by Nadia Gamboa 2007)

The Gallito Ciego Reservoir (Fig. 4), inaugurated in 1988, has 400 million m^3 of useful volume and 50 years of estimated lifetime. Water from the Jequetepeque River has been deviated in San Pedro Catchment giving rise to a complicated network of irrigation canals (Fig. 5). Studies were carried out to gather information on water, soil, biota, agricultural production, among others, as part of this project.

ONERN (1988) reported that 20.6% of agricultural land in San Pedro had salinity and poor drainage problems. These drainage problems are caused by the restriction of the underground flow due to impermeable barriers, rocky outcrops within the valley, and lithology with a predominance of medium to fine soils, drainage of canals and, largely, by rice crops. Likewise, the analysis reported that surface waters suitable for irrigation, had low Sodium Adsorption Ratio (SAR) values, although the salinity of well waters varied from medium to highly saline.

In addition, ONERN (1988) reported that the slight salinity problems identified in soils were due to the capillary rise of water from the water table and its subsequent evaporation on the surface. Besides, it was determined that 37.1% of the agricultural area in San Pedro was affected by salinity in 1969, 47.1% in 1978, 20.9% in 1984, 40.6% in 1986, and it was estimated 60% by 1995 (see Tables 1 and 2).

Table 1 Evolution of soil salinity in the San Pedro agricultural area (adapted from ONERN 1988)

Soil salinity level	Salzgitter (1969)		Puiggros (1984)		ONERN (1986)	
	ha	%	ha	%	ha	%
Normal	34,766	73.5	37,663	80.6	33,840	71.5
Slightly saline	5622	11.9	1909	4.1	7200	15.2
Moderately saline	3388	7.2	3211	6.9	3030	6.4
Strongly saline	3524	7.4	3950	8.4	13,460	6.9

Table 2 Evolution of saline and sodium-saline soils in the San Pedro agricultural area (adapted from ONERN 1988)

Type of soils	Puiggros (1984)		ONERN (1986)	
	ha	%	ha	%
Normal	37,663	79.7	33,840	71.5
Saline	5513	11.6	8600	18.2
Sodium-saline	4116	8.7	4860	10.3

4 Water and Soil Monitoring in San Pedro de Lloc

The salinization of soils is due to aridity conditions: superficial and underground water quality and pollution, marine influence, excessive use of fertilizers, among others. In the study area, some studies of water and soil quality have been carried out in recent decades. However, the collected studies contribute to the understanding of the current condition of soils in San Pedro de Lloc.

4.1 Surface Water Quality Monitoring

Quispe (2009) took superficial water samples in selected sites of the irrigation canals network, from the intake in Jequetepeque River (San Pedro Catchment, denoted Q1 in Fig. 6) to the drainage in Santa Elena Beach (Q12), where it pours directly into the sea. The main agricultural campaign starts in September, concludes in April, and the main crop is rice. The complementary agricultural campaign runs from February to July, and yellow corn, onion, and other minor crops are grown. Finally, the small agricultural campaign begins in April and ends in August, and the predominant crop is yellow corn. Therefore, the sampling dates selected were April, June, and September, 2007, considering local agricultural campaigns.

Water samples showed alkaline pH; values oscillated around 8.3, with a minimum of 7.62 (Q10 in September), and a maximum of 8.7 (Q2 in April). At Q9, the highest electrical conductivity and total dissolved solids content were reported in June 1794 μS/cm and 1376 mg/L, respectively. At the Q2 sampling site, the lowest electrical

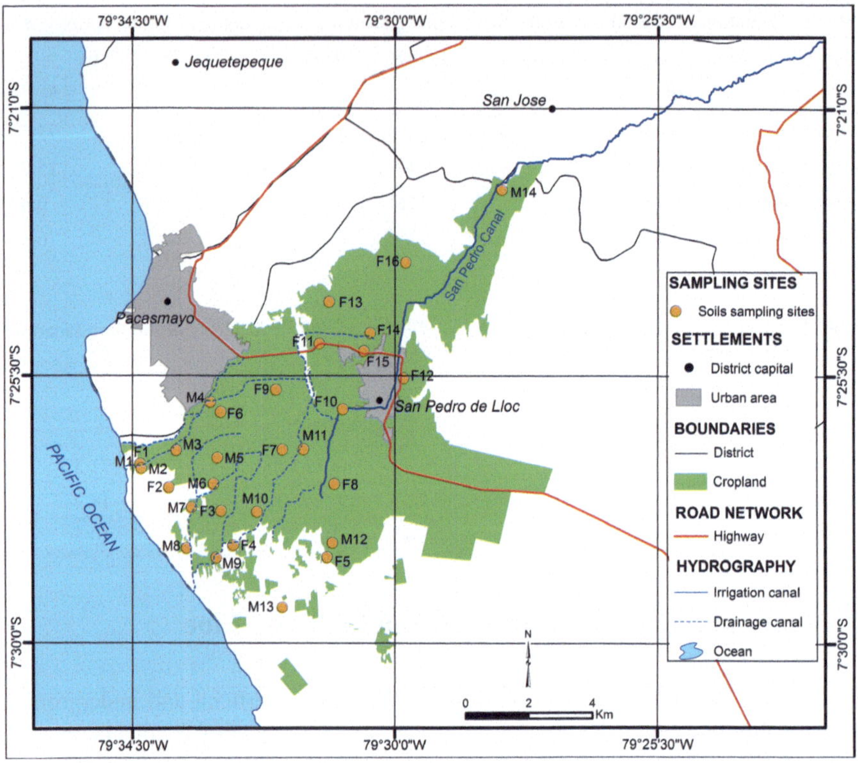

Fig. 6 Soil sampling sites. F or M denotes samples taken by Fox or Marchese, respectively (Tavares Corrêa, for this publication)

conductivity and TDS were reported in April, 191 μS/cm and 158 mg/L, respectively. Total dissolved solids levels increased as the sampling sites approached the coast. This correlation suggested solids input due to the erosion of soils without vegetation cover.

The maximum chemical oxygen demand COD (20.55 mg/L) at the Q7 sampling site in September was attributed to urban influence. Both the hardness and the chloride concentration increased as the sampling site came near the coast and for samples collected in September.

Nitrate concentrations reached minimum value at Q10 (0.4 mg/L in September) and maximum at Q5 (1.41 mg/L, April), the site located in an urban area. In the case of sulfate, it was observed that concentrations from samples collected from Q1 to Q6 sites in the three dates were below 70 mg/L. The samples collected at Q9 (447 mg/L), Q11 (423 mg/L), and Q12 in September were concordant with the high sulfate values expected in environments close to the coastline.

By comparison, where such information is available, dissolved phosphorus levels were higher in the samples collected in those stations nearest to the San Pedro Catchment during April. The highest value presented at Q7 site (3220 mg/L) at the exit of the district in September is ten times higher than at Q4, clearly showing the influence of the discharge of untreated domestic effluents to irrigation canals.

Quispe (2009) concluded that anthropogenic activities affected the quality of water used for irrigation, particularly in the upper part of the study area, and largely in the middle part of the study area. The urban wastewater would contaminate crops through polluted water distributed by the irrigation canals. Nitrates, chlorides and organic matter contents relate directly to organic and agricultural pollution along the San Pedro Canal. Nitrates levels are above legally permissible limits and could be attributed to bad agricultural practices because of an overdose of nitrogen fertilizers that can salinize the soil and alter the quality of groundwater.

4.2 Groundwater Quality Monitoring

The intensive rice harvesting with flood irrigation and use of nitrogen fertilizers motivated the study of nitrate content in groundwater due to its relationship with the syndrome of Methemoglobinemia, or Blue Child Disease. High levels of nitrates in drinking water or food oxidize the iron atom in the hemoglobin molecule, whereby this hemoprotein is unable to bind oxygen reversibly. Thus, nitrate concentration is a chemical parameter that must be under control in superficial and underground water (de Fernícola 1989; Arumi et al. 2006).

Vinelli (2012) determined physicochemical parameters in underground water in this area in 2007. Underground water from wells located in harvesting, livestock farming, and rural areas was sampled in April when the main campaign was closing, and June while small and complementary campaigns were still being carried out. While this study was carried out, it was known that wells located in V4, V5, V7, and V9 (in Fig. 6) were no longer used by any purpose, although those were previously used for irrigation, watering cattle or human consumption. The sampling stations were georeferenced. The pH values varied from neutral to alkaline, and a clear correlation was observed between CE and TDS.

Vinelli (2012) concluded that nitrate concentrations were below the legal limits in the country. Conductivity, pH, and chloride concentration were within legal limits, but chloride was high in abandoned wells or near the sea, as expected. Most of the samples were very hard waters, that is, $CaCO_3$ concentrations higher than 300 mg/L A well in an area of intense agricultural activity had higher levels of phosphate during the second sampling, which showed the impact of agricultural activity in water quality. On the other hand, the content of this anion in abandoned wells was lower probably due to fertilizer disuse or nearby agricultural activities.

4.3 Soil Monitoring

Fox (2013) made a diagnosis of soil loss by salinization in San Pedro de Lloc performing a space–time analysis of variation in agricultural use from 1980 to 2003. As part of this study, samples of subsurface soil were collected at a depth of 15 cm in August 2011 and August 2012. The sampling sites were georeferenced (see Fig. 6). Texture, pH, EC, and organic matter content were determined in these samples.

Samples collected in the transect nearest to the coastline ranged from strongly to extremely saline based on EC and pH values obtained. Those results are consistent with field observations because salt accumulation was observed on the surface of the cropland. Samples collected in the study area showed a sandy texture. However, samples with high alkaline content (i.e., samples collected nearest to the coastline) were suitable for plant growth.

Organic matter was higher than 10%; results suggested a more advanced soil salinization process, especially near the coastline. Some plots located in that zone were abandoned and covered with salt grass and crystallized salts. This situation and the loss of crop areas may be related. Marine influence is still possible due to the proximity of those plots with high salinity contents. The soil in the area closest to the coast is between strongly saline and extremely saline, according to the analysis performed. Moving away from the coast, soils vary from slightly saline to moderately saline at the coast, and from slightly saline to non-saline in the innermost remote areas. During fieldwork, some abandoned plots covered with saltgrass were observed (see Fig. 7).

Marchese (2015) studied salinization in agricultural soils in San Pedro de Lloc. Soil samples were collected in August 2013 and August 2014, in 14 culture plots in various agricultural conditions, considering existing crops, rain precipitation, and irrigation water availability. Grid sampling was used for selecting georeferenced

Fig. 7 Saltgrass over San Pedro de Lloc fields (Photograph by Adolfo Marchese, 2014)

sample sites describing three transects parallel to and four transects perpendicular to the coastline (see Fig. 6).

The soils sampled near the coastline were mostly sandy (>50%), tending to sandy loam in the middle and the amount of clay progressively increased. This tendency throughout all transects that were perpendicular to the coast prevailed due to the winds from the south. Sand advance was strong, due to the winds from the south.

The abandoned plots in the low and middle sectors had very low contents of organic matter due to the burning of crop residues and plowing, among other activities. Consequently, organic matter decreased below 1%. An important sign of degradation is the disappearance of the organic matter layer that covers and protects the soil.

The concentration of nitrate was high in the three sectors of the study area (between 30 and 100 mg/kg). Similar nitrate levels were expected in the samples in August 2013 and August 2014 but it was not so because of the severe drought between January and February 2014. In the absence of water or because of soil degradation, farmers use to change crops. In the M9 sampling site, rice was changed for quinoa (*Chenopodium quinoa*). Quinoa is a pseudo-cereal domesticated by the ancient Peruvian people, rich in proteins and minerals. It is known as an alternative crop due to its low water consumption and with a more flexible tolerance range than rice.

Light salinization was determined in the middle sector in August 2013 and verified by the presence of halophilic vegetation (see Fig. 7) in some sampling stations. In this study, some water samples collected in the irrigation canals and drainages had EC less than 4 dS/m, which correspond to non-saline soils (see Fig. 8). However, it is possible a cumulative effect of salt contribution to the soil, so a subsequent evaluation of this water was recommended.

When comparing the results obtained here with those reported in the 60s, 70s, and 80s (see Fig. 9), Marchese (2015) concluded that the salinization of these soils has accelerated since the number of non-saline crops has been reduced to one in the fourteen points of the selected sampling sites of this study. A possible cause could be the construction of the Gallito Ciego dam and the expansion of irrigation canals.

The incidence of sodium in the low sector is due to its proximity to the sea and the high water table. The variation of SAR over a year (from August 2013 to August 2014) increased both in the low and middle sectors. This phenomenon is linked to the increase in the degree of salinization at these same points and the loss of density and structure thereof. The presence of sodium in the soil profile concerns because this element can be exchanged for calcium or magnesium. The soils with this property lose structural stability since Ca and Mg are needed for stabilizing colloids and clays.

The water from the irrigation canals and drainages contributes to the sodification of the soil in a cumulative manner by the contribution of sodium in small amounts over time. According to the results of SAR, calcium and magnesium were in constant exchange for sodium. Thus, the structure of the soil was rapidly degrading since the exchanged sodium made unstable clays and colloids of the soil. On the other hand, the exchanged elements traveled to the macropores and were constantly being eliminated by leaching, making the groundwater hard.

Marchese (2015) concluded that textures of soils collected from the nearest transect to the coastline varied from slightly saline to extremely saline, and with a pH

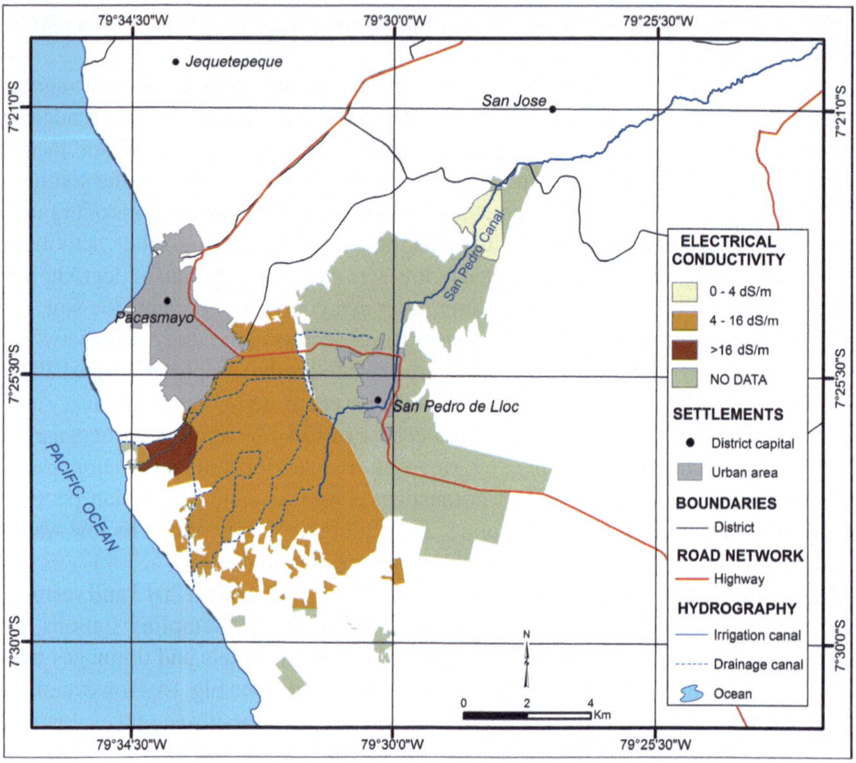

Fig. 8 Electrical conductivity of soils in San Pedro de Lloc, according to Marchese (2015) (Tavares Corrêa, for this publication)

between 7.5 and 8.5, which shows serious salinization and increased sodium content. It should be caused by low water retention capacity (less than 2%), excessive water evaporation, high water table, its proximity to the coast and excessive use of fertilizers based on sulfates and nitrates. However, high organic matter contents (>2%) are explained because rice is still harvested in those soils and irrigation water provides salts in excess. Sodium soils with a high SAR correspond to 45% of soils collected and exhibit medium to low water retention capacity due to predominant sand texture. This research concluded that the soils studied in San Pedro de Lloc are running through serious salinization and increased sodium content due to bad agricultural practices.

5 Effects of Intensive Agriculture on Soil Properties in San Pedro de Lloc

Angulo (2002) identified agricultural areas in the district of San Pedro de Lloc threatened by the soil salinization process associated with a rise in the water table. A map of apparent land use was drawn up based on aerial photographs from 1997 with three

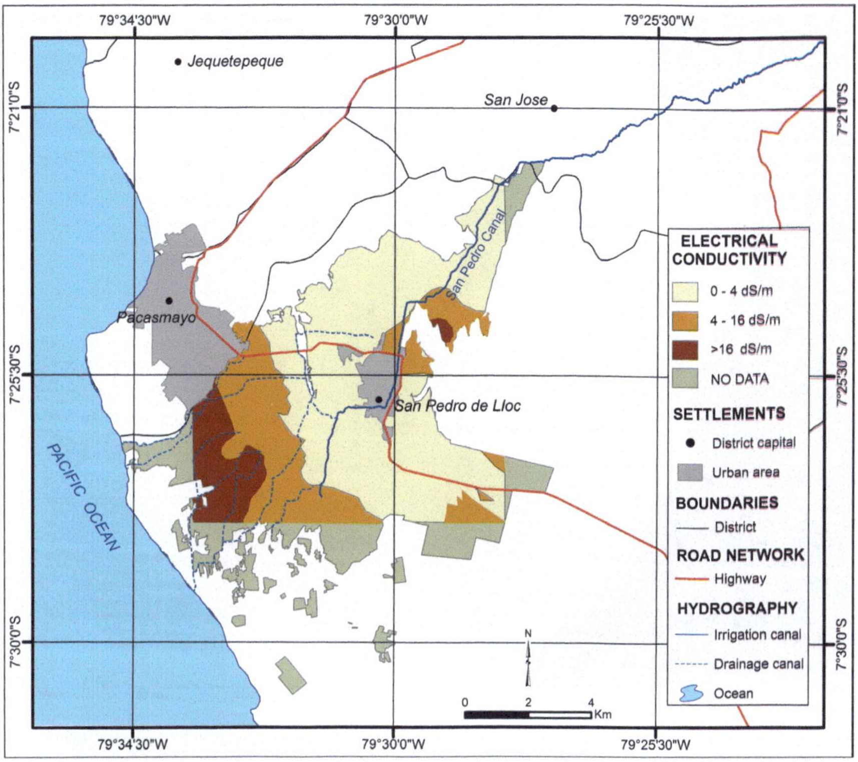

Fig. 9 Electrical conductivity of soils in San Pedro de Lloc, according to Salzgitter (1969) (Tavares Corrêa, for this publication)

kinds of uses (rice, polyculture or fallow crops, and non-agricultural use) confirmed in the field. The measure of the depth of 23 wells distributed evenly in the study area showed variations in the water table. A buffer zone of two thousand meters was designed around each well to delimitate the area before interpolating the depth values by regularized SPLINE. An index of representativeness of rice cultivation was proposed with the total area of each unit of analysis and the rice cultivation area related. Spearman's rank correlation coefficient was applied to the depth of the aquifer for each unit of spatial analysis (given by the depth of each well) and to the rice crop representativeness index.

Higher water table areas were located close to both the coast and the main irrigation canal (Fig. 10). According to Autoridad Administrativa del Agua Jequetepeque Zarumilla (2015), this trend is maintained in the current days. Therefore, this fact indicates that the canal is still the largest recharge agent for the aquifer inside the district. Of the 12,274.6 total ha in the area studied by Angulo (2002), 4971 ha presented a very shallow water table (Fig. 11), being threatened by soil salinization processes due to capillarity and evaporation, mainly in areas near the coast with seawater intrusion. Measurements made by Angulo in mid-May 2002, when the rice

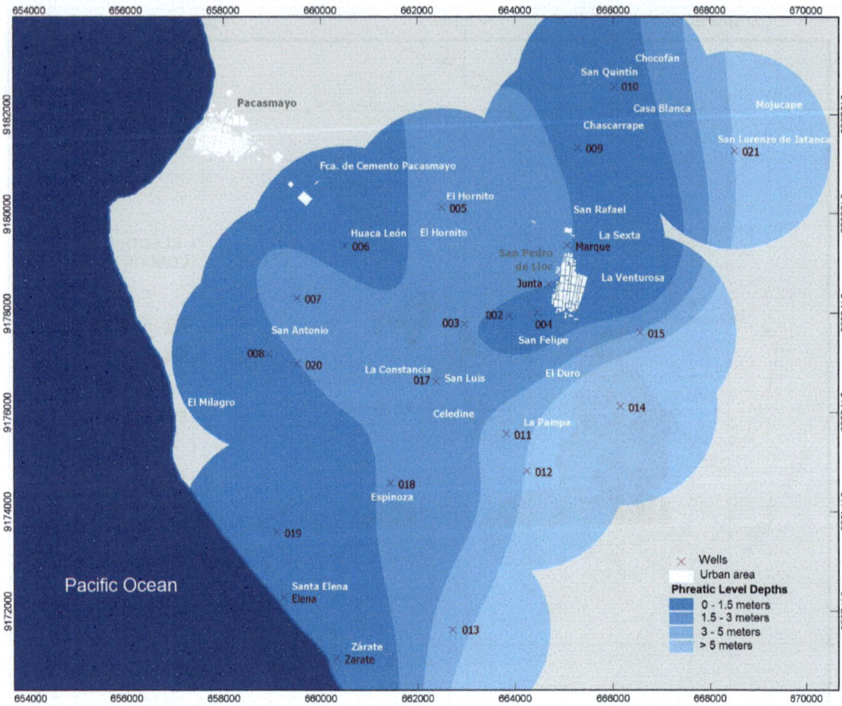

Fig. 10 Isodepth curves of the water table in San Pedro de Lloc in May 2002 (Angulo 2002)

harvest had already been carried out, showed that rice agriculture did not cause the aquifer to rise throughout the year. Assuming that the water table rises during the large campaign, it should be accepted that the soil would be well-drained (Angulo 2002). The water table on the low part of the study area between 2004 and 2015 tended to ascend at rates of 0.012–0.58 m/year (Autoridad Administrativa del Agua 2015).

Fox (2013) compared aerial photographs and satellite images for determining the spatial variation of the cultivated area in the study area between 1980 and 2003. The photographs of 1980 are relevant because they describe the most recent status available before the Gallito Ciego dam. The crop extent that was exclusively dependent on the seasonal flows of the Jequetepeque River was determined. Due to the construction of the dam and the resulting regulated flow, more water for irrigation activities was available throughout the valley all over the year. Therefore, more than one crop could develop in one year-round; more intensive would be the use of the soil; more water used that raises the water table and, then, an increased risk of soil salinization.

According to Fox (2013), the crops covered 36.63 km^2 in 1980 and 37.02 km^2 in 2003 in San Pedro de Lloc. The author pointed out agricultural areas (4.87 km^2) no

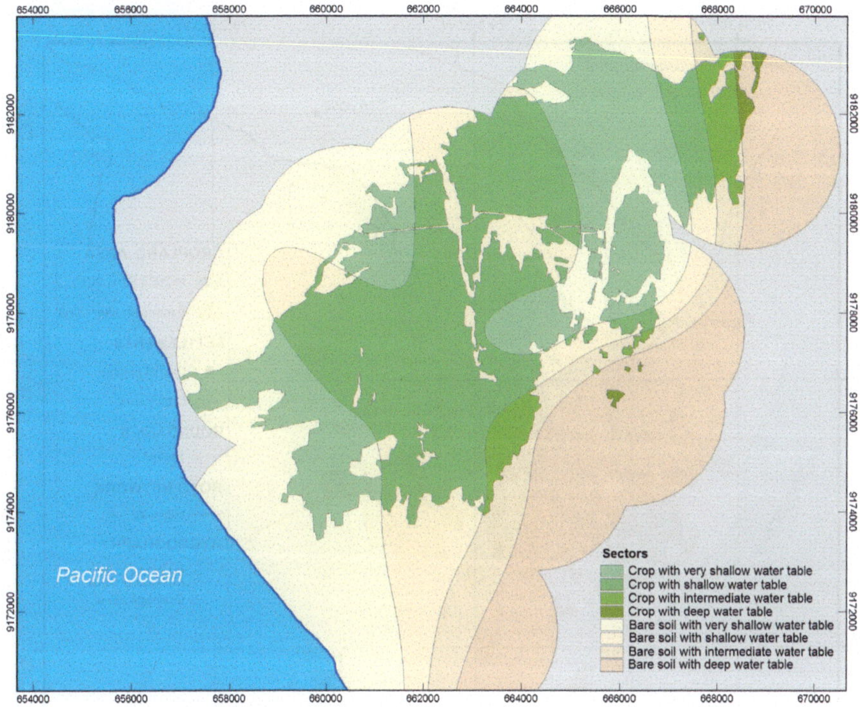

Fig. 11 Land use and relationship with water table depth (Angulo 2002)

longer planted and new agricultural areas (5.26 km^2) were included. Field observations served to identify the state of abandonment of agricultural plots. A broad part of these was located near the coast and covered with salt grass (*Distichlis spicata* (L.) Greene). On the other hand, most of the new farmland was located near the coastal strip. The transitional crops, such as chili pepper, and corn, were technically irrigated using underground pumping because the irrigation canals did not reach this area. The soils of these coastal sectors are typically sand texture and ideal for drip irrigation (USSLA-USDA 1954).

The extension of the crops of San Pedro de Lloc was again delimited to compare the results of Fox (2013) with the current conditions. To delimit a recent cropland extent in the study area, a georeferenced image of Google Earth dated April 17, 2019, was applied. In the same way, Sentinel 2 was used to ensure a good precision of the crop limits. This image, dated February 11, 2019, represents the moment at which all crops, and especially rice, reached their highest state of growth. The crop limits were digitalized in Arcmap v.10.5. Symmetrical Difference Analysis in ArcToolbox was used to compare crop areas of 2003 and 2019 (Fig. 12). The results confirmed that the cultivated areas increased by 63.6%, i.e., from 37.05 km^2 in 2003 (Fox 2013) to 60.56 km^2 in 2019. Most of this expansion results from the export-oriented agribusiness

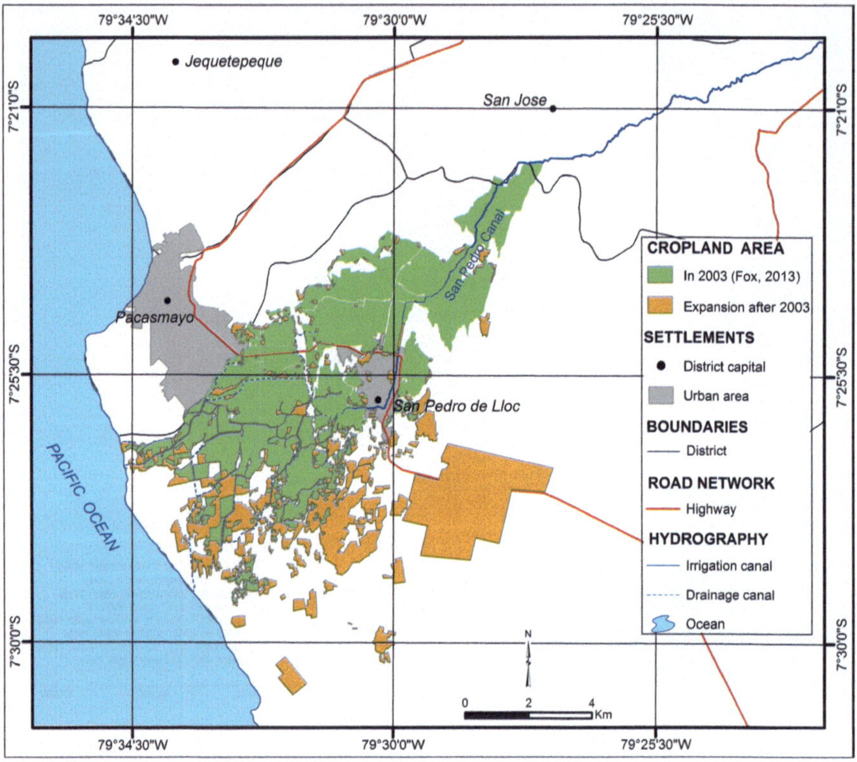

Fig. 12 Expansion of the agricultural area in San Pedro de Lloc from 2003 (Tavares Corrêa, for this publication)

located south of the traditional cropland. These crops employ a technified irrigation system supplied by groundwater. It is important to mention that only 0.87 km^2 of the total cultivated area in 2003 has ceased to be used for agriculture.

6 Measures of Soil Salinization Control

An adequate drainage system is the main salinization control of agricultural soils in San Pedro de Lloc. According to PEJEZA (2012), Santa Elena (17.92 km) and El Milagro (8.62 km) drainage channels formed a network that collects the irrigation waters leaving the rice crops and drive them toward the ocean. However, it is important to note that frequent washout of soils is also a control measure. Since the construction of the Gallito Ciego dam, soils throughout the valley can support two or even three agricultural campaigns due to the availability of irrigation water year-round, which prevents or reduces the accumulation of salts in the profile. Finally,

Alva et al. (1976), suggested that soils can reduce salinity by eliminating rice agriculture in sectors with light salinity and moderate drainage. Probably, farmers would not abandon rice cultivation since the high water table inhibits any other crop in these currently rice-dominated sectors.

7 Conclusions

After the construction of the Gallito Ciego dam, the flow regulation of the Jequetepeque River allowed a second agricultural campaign throughout the valley. The increase in the volume of water used for irrigation raised the water table, mainly in the cultivation areas near the San Pedro canal and those closest to the coastline. Due to this increase in the water table, a more intensive salinization process was expected. However, only a few cultivated areas were abandoned between 2003 and 2019. Although soils tend from light-to-moderate salinity, salt-tolerant crops are still cultivated in the study area without major restrictions.

Acknowledgements Nadia Gamboa and Carlos Tavares Corrêa thank the Dirección de Gestión de la Investigación (Research Management Office) of the PUCP for funding the research projects DGI 2011-0110 and DGI 2013-0097. Adolfo Marchese thanks the same office for the prize of the Programa de Apoyo a la Iniciación en la Investigación—PAIN (Support Program for Emerging Research).

References

Alva CA, Van Alphen JG, De La Torre A, Manrique L (1976) Problemas de drenaje y salinidad en la costa peruana. Bulletin 16 International Institute for Land Reclamation and Improvement. ILRI, Wageningen, pp 3–37

Angulo E (2002) Estudio de la profundidad del acuífero en un valle costero sometido a riegos intensivos. Academic Student Report, School of Arts and Human Science, Pontificia Universidad Católica del Perú. Lima, Perú

Arumi J, Núñez J, Salgado L, Claret M (2006) Evaluación del riesgo de contaminación de nitrato de pozos de suministro de agua potable rural en Chile. Rev Panam Salud Publ 20:385–392 https://iris.paho.org/xmlui/bitstream/handle/123456789/7904/a04v20n6.pdf?sequence=1&isAllowed=y

Autoridad Administrativa del Agua—Jequetepeque Zarumilla (2015) Evaluación de los recursos hídricos en la Cuenca del río Jequetepeque: Aguas Subterráneas. Lima, Diciembre 2015, Ministerio de Agricultura y Riego, Autoridad Nacional del Agua, Dirección de Conservación y Planeamiento de Recursos Hídricos, Autoridad Administrativa del Agua Jequetepeque Zarumilla, Administración Local del Agua Jequetepeque. 250pp. https://siea.minagri.gob.pe/siea/sites/default/files/Informe-coyuntura-arroz-280818_0.pdf

Burga M (1976) La encomienda a la hacienda capitalista. El valle del Jequetepeque del Siglo XVI al XX. Instituto de Estudios Peruanos. Serie: Estudios de la Sociedad Rural, vol 4. 1st edn. Lima

de Fernícola N (1989) Metahemoglobinemia Infantil Causada por Nitratos. Bol Sanit Panam 106:32–40. https://iris.paho.org/xmlui/bitstream/handle/123456789/17754/v106n1p32.pdf?sequence=1&isAllowed=y

Dillehay T, Kolata AL, Pino M (2004) Pre-industrial human and environment iteractions in northern Peru during the Late Holocene. Holocene 14:272–281. https://pdfs.semanticscholar.org/67c7/b8d9d413799ac1533465d265bf0ae142c6ff.pdf
Fox E (2013) Evaluación de pérdida de suelo por salinización en la parte baja de la cuenca del Jequetepeque: San Pedro de Lloc (1980–2003). Licentiate Thesis in Geography and Environment, School of Arts and Human Science, Pontificia Universidad Católica del Perú. Lima, Peru. https://tesis.pucp.edu.pe/repositorio/handle/20.500.12404/4809
INEI Instituto Nacional de Estadística e Informática (2017) Censos Nacionales 2017: XII de Población VII de Vivienda y III de Comunidades Indígenas. INEI Lima, Perú
Lostao J (1971) Recuperación de tierras mediante el riego y drenaje en la costa del Perú. 1er Seminario Latinoamericano Sobre la Evaluación Sistemática de Tierras y Aguas, México https://repositorio.ana.gob.pe/bitstream/handle/20.500.12543/1529/ANA0000310.pdf?sequence=1&isAllowed=y
Marchese A (2015) Estudio físico y químico de suelos agrícolas para la estimación del nivel de salinización en el sector Bajo de San Pedro de Lloc. Licentiate Tesis in Chemistry, School of Science and Engineering, Pontificia Universidad Católica del Perú, Lima, Peru. https://tesis.pucp.edu.pe/repositorio/handle/20.500.12404/6442
MINAGRI Ministerio de Agricultura y Riego (2015) Evaluación de recursos hídricos de la Cuenca del Jequetepeque—Aguas superficiales Technical Report, Lima
MINAGRI and SENAMHI (2018) Ficha técnica 09. Requerimientos Agroclimáticos del Cultivo de Arroz. https://www.minagri.gob.pe/portal/informacion-agroclimatica/fichas-tecnicas-2018?download=1355l:ficha-tecnica-cultivo-de-arroz
ONERN—Oficina Nacional de Evaluación de Recursos Naturales (1973) Evaluación de la Salinidad en el Perú. Contribución al "Grupo de Trabajo para la Evaluación y Control de Degradación de Tierras Áridas de América Latina" Proyecto Regional FAO/PNUD RLA 70/457. Ministerio de Agricultura—DGA/DIPRECO, Santiago, Chile
ONERN (1986) Perfil ambiental del Perú. Lima, 275 p
ONERN (1988) Plan de ordenamiento ambiental de la cuenca del río Jequetepeque para la protección del reservorio Gallito Ciego y del valle agrícola. ANA0000077_pdf
PEJEZA—Proyecto Especial Jequetepeque—Zaña (2012) Inventario de la Infraestructura hidráulica mayor de riego y drenaje. Yonan, enero, 2012 Ministerio de Agricultura PEJEZA Gerencia de Operación y Mantenimiento
Pulgar Vidal J (1987) Geografía del Perú, las ocho regiones naturales, 9th edn. PEISA, Barcelona
Puiggros Ingenieros Consultores (1984) Estudio a nivel definitivo de desarrollo agrícola del valle Jequetepeque, Yonán, Perú, as cited by ONERN (1988)
Quispe E (2009) Diseño de un programa de monitoreo de agua superficial en San Pedro de Lloc. Master Thesis in Chemistry, Graduate School, Pontificia Universidad Católica del Perú, Lima
Salzgitter Industriebau Gesellschaft MBH (1969) Proyecto Jequetepeque: estudio semidetallado de suelos. Gobierno de la República del Perú, Comité Especial del Valle del Río Jequetepeque. Tomo I (Informe General). https://repositorio.ana.gob.pe/bitstream/handle/20.500.12543/4052/ANA0002456_1.pdf?sequence=1&isAllowed=y
Schulz N, Boisier P, Aceituno P (2011) Climate change along the arid coast of Northern Chile. Int J Climatol 32:1803–1814
SENAMHI—Servicio Nacional de Meteorología e Hidrología (1988) Mapa de Clasificación Climática del Perú Método de Thornthwaite SENAMHI, Lima
Tavares Corrêa C (2004) El *Continuun Dunario* y el manejo de dunas litorales en el norte del Perú. B Soc Geog Lima 117:171–180
Trenberth KE (1997) The definition of "El Niño." Bull Am Meteorol Soc 78:2771–2778. https://doi.org/10.1175/1520-0477(1997)078%3C2771:TDOENO%3E2.0.CO;2
USSL-USDA United States Salinity Laboratory Staff (1954) Diagnosis and improvement of saline and Alkali soils. Handbook, vol 60. USDA, Washington, DC

Vinelli N (2012) Estudio analítico de nitratos en aguas subterráneas en el Distrito San Pedro de Lloc. Licentiate Thesis in Chemistry, School of Science and Engineering, Pontificia Universidad Católica del Perú, Lima https://tesis.pucp.edu.pe/repositorio/handle/20.500.12404/1463

Wilson J (1984) Geología de los cuadrángulos de Jayanca (13-d), Incahuasi (13-e), Cutervo (13-f), Chiclayo (14-d), Chongoyape (14-e), Chota (14-f), Celendín (14-g), Pacasmayo (15-d), Chepén (15-e). INGEMMET. Boletín. Serie A: Carta Geológica Nacional, vol 38. https://repositorio.ingemmet.gob.pe/bitstream/20.500.12544/157/2/A-038-Boletin_Jayanca-13d_Incahuasi-13e_Cutervo-13f_Chiclayo-14d_Chongoyape-14e_Chota-14f_Celendin-14g_Pacasmayo-15d_Chepen-15e.pdf

World Bank (2017) Gaining momentum in Peruvian agriculture: opportunities to increase productivity and enhance competitiveness. Lima

Effects of Salinity on Vineyards and Wines from Mendoza, Argentina

Rosana C. Vallone, Laura E. Martínez, Federico G. Olmedo, and Santiago E. Sari

Abstract More than 160,000 ha in Mendoza Province, Argentina, are cultured with vineyards, and a large percentage is grown in areas with high salinity. The correlations of soil salinity with the ion concentration in wines, in vineyards located in saline areas of the Mendoza River Basin were examined. Some relevant conclusions are: (i) chloride concentrations in petioles at harvest were in general lower than 1%, except for Cabernet Sauvignon and Malbec in high soil salinity sites; (ii) chlorides in musts exceeded legal limits in Cabernet Sauvignon, Tempranillo, and Pedro Giménez; (iii) chlorides in musts and in wines were not clearly related; (iv) the correlation between chlorides in soil and wine was $r = 0.50$; (v) there were significant correlations between soil pH and must pH and between wine K concentration and wine pH; (vi) the K concentration in musts and wines was not related to salinity variables or exchangeable soil K. The 2017 vintage wines from the area were tested by a wine tasting panel. The experts perceived salty flavors that were positively associated with bitterness in wines from districts with saline soil. Boron excess in vines is not frequently observed in the northern oasis of Mendoza.

Keywords Grapevine · Chloride · Potassium · Wine tasting · Musts

---

R. C. Vallone (✉) · L. E. Martínez · F. G. Olmedo · S. E. Sari
E.E.A. INTA Mendoza, Mendoza, Argentina
e-mail: rosanavallone@gmail.com

L. E. Martínez
e-mail: martinez.laura@inta.gob.ar

F. G. Olmedo
e-mail: olmedo.guillermo@inta.gob.ar

S. E. Sari
e-mail: sari.santiago@inta.gob.ar

R. C. Vallone
Soil Science Department, National University of Cuyo (UNCuyo), Mendoza, Argentina

E. Taleisnik and R. S. Lavado (eds.), *Saline and Alkaline Soils in Latin America*, https://doi.org/10.1007/978-3-030-52592-7_8

1 Salinity in Vineyard Soils of the Mendoza Province

Argentina is the fifth largest wine producer in the world. The main wine regions of the country are located in the provinces of Mendoza, San Juan, La Rioja. Salta, Catamarca, and Río Negro. The province of Mendoza is located in the arid region of Argentina, and almost all crops receive full irrigation. Mendoza produces more than 60% of the Argentine wines, and it is the source of an even higher percentage of the total exports. A large proportion is grown in areas with high salinity. The quality of the current irrigation water, both for its total degree of salinity (electrical conductivity, ECe) and its concentration of hazardous elements such as chlorides and sodium, and poorly drained soils are the most important factors that quantitatively and qualitatively characterize salinity problems in most areas of the province (Vallone and Nijensohn 2002).

The main aquifers in Mendoza are located in two major regions: one is in the north, irrigated by the Mendoza and Lower Tunuyán Rivers, and the second is in the south, belonging to the Oasis of the Diamante and Atuel Rivers. Most of the population of the province lives in the northern oasis. It has a large industrial expansion with activities that involve a variety of water uses: drinking water, agriculture, recreation, power, etc. The main crops in the area are grapevines, stone-fruit trees (peach, plum, apricot, and cherry), pear and quince trees and vegetables (mainly garlic, onion, tomato, and potato). According to the Regional Center of Aquifers of Mendoza, there is an annual underground water storage of about 15,000 hm^3 (Vallone 2008). According to the FAO (2015), 26.4% of effectively irrigated soils in the entire Mendoza Province are affected by some degree of salinity, particularly in the North Oasis. Irrigation areas concentrate in certain regions of the Mendoza River Basin and are less dense in the Tunuyán River.

Vallone et al. (2007) studied soil salinity at a 1:200,000 scale in all irrigated oases in the province and determined that 33.1% of the surface presented higher than 4 dS/m ECe (electrical conductivity values of the saturated soil extract). Of this, 25% was located in the Lower Tunuyán Basin, and 51% in the Mendoza River (Fig. 1).

In the North Oasis, where the city of Mendoza, (capital of the Province) is located, poor management of irrigation water together with agricultural, municipal, and industrial pollution have deteriorated the upper layers of underground water that are currently highly saline.

Morábito et al. (2004) found high salinity levels linked to subsurface drainage problems in the Central and North districts of the irrigation area of the Mendoza River. The authors observed similar problems in the area where the Mendoza and Tunuyán rivers are closest (Fig. 1, site a).

Considering the soil condition (cultivated or non-cultivated), the authors concluded that in all cases, the ECe of non-cultivated soil was significantly higher than that of cultivated soil. When analyzing the entire area under study (372,610 ha), it was observed that 63% of the basin showed soil salinity levels below 8 dS/m in the first half meter of soil. There was a close relationship between land use and the

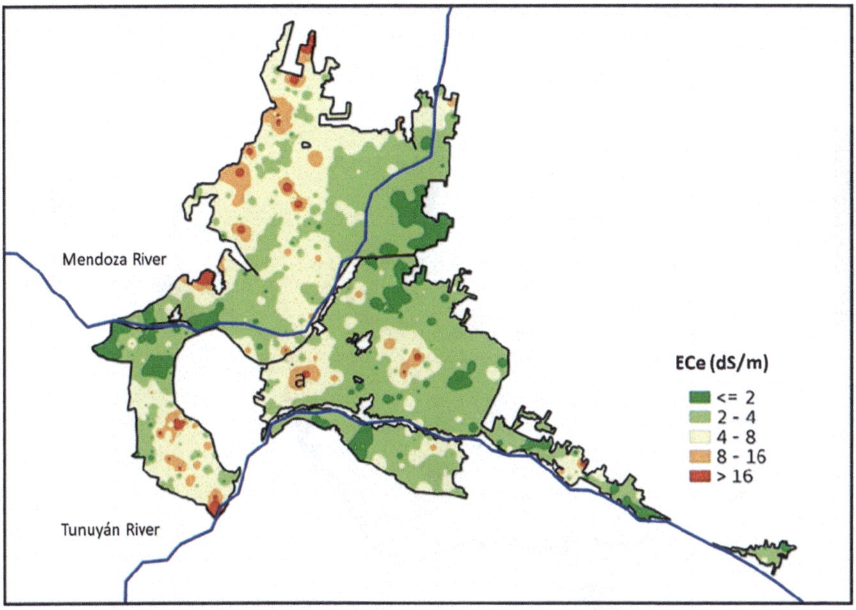

Fig. 1 Average electrical conductivity of the saturated soil extract up to 2 m of depth in the Mendoza and lower Tunuyán River Basin (Vallone et al. 2007)

corresponding salinity levels: the cultivated area (51% of the total) decreased with increasing soil salinity and practically became zero at levels exceeding 8 dS/m.

Morábito et al. (2010) determined temporal and spatial variation in soil salinity in the oasis of the Mendoza River in 1973 (127 sites) and in 2002 (173 sites) on a 1:50,000 scale. In all cases the ECe of non-cultivated soil was again, like in previous years, significantly higher than that of cultivated soil. Average ECe values in cultivated soil ranged between 2.4 and 2.8 dS/m and between 24 and 33 dS/m in non-cultivated soil. Yet there was considerable spatial variation and 59% of the cultivated area (94,533 ha) presented elevated salinity levels. Superposition of the two generated images (cultivated and non-cultivated) with the respective salinity maps revealed that 97% of the cultivated area had lower than 6 dS/m ECe values down to a meter depth and the whole cultivated area had values lower than 8 dS/m on the average (Fig. 2).

The Lower Tunuyán River is located in a plain that decreases in height from 800 to 500 m in West–East direction. Its climate and soil make this area favorable for high-quality vineyards, with adequate productivity. The basin has about 7800 agricultural establishments with approximately 100,000 surface-irrigated has. Viticulture is the most prominent activity (64% of the irrigated area, some 50,000 ha), followed by fruit production (17%): olives (7%), vegetables (6%), and pasture and forestry activities (3%) (Tozzi et al. 2017).

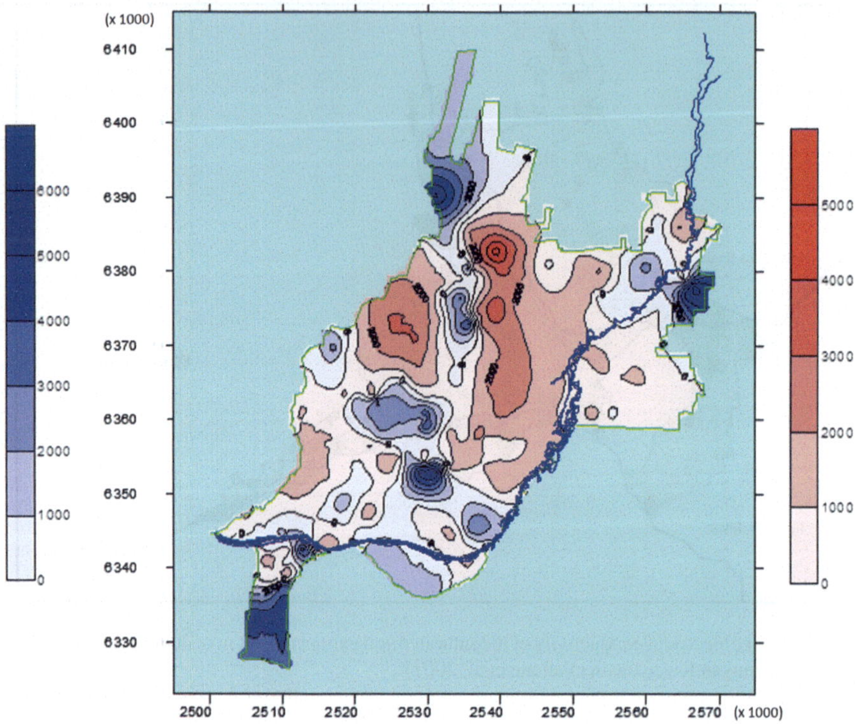

Fig. 2 Iso-difference map of salinity of cultivated soil between 0 and 50 cm for the Mendoza River Basin (1972 and 2002). Red and blue areas show the increase and decrease, respectively, in salinity for 2002 (Morábito et al. 2010)

Studies on soil salinity in the area evidenced that farmers tend to apply irrigation sheets that generally meet the leaching requirements and that soil salinity problems are due to the existence of sectors with high water tables. Tozzi et al. (2017) assessed the development of soil salinity, based on measurements of ECe, covering two periods, 2001–2002 and 2007–2009. Although the authors concluded that several factors explained the variations observed: phreatic levels, canal lining, rain, evapotranspiration and irrigation volume, they attributed this change mainly to a decrease in groundwater levels, caused by canal lining. Isosalinity curves of the surface layer revealed that 49% of the area in 2002 presented moderate saline soils, while in 2009, 60% of the surface showed a reduction in salinity. Mirábile et al. (2006) (Fig. 3), comparing the 1985–2002 period, also noticed a tendency toward a decreased salinity in approximately 60% of the surface analyzed.

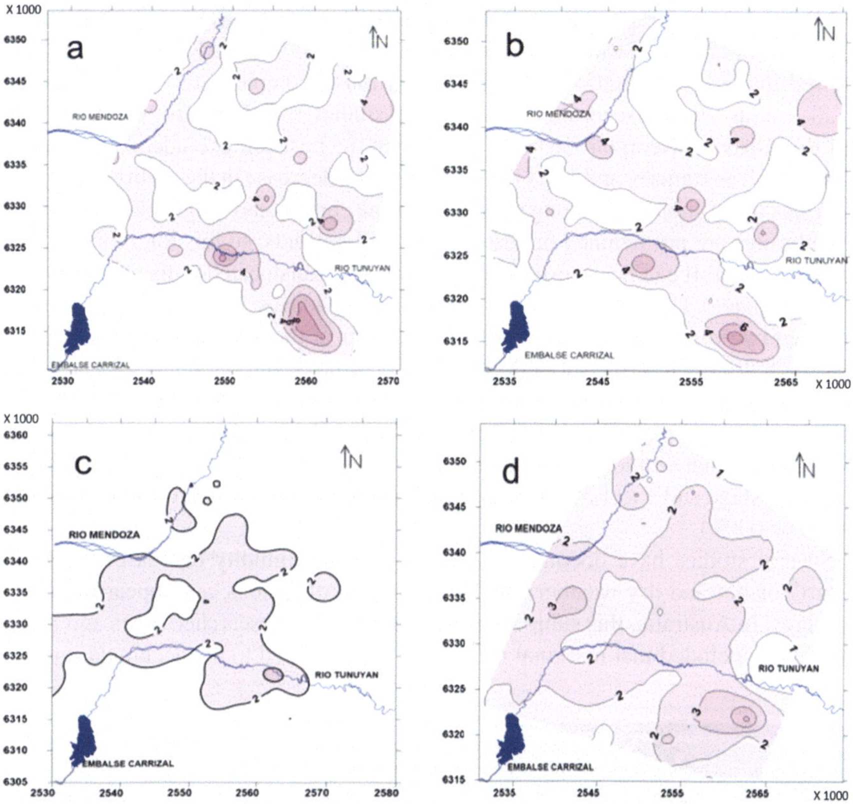

Fig. 3 Soil isosalinity maps of the LT River Basin for superficial and deeper layers, corresponding to agricultural campaign 2002 and 2009 (Tozzi et al. 2017). Reference 2002 campaign: **a** layer 0–0.5 m. **b** layer 0.5–1 m. 2009 campaign: **c** layer 0–0.5 m. **d** layer 0.5–1 m

2 Effect of Excess of Chloride, Sodium, and Boron on Grapevines

An increase in chloride toxicity has been noted in vineyards in the Mendoza River Basin. Possible reasons for this situation are the use of chloride-sensitive rootstocks, an increase in the drip irrigation area, irrigation management strategies with water deficit, deterioration in the groundwater quality, and soil physical problems (compaction, cementing, poor internal drainage, and poor infiltration (Vallone 2008, Vallone et al. 2017).

Chloride (Cl) and sodium (Na) concentrations increase progressively during berry development. It has been known for several years that Cl accumulation rate tends to stop near the veraison stage before growth is resumed, while Na showed a slightly lower increase during the second fruit growth stage compared with the first (Walker et al. 2000). Both varietal and, especially, rootstocks can affect the chloride and sodium concentration accumulation in leaves and fruits (Bernstein et al.1969). Ungrafted or rootstock Chardonnay and Syrah varietals such as K51–40 and 1202C showed higher concentrations of chlorides in must, compared to other rootstocks (Walker et al. 2010). According to Australian studies, 140 Ruggeri, Schwarzmann, and Rupestris St. George rootstocks are among the best salt excluders. Other rootstocks, such as Ramsey and 1103 Paulsen, exhibit a decrease in the ability to exclude chlorides when subjected to long periods of saline conditions (Tregeagle et al. 2006).

Chlorides are not retained or adsorbed by soil colloids, so this ion moves freely with soil water. It can be absorbed by the roots and translocated to the leaves where it accumulates. The visible toxicity symptoms in leaves are marginal chlorosis, followed by progressive necrosis toward the center of the leaf. Internodes are shortened, radial growth of the shoot decreases and anthocyanin pigments appear in red varieties (Fig. 4). According to the varietal tolerance, the vine is affected in soils with 10 to 25 $mmol_c Cl/L$ in the saturated extract. Concentrations in plant tissues, like petioles, that would indicate a possible toxicity would be higher than 0.5% in flowering stage and 1 to 1.5% or more in midsummer (in sensitive varietals the limit would be 0.8%).

Several studies have documented that if relative humidity decreases suddenly ("start" of hot and dry weather), the development of lesions can appear in only a few days. In Australia, this subject has been extensively researched years ago and it has been concluded that marginal necrosis is more related to the accumulation rate

Fig. 4 Salinity symptoms in vineyards. Above: typical view of necrosis of the canopy on trellises, a decrease in bud diameter and internode length. Below: anthocyanin pigmentation in cv. Merlot (left) and leaf centripetal necrosis (right)

than to the concentration of chlorides or sodium in blades (Downton and Hawker 1980; Walker 1994). Before the toxicity symptoms appear, there is a reduction in photosynthesis and growth due to a decrease in stomatal conductivity.

Stevens et al. (2011) studied the effects of irrigation with saline water for six years and at different times of the crop cycle (on average the EC of the saline water was 3.16 dS/m) on the composition of grape must of Colombard vines grafted on Ramsey. Potassium concentrations in must increased above those of controls irrigated with EC 0.5 dS/m water between the veraison and harvest stage. The pH also increased.

The pH of the grape juice depends on the total acidity (concentration of organic acids), the tartaric acid:malic acid ratio and the exchange of protons through the berry membrane (Boulton 1980). Therefore, the increase in Na and K concentrations in juice from saline treatments is indicative of proton exchange and tends to increase the pH.

Excess of boron (B) in grapevines in the northern oasis of Mendoza is not frequent, unlike in areas the nearby provinces of San Juan and La Rioja, where water and soil sometimes present high concentrations of this element. The characteristic symptoms are that blades become less lobed and bend down along the edges. Research carried out several years ago on vines in Australia, showed that there is a strong interaction between boron and chlorides on the growth and mineral composition of Cabernet Sauvignon and that the response of the plant cannot be predicted by only analyzing individual effects (Downton and Hawker 1980). The combination of elements is not more harmful than their individual effect. In fact, plants that received both treatments (B and Cl) did not develop symptoms of excess boron which were evident on the leaves of plants that only received boron treatment, although the leaf B concentration was similar. It seems that chloride offers some kind of "protection" against the development of excess boron symptoms. It is known that boron preferentially accumulates at very high levels in leaves in grapevines, and toxicity symptoms appear around 300 ppm (petioles 170 ppm approx.). This interaction has been proven in vineyards in La Rioja Province (Vallone 2004) (Fig. 5). Boron concentrations in irrigation water that exceed 1 mg/L are associated with symptoms of grapevine toxicity (Hart 1974).

Orientation values for boron in plant tissues are shown in Table 1.

3 Effect of Salinity on Must and Wine Quality

In addition to the damage of high salinity to the vine productivity, there is also a decrease in wine quality. Consumption of wine with high Cl concentrations can be unpleasant. According to the ability of connoisseurs to perceive salty taste close to the legal limits, sensory thresholds for the perception of sodium chloride in musts and wines have been established as a tool to make swift decisions about the suitability to vinify grapes (Levy et al. 2014). In Argentina, the limit of Cl in wines is 0.6 g/L expressed as potassium chloride (KCl) and 0.8 g/L as NaCl (Res C.35/2000, National Viticulture Institute regulations). Legal limits vary widely among countries:

Fig. 5 Symptoms of protection against boron excess by chloride in Cabernet Sauvignon (upper panel) and typical symptoms of boron excess in Petit Verdot (lower panel). Studies carried out in La Rioja Province, Argentina (Vallone 2004)

Table 1 Boron deficiency and toxicity levels in petioles of flowering grapevines

Level concentration	(mg/kg)
Deficient	25
Questionable	26–30
Adequate	>30
Possibly toxic	100–150 (>300 in leaves)

in Australia, the legal limit is 1 g/L expressed as NaCl, and wine should not contain more than 606 mg/L Cl.

In South Africa, the limit is 100 mg/L and in Sweden 60 mg/L of Na, whereas in Turkey the limit is 500 mg/L of chlorides measured as NaCl (Stockley and Looyd-Davis 2001).

In Australia, Walker et al. (2010) showed that Syrah wines from certain sites with soils presenting an EC between 2.1 dS/m and 3.3 dS/m, with ungrafted or grafted vines such as K51–40, exceeded the limits. The authors also observed that the Cl content in these Syrah wines increased after fermentation, presenting a ratio of 1.7:1 between the Cl content of the wine and the grapes. In the same study and unlike what was observed for Syrah, Chardonnay, a white wine, presented a different Cl extraction rate, with a 1:1 ratio between the Cl content of the wine and the grape. Tregeagle et al. (2006) and Gong et al. (2010) observed lower chloride concentrations in pulp and skin of Chardonnay compared to Syrah grafted on Ramsey and 1103 Paulsen.

4 Survey of the Mendoza River Basin

Vallone et al. (2017, 2018) studied from 2014 to 2017) vineyards cultivated on saline soils of the Mendoza River Basin, which had been selected with geostatistical techniques from a land utility map by Vallone et al. (2007). The study evaluated the relationship between soil salinity, pH and potassium on the composition and quality of regional musts and wines. The main results are as follows.

4.1 Soil Salinity of the Basin

The ECe values for the 53 sites assayed during the 4 years indicated that the problems of soil salinity in the first meter of depth are more marked in the “central” zone of the basin which presents poor drainage and groundwater near the surface (Fig. 6).

Fig. 6 Average electrical conductivity of the saturated soil extract (0–100 cm) corresponding to subzones within the Mendoza River Basin (Vallone et al., personal communication)

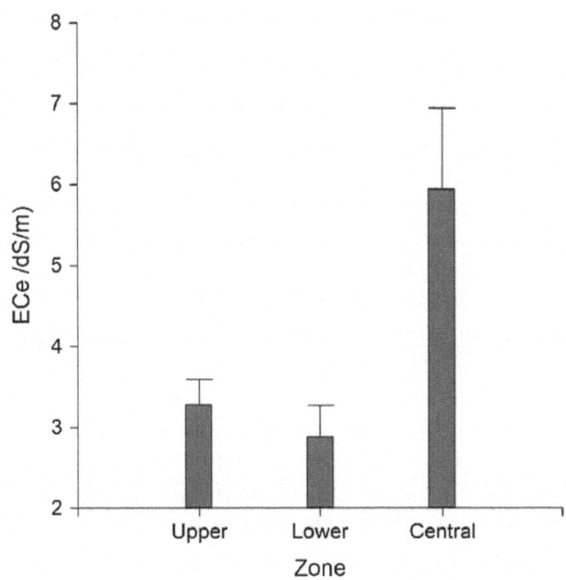

Table 2 Average (2014–2017) chloride concentration (%) and standard error (SE) per varietal in vineyards surveyed in the Mendoza River Basin, measured in petioles extracted at harvesting

Varietal Means	(%) ± SE
Cabernet sauvignon	0.95 ± 0.32
Pedro giménez	0.86 ± 0.11
Tempranillo	0.63 ± 0.04
"Criollas" (local wine varietal)	0.61 ± 0.06
Malbec	0.61 ± 0.06
Bonarda	0.60 ± 0.08
Syrah	0.54 ± 0.01
Torrontés	0.50 ± 0.02

4.2 Chlorides in Petioles

The average Cl concentrations at the time of harvest in petioles are relatively high, but generally lower than 1%, which is considered toxic in late summer. In a saline area with drainage problems, Malbec and Cabernet Sauvignon varietals exceeded this level. Varietals with the lowest concentrations were Bonarda, Syrah, and Torrontés (Table 2). The same chloride concentration is generally maintained in musts and wines.

4.3 Chlorides in Musts and Wines

There was a marked varietal effect on the concentration of chlorides in must, which varied according to the site. Cl concentrations in Tempranillo, Pedro Giménez and Cabernet Sauvignon exceeded the established limits, whereas Malbec, Bonarda, Syrah and "Criollas," a local wine varietal, evidenced lower concentrations (Fig. 7).

The chloride concentrations in musts are not clearly related to the values in wines (Fig. 8), because they would also depend on the extraction capacity during the fermentation process of the varietal, as Walker et al. (2010) postulated. During the vinification process, Cl increased mainly in the Syrah varietal, similar to data observed by Walker et al. (2010), whereas in Tempranillo and Pedro Giménez Cl concentration decreased and in Cabernet Sauvignon, Bonarda, Malbec, and Criollos the concentration remained almost the same (Table 3).

The skin:pulp ratio was evaluated during vinification of red wine varietals. The variability observed in both skin and pulp would be mainly due to the varietal and, to a lesser extent, to the water status of the plant that would affect the volume of the berry. The chloride concentration in skin (1.51 ± 0.18 g%g) was twice that in pulp (0.69 ± 0.09 g%g) and lower than published by Gong et al. (2010), for whom chloride concentrations in grape skin are 7 to 14 times higher than those in pulp. In addition, Cl concentrations in skin are 3–8 times higher than Na concentrations.

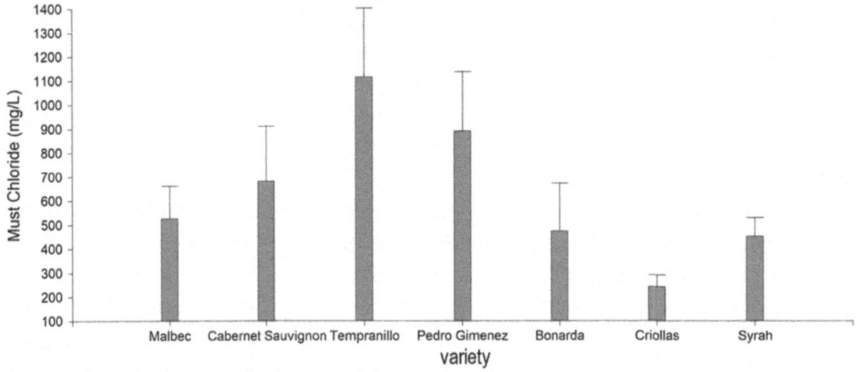

Fig. 7 Chloride concentrations expressed as KCl in musts of different varietals from saline soils of the Mendoza River Basin (vertical bar indicates standard error) between 2014 and 2016. (Vallone et al. personal communication)

Fig. 8 Cl in musts and Cl in wines relationship (expressed as ClK)

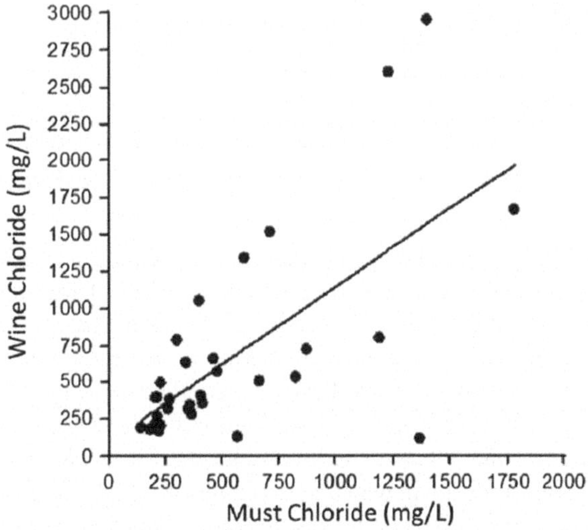

Table 3 Chloride concentration (expressed as KCl (mg/l)) in wines of different red and white grape varietals studied in the Mendoza River Basin between 2014 and 2016

Varietal	Means	SE	Min.	Max.
Syrah	1001.6	203.7	629.2	1330.8
Tempranillo	705.8	179.2	526.6	884.9
Pedro giménez	698.7	581.0	114.4	1860.7
Cabernet sauvignon	672.9	232.9	159.8	1667.3
Bonarda	470.5	125.2	316.3	718.4
Malbec	464.3	71.2	175.8	794.1
Criollas	281.4	40.7	191.8	388.6

During fermentation of red wine must, which involves a period of contact with the skin, the Cl concentration significantly increases during the first few days while the increase in Na concentration is only marginal. The increase in Cl concentration in Chardonnay vinification was significantly lower than in Syrah because of the lack of contact with the skin in the former varietal (Tregeagle et al. 2006; Walker et al. 2010).

During the 2014 harvest, extraction of chlorides was further studied in two red varietals, Malbec and Cabernet Sauvignon, from the Upper Mendoza River. Malbec showed higher chloride concentrations in skin and pulp and a larger proportion of skin with respect to the total berry weight compared with Cabernet Sauvignon. However, the resulting Malbec wine had less chloride concentration (464 mg/L) than Cabernet Sauvignon (673 mg/L). This would indicate that extraction of the anion during vinification is greater for Cabernet Sauvignon but the physicochemical causes have not been elucidated yet. The skin:pulp ratio (0.25 in Malbec and 0.47 in Cabernet Sauvignon) and the capacity to accumulate and extract Cl according to the varietal would probably better explain the differential behavior (Walker et al. 2003), but these aspects need to be further studied.

4.4 Relationship Between Soil Variables and Chloride and Potassium Concentrations in Musts and Wines

The correlation between Cl in must, wine, and soil ECe was between 0.41 and 0.52 according to the Pearson coefficients. One of the most relevant associations was whether there exists an association between soil chlorides and wine chlorides (and thus discards possible wine acidity corrections with hydrochloric acid, an illegal practice in Argentina). Pearson's coefficient was 0.50. The legal limit in wine for chlorides (expressed as potassium chloride, KCl) was exceeded with ECe values greater than 3.7 dS/m and soil chloride concentration of 13.9 $mmol_c$/L. The pH and K concentration in wines were highly correlated, like the soil pH and must pH, which would explain the high pH values found in Mendoza Province, compared with other viticultural areas in the world. This is related with the occurrence of regional alkaline soils with high base saturation.

The results also indicated that the chloride concentrations in wines and musts directly correlated with the ECe, with $r = 0.55$ ($p \leq 0.0015$) and $r = 0.41$ ($p \leq 0.023$), respectively, and indirectly with chloride and potassium ions in saline soils. K contents in musts and wines were not associated with any of the salinity variables or with exchangeable K in soil. Besides, no consistent relationship between K concentration in soil and musts was found in other viticultural areas.

Our data are in agreement with Dundon et al. (1984), while Walker and Blackmore (2012) observed a relationship between K in musts and soils. Varietal differences and their influence on the redistribution and mobilization of K within the plant should possibly be considered.

However, this is difficult to find in the potassium-rich soils in the oases of Mendoza because of the parent material mineralogy. In some regional studies with potassium fertilization for several years, no variations in K, pH and total acidity of the musts were observed (Vallone, personal communication). However, under K deficiency, as was the case of Concord grapes, modifications in K, pH or acidity were detected (Mattick et al.1972).

The pH of the musts ranged between 3.28 and 4.14 and was directly correlated with the soil pH ($r = 0.38$, $p \leq 0.041$) and indirectly with exchangeable soil K. The pH of the wine ranged between 3.1 and 4.5 and was inversely correlated with soil chlorides ($r = -0.29$, $p \leq 0.044$). This can also be analyzed in relation to the location in the river basin, since wines with the highest pH values were cultured in the central area (3.92 ± 0.06), showing highest salinity values, as previously mentioned. Somers (1975) found that high K concentrations in musts corresponded to high musts pH values. Rankine (1977) observed that Australian red wines had a wider pH range (up to 4.3) than French wines (up to 3.9), which was attributed to higher K concentrations in musts. Walker and Blackmore (2012) found higher pH values in wines that contained higher concentrations of soluble solids and K in musts. In agreement, in Mendoza Province a correlation between soluble solids and K in musts ($r = 0.43$, $p \leq 0.0227$) was found. The grape varietal affected the K concentration and pH of wines, Pedro Giménez, showed the lowest pH and Cabernet Sauvignon and Malbec the highest pH, which was inversely related to the varietal chloride concentrations in must (Table 4).

In summary, higher chloride concentrations in soil coincided with lower potassium absorption (evidenced by lower K concentrations in petioles), which resulted in lower pH and K concentrations in wines. The general decrease in potassium absorption with increasing salinity is partly attributable to the replacement of K by Na (Downton 1985). Figure 9 shows the spatial distribution and the relation between soil and wine K and between soil and wine Cl in the Mendoza River Basin.

Table 4 K concentratios in musts (mg/L) and pH of wines of different white and red varietals in the Mendoza River Basin

Varietal	K musts	pH wine
Pedro giménez	1.373	3.27
Tempranillo	1.467	3.43
Syrah	1.655	3.67
Criollas	1.665	3.74
Bonarda	1.885	4.03
Cabernet Sauvignon	1.892	4.06
Malbec	1.917	3.99

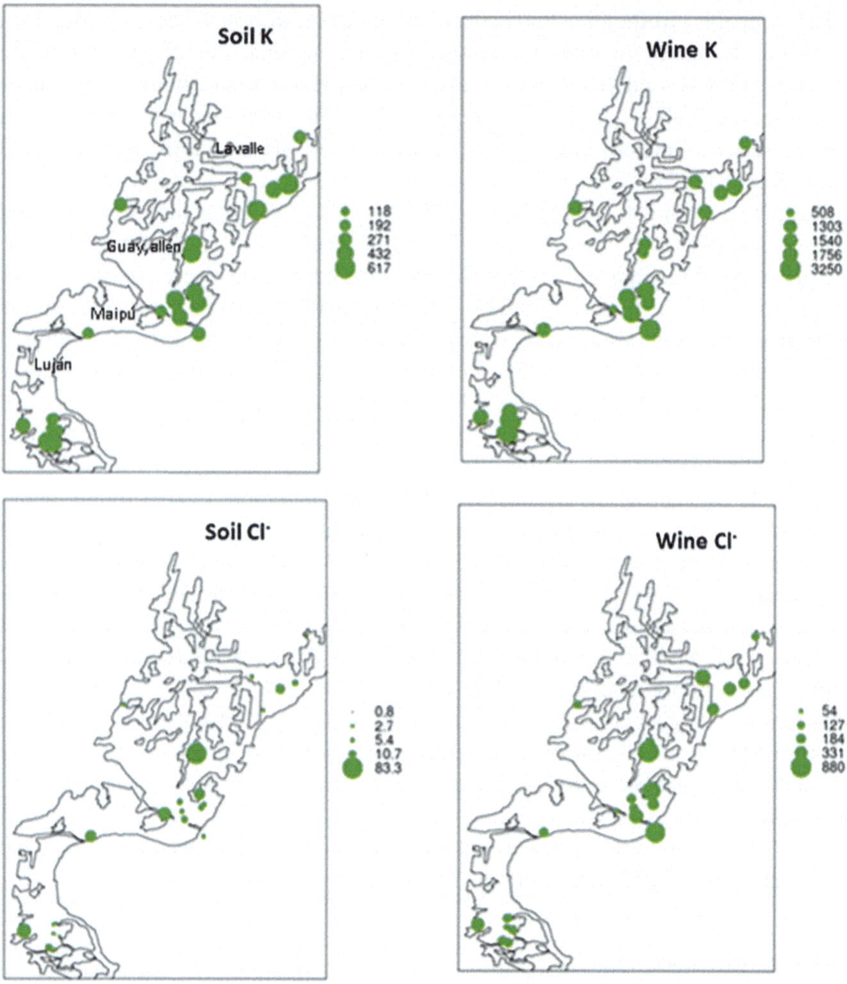

Fig. 9 Spatial distribution of exchangeable K concentration in soil (mg/kg), K in wine (mg/L), chlorides in soil (mmolc/L) and chlorides in wine (as KCl in mg/L) of vineyards located in the Mendoza River Basin (The size of the circles represent relative average contents)

4.5 *Effect of Soil Salinity on Organoleptic Characteristics of Wine*

To assess whether the chloride concentration in wines can be perceived by the consumers, 2017 vintage vinifications were tested by a tasting panel of experts. Wines were examined by 10 judges using descriptive sensory analysis. The following descriptors were used: sweet, salty, sour, and bitter in the mouth.

Table 5 Correlations between descriptors and Na and Cl concentrations in wines

Variable 1	Variable 2	Pearson Coef.	*p*-value
Salty	Bitter	0.81	0.0002
Salty	Sour	−0.66	0.0057
Salty Na	Na in wine	0.62	0.0106
Salty Cl	Cl in wine	0.27	0.3111

A positive relationship was observed between salty and bitter tastes and a negative relationship between salty and sour. The salty descriptor was associated with Na concentrations in wine (Table 5).

Analysis of the principal components and ANOVA (data not shown) identified wines from different localities as wines with saltier notes and these wines were mostly Syrah and Criollas varietals.

In summary, the ion composition of plants, musts and the wines of vineyards for the different sub-basins of the Mendoza Province, Argentina, and their relationship with soil salinity were examined. A panel of experts perceived salty flavors that were positively associated with bitterness in wines from districts with saline soil.

References

Bernstein L, Ehlig CF, Clark RA (1969) Effect of grape rootstock on chloride accumulation in leaves. J Am Soc Hortic Sci 94:584–590

Boulton R (1980) The relationship between total acidity, titratable acidity and pH in grape tissue. Vitis 19:113–120

Downton WJS (1985) Growth and mineral composition of the sultana grapevine as influenced by salinity and rootstock. Aust J Agric Res 36:425–434

Downton WJS, Hawker JS (1980) Interaction of boron and chloride on growth and mineral composition of cabernet sauvignon vines. Am J Enol Vitic 31(3):277–282

Dundon CG, Smart RE, McCarthy MG (1984) The effect of potassium fertilizer on must and wine potassium levels of shiraz grapevines. Am J of Enol Vitic 35:200–205

FAO (2015) Estudio del potencial de ampliación del riego en Argentina. UTF/ARG/017 desarrollo institucional para la inversión. UCAR—PROSAP. Roma, https://www.fao.org

Gong H, Blackmore DH, Walker RR (2010) Organic and inorganic anions in shiraz and chardonnay grape berries and wine as affected by rootstock under saline conditions. Aust J Grape and Wine Res 16:227–236

Hart BT (1974) A compilation of Australian water quality criteria. Aust. water resources technical paper #7, Aust govt publishing service, Canberra

Levy C, Petrie PR, Hasted AM, Johnson TE, Collins C, Bastian SEP (2014) Evaluation of sensory thresholds and perception of sodium chloride in grape juice and wine. Am J Enol Viti 65(1):124–133

Mattick LR, Shaulis NJ, Moyer JC (1972) The effect of potassium fertilisation on the acid content of 'Concord' grapejuice. Am J Enol Vitic 23:26–30

Mirábile C, Morábito J, Manzanera M, Tozzi D (2006) Dinámica de la salinidad del suelo en el oasis del río tunuyán inferior comparación 1985–2002. https://www.ina.gob.ar/legacy/pdf/CRA-IIIFERTI/CRA-RYD-10-Mirabile.pdf

Morábito J, Mirábile C, Pizzuolo P, Tozzi D, Manzanera M, Mastrantonio L (2004) Salinidad de suelos regadíos e incultos en el oasis norte de Mendoza—Argentina. XIX Congreso Argentino de la Ciencia del Suelo, Paraná, Entre Ríos–Argentina

Morábito J, Mirábile C, Manzanera M, Cappé O, Tozzi D, Mastrantonio L (2010) Evolución de la salinidad de suelos regadíos e incultos en el área del Río Mendoza. XX congreso nacional del agua—III simposio de recursos hídricos del cono sur. Conagua 2005. Argentina

Rankine BC (1977) Developments in malolactic fermentation in australian red tables wines. Am J of Enol Vitic 28:27–33

Somers TC (1975) In search of quality for red wines. Food Technology in Australia 27:49–56

Stevens RM, Harvey G, Partington DL (2011) Irrigation of grapevines with saline water at different growth stages: effect on leaf, wood, and juice composition. Aust J Grape Wine Res 17:239–248

Stockley C, Looyds-Davies S (2001) Analytical specifications for the export Australian wine. Aust Wine Research Institute, Adelaide

Tozzi F, Mariani A, Vallone R, Morábito J (2017) Evolución de la salinidad de los suelos regadíos del río tunuyán inferior (Mendoza—Argentina). Rev FCA UNCuyo 49(1):79–93

Tregeagle JM, Tisdall JM, Blackmore DH, Walker RR (2006) A diminished capacity for chloride exclusion by grapevine rootstocks following long-term saline irrigation in an inland versus a coastal region of Australia. Aus J Grape and Wine Res 12:178–191

Vallone RC (2004) Diagnóstico nutricional y manejo del problema de salinidad en suelos. Cultivos: vid y durazno, La rioja, Argentina. Informe servicio externo cátedra de edafología, facultad de Ciencias Agrarias, UNCuyo, 21 p

Vallone RC (2008) Situación de la salinidad en Mendoza. In: Taleisnik E, Grunberg K, Santa María G (eds) La salinización de suelos en la Argentina. EDUCC, Córdoba, pp 63–80

Vallone RC, Maffei JA, Olmedo GF, Morábito JA, Mastrantonio LE, Lipinski V, Filippini MF (2007) Mapa de aptitud de suelos con fines de riego y de riesgo de contaminación edáfica de los oasis irrigados de la provincia de Mendoza. Informe final al dep. General de irrigación, convenio con OEA. Más Anexos Mapas y Fichas de perfiles modales. Mendoza p 65

Vallone RC, Martínez LE, Olmedo GF, Sari SE (2017) Influencia de la salinidad edáfica en el contenido de cloruros y potasio de mostos y vinos en la cuenca del río Mendoza, Argentina. V Reunión de la Red Argentina de salinidad, villa mercedes, San Luis. https://redsalinidad.com.ar/wp-content/uploads/2018/11/Trabajos-V-RAS.pdf

Vallone RC, Martínez LE, Olmedo GF, Sari SE (2018) ¿Responden el pH, cloruros y potasio de mostos y vinos a la salinidad edáfica? XXVI congreso argentino de la ciencia del suelo, Tucumán. https://www.suelos.org.ar/sitio/cacs-2018-descarga-de-contribuciones/

Vallone RC, Nijensohn L (2002) Guía de orientación para regantes de zonas áridas con énfasis en áreas salinas. Ed. Tintar, Mendoza. p 150

Walker RR (1994) Grapevine responses to salinity. Bulletin de L'OIV 761–762:634–661

Walker RR, Blackmore DH, Clingeleffer PR, Godden P, Francis L, Valente P, Robinson E (2003) Salinity effects on vines and wines. Bulletin de l'OIV 76:200–227

Walker RR, Blackmore DH, Clingeleffer PR (2010) Impact of rootstock on yield and ion concentrations I petioles juice and wine of shiraz and chardonnay in different viticultural environments with different irrigation water salinity. Aust J Grape and Wine Res 16(1):243–325

Walker RR, Blackmore DH (2012) Potassium concentration and pH inter-relationships in grape juice and wine of chardonnay and shiraz from a range of rootstocks in different environments. Aust J Grape Wine Res 18:183–193

Walker RR, Read PE, Blackmore DH (2000) Rootstock and salinity effects on rates of berry maturation, ion accumulation and colour development in Shiraz grapes. Aust J Grape and Wine Res 6:227–239

Causes, Effects, and Management of Salinity Problems in Pecan Production in North Mexico

Dámaris Ojeda-Barrios, Adalberto Benavides-Mendoza, Adriana Hernández-Rodríguez, Laura Raquel Orozco-Meléndez, and Esteban Sanchez

Abstract Salinity and sodicity are critical problems for arid and semi-arid pecan (*Carya illinoensis*) producing regions of northern Mexico, causing decreases in yield and quality of the pecan nut. Wells are sources of irrigation water, but the quality is below the standards for agricultural use as it is high in salt content, and it gradually salinizes soils. These saline soils compromise the profitability of pecan production. The objective of this chapter is to provide insight into the causes, effects, and management of saline soils, focusing on ecological and economical solutions for the pecan orchards of northern Mexico.

Keywords Pecan nut · Antioxidant capacity · Chlorophyll · Foliar symptoms · Irrigated crops

D. Ojeda-Barrios (✉) · A. Hernández-Rodríguez · L. R. Orozco-Meléndez
Facultad de Ciencia Agrotecnológicas, Universidad Autónoma de Chihuahua, Chihuahua, Chihuahua, México
e-mail: dojeda@uach.mx

A. Hernández-Rodríguez
e-mail: aernande@uach.mx

L. R. Orozco-Meléndez
e-mail: laurao_14@hotmail.com

A. Benavides-Mendoza
Departamento de Horticultura, Universidad Autónoma Agraria Antonio Narro, Saltillo, Mexico
e-mail: adalberto.benavides@uaaan.edu.mx

E. Sanchez
CONACYT-Centro de Investigación en Alimentación Y Desarrollo A.C. Coordinación Delicias, Fraccionamiento Vencedores del Desierto, Delicias, Chihuahua, México
e-mail: esteban@ciad.mx

E. Taleisnik and R. S. Lavado (eds.), *Saline and Alkaline Soils in Latin America*, https://doi.org/10.1007/978-3-030-52592-7_9

1 Introduction

Currently, Mexico is the seventh-largest fruit and vegetable producer in the world, with a total estimated annual production of 32 million tons. The national output represents 1.7% of the global production, only preceded by China, 40.9%, India, 11.6%, USA, 3.4%, Brazil, 2.6%, Turkey, 2.3%, and Iran, 1.8% (FAO-FAOSTAT 2018).

Mexican horticulture production represents 20.67% of the gross agricultural domestic product. The area of harvested fruit trees represents 6.44% of the national surface. In 2018, a hectare of fruit trees was three times more profitable than a hectare of any other crop (SAGARPA 2018). The volume of exported fruits from 1961 to 2010 was 4.7 times higher than that of imported fruits (FAO-FAOSTAT 2018). The production volume of horticultural crops in Mexico has almost doubled from 1980 to 2011, going from approximately 8.5 to 15.7 million tons. This productive upswing is due to the rise in profitability, social equity, and contribution to human and environmental health linked to crop production (SAGARPA 2018). The climate of Mexico is advantageous to produce tropical fruits (FAO-FAOSTAT 2018). The state of Chihuahua has potential for the cultivation of warm zone horticultural crops. Currently, Chihuahua has 30,000 ha of apple trees, 84,928 ha of the pecan trees (*Carya illinoensis*), 2,500 ha of the peach trees, and 2,000 ha dedicated to viticulture.

One problem that global agriculture is facing is soil salinization. Saline stress is an environmental problem that affects the productivity of horticultural crops in arid and semi-arid regions. In Mexico, it affects 3.2% of the national territory, 600,000 ha. Soil salinization is observed in the states Sonora, Sinaloa, Tamaulipas, San Luis Potosí, Chiapas, Nuevo Leon, Oaxaca, Veracruz, Chihuahua, and Zacatecas (Fig. 1) (SAGARPA 2018).

Saline soils are common in arid and semi-arid climates. The use of saline water for crop irrigation and inadequate soil management cause a progressive deterioration of the soil and decrease in production and crop quality (Follet et al. 2005). Most crops in Mexico are sensitive to salinity, and when crops are cultivated on saline soils, the production may be severely reduced (Trasviña-Barriga et al. 2018). All regions in Mexico where pecans are grown (Fig. 1) have some degree of soil salinity.

The pecan tree is a crop of great importance worldwide due to its profitability (Ojeda-Barrios et al. 2012). It is cultivated in several countries: USA, Mexico, Australia, Brazil, Canada, Israel, South Africa, and China (Casales et al. 2018; Zhang et al. 2015). In Mexico, the total surface of planted pecan tree is 141,298 ha, and in 2018, the production reached 154,808 t, with a yield of 1.598 t ha^{-1}. Chihuahua has a planted surface of 84,928 ha; it corresponds to 62.8% of the national total, with a production of 96,926 t and an average yield of 1,691 t ha^{-1} (SAGARPA 2018).

Fig. 1 States of Mexico and the distribution of the main pecan nut-producing areas with saline soil. *Author* Dámaris Ojeda-Barrios

2 Irrigation Water Quality in Northern Mexico

In northwestern Mexico, groundwater extracted from deep aquifers complements the supply of water from dams, when the water demand for crop irrigation is not satisfied due to drought. In the pecan producing areas of the state of Chihuahua (Fig. 2), 40% of analyzed irrigation water samples presented values of electrical conductivity higher than 1.0 dS m^{-1} (Jasso et al. 2010). This indicates a plausible risk of soil salinization in all orchards with clay soils or that have soil compaction. Soil samples from that region presented sodium adsorption ratios (SAR) of 4.0 or lower, which indicates a low risk of soil sodification, and sulfates were low. With these concentrations, it is possible to find a deficiency of zinc (Zn), manganese (Mn), magnesium (Mg), and iron (Fe), which can cause damage to the vegetative growth, yield, and quality of the nut (Jasso et al. 2010; Ojeda-Barrios et al. 2012).

Table 1 shows the pH, electrical conductivity (EC), SO_4 concentration, and SAR in the pecan producing regions of the state of Chihuahua. Samples from Delicias had pH values higher than 7.5. In Jiménez, the mean concentration of SO_4 was 998 ppm, more than double that of Delicias. The highest SAR was observed in Chihuahua (Jasso et al. 2010).

In the nut-producing regions of Chihuahua (Fig. 2), irrigation water contains 500–1500 mg L^{-1} of dissolved salts, an annual equivalent of 5–15 tons of salt per hectare. When the concentration of soluble salts in the root zone reaches the aforementioned concentrations, it causes weak growth and suboptimal production (Miyamoto and Story 1995).

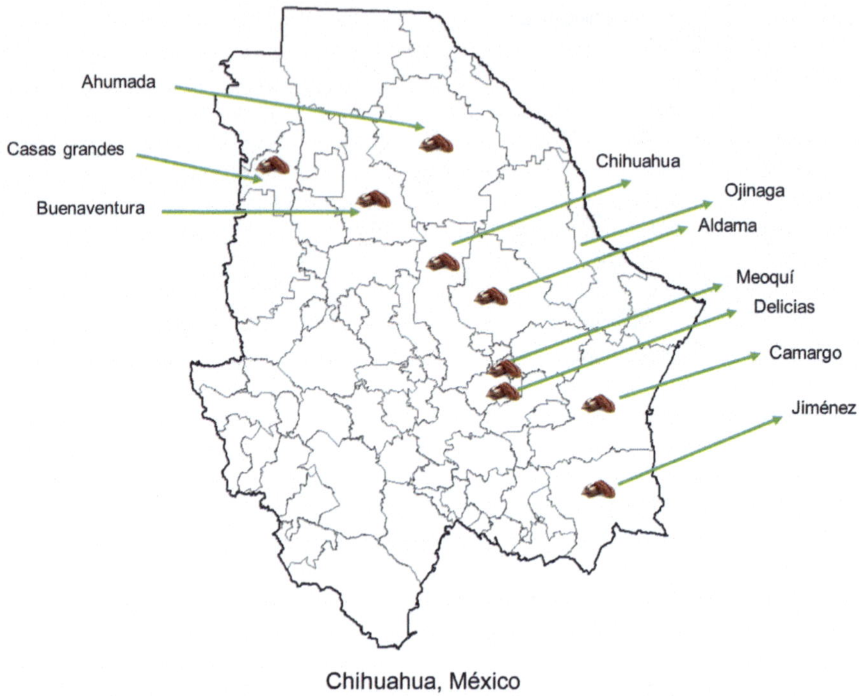

Fig. 2 Pecan nut-producing regions in the state of Chihuahua, Mexico. *Author* Dámaris Ojeda-Barrios

Table 1 Comparison of parameters evaluated in irrigation water by region: Mean values (Jasso et al. 2010)

Parameter	Jiménez-camargo	Delicias	Chihuahua
pH	7.45	7.81	7.21
EC, dS m^{-1}	1.97	0.90	0.64
SAR	1.07	1.99	13.15
SO_4, ppm	988	441	36

Low infiltration rate of a soil causes a reduction of salt leaching. If drainage is adequate and the salinity of the irrigation water is lower than 2.0 dS m^{-1} EC, there will be no damage to the pecan trees. When the irrigation water has a salt concentration of 1.5–1000 mg L^{-1}, pecan orchards established in clay soils with low infiltration rates are more likely to face salinity problems than orchards established in soil of sandy texture, with good drainage. When the concentration of soluble salts is low in the irrigation water, but it presents a high SAR, infiltration rate of the soil is reduced. This can be corrected using soil amendments (Grajeda et al. 2012).

3 Irrigation Requirements and Salinity Tolerance in Pecan Trees

Pecan orchards have a high demand for irrigation water because the crop requires the soil to be at field capacity (Brown 2010; Grajeda et al. 2012). In 2020, the state of Chihuahua will need 150 million cubic meters of irrigation water for the pecan crop. In Chihuahua, the area of pecan crop is expanding at a fast rate, which is putting the region in a critical water situation. Proper management of irrigation water and soil is required to avoid adverse effects on the soil. It is imperative to develop technology for more efficient use of irrigation water with an integrated vision. (Montes-Renteria et al. 2011). Some technological improvements have been made by changing surface irrigation to pressurized methods in the states of Chihuahua and Coahuila, where sprinkler and micro-sprinkler irrigation prevail.

Due to the lacustrine nature of Chihuahua, the aquifers have water with a high salt concentration, including Na, bicarbonates, and Mg ions. The presence of salts coupled with the tendency to apply controlled irrigation, such as sprinkling, micro sprinkling, or dripping, and the passage of machinery can affect the soil structure and lead to soil compaction (Pulido and González 2010).

When the irrigation water exceeds 1.0 dS/m and is used in low infiltration soils, such as clay, clay silt texture, or clay silt crumb, it is likely that after a few years, salinity damage will be observed (Nuñez-Colima et al. 2019). When the soil exceeds 2.0 dS m^{-1} EC, there are associated problems such as root rot and soil compaction, along with a reduction of productivity. At 3.0 dS m^{-1} EC, the pecan crop loses 40% of its productive potential, due to a reduction in the number of nuts, and incomplete filling of the kernel (Reyes et al. 2010).

In the Lagunera region of central-northern Mexico, between the latitudes of 24°37′00″ and 26°06′31″, a research was carried out in 18 pecan orchards, of which 75% had irrigation water with 0.7 dS/m CE or more, while SAR fluctuated from 0.7 to 13.8. From the orchards evaluated in 2015–2017, 28% had salinity and/or sodicity problems, caused in large part by water quality (Nuñez-Colima et al. 2019).

In the nut-producing region of Jiménez, Chihuahua, irrigation water samples were taken in 2015 in ten orchards, and the chemical composition was contrasted with current Mexican standards: NOM 127, for the standards of human use and consumption, and Federal Law of Rights—Provisions Applicable to Matters of National Waters, for agricultural irrigation.

Figure 3 shows the locations sampled. The parameters analyzed were pH, chlorides, Fe, Mn, Zn, Cu, As, B, Pb, Cd, Ni, Na, SDT (total dissolved solids), nitrates, hardness, and alkalinity. All values obtained from all samples were within the parameters established for both human consumption and agricultural use. Future periodic monitoring of Pb, Ni, As, and B will help to avoid the possibility of consequences for the population and negative impacts from farming activities (Valles-Aragón et al. 2017).

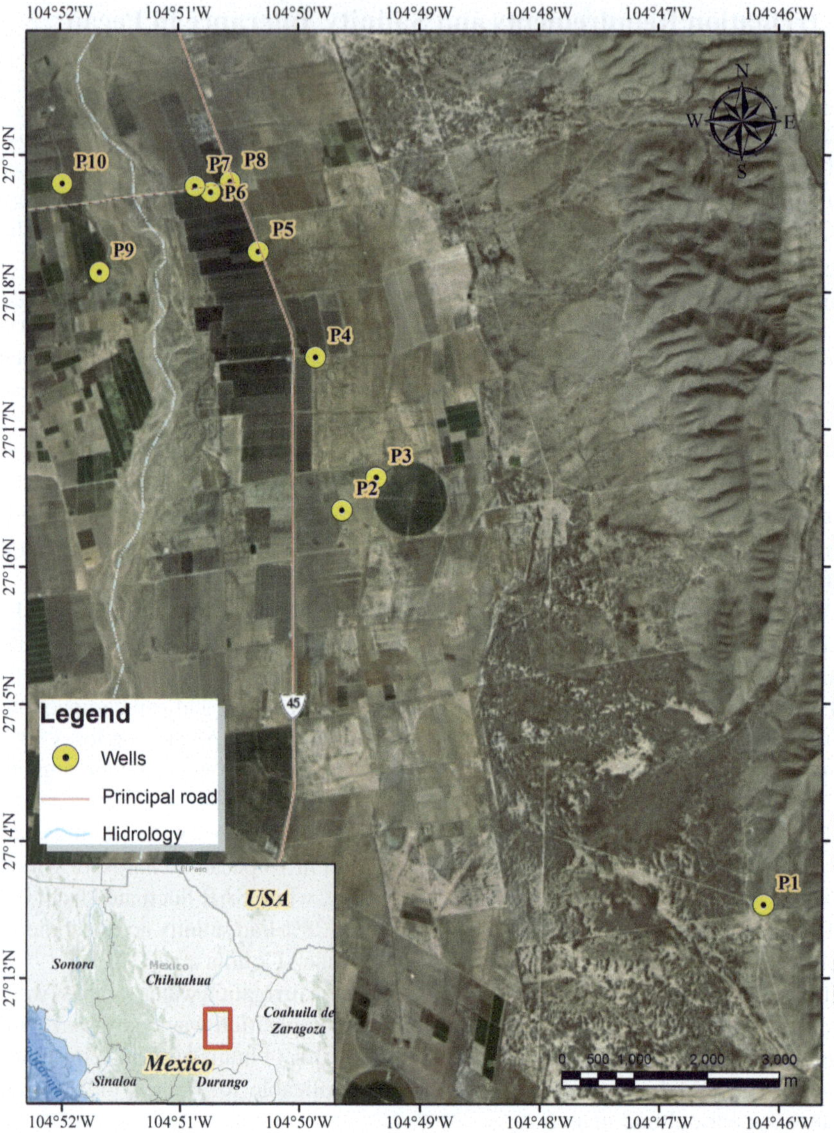

Fig. 3 Geo-position of the irrigation water wells in Chihuahua, México, sampled by the Jiménez-Las Pampas, 2014. Map data, Google, Maxar technologies Jiménez, Chih. (27 Apr. 2020). Google Maps. Google. Retrieved from: https://www.google.com.mx/maps/@27.2754158,-104.81917,16389m/data=!3m1!1e3

4 Salinity Stress in Pecan Trees

The pecan tree should be cultivated in deep, fertile, alluvial soil, with a high moisture retention capacity and good drainage. Unfortunately, many pecan orchards have poor soil quality, distinct from the preferred ecological niche (Campos-Villareal et al. 2017). *Carya illinoensis* is susceptible to high levels of salt, both in irrigation water and in the soil; its tolerance threshold is 2–3 dS m^{-1} (Miyamoto and Nesbitt 2011).

Saline soils have a reduced water potential and can cause an excessive absorption of ions, mainly Na and Cl, which results in nutritional imbalances and interferes with various metabolic processes (Trasviña-Barriga et al. 2018). Saline stress causes adverse effects on proline accumulation, the activity of antioxidant enzymes catalase (CAT) and superoxide dismutase (SOD), the concentration of photo synthetic pigments, and other factors that influence the growth, development, and productivity of plants (Meguekam et al. 2014). Superoxide dismutase and catalase are among the most effective enzymes in the elimination of compounds that cause oxidative stress to plants (Hand et al. 2017).

The study by Castillo-Gonzalez et al. (2014) measured some physiological and biochemical responses to salt stress in pecan trees (Fig. 4). The parameters considered included: concentration of chlorophyll, carotenes, proline, sulfates, total soluble proteins, hydrogen peroxide, the activity of SOD and CAT enzymes, antioxidant capacity, and relative content of leaf water and foliar area. The experimental design was at random with four categories of visual leaf damage: no visual damage, scarce, moderate, and severe damage. Data for all four categories were similar for

Fig. 4 Pecan trees with severe salinity damage in Jiménez, Chihuahua. *Photos* Dámaris Ojeda-Barrios

both the 2013 and 2014 production years. Leaves without visual damage had the highest concentration of chlorophyll, relative water content, and leaf area index, with a minimum concentration of carotenes, sulfate, and H_2O_2, and low CAT and SOD activity. Scarcely damaged leaves had intermediate values. Leaves with moderate damage had the highest values for SOD activity and DPPH (antioxidant capacity assay) values. Leaves with severe damage had the lowest concentration of chlorophyll and the maximum concentration of carotenes, proline, and sulfates. (Castillo-Gonzalez et al. 2014).

Another study in Chihuahua, Mexico, also showed how salinity adversely affects both pecan nut quality and yield. Pecan seedlings showed physiological changes after being irrigated with four concentrations of sodium sulfate over 70 days: 1000, 2000, 3000, and 4000 mg/l applied twice weekly. The aim was to identify and quantify how saline stress affected physiological parameters. Seedlings exposed to Na_2SO_4 had reduced height and stem diameter. At the highest Na_2SO_4 exposure level, proline concentration in the leaflets was 820% higher, 2.63 mg/g, than in the controls, 0.32 mg/g and chlorophyll was 35% lower, 23.4 mg/l, than in the controls, 36 mg/l. Meanwhile, the sulfate concentration increased 104%, from 84.47 to 172.5 mg/g. The root biomass decreased from 30.5 to 9.5 g, and leaf biomass decreased from 26.7 to 10 g. No disease symptoms were apparent in any seedling, suggesting that these changes were induced by Na_2SO_4 stress alone. Proline, chlorophyll, and sulfate concentration, as well as root and shoot biomasses, were strongly affected by irrigation with Na_2SO_4 at higher concentrations than 2000 mg/l (Moreno-Izaguirre et al. 2016).

The pecan tree is susceptible to damage by Na (Miyamoto et al. 1986). High concentrations of Na cause a reduction in the growth and production of the tree in two ways: by producing an osmotic effect and by ion-specific toxicity (SAGARPA 2018). The osmotic effect causes a decrease in the availability of water for the plant. The tree cells undergo an osmotic adjustment through the accumulation of salts and synthesize organic compounds to counteract the osmotic effect. That process uses energy that would otherwise be used for tree growth (Maas 1990). The result is a small, low-vigor tree with small leaves that otherwise have a healthy appearance. When soil salinity in the root zone, at 120 cm depth, exceeds 2.5 dS m^{-1}, the trees suffer a decline in growth rate, a decrease in the size of the nut, and production is affected (SAGARPA 2018). When EC is above 6 dS m^{-1}, tree death occurs. Ion-specific toxicity caused by Cl, Na, and B ions affects the yield and growth of the tree.

Salt ion toxicity causes burns on the tips and margins of the leaves. These signs appear at the end of July, and they continue their development until August and September, where the symptoms appear strongly. At this stage, tree growth is considerably reduced. The detection of the symptomatology is challenging since the symptoms can also be caused by insecticide and herbicide applications or water deficiencies. The burns are small initially, only measure 0.5 cm, so it is advisable to check the leaves on the underside. It is recommended to perform soil and leaf analysis to rule out other damage that is not caused by salinity (Miyamoto et al. 1986; Maas 1985).

High concentrations of Na and Cl in the soil also cause nutritional deficiencies in the plant. The leaves absorb Ca, B, and Na, that is why aspersion irrigation with water that has high concentrations of B or Na can cause damage to the foliage (Maas 1985).

In El Paso, Texas USA Miyamoto et al. (1986) demonstrated that pecan tree trunk diameter decreased by 12% for each unit of EC that increased in the soil from a tolerance threshold of 2.5 dS m^{-1}. In the region of Comarca Lagunera, Mexico, a study revealed that the yield of pecan decreased when soil salinity presented levels of 3 dS m^{-1} or more or when soil sodicity was a PSI 3% or more (Santamaría and Medina 2005).

In Hermosillo, Sonora, Mexico, a study was carried out in several adult pecan orchards with several irrigation systems: rolled, drip, and spray. It was found that clay texture orchards with superficially buried drip irrigation presented a higher risk of sodicity and accumulated soluble Na above the permissible level of 486 mg kg^{-1}. A significant accumulation of salts was observed between soil sampling dates, exceeding the permissible threshold of 1.5 dS m^{-1}. That study showed the importance of performing heavy watering before the sprouting of trees to leach salt from the soil (Grajeda et al. 2012).

5 Management of Saline and Sodic Soils in Pecan Orchards

The methods for the improvement of saline and sodic soils are grouped into physical, chemical, agro-technical, and biological approaches. Due to the complexity of saline soils, a combination of methods is advised, to avoid the adverse effects on pecan production associated with salinity. The current trend in management of saline soils has been directed toward the search for ecological and economical techniques (Miyamoto and Nesbitt 2011).

Drainage construction and soil washing are physical methods. The hydrogeological conditions of the soils will determine the type of drainage, horizontal, vertical or combined, that is appropriate, but the fundamental objective is to keep the level of groundwater below critical levels to avoid contamination and affecting the crops (Jasso et al. 2010). Washing can eliminate excess soluble salts from the soil profile and restore favorable soil conditions for plant development. Soil washing should not be considered if the quality of the water is saline or of poor quality. The chemical methods used in saline-alkaline soil, fundamentally function by replacing the sodium and magnesium contained in the soil adsorption complex, with calcium. In some cases, incorporating gypsum to sodic soils can improve Ca absorption by trees (Esquer 2013). The main function of agro-technical methods is to improve the physical properties of soils affected by salts, with organic products, such as compost, semicompost, or vermicompost, among others. The organic substances neutralize alkalinity and favor the formation of humus, which improves soil structure. Organic fertilizers contain a large amount of nutrients and beneficial microorganisms. Biological methods can include the seeding salt tolerant plants in the orchard, such as forages, to extract salts from the soil.

6 Conclusions

In northern Mexico, high saline irrigation water coupled with poor quality soil conditions is compromising the economic viability of pecan production. This chapter highlights the salinity problem in pecan orchards of arid and semi-arid climates and suggests solutions to amend soil salinity and sodicity. Pecan cultivation should be avoided in areas where water is limited, and it is necessary the conservation of resources, the restoration of areas affected by salinity and the sustainable management of existing pecan orchards. To do that, information that promotes water and soil conservation and presents efficient ecological solutions for pecan producers is needed.

References

Brown PW (2010) Pecan water use estimates for Southern Arizona and Northern México. In: XI simposio internacional de nogal pecanero. Memoria científica no. 1 hermosillo, sonora, 12 de septiembre 2010

Campos-Villarreal GA, Arreola-Avila JG, Chavez-Simental A, Trejo Calzada R, Borja de la Rosa A, Lopez Santos A, Hernández Salgado JR (2017) Physiological response, ion accumulation and dry weight of pecan (*Carya illinoinensis* (Wangenh) K. Koch) rootstocks grown under saline stress conditions. Interciencia 42(11):744–749

Casales FG, Van der Watt E, Coetzer GM (2018) Propagation of pecan (*Carya illinoensis*): a review. African J Biotech 17(18):586–605

Castillo-Gonzalez J, Ojeda-Barrios DL, Sanchez-Chávez E, Ruíz-Anchondo T, Parra-Quezada R, Guerrero-Prieto V (2014) Indicadores bioquímicos bajo condiciones de salinidad en nogal pecanero. Proceedings XXXIX congreso nacional de la ciencia del suelo. Chihuahua. México. 24 October 2014

Esquer PLA (2013) Factores que influyen en la productividad del nogal pecanero. In: XI Simposio internacional de nogal pecanero. Memoria científica no. 1 hermosillo, sonora, 12 de septiembre 2013

FAO-FAOSTAT (2018) Metodología provisional para la evaluación y representación cartográfica de la desertificación. https://www.fao.org/faostat/es/#home

Follet RF, Castellanos JZ, Buenger ED (2005) Carbon dynamics and sequestration in an irrigated vertisol in central Mexico. Soil Till Res 83(1):148–158

Grajeda GJ, Sabori PR, Valenzuela MA, Quijada FJ, Nuñez MH, Rodríguez JC (2012) Salinidad del suelo en huertas de nogal pecanero *Carya illinoinensis* (Wangenh.) K Koch. Rev Ciencias Biológicas y de la Salud de la Univ de Sonora 13(3):22–27

Hand MJ, Taffouo VD, Nouck AE, Nyemene K, Tonfack B, Tl M, Youmbi E (2017) Effects of salt stress on plant growth, nutrient partitioning, chlorophyll content, leaf relative water content, accumulation of osmolytes and antioxidant compounds in pepper (*Capsicum annuum* L.) cultivars. Notulae Botanicae Horti Agrobotanici Cluj-Napoca 45:481–490

Jasso IR, Chávez SN, Figueroa VU, Rivera GM, Sabori PR (2010) Salinidad del agua de riego y su efecto en la productividad del nogal pecanero. In: XI Simposio internacional de nogal pecanero. Memoria científica no. 1 hermosillo, sonora, 12 de septiembre 2010

Maas EV (1985) Crop tolerance to saline sprinkling water. Plant Soil 89:273–284

Maas EV (1990) Crop salt tolerance. In: Tanji KK (ed) Agricultural salinity assessment and management. America social of civil engineering practices. no. 17 pp 262–236

Meguekam TL, Taffouo VD, Marius-Nicusor G, Maria ZM, Emmanuel Y, Akoa A (2014) Differential responses of growth, chlorophyll content, lipid peroxidation and accumulation of compatible solutes to salt stress in peanut (*Arachis hypogaea* L.) cultivars. Afric J Biotech 13:4577–4585

Miyamoto S, Nesbitt M (2011) Effectiveness of soil salinity management practices in basin-irrigated pecan orchards. HortTechnology 21(5):569–576

Miyamoto S, Storey BJ (1995) Soil management in irrigated pecan orchards in the southwestern United States. HortTechnology 5(3):219–222

Miyamoto S, Riley T, Gobran G, Petticrew J (1986) Effects of saline water irrigation on soil salinity, pecan tree growth and nut production. Irrig Sci 7:83–95

Montes-Rentería GD, Arreola-Ávila JG, Trejo-Calzada R, Rodríguez-López JS (2011) Acumulación de iones en nogal pecanero [*Carya illinoensis* (Wangenh.) K. Koch] de maduración temprana, sometidos a diferentes condiciones de salinidad. Revista Chapingo Serie Zonas Áridas 10(2):131–139

Moreno-Izaguirre E, Ojeda-Barrios D, Avila-Quezada G, Guerrero-Prieto V, Parra-Quezada R, Ruíz-Anchondo T (2016) Sodium sulfate exposure slows growth of native pecan seedlings. Phyton 84(1):80–85

Núñez Colima JA, Moreno Reséndez A, Valenzuela Núñez LM, Rodríguez MR, González TA, García PC, Esparza RJ, Molina OJ (2019) Influencia de variables climáticas en el contenido de N en *Carya illinoensis* Koch. Nova Scientia 11(22):207–223

Ojeda-Barrios DL, Abadía J, Lombardini L, Abadía A, Vázquez S (2012) Zinc deficiency in field grown pecan trees: changes in leaf nutrient concentrations and structure. J Sci Food Agr 92:1672–1678

Pulido ML, González MJ (2010) Salinidad de suelos en distritos de riego, sensores remotos y cambio climático. Terra Latinoamericana 28(1):15–26

Reyes FS, Ávila JA, Murrieta AL, Calzada RT, Arriaga OG, Herrera GG (2010) Efecto de niveles de NaCl sobre fotosíntesis y conductancia estomática en nogal pecanero (*Carya illinoinensis* (Wangeh.) K. Koch). Revista Chapingo Serie Zonas Áridas 9(2):135–141

SAGARPA (2018) Crecimiento en producción de nuez, favorece a exportación a Norteamérica In: Secretaria de agricultura, ganadería, desarrollo rural, pesca y alimentación 74:06. https://www.gob.mx/agricultura/prensa/se-incrementa-83-por-ciento-la-produccion-de-nuez-en-mexico. Accessed 20 Dec 2019

Santamaría CJ, Medina MC (2005) Salinidad de suelo y agua, producción de nuez y áreas de riesgo en la comarca lagunera. Agrofaz 5(3):105–112

Trasviña-Barriga A, Bórquez-Olguín R, Leal-Almanza J, Castro EL, Gutiérrez CM (2018) Rehabilitación de un suelo salino con yeso agrícola en un cultivo de nogal en el valle del yaqui. Terra Latinoamericana 36(1):85–90

Valles-Aragón MC, Ojeda-Barrios DL, Guerrero-Prieto VM, Prieto-Amparán JA, Sánchez-Chávez E (2017) Calidad del agua para riego en una zona nogalera del estado de Chihuahua. Rev Int Contam Ambient 33(1):85–97

Zhang R, Peng F, Li Y (2015) Pecan production in China. Sci Hortic 197:719–727

Salinity in Humid, Waterlogged and Flooded Environments

Genesis, Properties and Management of Salt-Affected Soils in the Flooding Pampas, Argentina

Perla A. Imbellone, Miguel A. Taboada, Francisco Damiano, and Raúl S. Lavado

Abstract The Flooding Pampa is a plain, covering around 90,000 km^2 composed of eolian and alluvial deposits. It is characterized by its geomorphology, with low (0.1–0.01%) slopes and elevations, including the coastal flat. A significant fraction of the region has permanent or temporary lagoons, generally connected with the underground water. Two kinds of water excesses are distinguished: (i) waterlogging of low duration and intensity and (ii) floods, intense and prolonged. Most soils have developed from loessial and loessoid sediments. They are composed mainly of a pyroclastic volcanic association, with a variable proportion of plagioclases, quartz, volcanic glass, lithoclasts and heavy minerals. However, the soils of the East sector of the region were generated by the influence of the sediments from the Rio de la Plata. They show a higher proportion of kaolinitic, smectites and interstratified expansible clays, than other soils of the region. Most salt-affected soils of the region belong to the US Soil Taxonomy Orders of Mollisols, Alfisols, Entisols and Vertisols. Their limitations are not easy to reverse technically and from the economic and ecological point of view. Conversely, it is reasonable to introduce management and reclamation technologies adapted to those limiting conditions.

P. A. Imbellone
Instituto de Geomorfología y Suelos, Universidad Nacional de La Plata, La Plata, Argentina
e-mail: micromorfologia@igs.edu.ar

M. A. Taboada
Instituto de Suelos, Instituto Nacional de Tecnología Agropecuaria (INTA), CONICET, Hurlingham, Buenos Aires, Argentina
e-mail: taboada.miguel@inta.gob.ar

F. Damiano
Instituto de Clima y Agua, Instituto Nacional de Tecnología Agropecuaria (INTA), Hurlingham, Buenos Aires, Argentina
e-mail: fdamiano52@hotmail.com

R. S. Lavado (✉)
Facultad de Agronomía, Universidad de Buenos Aires and INBA—CONICET/UBA, Buenos Aires, Argentina
e-mail: lavado@agro.uba.ar

E. Taleisnik and R. S. Lavado (eds.), *Saline and Alkaline Soils in Latin America*, https://doi.org/10.1007/978-3-030-52592-7_10

Keywords Flooding pampas · Lowlands · Endorheic drainage · Loessial sediments · Natric horizons

1 Introduction

For more than a century, geologists and geographers who regionalized the Pampas (Tapia 1937; Daus 1946; Frengüelli 1950; Fidalgo et al. 1975; INTA 1977; Sala et al. 1983; Fucks et al. 2012, among others), described a particular area because of its geological, physiographic and geomorphological characteristics. However, they did not agree in its spatial representation. Tricart (1973) carried out a unified stratigraphic, pedological and geomorphological study, showing the relationship between geoforms and geological processes occurred from the Pleistocene.

Recently, Zárate and Rabassa (2005) determined the tectonic basin based on morpho-structural elements and the subsoil structure and geology and its influence on the superficial morphological features. The Flooding Pampas, including the basin of the Salado River, occupies the Center and East of Buenos Aires province and peripheral areas. Figure 1 shows the location of the region, including other regions of the whole Pampas.

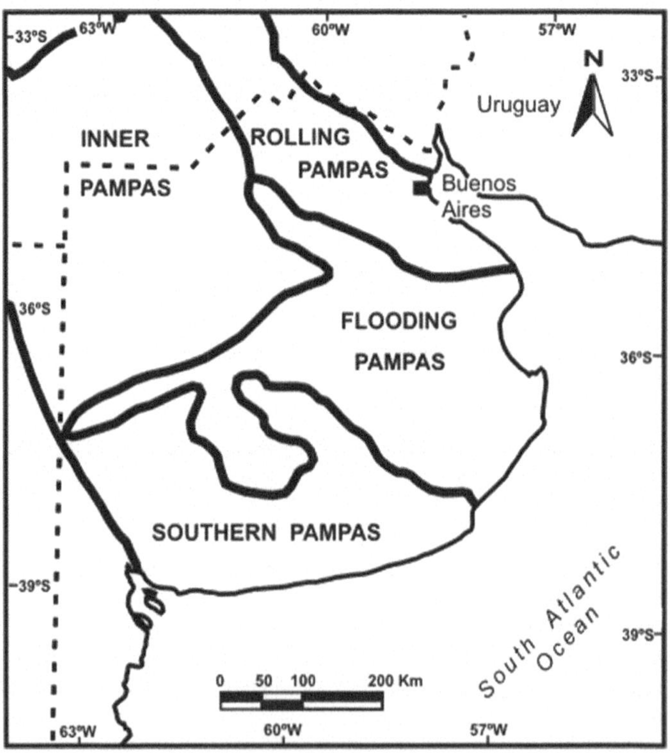

Fig. 1 Location of the flooding Pampas, in the Buenos Aires province, Argentina

Soils show some differences when compared with soils of other regions of the Pampas. Early soil maps registered gley and halo-hydromorphic soils (Bonfils 1966). Contrary to other agriculturized areas of the Pampas, the region, devoted mainly to cattle husbandry, is covered mainly by different kinds of grass communities' especially tall grasses (Soriano et al. 1991). The floristic composition of these grasslands varies in accordance with the local soil heterogeneity. Some patches of the original grassland still remain, but the area shows a degraded grassland invaded by several exotic species (Soriano et al. 1991). Recently it was invaded by the legume *Lotustenuis*, which increased its forage quality (see also the chapter by Nieva and Ruiz, in this book: "*Lotus* spp.—a foreigner that came to stay forever: economic and environmental changes caused by its naturalization in the Salado River Basin (Argentina)").

2 Geomorphological and Hydrological Characteristics

The area is an aggradational plain, covering around 90,000 km^2 composed of eolian and alluvial deposits of the Pleistocene. It is characterized by its geomorphology, with low (0.1–0.01%) slopes (Zárate and Rabassa 2005) and elevations, including the coastal flat, from about 0 to 100 m above sea level. The hydrology and the drainage system developed on a landscape of flat lowlands with a very low gradient. No drainage network is observed and the stream density is near 0 km^{-2} (Kuppel et al. 2015; Taboada and Damiano 2017). As an example, in the center of the region, covering 50,000 km^2, the Salado River is the only drainage way. The source of the river is located in another province and flows along the axis of the tectonic basin and receives waters from several tributaries in the periphery. It is a typical plain river, with a permanent regime and extremely variable flow draining the superficial and subsurface water excesses. The excess of water in the lowlands is evacuated to the ocean with extreme slowness, causing waterlogging and even floods. These events occur despite the large number of canals built at the beginning of the twentieth century to discharge the excess of water.

According to González (2005) the recharge of the hydraulic system is direct autochthonous but the discharge is mainly consumptive; it is located in courses and lagoons, and the outflows move toward the Samborombón Bay. The underground water moves very slowly, with gradients in the order of cm km^{-1}. From the chemical point of view, there is a very marked amplitude in the upper aquifers located in the so-called Postpampean (saturated zone, soil/phreatic aquifer) where the concentration ranges from less than 500 mg l^{-1} to 1,500 mg l^{-1}, while the saline content in aquifers located in the so-called Pampean, ranges from 500 mg l^{-1} to 20,000 mg l^{-1}. There is a tendency to increase the saline content toward the axis in the Salado River. The deeper semi confined aquifer, so-called Puelche, shows saline concentrations between 2000 and 10,000 mg l^{-1}. In the deeper aquifers salinity reaches concentrations of 60,000 mg l^{-1}. Recently, Kuppel et al. (2015) stated that the poor connection between the underground and superficial waters promote fast floods in the Flooding Pampa.

A significant fraction of the region has permanent or temporary lagoons of endorheic drainage, generally connected with the underground water. As a consequence, phreatic water is frequently near the superficial soil and it is highly saline, as well as the stream water (Dangavs 2005). The water bodies in the lowlands, either temporary or permanent, are shallow. The water outcrops or derives from shallow phreatic water, rainfall and, exceptionally, runoff. This is attributed to the almost null slopes and the coalescence of nearby water bodies. The results of the several canals built to drain those waters to the Samborombón Bay are debated for several reasons (inefficiency to conduct water excesses, soil salinization) (Bazzuri et al. 2018).

The present climate is humid and temperate, but the records indicate that climate conditions have changed from dry to wet alternatively, with variable intensity and recurrences (Ameghino 1884; Deschamps et al. 2017; Krause and Laurencena 2005; Fucks et al. 2012; among others). Periodical water excesses are a typical environmental event in the region, determined by climatic, edaphic and topographic factors, and their duration and intensity vary from year to year. Water excesses are partially associated to the seasonal cyclic behavior of the water table and to the occurrence of intense precipitations. Two kinds of water excesses are distinguished. One of them is waterlogging, marked by a lower duration and intensity; it takes place regularly at the end of winter and the onset of spring (Fig. 2). The others, true floods, are more intense and prolonged, but do not occur in a precise period of the year (Taboada and Damiano 2017; Kuppel et al. 2015). Roads and railroads embankments usually act as true dams.

There is a relationship between underground water and waterlogging; the response of phreatic levels depends on the thickness of the unsaturated zone and the magnitude of the phreatic movement and its relation with infiltrated rainfall. Waterlogging occurs also due to the clogging of lowlands and swamps after successive wet years (Fig. 3). In a regional view, waterlogging is related to the underground water outcrop when the soil storage capacity is surpassed, then this water is transferred to the soil surface (Hernández 2001; Krause and Laurencena 2005).

a b

Fig. 2 **a** Waterlogged landscape and **b** Flooded landscape, both in the center of the Flooding Pampas. *Photos* M. Taboada and R. S. Lavado

Fig. 3 **a** Phreatic water and capillary rise in depressions with seasonal vegetation ("espadaña", *Zizaniopsis bonariensis*). Note thickening of surface horizon which favors sediment filling in winter. **b** similar to (**a**) under vegetation of waxyleaf nightshade or "duraznillo" (*Solanum glaucophyllum*) in non-halomorphic low-lying areas (**c**) and (**d**) in floodplains with grassy vegetation (saltgrass, *Distichlis sp.*). Salado river basin. Salt efflorescences (to the right of the scale) which appeared one week after pit digging. *Photos* P. A. Imbellone

The effects of floods on the soils are closely related to the quality of the flooding water, its salt content and the prevalent ions. Water quality depends on the flooding water origin: rainfall (salt-free) or subsuperficial (saline). The key factor for each alternative depends on the soil profile characteristics, in particular the occurrence of natric horizons of low permeability, which control the up and down movements of water. These horizons are common in the Flooding Pampas. Figure 4 shows two cases, one with free rise from the phreatic water and the other a soil with the rise of phreatic water confined by the bottom of the natric B horizon, which limits the capillary ascent. In this case there are two different water types: above (rainfall water) and below (phreatic water) the B horizon. The soil undergoes both hydromorphisms (shallow and deep) simultaneously (Imbellone 2018).

When the water balance is positive, the phreatic water rises in soils without B horizon or other impervious layers and can reach the surface. Dissolved salts move

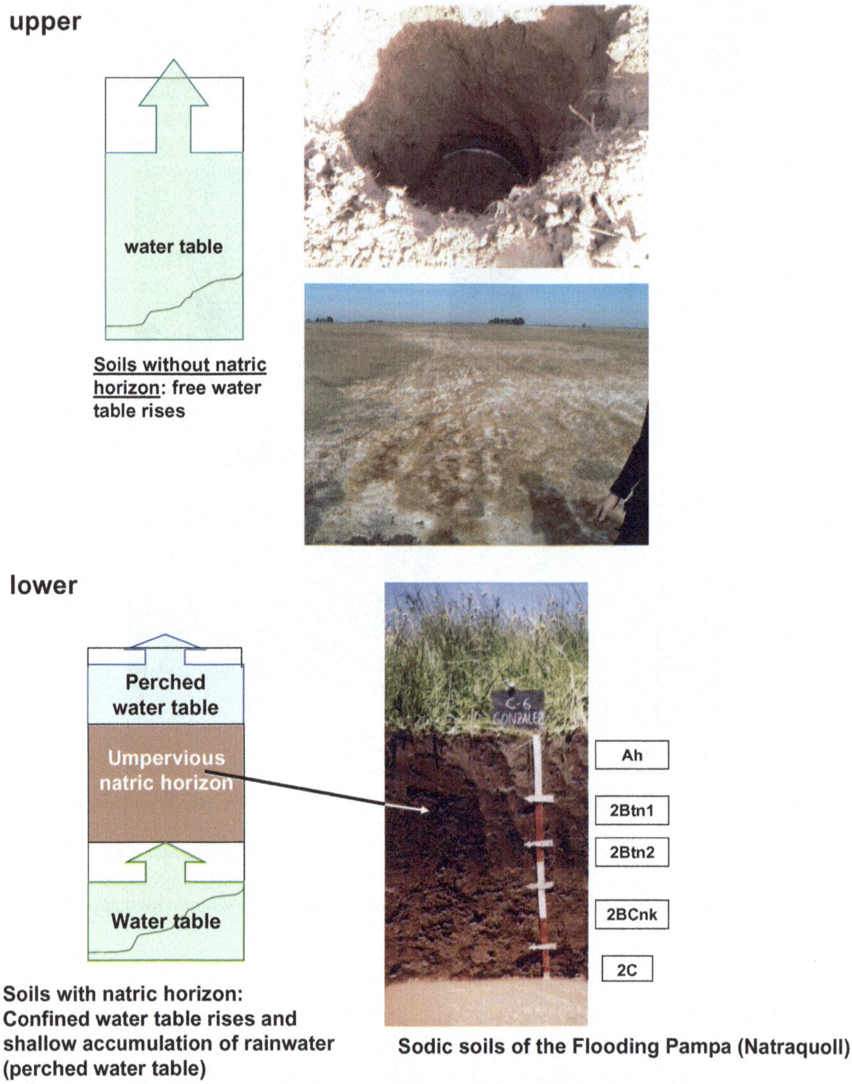

Fig. 4 Conceptual model of water (and salts) movement in soil profiles without B horizon (upper panel) and with 2Btn horizon (lower panel); soil profiles are shown on the right. *Photos* F. Damiano

with water by mass flow; they reach the soil surface when the soil dries. In soils with natric B horizon, salinization of top horizons also occurs, but following a complex pathway. Rainfall water is checked at the top of the B horizon and phreatic water is checked at the bottom of the same horizon. Salts rise through the B horizon by diffusion and when the soil dries, they finally reach the soil surface, this time by mass flow (Lavado and Taboada 1987). Salts, mainly sodic, finally form saline

Fig. 5 Saline efflorescences within the soil profile of the region (left); crystallization (scanning electron micrograph—SEM) of sodium sulfate (thenardite) on the soil surface (center) and crystallization (SEM) of calcium sulfate (gypsum) on the soil surface (right). *Photos* P. A. Imbellone

efflorescences of different thickness when the soil dries, either in the soil profile or on the soil surface (Fig. 5).

The salinization process is usually accompanied by alkalinization or sodification processes. The excess of salts tends to flocculate clays and other soil components but in humid climates, exchangeable sodium predominates. As a consequence, colloids become destabilized and dispersed; the pore system collapses and water movement is strongly affected.

3 Characteristics of the Sediments

The soils of the Pampas have developed from loessial and loessoid sediments, deposited in the late Pleistocene-Holocene. They are composed mainly of a pyroclastic volcanic association, with a variable proportion of plagioclases, quartz, volcanic glass, lithoclasts and heavy minerals (Imbellone and Teruggi 1993, Etchichuri and Tófalo 2004, among others). Most clay in those soils is considered inherited, due to the climatic characteristics (humid/semi arid and temperate/cold) of the late Pleistocene-Holocene, although soils contain abundant sodic plagioclases in the coarse fraction. The alteration of primary minerals and formation of secondary ones have been moderate and barely evident (Morrás 2017).

Although many soils of the Pampas possess argillic and sometimes natric horizons, the content of clay is less than 50% in soil profiles derived from loess. As an example, Table 1 presents the clay mineralogy of sodic soils in the center of the region: Typic Natrudoll, pedon 1; Abruptic Natrudoll, Pedon 2; Thapto-natric Epiaquoll, Pedon 3 (Subgroup originated and used in Argentina but not included in the original US Soil Taxonomy; Vertic Natraqualf, Pedon 4.

A different picture is observed to the East of the Salado River basin, where natric horizons were generated by the influence of the sediments from the Rio de la Plata marginal lagoon, the deposition of salt-laden aerosols from the sea and past sea ingressions or present marine sediments. The soils show a higher proportion of smectites and interstratified expansible clays than the soils of the central area. Also a

Table 1 Semi-quantitative estimate of clay minerals of sodic soils in the center of the Flooding Pampas region (After Imbellone et al. 2018)

Estimated clay mineralogy								
		Fraction <2 μm			Fraction <2 mm			
Horizon	Depth (cm)	Illite (%)	S + I (%)	K (%)	Illite (%)	S + I (%)	K (%)	Clay(%)
Pedon 1—Typic Natrudoll								
A	0–18	80	15	5	9	2	1	11.2
2Btnb	18–38	67	30	3	20	9	1	29.2
2BCnb1	38–53	65	32	3	15	8	1	23.7
2BCnb2	53–67	83	10	3	15	2	1	17.5
3Btnkxb	67–80	40	59	1	10	15	Tr	26.0
3BCkxb	80–123+	58	40	2	15	10	1	25.2
Pedon 2—Abruptic Natrudoll								
A	0–16	80	10	10	11	1	1	13.4
2Ab	16–30	85	10	5	16	2	1	18.2
2ACb	30–40	75	20	5	10	2	1	12.6
3Btnb	40–52	71	27	2	20	8	1	28.8
3BCnb	52–64	57	40	3	7	5	Tr	11.5
4Btnkxb	64–96	58	40	2	12	8	Tr	21.1
4BCnkxb	96–118	48	50	2	12	12	Tr	24.5
5C	118–137	49	47	4	6	5	Tr	11.5
5Ckm	137–140+	–	–	–	–	–	–	–
Pedon 3—Thapto Natric Epiaquoll								
A	0–25	82	17	1	7	1	Tr	8.9
AC	25–45	64	32	4	5	2	Tr	7.4
C	45–60	62	33	5	2	1	Tr	3.1
2Btnb	60–85	62	35	3	20	11	1	31.7
2BCb	85–119	60	38	2	7	4	Tr	11.0
2BCkxb	119–156	64	32	4	2	1	Tr	3.3
2Ck	156–171+	62	37	1	3	2	Tr	5.0
Pedon 4—Vertic Natraqualf								
A1	0–15	75	17	8	7	2	1	9.2
A2	15–28	85	11	4	9	1	Tr	10.1
Btnssb	28–52	41	57	1	18	26	1	44.7
2BCkxb	52–95	58	40	2	17	12	1	28.9
2Ck	95–110+	73	26	1	14	5	Tr	19.6

Note S, Smectite; K, Kaolinite; I: Interstratified. Tr: Traces

higher proportion of kaolinitic clays is found. These aspects were studied some years ago by Lavado and Camilión (1985) and more recently by Imbellone and Mormeneo (2011).

Because of the topographic characteristics and the nature of the sediments, this coastal plain is an adequate environment for the development of hydromorphic and alkaline soils. This littoral environment was formed in the recent evolutionary stage of the Río de la Plata between the glaciation at the end of Pleistocene and the Holocene transgression. This area limits to the west with the loessial plain, separated by a topographic step, carved during the maximal transgression and by the subaqueous fraction of the Paraná delta. It is a level to concave strip located at altitudes lower than 5 m above sea level. The mudflat, where halo-hydromorphic soils occur, is located between 1.25 and 2.50 m above sea level. Soil developed in the Villa Elisa facies of Las Escobas formation, ranging in age from 3500 to 2500 years BP, composed of the sediments deposited in the regressive stage of the sea. Those sediments contained abundant silt and clay, which flocculated in an environment subjected to tides mixed with fresh and salt waters (Cavalloto and Violante 2005). Then, the nature of the original materials, landforms and climate favored the development of alkaline and hydromorphic soils (Imbéllone et al. 2009).

The alkaline and hydromorphic soils in the littoral plain of the Río de la Plata (Fig. 6), have as much as 80% total clay and up to 50% expansible clay. The mineralogy indicates a maximum of 70% illite and traces to 20% kaolinite in the clay fraction (<2 μm). This means that the bulk soil contains 30% expansible minerals or more. The soils are very expansible due to the mineralogical composition of the fine fraction, besides, soils undergo a short period of water deficit in summer, which causes bypass flow along cracks, and the distribution of an alkaline solution in the total

Fig. 6 Natural environment of the coastal plain of the Rio de la Plata. **a** halo-hydrophilous vegetation (*Distichlis spicata, Cynodon dactylon, Solanum glaucophyllum, Cyperus prolixus*). **b** Natraquerts of the coastal plain of the Rio de la Plata: A, Bssng 1, Bssng 2, Bssng 3, 2Cgn. *Photos* P. A. Imbellone

mass of the soil because soils are alkaline from the surface. Clay minerals belong to the sediments formed during the Holocene transgression which accompanied the sea level decline. In this environment the materials would come in part from the contribution of "Pampean sediments" with abundant illite and in part from discharges from the upper and middle Paraná River, and upper Paraguay River, carrying smectites and kaolinite (Cavallotto and Violante 2005). However, the expansibility is also linked to swelling processes due to entrapped air during floods (Taboada et al. 2001).

4 Halomorphic Soils, Genesis and Properties

Soils are the consequence of the pedogenetic processes, which consist in a set of reactions acting simultaneously and/or sequentially, causing the horizonation and other soil characteristics. The five soil forming factors (climate, relief, parent material, biota, and time) bring about elemental processes like additions, removals, translocations and transformations of substances from the pedosphere (Buol et al. 1997). Soil salinization is promoted by specific factors like negative water balance, hydrogeological geomorphology, shallow water table; highly conductive sediments, high evapotranspiration capacity, salt-tolerant species, etc. (Nossetto et al. 2008). Halo-hydromorphic soils is a group of soils with different properties, whose common feature is that they have been or are affected by salts, or will be potentially affected in the near future. Another common feature is their limitations for agriculture. They are often called "salt-affected soils" and include, mainly but not exclusively, saline soils *sensu stricto* and sodic and alkaline soils. More than a century ago an evolutionary ideal outline was established for them (Fig. 7), which is applicable to the halo-hydromorphic soils of the Flooding Pampa.

As an example, the classic Solonchak soils are equivalent to Salids, according "US Soil Taxonomy" and they are not frequent in the Pampas, where the predominant water regimes are Udic, Ustic and Aquic. The Solonetz are equivalent (according to their evolutionary stage) to Natraqualfs, Natrudolls or Natraquolls, characterized by

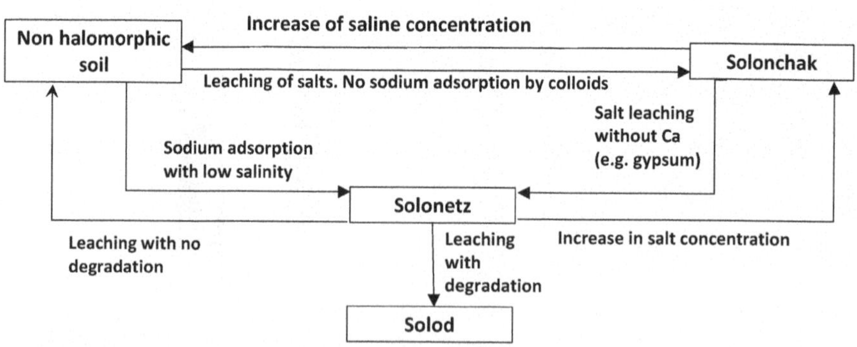

Fig. 7 Halomorphic soil genesis, according to an evolutionary outline (From Imbellone et al. 2010)

the occurrence of Natric horizons, high pH and dispersion of organic and inorganic colloids. The Solod soils are mainly equivalent to Natralbolls, resulting from the continual leaching of the soil, with pH decrease in the A horizon, and the development of an E horizon, with a marked whitening at the top of the columnar structure due to degradation of inorganic colloids and accumulation of silica (Fig. 8).

One characteristic of those soils is the loss of organic substances in the runoff and flood waters. Several research works carried out during the 1980 and 1990 decades were summarized by Lavado (2006). Organic matter -highly soluble in these soils due to sodicity and high pH- moves within the soil and on the surface (runoff water), accumulating in the so-called "black alkali" spots (Fig. 9). This organic matter has a low polymerization degree, with a predominance of dispersible fulvic acids. These and the organo-minerals complexes move with the water, reaching lagoons, water courses or the moving flooding water and finally reach the Samborombon Bay. This process involves a huge loss of humic substances from the soils of the basin, which end in the organic detritus which is a significant support for the natural life of the Ramsar site located in the Samborombón Bay.

Fig. 8 **a** Natraqualf. **b** Natralboll. **c** Detail of the columnar structure after manual removal of the A horizon (Natraqualf), in the Flooding Pampas region. *Photos* P. A. Imbellone

Fig. 9 **a** Black alkali spots in soils of the Flooding Pampas. **b** Soil showing phreatic water at 65 cm, darkened with dispersed humic substances and saline efflorescences in the dry left part of the soil profile. *Photos* P. A. Imbellone

5 Halomorphic Soil Distribution and Classification

According to different soil surveys (INTA 1990, 2018), more than 60% of the soils of the Flooding Pampa are affected by water excess, together with an excess of salts and exchangeable sodium. Most soils have clayey, sodic and low permeability horizons (Natric horizons Btn and BCn). The mapping units are halo-hydromorphic complexes and associations, composed mainly by the Great Groups Natraquolls, Natraqualfs, Natralbolls, Argialbolls, Thapto-natric Hapludolls and other soils rich in exchangeable sodium (Soil Survey Staff 2014). Those soils not only occupy small patches but also large extensions.

Unpublished soil maps elaborated by Moscatelli and Cruzate, grouped soils with problems of alkalinity and/or high pH, in an area exceeding the Pampean Region. Figure 10 shows the geographic distribution, showing the depth where the alkaline horizon appears. The total area of soils affected by some degree of alkalinity is about 116,000 km^2.

The more common soils at subgroup level according to Soil Taxonomy (Soil Survey Staff 1999) are Typic Natraqualfs (32,415 km^2), Typic Natraquolls (25,561 km^2), Typic Natralbolls (5335 km^2), Mollic Natraqualfs (3501 km^2) (Morrás 2017). The soils affected by sodification processes in the coastal plain of the River Plate (Fig. 11) were classified as Entisols (Aquents) and Vertisols (Aquerts) (INTA 2018).

Soil profiles with natric horizons are shown in Figs. 12 and 13.

6 Soil Utilization and Management

The Natraqualfs are the regional soils with the poorest quality for agriculture. The Natraquolls, instead, usually have better agronomic quality, which enables the implantation of forage species, like tall fescue, *Lotus*, and others. At present, many soils are devoted to summer crops (soybean, sunflower), although with low yields and risks of severe water stress in dry summers or harvest problems in wet autumns.

Diverse technologies are available to manage these soils, which differ in its effectiveness; they must be tested in each particular case. The different management and reclamation technologies vary according to the objectives. The basic principles which guide most of them are: (i) to control the capillary rise from the phreatic water; (ii) to increase the infiltration rate; (iii) to restore or create a plant cover and (iv) to improve the physical and chemical soil condition. The technologies differ also in their complexity degree, because they include from simple agronomic practices, like grazing management (Lavado and Taboada 1987; Di Bella et al. 2015; Vecchio et al. 2018); use of plant covers or mulches, interseeding, introduction of forage species, such as Leguminosae (Fabaceae) or megathermal graminoids (Otondo et al. 2015; Taboada et al. 1998); amendments utilization (Torres Duggan and Rodríguez 2017);

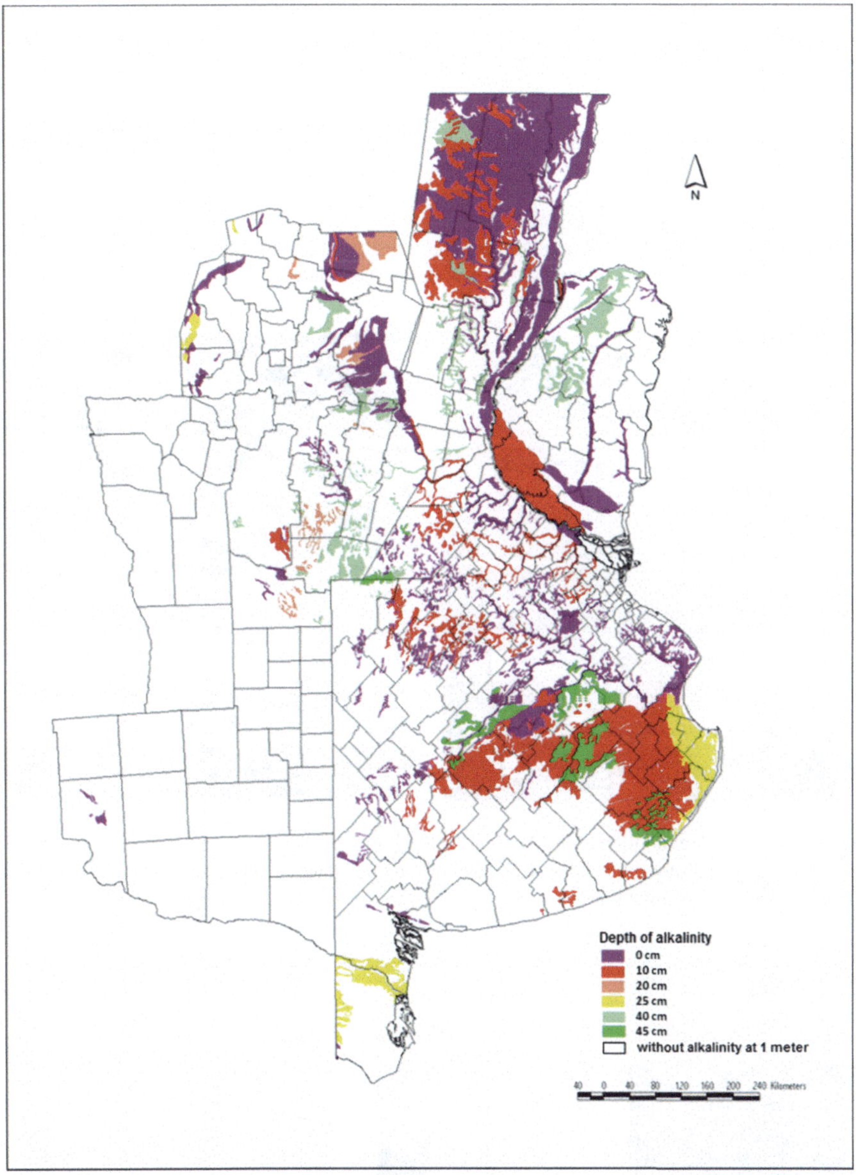

Fig. 10 Geographic distribution of alkaline soils in the Pampas, Chaco and adjacent regions (Moscatelli and Cruzate unpublished). The figure shows the depth of alkaline horizons

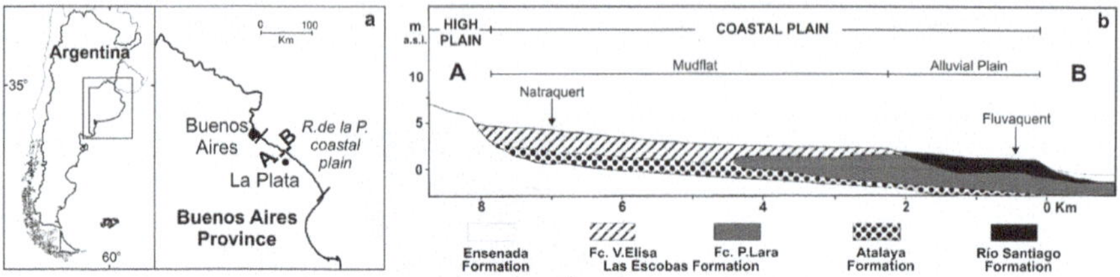

Fig. 11 **a** Location of the coastal plain and the transection at the right. **b** Schematic location Natraquerts and Fluvaquents in the coastal plain of the river plate. (from Imbellone et al. 2009)

Fig. 12 **a** Typic Argialboll profile (A, E, Bt, BCg horizons. **b** Details at the base of the same soil with strong hydromorphism due to the rise of phreatic water, Cañuelas county, Buenos Aires province. *Photos* P. A. Imbellone

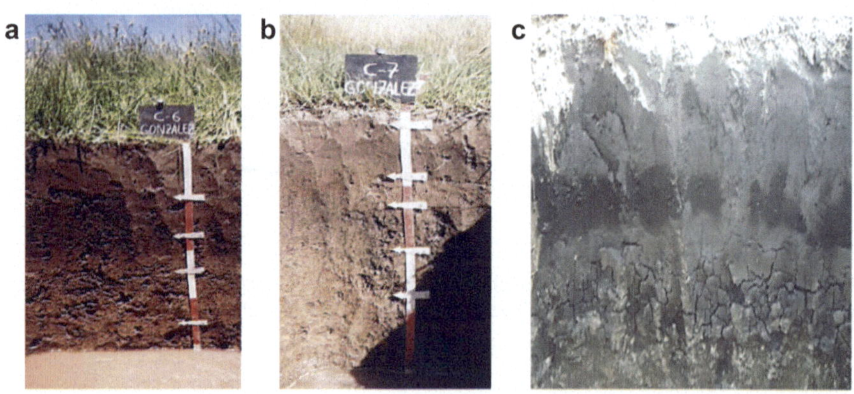

Fig. 13 Representative soils **a** Typic Natraquoll y **b** Typic Natraqualf (Cañuelas county), and **c** Typic Natrudoll (Lobos county), Buenos Aires province. *Photos* F. Damiano

deep tillage and land systematization for surface water management and improvement of drainage (Alconada et al. 1993; Damiano and Parodi 2015).

There are other potential technologies applicable to those soils. However, there are technical limitations, like groundwater as a source of salts, or economical restraints. Some technologies have been studied locally like gypsum application, whereas others need more local experimentation, like organic products and biological improvements through the use of adapted forage species (Taboada et al. 2017; Casas 2018). Fertilization is not a measure for the reclamation of salt-affected soils, but it can contribute to such objective, as a support of other technologies (Rubio et al. 2010). Many of the mentioned technologies only increase livestock productivity, but without modifying permanently soil properties. In those cases, a new unstable water and saline equilibrium is established, which is more productive, but the basic status of the soil does not change.

7 Concluding Remarks

The Flooding Pampa is an aggradational basin, filled with silty eolian (loessoid), and fluvial sediments during the Pleistocene and Holocene. The geomorphological and hydrological expression is their low slopes and low speed of the surface (runoff) and subsurface (phreatic water) fluxes. Conspicuous morphological features of the region are the numerous lagoons with associated dunes linked to deflation processes. The lagoons act as (a) bodies of endorheic drainage in normal rainfall periods or (b) as connectors of masses of overflow water during abundant rainfall periods, integrated to the drainage network.

Water excess is reflected in the soils in different ways: formation of perched water over the horizons acting as aquitards and/or rise of phreatic water at the base of the aquitards and/or reaching the topsoil when those horizons are lacking. The different behavior is determined by the subsoil particle size and is related to the different types of floods and the quality of that water, related to the type of soils.

The limitations imposed by landscape and soils characteristics are not easy to reverse technically and are not affordable from the economic and ecological point of view. Conversely, it is reasonable to introduce practices adapted to those limiting conditions, like grazing management, maintaining the plant cover, complemented by fertilization and limited quantities of mineral and organic amendments. Other technologies are economically limited.

References

Alconada M, Ansin OE, Lavado RS, Deregibus VA, Rubio G, Gutiérrez FH (1993) Effect of run-off water retention and grazing on soil and vegetation characteristics of a temperate humid grassland. Agric Water Manage 23:233–246

Ameghino F (1884) Las secas y las inundaciones en la provincia de Buenos Aires. Obras de retención y no de desagüe. MAAPBA, 4th reprint (1978). La Plata
Bazzuri ME, Gabellone NA, Solari LC (2018) The effects of hydraulic works and wetlands function in the Salado-river basin (Buenos Aires, Argentina). Environ Monit Assess 190(4):99. https://doi.org/10.1007/s10661-017-6448
Bonfils CG (1966) Rasgos principales de los suelos pampeanos. INTA, Buenos Aires
Buol SW, Hole FD, McCracken RJ, Southard RJ (1997) Soil genesis and classification, 4th edn. Iowa State University Press, Ames
Casas RR (2018) Fitorremediación de suelos salinos. In: Brutti LN, Beltran MJ, de Salamone IG (eds) Biorremediación de los recursos naturales. Ediciones INTA, Buenos Aires
Cavallotto JL, Violante RA (2005) Geología y geomorfología del río de la Plata. In: de Barrio R, Etcheverry RO, Caballé MF, Llambías E (eds) Geología y recursos minerales de la provincia de Buenos Aires. La Plata. pp 237–254
Damiano F, Parodi GN (2015) Manejo agrohidrológico de áreas deprimidas. El agua en la producción agropecuaria. In: Casas RR, Albarracín GF (eds) El deterioro del suelo y del ambiente en la Argentina. Prosa. Buenos Aires. pp 301–317
Dangavs NV (2005) Los ambientes acuáticos de la província de Buenos Aires. In: de Barrio R, Etcheverry RO, Caballé MF, Llambías E (eds) Geología y recursos minerales de la provincia de Buenos Aires. La Plata. pp 219–236
Daus F (1946) Morfología general de las llanuras argentinas. Geografía de la República Argentina GAEA. Buenos Aires. pp 115–196
Deschamps JR, Otero O, Tonni EP (2017) Cambio climático en la pampa bonaerense: las precipitaciones desde los siglos XVIII al XX. In: Waldman SM (ed) Inundaciones y manejo de cuencas. CADIA, Buenos Aires, pp 67–84
Di Bella C, Rodríguez AM, Jacobo EJ, Golluscio RA, Taboada MA (2015) Impact of cattle grazing on temperate coastal salt marsh soils. Soil Use Manage 31:299–307
Etchichuri MC, Tófalo OR (2004) Mineralogía de arenas y limos en suelos, sedimentos fluviales y eólicos actuales del sector austral de la cuenca chacoparanense. Regionalización y áreas de aporte. Revista Asoc Geológica Arg 59:317–329
Fidalgo F, De Francesco F, Pascual R (1975) Geología superficial de la llanura bonaerense. In: Geología de la provincia de Buenos Aires. VI congreso geologico Argentino. Bahía Blanca. pp 103–138
Fucks E, Carbonari J, Pisano F, Huarte R (2012) Aspectos geomorfológicos del sector medio e inferior de la Pampa Deprimida, provincia de Buenos Aires. Rev Soc Geológica España 25(1–2):107–118
Frengüelli J (1950) Rasgos generales de la morfología y la geología de la provincia de Buenos Aires. Ministerio de Obras Públicas de la provincia de Buenos Aires. Lab. ensayos materiales invest. Tecnol (LEMIT) Ser II 33:1–72
González N (2005) Los ambientes hidrogeológicos de la provincia de Buenos Aires. In: de Barrio R, Etcheverry RO, Caballé, MF, Llambías E (eds) Geología y Recursos minerales de la provincia de Buenos Aires. La Plata. pp 359–374
Hernández M (2001) La importancia de la zona no saturada en la hidrología de las llanuras. Anales Academia Nac Cs Exactas, Físicas y Nat 53:73–82
Imbellone PA, Guichon B, Giménez J (2009) Hydromorphic processes in soils. North eastern Buenos Aires province. Argentina. Latin Amer J SediBasin Anal 16:3–18
Imbellone PA (2018) Guía de descripción e interpretación de suelos hidromórficos. In: Álvarez C, Quiroga A, Fernández R (eds). Análisis y evaluación de propiedades físico hídricas de los suelos. INTA, Anguil, Argentina, pp 9–20
Imbellone PA, Giménez JE, Panigatti JL (2010) Procesos de sodificación y salinización. Suelos de la región pampeana. Procesos de Formación. INTA, Buenos Aires, pp 261–288
Imbellone PA, Giménez JE, Mormeneo ML (2018) Suelos con propiedades frágicas del centro y oeste de la provincia de Buenos Aires Argentina. In: Imbellone PA, Álvarez CR (eds) Compactaciones naturales y antrópicas en suelos argentinos. AACS, Buenos Aires, pp 180–244

Imbellone PA, Mormeneo ML (2011) Vertisoles hidromórficos de la planicie costera del río de La Plata, Argentina. C Suelo 29:107–127
Imbellone PA, Teruggi ME (1993) Paleosols in loess deposits of the Argentine Pampas. Quat Int 17:49–55
INTA (1977) La Pampa Deprimida. Condiciones de drenaje de sus suelos. Departamento de suelos. Publicación 152. Buenos Aires
INTA (1990) Atlas de suelos de la República Argentina. SAGyP-INTA. Proyecto PNUD Argentina 85/019. Buenos Aires
INTA (2018) Mapa de suelos de la provincia de Buenos Aires. Cartografía Digital: Mapa de suelos 1:50.000. https://www.geointa.inta.gob.ar/2013/05/23/cartografia-digital-de-suelos-y-fotomosaicos-georeferenciados-de-buenos-aires
Krause E, Laurencena P (2005) Aguas superficiales, relación con el régimen subterráneo y fenómenos de anegamiento. In: de Barrio R, Etcheverry RO, Caballé MF, Llambías E (eds) Geología y Recursos minerales de la provincia de Buenos Aires. La Plata. pp 313–326
Kuppel S, Houspanossian J, Nosetto MD, Jobbagy EJ (2015) What does it take to flood the pampas?: Lessons from a decade of strong hydrological fluctuations. Water Resour Res 51:2937–2950
Lavado RS (2006) Materia orgánica, nitrógeno, fósforo y azufre en los suelos afectados por sales de la región pampeana. In: Alvarez R (ed) Materia orgánica. Valor agronómico y dinámica en suelos pampeanos. EFA, Buenos Aires, pp 195–206
Lavado RS, Camilión MC (1985) Clay minerals in Argentinean salt affected soils. Clay Res 3:68–74
Lavado RS, Taboada MA (1987) Soil salinization as an effect of grazing in a native grassland soil. Soil Use Manage 3:143–148
Morrás HJM (2017) Propiedades químicas y físicas de suelos halomórficos de la fracción norte de los Bajos Submeridionales. In: Taleisnik E, Lavado RS (eds) Ambientes salinos y alcalinos de la Argentina. Orientación Gráfica Editora, Universidad Católica de Córdoba, Buenos Aires, Recursos y aprovechamiento productivo, pp 29–54
Nosetto MD, Jobbagy FG, Tóth T, Jackson RB (2008) Regional patterns and control of ecosystem salinization with grassland afforestation along a rainfall gradient. Global biogeochem cycles, 22. GB 2015, https://doi.org/10.1029/2007GB003000
Otondo J, Jacobo EJ, Taboada MA (2015) Mejoramiento de propiedades físicas por el uso de especies megatérmicas en un suelo sódico templado. C Suelo 33:119–213
Rubio G, Gutiérrez Boem FH, Lavado RS (2010) Responses of C_3 and C_4 grasses to application of nitrogen and phosphorus fertilizer at two dates in the spring. Grass Forage Sci 65:102–109
Sala JM, González N, Kruse E (1983) Generalización hidrológica de la provincia de Buenos Aires, In: Fuschini Mejía MC (ed) Hidrologia de las grandes llanuras: actas del coloquio de Olavarria abril 1983. UNESCO, comite nacional argentino para el programa hidrológico internacional. Buenos Aires pp 973–1009
Soil Survey Staff (1999) Soil taxonomy. A basic system of soil classification for making and interpreting soil surveys. 2nd edn. NRCS, U.S. Department of Agriculture. Washington
Soil Survey Staff (2014) Keys to soil taxonomy, 12th edn. United State Departament of Agriculture, Washington
Soriano A, León RJC, Sala,OE, Lavado RS, Deregibus VA, Cauhépé MA, Scaglia OA, Velázquez CA, Lemcoff JH (1991) Rio de la Plata grasslands. In: Coupland RT (ed) Temperate subhumid grasslands. Ecosystems of the World, vol 8. Natural grasslands A. Elsevier, Amsterdam, pp 367–407
Taboada MA, Damiano F, Lavado RS (2017) Tecnologías de manejo y rehabilitación de suelos halomórficos de la Pampa Deprimida y el Oeste Bonaerense. In: Taleisnik E, Lavado RS (eds) Ambientes salinos y alcalinos de la Argentina. Recursos y aprovechamiento productivo. Orientación Gráfica Editora, Universidad Católica de Córdoba, Buenos Aires, pp 55–88
Taboada MA, Damiano F (2017) Inundación y manejo de suelos en la Argentina. In: Waldman S (ed) Inundaciones y manejo de cuencas. CADIA, Buenos Aires, pp 145–169
Taboada MA, Lavado RS, Rubio G, Cosentino DJ (2001) Soil volumetric changes in natric soils caused by air entrapment following seasonal ponding and water table rises. Geoderma 101:49–64

Taboada MA, Rubio G, Lavado R (1998) The deterioration of tall wheatgrass pastures in saline sodic soils. J Range Manage 51:241–246

Tapia A (1937) Datos geológicos In: Aguas minerales de la República Argentina II. Provincia de Buenos Aires, Ministerio del Interior. Comisión nacional de aguas minerales. Buenos Aires, pp 23–90

Torres Duggan M, Rodríguez MB (2017) Bases conceptuales y pautas para el manejo de yeso en suelos afectados por sales. In: Taleisnik E, Lavado RS (eds) Ambientes salinos y alcalinos de la Argentina. Recursos y aprovechamiento productivo. Orientación Gráfica Editora, Universidad Católica de Córdoba, Buenos Aires, pp 251–274

Tricart JLF (1973) Geomorfología de la Pampa Deprimida. INTA, Buenos Aires

Vecchio MC, Golluscio RA, Rodríguez AM, Taboada MA (2018) Improvement of saline-sodic grassland soils propertssies by rotational grazing in Argentina. Rangel Ecol Manag 71:807–814

Zárate MA, Rabassa J (2005) Geomorfología de la provincia de Buenos Aires. In: de Barrio R, Etcheverry RO, Caballé MF, Llambías E (eds) Geología y recursos minerales de la provincia de Buenos Aires. La Plata. pp 119–138

Origin, Management and Reclamation Technologies of Salt-Affected and Flooded Soils in the Inland Pampas of Argentina

Miguel A. Taboada, Francisco Damiano, José M. Cisneros, and Raúl S. Lavado

Abstract The halo-hydromorphic soils in the Inland Pampas, located in subhumid and semiarid environments of central Argentina, are governed by the depth and salinity of water table, salt content and sodium in the profile, and the excess of water. Its aptitude of use is ruled by the A horizon thickness, cemented horizons (hardpans or fragipans) and occurrence/absence of plant cover. The occurrence of strong natric horizons interacts with heavy rainfall or underground water rise and causes floods or waterlogging. In recent years, there has been a generalized rise of underground water, causing waterlogging and salinity problems, even in non-saline cropped soils. Tested soil management and reclamation techniques of those halo-hydromorphic soils are summarily analyzed. The technologies were grouped according to their effects on soils, vegetation, and surface and underground water. They include chemical and biological treatments, grazing management, fertilization, measures to control the salt rising from phreatic water, runoff control, tree "biodrainage" and others. The environmental and economic success and perdurability of such technologies could be relatively guaranteed only by a site-specific approach, on account of the high spatial and temporal variability of those environments.

M. A. Taboada (✉)
Instituto de Suelos, Instituto Nacional de Tecnología Agropecuaria (INTA), Hurlingham, Buenos Aires and CONICET, Argentina
e-mail: taboada.miguel@inta.gob.ar

F. Damiano
Instituto de Clima y Agua, Instituto Nacional de Tecnología Agropecuaria (INTA), Hurlingham, Buenos Aires, Argentina
e-mail: fdamiano52@hotmail.com

J. M. Cisneros
Facultad de Agronomía y Veterinaria, Universidad Nacional de Río Cuarto, Río Cuarto, Argentina
e-mail: cisneros.pepe@gmail.com

R. S. Lavado
Facultad de Agronomía, Universidad de Buenos Aires and INBA—CONICET/UBA, Buenos Aires, Argentina
e-mail: lavado@agro.uba.ar

E. Taleisnik and R. S. Lavado (eds.), *Saline and Alkaline Soils in Latin America*, https://doi.org/10.1007/978-3-030-52592-7_11

Keywords Water excess · Soil technology · Pastures · Satellite images · Toposequence

1 Introduction

The Inland Pampas, also called Inner Pampas, cover the semiarid and subhumid areas located in several provinces: the northwest of Buenos Aires, northeast of La Pampa, southeast of Córdoba and southwest of Santa Fe. The location of this region is shown in Fig. 1 in the chapter by Imbellone et al., in this book: "Genesis, properties and management of salt-affected soils in the Flooding Pampas, Argentina." In this region there are some 15 million hectares in which the dominant saline and sodic soils are, simultaneously, affected by floods and waterlogging or droughts. In those areas plants are affected by cyclic stresses: salt, water, anoxia stress, nutritional, root confinement and others, acting either separately or in combination.

The remainder grassland ecosystems and the agrosystems of the region are distributed in dynamic patches, where groundwater acts as a common element (Alconada-Magliano et al. 2011).

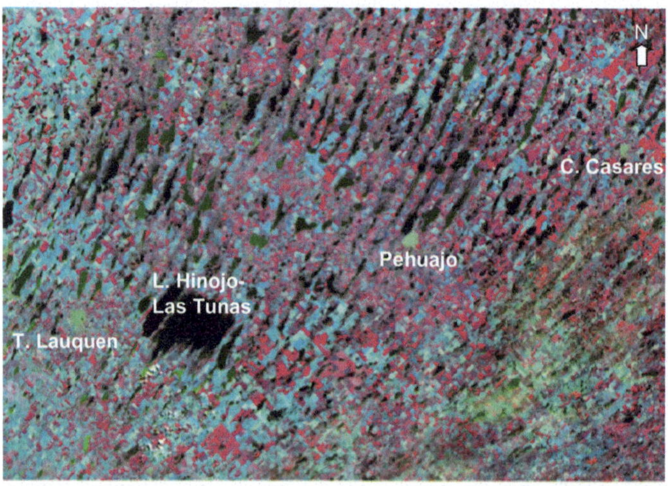

Fig. 1 Landsat satellite images showing longitudinal dunes and inter-dunes depressions in several counties of Buenos Aires Province (August 2014). Processed by F. Damiano

2 Water Excess Problems

Approximately 25% of the surface of Argentina is subject to various degrees of floods and waterlogging. Three levels of water excess are recognized: wetlands (12%), soils suffering temporary excess of water, waterlogging and/or flooding (9%) and soils with hydromorphic features (4%) (INTA 2019). The Inland Pampas also show those types of water excess. Their occurrence is related to factors such as regional relief, drainage network, the position in the landscape, and the impact of anthropic factors (changes in the land use, from pastures toward annual crops, poorly planned infrastructure works and even unauthorized channels and other works). Water excess affects pastures and crops, rural facilities, even settlements and small cities, with serious social and economical consequences (Taboada and Damiano 2017).

The Inland Pampas plain has a regional, gradual and gentle slope from west to east with a gradient averaging 25 cm/km, and poor development of watercourses to evacuate surface water excess in rainy periods. An additional and significant factor is that part of the region is furrowed by dunes that hinder surface drainage (Fig. 1). Those dunes interrupt the general slope and cause an arheic drainage, preventing water excesses to organize in courses; water can only be removed by deep drainage or evaporation. The whole region is dotted with swamps and lagoons, subject to surface expansion or contraction, in response to precipitation excess or deficit. The surface fluctuations in the water bodies are shown in Landsat satellite images (Fig. 2).

The soils of this area developed from recent sandy materials of variable thickness laid over clay sediments which support the water table, which depth is related to the thickness of the sandy mantle: it is deep in the dunes crest but shallow and even emerging in inter-dunes areas.

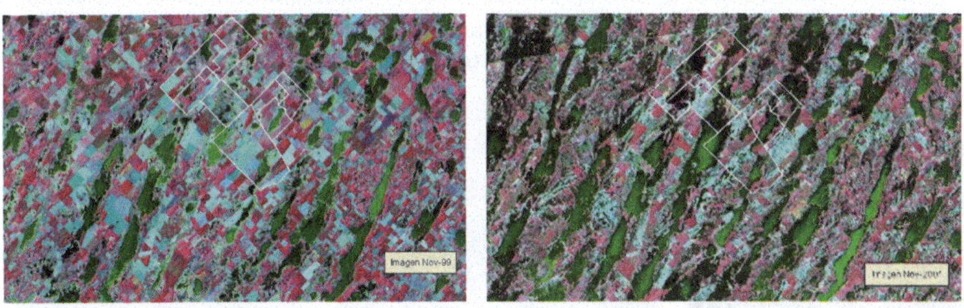

Fig. 2 Landsat satellite images showing longitudinal dunes (blue and pink colors, according to land use) and inter-dunes areas (green color) under water deficit, November 1999 (left) and water excess, November 2001 (right) in NW Buenos Aires Province (Lincoln County). Processed by F. Damiano

3 Soils

Figure 3 shows representative soil profiles of the Buenos Aires Province. In undulating relief, the soils are sandy loam, neutral to acid, with very good internal drainage and low water retention. Groundwater is deep and sometimes of good quality (Fig. 3a and b). In inter-dunes depressions, the soils show signs of hydromorphism, humic-clay films and lime concretions (Fig. 3c). These soils have a buried or thapto horizon, remnant of a paleoclimate. This horizon causes the development of temporary water-saturated layers, which originate gray colors, signs of hydromorphism.

In other areas of the Inland Pampas, soils without thapto horizons occupy the lower positions in the landscape. An example is the soil as shown in Fig. 4. In those soils, the water table is close to the soil surface and moves freely in the profile, causing soil salinization and alkalinization.

In the Southeast of Córdoba Province, some 1.8 million ha affected by halo-hydromorphism have been recorded. Most of those soils range from undifferentiated saline-sodic sandy profiles to soils with strong argillic-natric horizons (Fig. 5).

Soil behavior and management in this part of the Inland Pampas depends on water table depth and its salinity. Taking into account water table depth, three different soil conditions are found:

a. Alkaline (sodic) non-saline soils, with phreatic levels below 2 m, well drained but with a marked natric horizon, classified as Typic Natrustolls. They are characterized by the occurrence of patches of crops and pastures of diverse productivity.
b. Saline-sodic soils with phreatic water between 0.5 to 1.0 m. This is a critical water table depth (Fig. 6) which causes salt accumulation on the soil surface and limits plant growth (Yang et al. 2011). For local sandy loam and loam soils, the critical water table depth was established around 1.0–1.3 m, while salinity ranges from 2 to over 30 dS m^{-1} (Cisneros et al. 1999). These soils are classified as Typic

Fig. 3 Representative soils of the dunes area in Buenos Aires Province: **a** Entic Hapludoll (25 de Mayo County), **b** Thapto-Argillic Hapludoll (Saladillo County) and **c** Thapto-Natric Hapludoll (Lincoln County). Photos **a, b** and **c** F. Damiano

Fig. 4 Left, saline soils without thapto horizon; Right, eroded landscape with salts in surface, and plants of *Salicornia* sp. (General Villegas County, Buenos Aires Province). *Photos* F. Damiano

Fig. 5 Left: saline-sodic soils in SE Cordoba with hardpan. Right, poorly developed sandy-loam soil with saline groundwater showing a high capillary fringe. *Photo* J. Cisneros

Natraquolls and Typic Natraqualfs, very often with cemented horizons (fragipans and hardpans). Halophytic vegetation predominates, composed by *Distichlis spicata, Paspalum vaginatum, Cynodon dactylon* and *Atriplexundulata,* among other species.

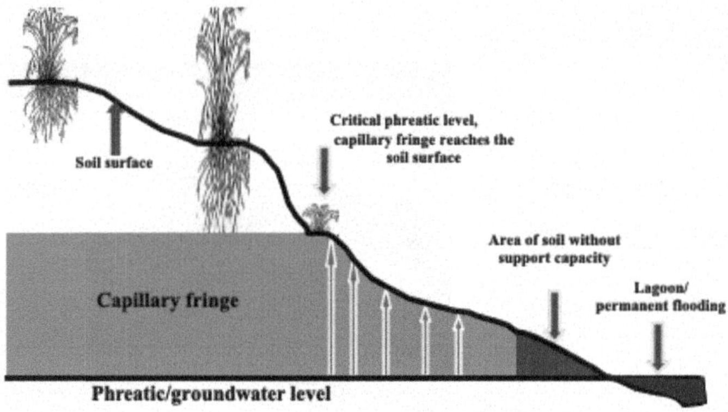

Fig. 6 Toposequence scheme with different environments related to phreatic water depth (vertical and horizontal scales not real)

c. Soils with 0.5 m deep to surface phreatic water. They are located in highly susceptible to waterlogging sites or in swamps and lagoons expansions. In these extreme conditions, soils lose the capacity to support agricultural equipments and cattle, and suffer intense compaction from trampling. These areas are highly variable, depending on the period of surface water permanence and its salinity, and can be distinguished by the dominant vegetation. *Spartina* sp. communities predominate in the more saline and flooded sites. While *Thypa latifolia* communities predominate in less saline areas. These are areas of low productivity but are important for sustaining biodiversity and hydrological control.

4 Processes of Salinization

Soils are affected by water, either surface or below surface. Non-saline water produces redox features in the soil, although exclusively hydromorphic soils are not common in the area. When the water is saline, the effects depend on the nature of the anion accompanying sodium, the main cation. When chloride and/or sulfate ions predominate, soils become saline, and plants are subject to various degrees of osmotic stress and ion toxicities. Conversely, when the anions are mainly carbonates and bicarbonates, the soils show alkaline features (high pH, aggregate destabilization, water infiltration reductions and so on), making soil reclamation very expensive and hard to achieve (Lavado and Taboada 2017).

Figure 7a, b and c show a typical soil toposequence in the NW of the Buenos Aires Province affected by rising phreatic water: high position in the relief, Aquic Hapludoll; mid-slope, Thapto-natric Hapludoll and lower position, flooded Typic Natraquoll. This toposequence shows the close relationship between the chemical composition of phreatic water and that of the soils (Damiano et al. 2015). The phreatic

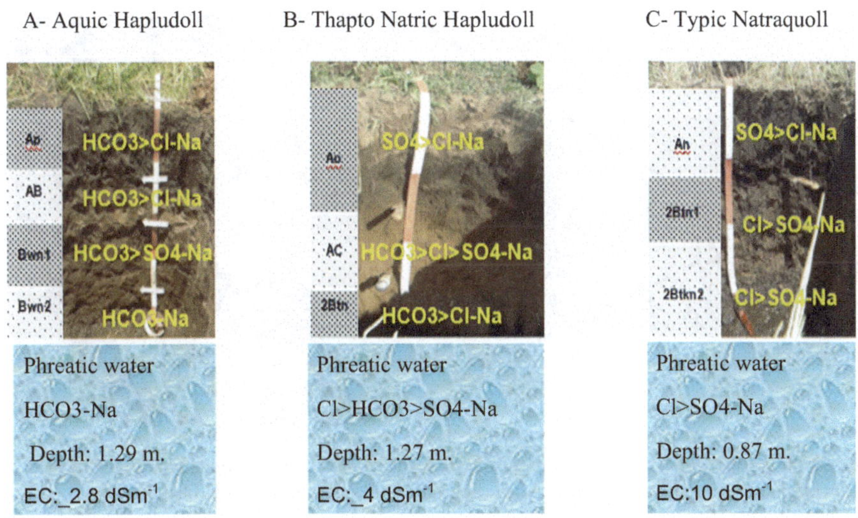

Fig. 7 Toposequence of soil profiles in Lincoln County, showing the close relationship between the chemical composition of the soil saturation extract and phreatic water. *Images* F. Damiano

water varies from sodic bicarbonate in the first to sodic sulfate/chloride in Typic Natraquoll, no seasonal changes due to rainfall were observed.

Remote sensors are a promising tool to study and to manage landscape units with risk of saline and hydric water problems. As an example, Spot 7 satellite images (resolution 10 m) were processed using a salinity index (B3 − B4/B3 + B4) on two dates: December 2012 and February 2013 (Damiano et al. 2015). In December 2012 (the beginning of summer in the Southern Hemisphere, Fig. 8), the rise of a very

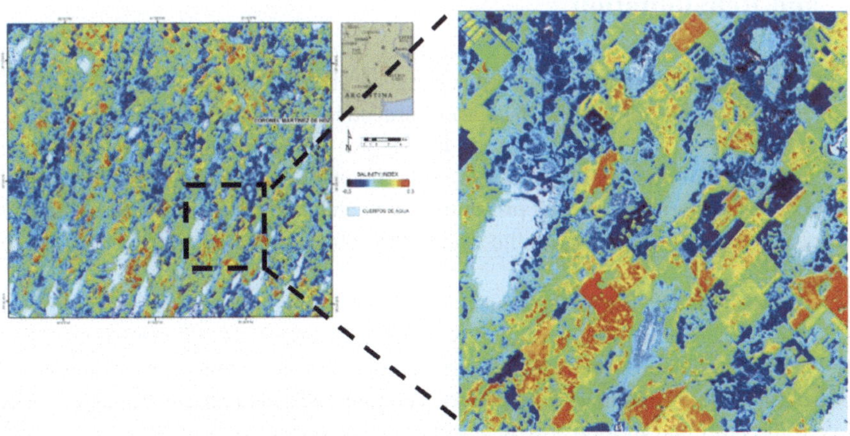

Fig. 8 Image spot processed using the salinity index (left) and the expanded area of the study area (right). Lincoln County, NW Buenos Aires, December 2012. Processed by F. Damiano

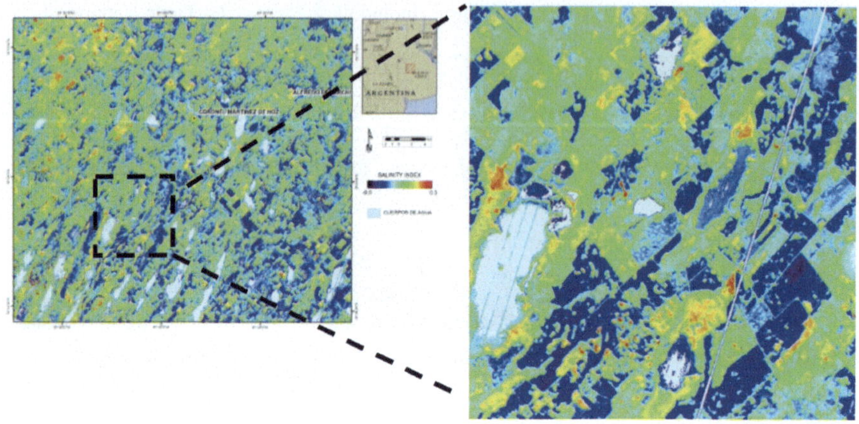

Fig. 9 Image spot processed using the salinity index (left) and the expanded area of the study area (right). Lincoln County, NW Buenos Aires, February 2013. Processed by F. Damiano

saline water table (>10 dS m^{-1}) reaching the surface of low areas, caused a positive salinity index (red colors). In the high positions of the relief, conversely, the phreatic water was more than 1 m depth and its salinity was lower than 3 dS m^{-1}. This area shows a negative salinity index (blue colors). Later, the phreatic water went down in the lowlands but soil salt content remained unchanged; however, heavy rainfall in December 2012 leached the salts from the soil surface and the saline index decreased (blue colors) as can be seen in the February 2013 image (Fig. 9).

5 Diagnosis and Principles of Management and Rehabilitation

In recent years, there were accelerated changes in land use in the studied region, as well as in Argentina in general. The agricultural advance even on halo-hydromorphic marginal soils (i.e., VI class of use aptitude, USDA 1961), devoted previously to cattle husbandry, meant more grazing pressure on remaining grasslands. This change was caused by the international high prices of grains and the technologies introduced in the local agriculture at the end of the twentieth century (Paruelo et al. 2005; Viglizzo et al. 2010).

To manage or rehabilitate those salt-affected soils, there are various technologies, differing in their objectives, complexity degree, effectiveness and persistence. The effect of any of them will be ephemeral if the impact of salts from groundwater is not controlled. To control underground water, economically impractical drainage works have to be built and in their absence, many technologies are only able to increase biomass productivity but do not change markedly soil properties (Cisneros et al. 2008; Lavado and Taboada 2017).

Due to the high variability of those soils and their space–time dynamics, site-specific diagnoses are necessary to determine the most adequate technologies. This is valid both at parcel or basin level. The diagnosis should define the productive potential and the risks of salinization/flooding in each site, to enable the design of site management schemes. Some basic diagnostic tools are: (i) Zoning and detailed mapping of the different environments present on the site, based on: phreatic water depth and salt content, flooding frequency and current vegetation communities (halophytes, hydrophytes, glycophytes). Water Table mapping and phreatimeter setting up in different environments are very useful tools to site-specific management in halo-hydromorphic areas; (ii) Monitoring soil degradation to determine the success of the management and reclamation strategies. Among the indicators are percentage of living cover in relation with saline patches, salt content of soil surface, occurrence of invasive species (e.g., *Cynodon dactylon*), runoff or overgrazing evidences, etc., and (iii) analyzing drainage possibilities of phreatic water, particularly, at the basin scale. This means the possibility to evacuate water toward channels, rivers, lagoons and evaluate the impact of such measures. The control of phreatic water depth is key for the viability of other techniques such as chemical corrections (e.g., gypsum) because it allows to evacuate salts and sodium excess.

It is important to know the soil use at regional scale. A strong relationship between the changes in soil use in the region, which took place since the end of the twentieth century, and the rise of the water table has been observed (Nosetto et al. 2015; Mercau et al. 2016). This salinization process could affect even areas of well-drained soils. Thus, the success of the various techniques will depend on the soil use dynamics, and on other factors affecting the phreatic water, at a regional scale. Table 1 lists techniques of soil management and rehabilitation, and their impact.

The technologies mentioned in Table 1 are considered in the following sections.

5.1 *Grazing Exclosure*

Grazing exclosure is a temporal livestock exclusion from areas of lowest productivity or with significant soil deterioration. This technique is proposed as an initial step to recover sites, including saline borders of lagoons or exposed beds of temporary ponds, where plant cover is below a threshold, around 20–30%. The grazing exclusion may be permanent or for a defined period, usually some months. A six-months grazing exclosure, in a degraded *Cynodon dactylon* pasture, showed positive changes in biodiversity, water infiltration and forage yield (Cisneros 1994).

5.2 *Rotatory Grazing*

Heavy grazing all year round could cause loss of soil cover and rise of the phreatic water (Lavado and Taboada 2017). It is possible to control soil surface salinization by

Table 1 Summary of soil, water and vegetation management techniques and their impacts on soil quality (Modified from Cisneros et al. 2008)

Management technique	Reduce capillary rise	Increase infiltration rate	Plant cover recovery	Lower phreatic level	Increase pasture production	Runoff control
Grazing exclosure	X	X	X		X	
Rotatory grazing		X	X		X	
Organic amendments	X	X	X		X	X
Soil loosening	X	X	X			
Interseeding, direct seeding, mulching	X	X	X		X	
Pasture seeding	X	X			X	X
Subsoiling—mole drainage		X		X		X
Saline patches revegetation	X	X	X		X	
Vegetative transplant			X		X	
Afforestation (biodrainage)	X		X	X		X
Calcium amendments		X	X		X	
Localized drainage	X			X		
Fertilization					X	
Agro-hydrological management		X	X	X	X	X

giving pastures periodic breaks or introducing rotatory grazing, which favors surface cover. Rotatory grazing is a technology for small plots, combining short times of intensive grazing with long resting periods. The system avoids gradients in grazing pressure and trampling around water sources, both common in large plots. It is based on the complete consumption of the pasture, avoiding species selection by animals. This grazing management favors organic matter and nutrient return to the soil as manure and urine as well as increasing standing dead plants and litter. However, the small increase in organic residues is not translated into significant accumulation of carbon in those soils, even in the long term. Rotary grazing management is very effective for increasing plot productivity and avoiding soil degradation but, like other technologies, it does not change main soil properties and characteristics.

5.3 Organic Amendments

Organic inputs are mostly diverse types of manure, some agroindustry residues and other by-products. A critical point to be considered is the distance between the source and the site of application, due to the large volumes of products to be used, their characteristics, transportation cost, storage, etc. These organic amendments are often wasted because of lack of knowledge or shortage of appropriate machinery. Manure adds nutrients to soils and also improves structure and other physical properties and, consequently, significantly increases forage productivity. Physical improvement is crucial in sodic soils, characterized by physical and hydrological limitations. However, organic matter alone is often inadequate to stabilize the structure in very alkaline soils and could even increase the dispersive potential of such soils. In those cases, it is necessary to flocculate the soil first (Lavado and Taboada 2017).

There are many industrial and urban by-products that could eventually be used for the same purpose, but they should be carefully analyzed as they could carry elements or substances that are toxic for people and the environment, as well as pathogens and parasites. The use of those products is usually subjected to regulations and intervention of different agencies.

5.4 Surface Soil Loosening

Grazing all year round on saline and flooded soils cause severe problems, among them surface compaction, which reduces water infiltration and increases capillary rise, leading in turn to increased surface salinization. Surface mechanical loosening is a technique that disrupts compacted horizons with vertical and/or horizontal cutting tools. Its success depends on the maintenance of a high proportion of vegetation cover, either live or dead. This technique is not recommended for bare soils, where any machinery intervention could be harmful.

Fig. 10 Subsoiler with narrow blades and wings to loosen soil, with minimum cover disturbance. *Photo* J. Cisneros

The best tools for loosening surface soil with minimum cover disturbance are rigid subsoilers with narrow blades (Fig. 10). The work depth is related to the compacted top horizon thickness, usually around 7–10 cm; tillage direction must be perpendicular to the slope, to make use of the surface water retention due to the microrelief caused by the tillage operations. This increases soil water retention and reduces runoff. The duration of the effect varies with subsequent land use and the grazing regime, but it usually lasts from 2 to 4 years.

5.5 Pasture Seeding

The idea to use perennial or annual pastures to improve the soils is not new. More than 60 years ago the improvement of sodic soils by seeding S*orghum technicum* (broom corn) was proposed (Sauberán and Molina 1963). It was thought that incorporating biomass on bare sodic soils would cause high biological activity, which would generate large quantities of CO_2, thus reducing soil pH, which would gradually change soil properties. However, although soil productivity increased, soil resilience and the occurrence of high water table make soil changes small and ephemeral.

Recently, promising results were obtained by the introduction of tropical C4 species, like *Chloris gayana* Kuhn (Gramma Rhodes) or *Panicumcoloratum* L. (panicgrass). Cisneros et al. (2008) in SE Córdoba found reductions in salt content when the soils after those forages were established. *Chloris gayana* showed high productivity with shallow phreatic water but low salinity in controlled conditions studies (Chiachera et al. 2016). Accordingly, in SE Córdoba, this species showed the best behavior in areas less affected by salinity. Similar results were shown by some varieties of *Festuca* sp., *Phalaris* sp., *Lotuscorniculatus* and *L. tenuis*. Some interesting native species are mentioned below. In NW Buenos Aires a pasture of *Thinopyrum ponticum* (tall wheatgrass) reduced soil electrical conductivity and pH, a year after seeding after replacing a degraded natural pasture. Improvements in biomass production were observed, as well as improvements in physical properties were measured (Casas 2018).

5.6 Interseeding, Direct Seeding, and Mulching

Interseeding, direct seeding and mulching are agricultural practices grouped in this section according to their results.

Thynopyrum ponticum, tall wheatgrass, is the species usually chosen for interseeding halophytic grasslands dominated by *Distichlis spicata* o *Spartina* sp., due to its tolerance to salinity and to short flooding periods. Other interesting species found in SE Córdoba Province are *Melilotusofficinalis* and *M. albus*. In a germination study under controlled conditions, with a maximum salinity level of 20 dS/m, the order was *Thinopyrton ponticum* > *Chloris gayana* > *Melilotus albus* > *Lotus tenuis* (Miñan et al. 2014).

The effect of direct seeding was determined in a study comparing wheat grown in a saline Entisol. Wheat grown conventionally showed an average of 6.8 dS m^{-1} at 0–60 cm depth, while wheat the grown in direct seeding (soil surface covered with stubble) the soil salinity level reduces to a half: 3.9 dS m^{-1} at 0–60 cm depth (Taboada and Lavado, personal communication).

The application of mulch can reduce surface salinity, improving seed germination. The mulch may be crop or pasture residues, unused forage rolls, etc. In degraded grasslands, the cover could be accomplished by cutting and chopping grass bushes and spreading them on bare surfaces or saline patches (Fig. 11).

Fig. 11 Interseed *Melilotusalbus* in saline-sodic soils in SE Cordoba. *Photo* RR Casas

5.7 Deep Subsoiling, Mole Drainage

Deep subsoiler is a tool causing underground channels to move water to the soil depth, and accumulate it there. In an experiment, in NW Buenos Aires Province, this technology improved physical properties like porosity, water infiltration and so on, leaching salts in response to rainfall and increased grassland productivity (Altamore et al. 1983).

Mole drainage is carried out with a tool which cuts the soil vertically and usually holds a bullet-shaped device. This tool can be applied in situations of poor drainage, with two main objectives: (i) To improve water infiltration and drainage in soil with internal drainage limited by a deep impervious layers (fragipan, sodic B horizon and others), frequent in some areas. In such cases, the tool must break the obstacle in order to allow better water circulation in depth. Usually, the depth of the technology is around 40 cm (Fig. 12). (ii) To depress phreatic water, in flooded lowlands occupied by *Spartina* sp. or *Distichlis spicata*. In such cases, this tool could cause drainage for the phreatic water toward lower positions in the basin, usually collecting the drained water in lagoons. A previous condition is that the salinity of drained water be similar to that of the lagoon, to avoid contaminating it. In those cases, the work is carried out at 1 m depth and 2–5 m distance, according to soil characteristics.

5.8 Species Transplant

This technology is based on vegetative propagation of plants at extremely saline/waterlogged sites. Native species are generally proposed for such sites, but there are usually no commercial seeds available, thus transplantation is an alternative.

Fig. 12 Subsoiler with mole drain to depress phreatic level. *Photo* RR Casas

In SE Córdoba Province, two promising species are mentioned: one is *Paspalum vaginatum*, which grows in lagoons borders and is highly tolerant to salinity and waterlogging, having high capacity for surface cover by stolons and rhizomes. Planting bushes properly totally covers the soil surface after 1–2 years.

The other species is the hygrophilous *Echinocloa helodes*, highly tolerant to waterlogging but not tolerant to high salinity (Cantero et al. 1991). It has a high potential as forage and was successfully used to recover flooded lowlands in SE Córdoba, transplanting stolon and rhizomes.

5.9 Afforestation or "Biodrainage"

The replacement of dry forest trees by crops causes phreatic water rise and soil salinization in some places in the world (Jobbágy et al. 2008, see also the chapter by Jobbaggy et al., in this book: "Salt accumulation and redistribution in the dry plains of Southern South America: Lessons from land use changes"). This process is widespread in the central-west part of Argentina, including parts of the Inland Pampas. Forest "biodrainage" is a technology aiming to restore partially the lost hydrological equilibrium caused by deforestation. It is based on the high evapotranspiration potential of forest species, in particular, evergreen ones and the consequent depression/regulation of groundwater levels, and mitigation of waterlogging phenomena.

Some research in SE Córdoba Province showed a potential for small forest masses to depress water tables by approximately 0.5 m, when compared to the surrounding agricultural areas (Angeli et al. 2006). A similar impact was observed in NW Buenos Aires (Alconada Magliano et al. 2009). This fact has motivated the Province of Córdoba to enact an Agroforestry Law that imposes the 2–5% afforestation of each farm (Ministerio de Agricultura y Ganadería de Córdoba 2018). Several species of *Eucalyptus* showed tolerance to salinity, waterlogging or both and were proposed for such purpose, among them, *E. amplifolia, E. astringens, E. botryoides, E. longifolia, E. occidentalis, E. sargentii, E. umbra, E. diptera, E. dundasii, E. gomphocephala, E. pilularis, E. robusta, E. rudis, E. camaldulensis* and *E. siderophloia* (Cisneros et al. 2008).

5.10 Calcic Amendments

Calcic amendments are minerals or industrial by-products of low economic value on their own, for which transport is the main cost in their application. That is why amendments are usually used locally and very often not by choice but for economic reasons. The distribution of gypsum is generalized and it is the most widely used amendment. There are no deposits of other sulfates, sulfides or sulfur, or other amendments of potential value, in areas near the Pampas Region. There are, instead, industries that produce by-products rich in sulfuric acid, for example. These materials can be used as amendments of soils affected by salts. However, toxicity or environmental disturbance, or the danger of handling these by-products limits their use.

The practice of mass application of gypsum, as it is applied in fully irrigated areas is technically and economically non-viable. Conversely, the application of small amounts of gypsum is recommended in the planting band to produce better aggregation and reducing soil pH, improving germination of forage species (see also the chapter by Torres Duggan and Rodríguez, in this book: "Conceptual and practical frameworks to address gypsum management in salt-affected soils").

5.11 Localized Drainage

The objective of this technique is to depress the water table in small areas, improving their productivity. The design of a project using this technique requires a detailed knowledge of the relief and of the direction of the water table flux, in order to guide the evacuation routes and the final disposition of the water surpluses (see also Agro-hydrologic management techniques, below). Studies carried out in NW Buenos Aires Province show the technical feasibility of a drainage system, with drains 150 cm deep (Vázquez Amabile et al. 2017). However, due to economic limitations, this type of drainage is not common in Argentine field production.

Other simple and economical technologies suitable for use in small areas are mole drainage (see above) or a combination of mole drainage and open drains. Polders or hydrological isolation and pumping systems in small areas are used for flood protection of rural infrastructure (houses, silo plants, pen, etc.). Those systems are used also to protect urban conglomerates on flood-prone areas.

5.12 Agro-Hydrologic Management

The basic objective of agro-hydrological systematization is to prevent, retain or delay the accumulation of water excesses, concentrating them in the less productive areas of the landscape. The idea is to remove the water to the atmosphere by evapotranspiration and to the water table by infiltration, and to retain water during periods of drought (Fig. 13). This practice is a realistic solution within the possible economic context, dealing with those prolonged and recurrent flooding of low hydraulic load that are common in plains. The technology could be applied in plains with a slope less than 0.5% (low kinetic energy of the water), humid-subhumid climate regime and hydromorphic and halomorphic soils with severe limitations of surface and subsurface drainage and susceptible to the rise of salts from the water table.

The works can be carried out with existing machinery in the farm and they are considered a necessary preliminary step for other soil improvement practices

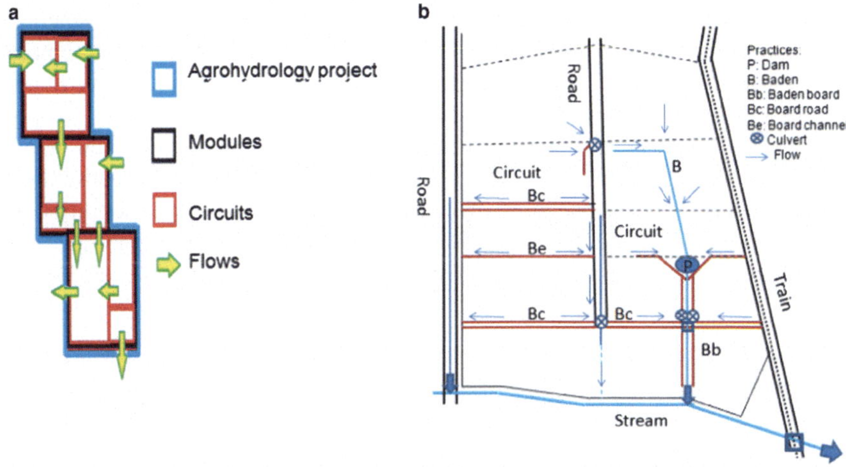

Figure 13 **a** Graphical relation of the glossary used in agro-hydrology, and **b** schematic design of a project, with hydrological circuits

(Damiano and Parodi 2015). As in any type of hydraulic water control design, the larger the volume of water to be controlled, the greater the effort and the cost of the works involved.

5.13 Fertilization

This technique is the last stage for the recovery of pastures in halo-hydromorphic soils, the phase of productivity optimization. The success of the practice is guaranteed only when the surface cover and the occupation of the volume of soil by the roots are large. Nitrogen fertilizers are exposed to heavy losses due to ammonia volatilization or denitrification.

Dry matter yield and seed production of *Thinopyrum scabrifolium* (Criollo tall wheatgrass) increased when N was applied. The maximum pasture response was found with the application of 200 kg N/ha (Cantero et al. 1985). There are few experiments regarding responses to fertilization in low productive saline-sodic soils, probably due to the low return on investment.

5.14 Revegetation of Extremely Saline Areas

The productivity recovery of forage in sites affected by extreme salinity, like the beds or borders of drained lagoons, overgrazed or cleared areas with highly saline water tables, is one of the greatest technical challenges. This recovery requires the joint implementation of several techniques discussed in this chapter, like fencing, mulch cover, amendments, high density interseeding, and species transplant in more severely affected areas and short-time grazing in the initial years of revegetation. Once the soil cover is recovered, in a medium-term program, it is possible to apply technologies to increase grassland productivity, like surface loosening, new interseeding and nitrogen fertilization. The agro-hydrology technologies, because their control of runoff gives those sites of extreme degradation more possibilities of recovery. At a basin scale, the vegetation cover in lowlands reduces the salt load of runoff, because surface soils remain with low salt content (Cisneros et al. 1999).

6 Conclusions

The halo-hydromorphic soils in the Inland Pampas show a varied gradient from sandy textures and semiarid climate in the west to sandy loam to sandy clay loam including buried horizons ("thapto" horizon) in the center-east of the region.

There are rehabilitation technologies for these saline-sodic soils. However, prior to their implementation it is important to consider the influence of groundwater as a source of salts and sodium on the soil surface. Therefore, to avoid the arrival of salts to the top soil, it is important to keep it covered by vegetation. Other technologies are application of organic products, biological improvement through planting of adapted species, gypsum application, direct sowing and others. Agro-hydrological systematization should be considered to handle excess surface water in the fields. Grassland fertilization is a tool that deserves more attention.

Acknowledgements The authors thank Instituto de Clima y Agua–INTA and Comisión Nacional de Actividades Espaciales (CONAE) for providing elements of satellite images.

References

Alconada Magliano MM, Bussoni A, Rosa R, Carrillo Rivera JJ (2009) El bio-drenaje para el control del exceso hídrico en Pampa Arenosa, Buenos Aires, Argentina. Invest Geog [online]. https://www.scielo.org.mx/scielo.php?script=sci_arttext&pid=S0188-46112009000100005&lng=es&nrm=iso

Alconada-Magliano MM, Fagundo-Castillo JR, Carrillo-Rivera JJ, Hernández PG (2011) Origin of flooding water through hydrogeochemical identification, the Buenos Aires plain, Argentina. Environ Earth Sci 64:57–71

Altamore R, Torres RF, Lavado RS, Giménez JE (1983) Efecto del subsolado sobre un suelo salino-alcalino del oeste bonaerense. C Suelo 1:45–51

Angeli AR, Cholaky CG, Cantero GA, Cisneros JM (2006) Biodrenaje forestal: efectos sobre la salinidad del suelo y el nivel freático en el sur cordobés. Paper presented at the XX Congreso Argentino de la Ciencia del Suelo. Salta, Argentina. September 2006

Cantero GA, Bonadeo E, Becerra V, Marcellino JR (1985) Influencia de la fertilización nitrogenada sobre el comportamiento de agropiro criollo (*Agropyron scabrifolium* (Doell) Parodi). I. Producción de materia seca, proteína bruta y semillas. Rev UNRC 5:5–17
Cantero JJ, Nuñez C, Gaich G, Cisneros JM (1991) *Echinochloa helodes* (Paniceae, Gramineae): Biología y regeneración vegetativa. Rev UNRC 11:79–103
Casas RR (2018) Fitorremediación de suelos salinos. In: Biorremediación de los recursos naturales. INTA-FAUBA. Buenos Aires
Chiacchiera S, Bertram N, Taleisnik E, Jobbágy E (2016) Effect of water table depth and salinity on growth dynamics of Rhodes grass (*Chloris gayana*). Crop Pasture Sci 67(8):881–887. https://doi.org/10.1071/CP15241
Cisneros JM (1994) Caracterización de la afectación hidrohalomórfica en ambientes representativos del centro-sur de Córdoba. M.Sc. thesis in Soil Science, Faculty of Agronomy, University of Buenos Aires (FAUBA)
Cisneros JM, Cantero JJ, Cantero GA (1999) Vegetation, soil hydrophysical properties, and grazing relationships in saline-sodic soils of Argentina. Can J Soil Sci 79:399–409
Cisneros JM, Degioanni A, Cantero JJ, Cantero A (2008) Caracterización y manejo de suelos salinos en el área pampeana. In: Taleisnik E, Grunberg K, Santa Maria G (eds) La salinización de suelos en la Argentina: su impacto en la producción agropecuaria. EDUCC, Córdoba, pp 17–46
Damiano F, Parodi GN (2015) Manejo agrohidrológico de áreas deprimidas. In: Casas RR, Albarracín GF (eds) El deterioro del suelo y del ambiente en la Argentina. Tomo 1. Prosa-FECIC, Buenos Aires, pp 301–317
Damiano F, Alconada Magliano MM, Fagundo Castillo JR (2015) Estimación de la composición química del agua subterránea a partir de su salinidad. https://www.academia.edu/15253656/Damiano_et_al_hidrogeoquimica_trabajo
INTA (2019) Sistema de Información de Suelos del INTA. https://sisinta.inta.gob.ar/
Jobbágy EG, Nosetto MD, Santoni CS, Baldi G (2008) El desafío ecohidrológico de las transiciones entre sistemas leñosos y herbáceos en la llanura Chaco-Pampeana. Ecol Austral 18:305–322
Lavado RS, Taboada MA (2017) Génesis y propiedades de los suelos halomórficos. In: Taleisnik E, Lavado RS (eds) Ambientes salinos y alcalinos en la Argentina. Orientación Gráfica Editora, Universidad Católica de Córdoba, Buenos Aires, Recursos y aprovechamiento productivo, pp 9–27
Mercau JL, Nosetto MD, Bert F, Giménez R, Jobbágy EG (2016) Shallow groundwater dynamics in the Pampas: climate, landscape and crop choice effects. Agric Water Manag 163:159–168
Ministerio de Agricultura y Ganadería de Córdoba (2018) Ley 20.880 Plan Provincial Agroforestal. https://magya.cba.gov.ar/upload/Ley_20880_Plan_Provincial_Agroforestal.pdf
Miñan JP, Cisneros JM, Petryna L (2014) Niveles de salinidad y germinación de cuatro especies forrajeras. Paper presented at the XXIV Congreso Argentino de la Ciencia del Suelo, Bahía Blanca, 5–9 mayo 2014
Nosetto MD, Paez R, Ballesteros SI, Jobbagy EG (2015) Higher water-table levels and flooding risk under grain versus livestock production systems in the subhumid plains of the Pampas. Agric Ecosyst Environ J 206:60–70
Paruelo JM, Guerschman JP, Verón SR (2005) Expansión agrícola y cambios en el uso del suelo. Cienc Hoy 15:14–23
Sauberan C, Molina JS (1963) Recuperación de terrenos "salitrosos" por métodos biológicos. C Investig 19:449–458
Taboada MA, Damiano F (2017) Inundación y manejo de suelos en la Argentina. In: Waldman S (ed) Inundaciones y manejo de cuencas. CADIA, Buenos Aires, pp 145–169
USDA (1961) Land Capability Classification. Agricultural Handbook N° 210. Disponible en: www.nrcs.usda.gov/Internet/FSE_DOCUMENTS/nrcs142p2_052290.pdf
Vázquez Amabile GG, Bosch N, Ricca AP, Rojas DE, Ortiz de Zárate ML, Lascombes J, Feiguín MF, Cristos D (2017) Napa freática: Dinámica, variables de control y contenido de nitratos en suelos de pampa arenosa. C Suelo 35:117–134

Viglizzo EF, Carreño LV, Pereyra H, Ricard F, Clatt J, Pincén D (2010) Dinámica de la frontera agropecuaria y cambio tecnológico. In: Viglizzo EF, Jobbágy E (eds) Expansión de la Frontera Agropecuaria en Argentina y su Impacto Ecológico-Ambiental. INTA, Buenos Aires, pp 17–22

Yang F, Zhang F, Yin Y, Liu Z, Huang Z (2011) Study on capillary rise from shallow groundwater and critical water table depth of a saline-sodic soil in western Songnen plain of China. Environ Earth Sci 64:2119–2126

Salt-Affected Soils of Pantanal Wetland

Sheila A. C. Furquim and Thiago T. Vidoca

Abstract The Pantanal, one of the largest wetlands of the world, has a high incidence of salt-affected soils due to favorable past and current conditions, such as high evapotranspiration rates and restriction of leaching. According to maps produced during the 1980s, these salt-affected soils are mainly classified as Gleyic Solonetz, Eutric Planosols, and Eutric Vertisols, the former being the dominant soil class in the wetland. More recent and detailed studies performed mainly in two Pantanal subregions, Nhecolândia and Barão de Melgaço, revealed important occurrence of salt-affected soils in areas that have not been previously mapped. In the Nhecolândia, Saline–Sodic soils, currently submitted to salinization and solonization, occur around hundreds of alkaline–saline lakes. In the Barão de Melgaço subregion, Sodic and Solodized Solonetz occur in the highest positions of the landscape, such as paleolevees and paleochannels, whereas Solods are encountered in the lower geomorphic units, specifically depressions and intermittent watercourses. Although the genesis of Saline–Sodic soil is still observed in Nhecolândia, in consonance with the favorable environmental conditions for the development of salt-affected soils in the wetland, the salinity of the soils in the highest areas of Barão the Melgaço seems to be mostly a heritage from more arid conditions in the past.

Keywords Alkaline–saline lakes · Saline–Sodic soil · Seasonal flooding · Solodization · Solodized solonetz · Solod

S. A. C. Furquim (✉)
Environmental Sciences Department, Universidade Federal de São Paulo (UNIFESP), Diadema-SP, Brazil
e-mail: sacfurquim@gmail.com; sfurquim@unifesp.br

T. T. Vidoca
Geography Department, Universidade de São Paulo (USP), São Paulo-SP, Brazil
e-mail: vidoca@gmail.com

E. Taleisnik and R. S. Lavado (eds.), *Saline and Alkaline Soils in Latin America*, https://doi.org/10.1007/978-3-030-52592-7_12

1 The Pantanal Wetland: An Overview

The Pantanal is one of the largest wetlands of the world, with an area of 150,501.84 km^2 (Padovani 2010). It is located in western Brazil (90%) and in small areas of eastern Bolivia and north-eastern Paraguay (10%), in the geographical center of South America (15°–23° S; 54°–60° W) (Padovani 2010; Assine et al. 2015) (Fig. 1).

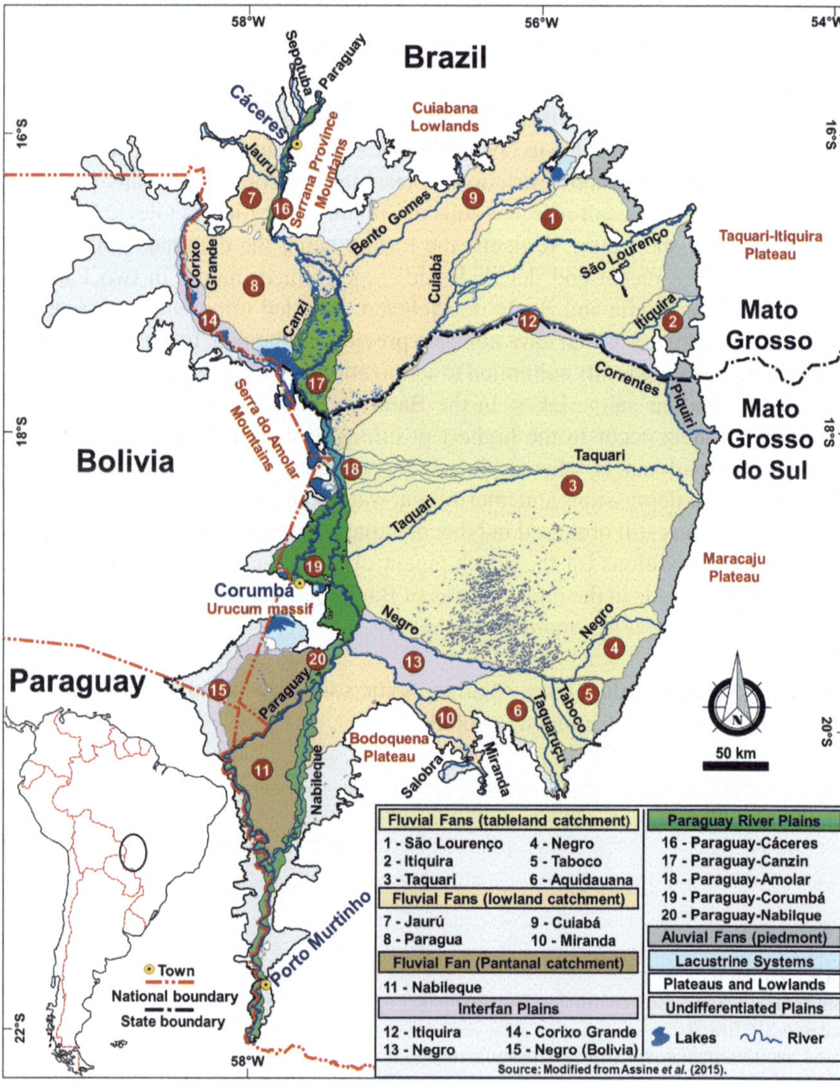

Fig. 1 Location of Pantanal in Brazil, Bolivia, and Paraguay and depositional systems tract of Pantanal (adapted from Assine et al. 2015)

The region is occupied by huge rural properties and the extensive animal husbandry, mainly based on native pasture, has been the dominant economic activity over the last two centuries. Commercial fishing, mining, and tourism are also present, but in minor proportions (Seidl et al. 2001). Although the areas being deforested have increased, mainly for implantation of cultivated pasture (Silva et al. 2011), the region still has approximately 80% of original natural vegetation, with low anthropic influence and many pristine ecosystems (Alho 2011; Pott and Silva 2016).

The wetland presents a unique combination of *Cerrado*, Amazon, Atlantic Forest, and *Chaco* biogeographical provinces. The main terrestrial vegetation type is the savanna or *Cerrado* (including *Cerrado* woodland, floodable monodominant savannas, and grasslands), followed by steppic savanna and pioneer woodlands (intermingled with monodominant savannas). Seasonal deciduous and semi-deciduous forests, scrub, and herbaceous vegetation are found in smaller areas (Pott and Silva 2016). Several studies have revealed that the Pantanal region harbors, for example, 900 species of herbaceous and 756 species of woody plants, 600–700 species of birds, 263 of fish and 60–80 of mammals, including jaguar, puma, tapir, ant-eater, and capybara (Junk et al. 2006). Because of its biodiversity, combined with its enormous territory range, the Pantanal is recognized as a National Heritage site by the Brazilian Constitution, a World Heritage Area (Pantanal National Park) and a Biosphere Reserve by UNESCO, and a wetland with international importance by the Ramsar Convention of Wetlands.

The climate of the Pantanal is Tropical Savanna-Aw, according to Koppen's classification (Por 1995). The mean annual precipitation varies between 850 and 1100 mm, reaching 1500 mm in the eastern and northern highlands (escarpments, plateaus, and hills). The wet season in the wetland occurs generally from December to March, whereas the dry season is from June to August (Alvarenga 1984; Alfonsi and Camargo 1986). The mean annual temperatures range from 22.4 to 25.6 °C, with October to December and November to January being the hottest months in the north (Cáceres) and south (Porto Murtinho), respectively (Tarifa 1986; Alfonsi and Camargo 1986). As the annual potential evapotranspiration is at least 1400 mm, the Pantanal experiences a high annual hydric deficit, greater than 300 mm (Alfonsi and Camargo 1986).

The upper course of the Paraguay river, a major tributary of the La Plata basin, is the main branch that drains the wetland, running from north to south in its western part (Collischonn et al. 2001; Berbery and Barros 2002). The Paraguay river and its tributaries and distributaries seasonally increase their discharge and occupy the floodplain in a complex pattern (Hamilton 1999). The inundation peak occurs from February to April in the north/east and from May to July in the southern part (Hamilton et al. 1996), triggered by the increase in precipitation in the wetland but, mainly, in the eastern and northern highlands, where are the sources of many watercourses (Adámoli 1986; Por 1995; Hamilton 1999). Thus, because of the time lag between the inundation flows from the north/east to south, the maximum area that is simultaneously flooded (110,000 km^2) is smaller than the total area that is flooded during the whole season (130,000 km^2) (Hamilton 2002).

The waters coming from the highlands (200–900 m), formed by Precambrian crystalline and Mesozoic sedimentary rocks, are the main responsible for carrying the sediments that have filled the Pantanal sedimentary basin probably since the Plio-Pleistocene (Del'Arco et al. 1982; Godoi Filho 1986). The modern wetland is a huge plain, with altitudes between 80 and 200 m and gradients lower than 1° (Assine et al. 2016; Silva 1986), where fluvial processes have built fans and megafans (e.g., São Lourenço, Taquari, Nabileque, Cuiabá), interfan floodplains (e.g., Itiquira, Negro) and the major floodplain of the Paraguay river (Assine et al. 2015) (Fig. 1). Lakes and ponds are also found in active floodplains and ancient lobes of megafans (Assine et al. 2015). The Pantanal formation, with an estimated maximum thickness of 500 m, is mostly composed of fine sandy and silty-clayey unconsolidated sediments and immature, angular to subrounded and fine to coarse grained sandstones, with frequent ferruginous cementation or concretions and a tendency to become finer toward the surface (Del'Arco et al. 1982; Godoi Filho 1986; Assine et al. 2005, 2016).

The soils of the wetland develop mainly from the unconsolidated sediments of the Pantanal formation. According to Couto and Oliveira (2010), based on Amaral Filho (1986) and using the WRB classification (IUSS Working Group WRB 2015), Gleyic Solonetzes (32,281 km^2), Dystric Plinthosols (32,279 km^2) and Ferric Podzols (28,767 km^2) are the dominant soils in the wetland (Fig. 2a). The former is common in many parts of Pantanal, whereas Dystric Plinthosols and Ferric Podzols are mostly found in the north and central parts, respectively. Eutric Planosols (18,073 km^2) and Dystric Gleysols (17,843 km^2) still present important spatial distribution, occurring mostly in the central part and adjacent to the main watercourses, respectively. Other soils classes have reduced distribution, such as Eutric Vertisols, Haplic Arenosols, Eutric Fluvisols, Luvic Chernozems, Haplic Acrisols, and Lithic Leptosols. The last three soil classes occupy mainly the higher zones located near the borders of Pantanal (Couto and Oliveira 2010).

The genesis of most of the Pantanal soils is intrinsically related to flooding dynamics, being important to consider factors such as floodable and not-floodable sites, annual period of water saturation and maximum/minimum depth of groundwater in the profile. Many soils are affected by gleyzation, especially in areas experiencing longer periods of water saturation, which imprints redoximorphic features on the different classes, such as Gleysol, Solonetz, Plinthosol, Planosol, Podzol, Arenosol, and Vertisol. Plinthization, with the consequent formation of plinthites and petroplinthites, is also common in many classes, especially where seasonal variation of the subsurface water level is more intense. This process is important mainly in the north, which is clearly expressed by the dominance of Plinthosols (Couto and Oliveira 2010; Couto et al. 2017).

Podzolization is also significant, occurring mainly in sandy soils that are characterized by good drainage, such as in the Taquari megafan. However, recent studies have revealed that the extension of Podzols in the south part of this megafan (Nhecolândia subregion, Fig. 3) is smaller than the area mapped in Fig. 2a (Couto et al. 2017). Genesis of soils with textural contrasts, such as Solonetzes, Planosols, and Acrisols may have an origin directly linked to pedological processes, like clay illuviation, and ferrolysis (Nascimento 2012; Nascimento et al. 2015; Oliveira Junior 2015), or be

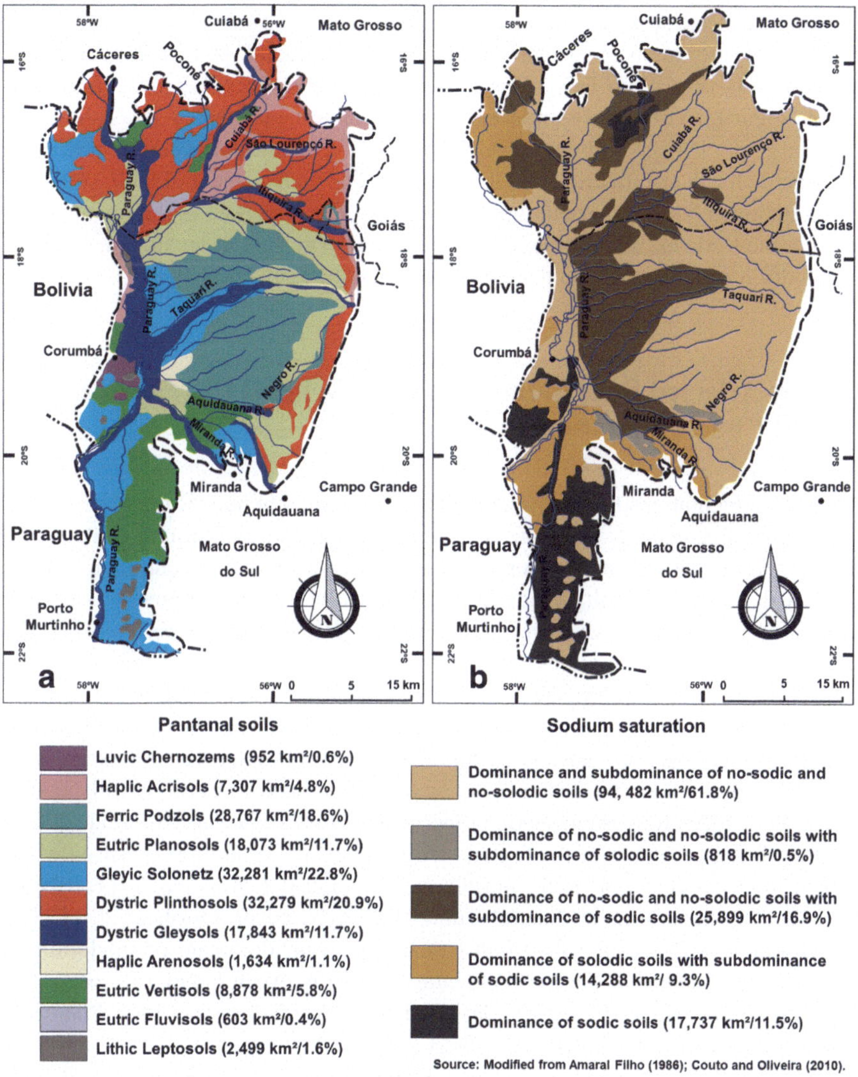

Fig. 2 Brazilian Pantanal with: **a** Distribution of soils; **b** Distribution of soils with sodium saturation (adapted from Amaral Filho 1986; Couto and Oliveira 2010)

strongly influenced by sedimentation processes, including flooding and river avulsion (Nascimento et al. 2013). In the last case, the textural differences are formed by the deposition of layers with different textures, driven by changes in the inundation dynamics, and in the position of the river channel (avulsion) (Nascimento et al. 2013, 2015; Couto et al. 2017).

Finally, processes such as salinization, solonization, and solodization were and/or are still very common in the Pantanal wetland, producing large areas with salt-affected soils (Fig. 2b). This chapter will focus on these soils, presenting their morphological, chemical, and mineralogical characteristics, their relationship with the local geological, geomorphological, and hydrological dynamics and the processes responsible for their formation and functioning. The management of these soils will not be included because agricultural activities are not important in the region, inasmuch as the wetland still has preserved ecosystems in most of its territory and animal husbandry is mainly based on the use of native pasture.

2 Salt-Affected Soils in the Pantanal: General Distribution and Characterization

The environmental conditions of the Pantanal, presented in item 1, favor the formation of salt-affected soils by salinization and/or solonization (Furquim et al. 2017). High evapotranspiration and seasonal hydric deficit permit the concentration of solutions, with consequent salt accumulation in the soils by mineral precipitation and increasing of ions in the exchangeable complex (Bohn et al. 1985). A nearly flat surface, contrasting soil textures with consequent lower hydraulic conductivity in the subsurface, and high levels of the subsurface water, at least during part of the year, are characteristics that restrict water flow and reduce salt leaching (Westin 1953; USSL Staff 1954). Besides, the seasonal accumulation of water improves evapotranspiration rates and facilitates the instability of aggregates followed by colloidal dispersion (USSL Staff 1954; Rengasamy and Sumner 1998; van Breemen and Buurman 2003).

Thus, Pantanal is one of the regions with highest concentration of salt-affected soils in Brazil, second only to the northeastern semi-arid (Gardi et al. 2014; Ribeiro et al. 2003). According to Amaral Filho (1986), based on Orioli et al. (1982), 58,518 km^2 (38.2%) of the Brazilian Pantanal present dominant or subdominant soils with Exchangeable Sodium Percentage (ESP) between 6 and 15% or $\geq$15% (Fig. 2b). These soils are mainly classified as Gleyic Solonetz, Eutric Planosols ($\geq$6% ESP $\leq$15%) and Eutric Vertisols ($\geq$6% ESP $\leq$15%) (Fig. 2a) (Couto and Oliveira 2010). As previously mentioned, Gleyic Solonetz is the dominant soil class of the wetland (Fig. 2a), occupying wide areas in the south, central-western, and north-western regions, especially in the fluvial fans of Nabileque, Taquari, Cuiabá, and Paraguay (Fig. 1). Eutric Planosols occur mostly in the Taquari and Negro fluvial fans (Fig. 1), located in the central part of the wetland (Fig. 2a), whereas Eutric Vertisols are mainly encountered in the south (Fig. 2a), especially in the Miranda fluvial fan and surrounding lowlands (Fig. 1) (Couto and Oliveira 2010; Assine et al. 2015).

Gleyic Solonetz tends to present A, E, Btn, and/or Btgn horizons, commonly showing an abrupt transition between E and B. The texture is generally sandy in the A and E and clayey in the B horizons. The B horizon presents an ESP $\geq$15% and an extremely hard consistency when dried. Salt-affected Planosols are generally formed by A, E, Bg, and Btg horizons. The E can be very thick and generally presents an abrupt transition to the underlying horizon. The texture is mainly sandy near the surface and loamy in the B, but loamy and clayey or sandy and clayey textures can occur, respectively. ESP ranges from 6 to 15% and low activity clays dominate. Finally, salt-affected Vertisols are formed by clayey A and C horizons, with hard to extremely hard consistencies. The ESP varies between 6 and 15% and, in some cases, the electrical conductivity (EC) reaches values $\geq$4 dS m^{-1} (Orioli et al. 1982).

However, more detailed studies have revealed an important presence of salt-affected soils in places not mapped by previous surveys, which were produced in general scales. The south part of the Taquari megafan (central Pantanal), especially the lowlands known as low Nhecolândia (Fernandes et al. 1999) (Fig. 3a), have hundreds of alkaline–saline lakes surrounded by Saline–Sodic soils (Sakamoto 1997; Furquim 2007; Furquim et al. 2008) and brackish lakes surrounded by Sodic and degraded Sodic soils (Solodized Solonetz and Solod), (Furquim et al. 2017). Also, in the São Lourenço fluvial fan (Assine et al. 2015), in a sub-region known as Barão de Melgaço (Silva and Abdon 1998) (Fig. 3), higher and lower geomorphic units present a significant presence of Sodic and degraded Sodic soils (Solodized Solonetz and Solod), respectively. These areas have been intensely studied by different research groups over the last two decades, focusing on soil morphological, chemical, and mineralogical characteristics in order to understand their genesis. Thus, further information about these soils will be given in the following sections.

3 Salt-Affected Soils of the Low Nhecolândia

3.1 *The Nhecolândia Subregion*

The Nhecolândia subregion (area ~26,921 km^2) (Silva and Abdon 1998), located in central Pantanal (Fig. 3), corresponds to the southern half of the Taquari fluvial megafan (Fig. 1) (Assine et al. 2015). Low Nhecolândia (Fig. 3b) is a zone with low altitudes that occupies abandoned lobes, defined as elongated areas formed by the ancient deposition of sediments by a drainage network (Zani et al. 2009; Fernandes et al. 1999). This zone is marked by the presence of 637 alkaline–saline lakes, located on the top of sand hills, and 8214 freshwater lakes, located within depressed and elongated intermittent channels. The alkaline–saline lakes (*salinas*) occur inside rounded depressions that generally have a diameter of 500–1000 m and a depth of 2–5 m. These lakes are mainly fed by rainwater and rarely dry up, presenting water with a pH >8, electrical conductivity (EC) >5 dS m^{-1}, and a high concentration of carbonates and sodium. The sand hills (*cordilheiras*) are narrow and elongated land strips that occupy the highest topographical position of the landscape, being 2–3 m higher than the surroundings. Due to this slight difference in altitude, the complex sand hills/alkaline–saline lakes is not typically reached by seasonal flooding, allowing the development of dense savanna vegetation in the hills and around the

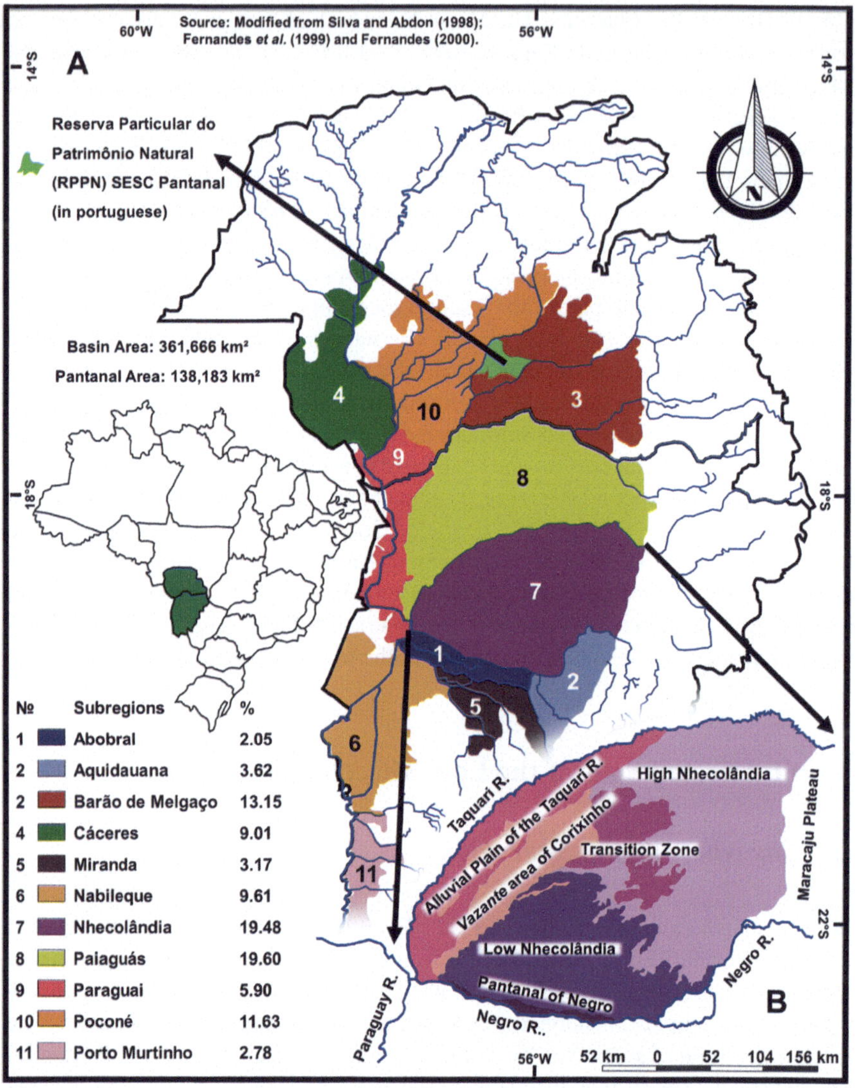

Fig. 3 A-Subregions of the Brazilian Pantanal (adapted from Silva and Abdon 1998), with the location of the RPPN in the Barão de Melgaço subregion; B-subregion of the Nhecolândia and its divisions, including the low Nhecolândia (adapted from Fernandes et al. 1999; Fernandes 2000)

lakes (Barbiero et al. 2002; Almeida et al. 2003; Evans and Costa 2013; Costa et al. 2015; Freitas et al. 2019).

The freshwater lakes, locally known as *baías*, are temporary lakes with different forms and dimensions and a depth of up to 2 m. They usually emerge during the flood period, constituting seasonal watercourses locally called *vazantes*, which are adjacent to the sand hills and can be several kilometers long and 10–40 m wide. The complex freshwater lake/intermittent watercourse receives the inundation seasonal water and, for that reason, are covered by open-grass savanna and swampy grasslands. In contrast to the alkaline–saline lakes, this surface freshwater, together with the subsurface freshwater of the sand hills, generally present a pH <7, EC between 0.01 and 0.6 dS m^{-1}, and a chemical profile dominated by carbonates/chloride and sodium/calcium/potassium (Barbiero et al. 2002, 2007; Almeida et al. 2003; Parizotto 2012; Evans and Costa 2013).

The saline waters of low Nhecolândia have currently arisen through the evaporative concentration of freshwater (Barbiero et al. 2002; Almeida et al. 2010; Furquim et al. 2010a). Both freshwater and alkaline–saline lakes act as recharge zones, in a process called depressed-zone recharge. Thus, freshwaters from precipitation, runoff, and shallow subsurface flows reach the alkaline–saline lakes during the wet season and the water retained throughout the year in the isolated depressions evaporates and concentrates, especially during the dry season (Freitas et al. 2019). However, concurrently with the formation of saline water, some of the alkaline–saline lakes have been degraded by the atypical entrance of the freshwater flood, as a result of the erosion of the fragile sand hills. Thus, new intermittent watercourses are created in detriment to some sand hills and the associated alkaline–saline lakes are being transformed into brackish lakes, which have intermediate chemical characteristics between saline and freshwater (Almeida et al. 2003; Fernandes 2007).

3.2 Salt-Affected Soils Associated with the Alkaline–Saline Lakes

The coexistence of surface and subsurface freshwater, alkaline–saline lakes, and brackish lakes over short distances (≥200 m) in low Nhecolândia imprints different morphological, physical, chemical, and mineralogical characteristics on the related soils, allowing greater soil diversity in the region. In the sand hills, freshwater lakes, and intermittent watercourses, Arenosols (A–C horizons) commonly occur, characterized by a fine sand texture (≥800 g kg^{-1} of sand), single grains and low pH (≤5.3). Podzols (A—Bh, Bs, Bhs—C) are also common, characterized by sand and loamy sand textures (≥850 g kg^{-1} of sand), single grains and pH generally between 5 and 7 (Cunha 1980; Sakamoto 1997; Silva and Sakamoto 2003; Fernandes 2000; Cardoso et al. 2016; Schiavo et al. 2012). However, some of the Arenosols and Podzols present different chemical characteristics, with a pH up to 10 and ESP ≥15% (Cardoso et al. 2016) and pH up to 7.5 and ESP between 6 and 10% (Schiavo et al. 2012), respectively.

Soil surveys carried out in transects around alkaline–saline lakes reveal soils with similar morphology, generally with A and/or H horizons on the surface, followed by a sequence of E, Bkg, and Bnx/Bnq in the subsurface (Fig. 4). A horizons present light (e.g., pale brown, very pale brown) or dark colors (e.g., very dark grayish brown), depending on the presence of vegetation around the lake, a sandy texture

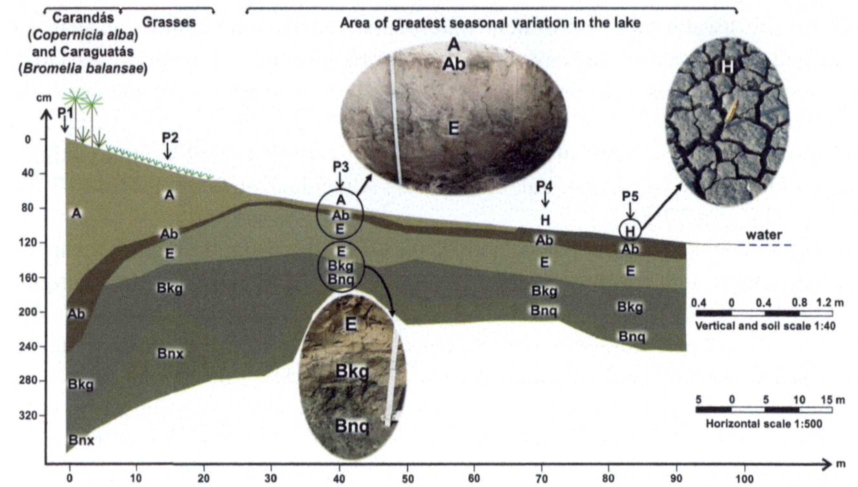

Fig. 4 Morphology of a transect with Saline Sodic soil around a representative alkaline—saline lake in the low Nhecolândia (adapted from Furquim 2007)

and are single grain in the highest zones of the depression. They are laterally replaced in the zone of variation at the lake level by the H horizon, which is black to dark gray, prismatic, loamy sand, and rich in decomposed algae. The E is usually light brownish gray, single grain and sand, with vertical and darkish volumes of colloidal eluviation. The Bkg horizon is grayish (e.g., gray, very dark grayish brown), loamy sand and massive, with abundant carbonate nodules. Finally, the Bnx and Bnq horizons are greenish (5Y hue) (e.g., olive gray, olive), massive, with more clay than the other horizons (sandy loam texture) and an extremely hard consistency. The former occurs in the highest zones around the lake and becomes much softer when saturated, whereas the latter occurs toward the lake (P3–P5, Fig. 4), maintaining its hardness even under saturation (Sakamoto 1997; Barbiero et al. 2000; Fernandes 2000; Silva and Sakamoto 2003; Silva et al. 2004; Furquim 2007; Furquim et al. 2010a; Martins 2012; Furian et al. 2013).

Only a few of these studies present chemical analyses of the studied soils. Thus, based on the available data, these soils are characterized by an alkaline pH, mostly $\geq$8 but up to 10.8, and high electrical conductivity (EC), in general $\geq$9 dS m^{-1} but up to 19.2 dS m^{-1}. Na^{+}, K^{+} and Ca^{2+} are the most abundant exchangeable cations, with exchangeable sodium saturation (ESP) between 20 and 50% in most of the horizons. The highest values of pH, EC, and ESP are found in the subsurface, in the Bnx and/or Bnq horizons. The values of organic carbon are low, generally $\geq$10 g kg^{-1}, but they can reach 22 g kg^{-1} in the H horizons (Sakamoto 1997; Furquim 2007; Martins 2012; Furquim et al. 2008).

Following the classification of the salt-affected soils, those associated with the alkaline–saline lakes are Saline–Sodic soils, as much as they present EC $\geq$4 dS m^{-1} and ESP $\geq$15% (USSL Staff 1954; Bohn et al. 2001; Sparks 2003). Saline Sodic soils are formed by the pedological processes of salinization and solonization. The former is the accumulation of salts more soluble than gypsum in the profiles, which provokes the increase of exchangeable cations and, consequently, high electrical conductivity of the soils (Gedroiz 1912; USSL Staff 1954; Fanning and Fanning 1989). The latter process provokes higher percentages of Na^+ in the exchangeable complex and high pH (USSL Staff 1954; Fanning and Fanning 1989; Schaetzl and Anderson 2005), mainly because of the dissolution of sodium carbonates, with the liberation of OH^- into the solution (McBride 1994) (Eqs. 1 and 2).

$$Na_2CO_3 \rightarrow 2Na + CO_3^{2-} \quad (1)$$

$$CO_3^{2-} + H_2O \rightarrow HCO_3^- + OH^- \quad (2)$$

Detailed studies on the mineralogy of these Saline–Sodic soils have revealed that the genesis of the H and B horizons are intrinsically linked to the authigenic formation of minerals and amorphous materials. In general, quartz dominates in the fine silt (0.02–0.002 mm), kaolinite and quartz in the coarse clay (0.002–0.0002 mm), and mica, smectite and kaolinite in the fine clay ($\leq$0.0002 mm) (Furquim 2007; Furquim et al. 2010a). Carbonates and amorphous silica also occur in minor amounts. According to these authors, kaolinite in both the coarse and fine clay are enriched with octahedral Fe^{3+} and is inherited from the Pantanal formation. These studies present the following set of evidence that corroborates the detrital origin: (i) kaolinite is ideally formed in well-drained soils, with low pH and K^+ concentration (Tardy et al. 1973; Bohn et al. 2001; Velde 1995), which are the opposite of the conditions encountered in the studied Saline–Sodic soils; (ii) it is the dominant mineral in coarse clay but occurs in smaller proportions in fine clay, which is the fraction that usually concentrates authigenic minerals (Allen and Hajek 1989); (iii) the source of the sediments and water of Nhecolândia (Maracaju Plateaux) is a tropical region covered mainly by acid soils, such as Ferralsols (Gardi et al. 2014), which usually contain high amounts of Fe-kaolinite.

On the other hand, minerals from the mica and smectite groups, carbonates and amorphous silica have an authigenic formation in the Saline–Sodic soils and are strictly related to the evaporative concentration of the salts in the water. Barbiero et al. (2002) and Barbiero et al. (2008) showed that fresh and saline water in the low Nhecolândia belong to the same chemical family and that most of the chemical compositional changes in its waters are related to precipitation of some elements (mainly Ca^{2+}, Mg^{2+}, K^+) into the soil by evaporative concentration of surface and subsurface water in and around the alkaline–saline lakes. Mineralogical analyses of the fine clay corroborated these conclusions, attesting an authigenic origin of large amounts of smectite in the H horizon (Furquim et al. 2008), carbonates in the Bkg's nodules (Furquim et al. 2010a) and large amounts of mica in Bnx and Bnq (Furquim et al. 2008; Furquim et al. 2010b; Andrade et al. 2020).

Trioctahedral Mg-smectite dominates the mineral assemblage of the H horizon, specifically stevensite and saponite interstratified with a dioctahedral mineral. These flake-shaped clay minerals are characterized by zero or low substitution of Al for Si in the tetrahedra and dominance of Mg^{2+} in the octahedra. Their occurrence exactly in the area of seasonal variation of the lake (H horizon) (Fig. 4) and the very low amounts of rare earth elements (REE) is evidence that these minerals are precipitating directly from the alkaline–saline lakes (Furquim et al. 2008). This precipitation is likely facilitated by Al and Fe bounds to organic colloids that migrate down in the profile, allowing the formation of Al-poor clay minerals on the surface (Barbiero et al. 2016). Precipitation of carbonates, mainly calcite ($CaCO_3$) and dolomite ($CaMg(CO_3)_2$), was also observed in nodules of the Bkg (Furquim et al. 2010a). Thus, neoformation of Mg-smectite, dolomite and calcite in the Saline–Sodic soils are the main responsible from the decrease of Mg^{2+} and Ca^{2+} from saline waters observed by Barbiero et al. (2002, 2008) and for important characteristics of the H and Bkg horizons.

The Bnx and the Bnq are dominated by dioctahedral Fe-phases from mica group, specifically Fe-illite and glauconite (73–76% of the layers), interstratified with minor amounts of kaolinite (called as K-I) and smectite (called as I-S). These minerals are lath-shaped, present Al^{3+} and Fe^{3+} as the main octahedral cations and have a green color causing the greenish matrix (5Y hue) of the natric horizons (Furquim et al. 2010b; Andrade et al. 2020). Furquim et al. (2010b) demonstrated that Fe-illite and glauconite form by precipitation from the soil solution, which is corroborated by the suitability conditions for this process, such as alkaline pH, negative Eh and appropriate concentrations of $Si(OH)_4$, K^+, Mg^{2+} and Fe in the surface and subsurface waters. Thus, the decrease of K^+ from the saline waters, observed by Barbiero et al. (2002, 2008), occurs due to the involvement of this ion in mica formation around the lake during the water evaporation. The first stage of mica neoformation seems to be a Si-rich amorphous material with very similar chemical composition of the Fe-mica, which is also probably responsible for the hardness of the Bnq, where this material was found in much higher amounts (Furquim et al. 2010a, b).

3.3 *Salt-Affected Soils Associated with Brackish Lakes*

The conversion of alkaline–saline lakes into brackish lakes due to the atypical input of flooding waters, as cited in item 3.1, have provoked the transformation of Saline–Sodic soils into Sodic and degraded Sodic soils (Solodized Solonetzes and Solods) in low Nhecolândia. Furquim et al. (2017) studied soil transects around three brackish lakes, with different progressive geochemical characteristics (Almeida et al. 2003): (i) Carandazal lake, with water more similar to that in alkaline–saline lakes (pH of 6.5; EC of 1.9 dS m^{-1}). (ii) Banhado lake, with waters more similar to that in freshwater lakes (pH of 6.2; EC of 0.8 dS m^{-1}); (iii) and Cerca lake, with intermediate pH (6.4) and EC (1.1 dS m^{-1}) values. The distribution of the soils around each brackish lake is strictly related to the water geochemistry.

Sodic soils, characterized by pH $\geq$ 8.5, EC $\leq$4 dS m^{-1}, ESP $\geq$ 15% and Sodium Adsorption Ratio (SAR) > 13 (USSL Staff 1954; Fanning and Fanning 1989; Schaetzl and Anderson 2005), occur mostly around the Carandazal lake, which is more similar to the alkaline–saline lakes. In general, these soils present an A horizon that is grayish brown, sandy to silt loam, and with platy structure. In the subsurface, there are two B natric horizons, which are greenish, mostly massive, sandy loam to sandy clay loam and extremely hard. The main morphological difference between them is the hue, which is 5Y in the Bnc1 and 2.5Y in the Bnc2. The EC tends to be <2 dS m^{-1} in the whole profiles and the pH and ESP are >8 and >15% in the B horizons, respectively, with higher, equal or smaller values in the overlaying horizons. SAR varies from 15 to 135 and exchangeable $Al^{3+} + H^{+}$ is zero or near zero (Furquim et al. 2017).

Solods are soils where the salt-affected characteristics are very discreet. They are characterized by low EC ($\leq$4 dS m^{-1}), ESP ($\leq$10%), and SAR (<13) and variable pH, tending to be acid to neutral near the surface and neutral to alkaline in the subsurface (Fanning and Fanning 1989; Schaetzl and Anderson 2005; Furquim et al. 2017). These soils are dominant around the Banhado Lake, which is more similar to freshwater, commonly presenting profiles with A, Bnc1, Bnc2, and Bc horizons. The first three horizons have similar characteristics than those described for the Sodic soil around the Carandazal lake. The Bc is also clay enriched, mostly massive and as hard as the other B horizons, but it is not greenish, presenting a matrix marked by a 10YR hue. EC, ESP and SAR are mostly $\leq$1 dS m^{-1}, $\leq$6% and $\leq$5, respectively, although some higher values are observed mainly in the B horizons. The pH varies from 4.9 to 5.8 near the surface and from 5.0 to 8.0 in the B horizons (Furquim et al. 2017). Values of exchangeable $Al^{3+} + H^{+}$ tend to be high through the whole profile, reaching 50.3 $mmol_c$ Kg^{-1}.

Finally, the Solodized Solonetzes typically have intermediate characteristics between Sodic and Solod soils, such as ESP $\geq$10%, SAR <13 and pH from 6 to 8.5 (Fanning and Fanning 1989; Schaetzl and Anderson 2005; Furquim et al. 2017). In contrast to Sodic soils, these parameters tend to be lower in the upper horizons. They are the dominant soils in the transect around Cerca lake, where the water also has intermediate geochemical characteristics (pH and EC), with values between the other two brackish lakes. A, Bnc1, and Bnc2 horizons also occur in the studied Solodized Solonetzes, with similar morphological characteristics to Sodic soils. EC values are $\leq$1 dS m^{-1}, both ESP and pH values are $\leq$5 in the A horizon and $\geq$10 in the B horizons, SAR is $\leq$10 and exchangeable Al^{3+} and H^{+} is high in the near-surface horizons (9–53 $mmol_c$ kg^{-1}) (Furquim et al. 2017).

The Bnc1, Bnc2, and Bc horizons also present a transformation sequence from the Bnq around the alkaline–saline lakes. They are all analogous in the subsurface position, clay-enrichment, and hardness. Bnc1 is the most similar to the Bnq, maintaining a more preserved green color (5Y), whereas Bnc2 presents more yellow-greenish colors (2.5Y). Bc, occurring only around the Banhado lake, is the most degraded, no longer presenting greenish colors (10YR). Although the chemical characteristics between Bnc1 and Bnc2 are not significantly different, Bc clearly has the lowest ESP, pH, and SAR and the highest contents of $Al^{3+} + H^{+}$. It is also important to highlight that all these parameters tend to decrease toward the lake, especially in Carandazal and Banhado lakes, showing that the more degraded salt-affected soils are in the lowest positions of the depressions.

These soil transformations are intrinsically associated with the geomorphological and hydrological processes observed in the Taquari megafan. As commented before, the erosion of the sand hills, caused by the inundation flows are responsible for the creation of new intermittent watercourses (*vazantes*) that are progressively transforming the alkaline–saline lakes into brackish lakes. Thus, the associated soils are transformed by this new hydrological dynamics, leaving the more preserved salt-affected soils in the higher positions in the landscape (Almeida et al. 2003; Fernandes 2007). Erosional processes commonly act in abandoned lobes (Assine 2003), forming degradational landforms, such as the intermittent watercourses, over aggradational features. These processes have been apparently more intense over the last 3000 years, when the climate probably become more humid in the region (Whitney et al. 2011; McGlue et al. 2012).

The observed sequence of soils in the low Nhecolândia is rare in the literature, being in accordance with the model of salt-affected soils' evolution proposed by Gedroiz (1912, 1917, 1925). The process of salinization is no longer active and only solonization is responsible for the formation of the Sodic soils associated with the Carandazal lake. Thus, these soils experienced leaching of soluble salts and consequent decrease of EC, with absolute or relative increase of Na^+ in the exchangeable complex, preserving high ESP, SAR and pH. Solodized Solonetzes and Solods, respectively, associated with Cerca and Banhado lakes, are formed by solodization, which reflects the progressive leaching caused by the intermittent watercourses. The first stage of solodization forms Solodized Solonetz, after the leaching of most of the basic cations, including Na^+, from the upper horizons, but preserving a high Na^+ content in the B horizons, which maintain relatively high values of ESP, pH, and SAR. The last stage, responsible for the formation of Solods, represents the leaching of mono- and divalent basic cations from the whole profile. Basic cations are generally replaced by exchangeable $Al^{3+} + H^+$ during both stages of solodization (Gedroiz 1912; Sumner et al. 1998; Sparks 2003).

Andrade et al. (2020) showed that mineralogical analyses of the clay fraction corroborate the evolutionary sequence around the brackish lakes, demonstrating that the Fe—mica in the Bnq related to the alkaline–saline lakes, is no longer stable and has been transformed into mixed-layered minerals (MLM) enriched in smectite and kaolinite under the leaching conditions. In the Sodic soils, illitic phases still dominate the samples as K-I and I-S, but the total percentage of illite layers in the MLM is lower than in the Saline–Sodic soils, varying between 52 and 68%. The total percentage of smectite (in K-S, I-S) and mainly kaolinite (in pure kaolinite, K-S, K-I) increases, reaching 7–21% and 19–35%, respectively. In the Solods, the total percentage of illite (in I-S phases) is significantly smaller, reaching 36–41%. Total kaolinite layers are maintained (21–31% in pure kaolinite, K-S and K-I), whereas the total smectite layers increase to 31–37% (in K-S and I-S). These progressive changes are also verified in the B horizons, inasmuch as the highest percentages of mica and the lowest percentages of kaolinite and smectite layers are observed in the Bnc1 horizon of the Sodic soils, whereas the lowest percentages of illite and the highest percentages of smectite layers are observed in the Bc horizon of the Solod.

The decrease in greenish Fe-enriched mica is probably responsible for 5Y hues in Bnc1 and 10YR in Bc.

Thus, these mineralogical trends represent the progressive morphological and chemical transformations from Bnq into Bnc and Bc horizons through the replacement of the salinization/solonization in the Saline–Sodic soil by solonization and solodization in Sodic and Solod soils, respectively. Illitic MLM, neoformed around the alkaline–saline lakes, has been transformed to mixed-layered smectite and kaolinite under more acid geochemical conditions, without the complete dissolution of the minerals. The transformation of illitic MLM into smectite occurs through K^+ loss, opening of the interlayer allowing the input of cations, and the transformation of smectite into kaolinite includes more complex changes in the structure, including removal or inversion of the tetrahedral sheet and increase in octahedral Al by isomorphic substitution (Karathanasis and Hajek 1983; Amouric and Olives 1998; Dudek et al. 2006; Ryan and Huertas 2009).

4 Salt-Affected Soils of the Barão de Melgaço Subregion

Barão de Melgaço (Fig. 3) is located in the western half of the São Lourenço megafan and smaller parts of the Cuiabá fan (Fig. 1). Important soil studies in this subregion were conducted in the Private Natural Heritage Reserve (RPPN—Reserva Particular do Patrimônio Natural), a property of the Pantanal Social Service of Commerce (Serviço Social do Comércio, SESC Pantanal). The RPPN, with an area of 1060 km^2, is partially located in the fluvial floodplain of the Cuiabá river (36%) and, mostly, in an abandoned Pleistocene lobe of the São Lourenço fan (64%) (Nascimento 2012; Assine et al. 2015). The Cuiabá floodplain is 10–12 km wide strip on the left bank of the river, covered with riparian forest. On average, it remains flooded for 140 days of the year, with the inundation reaching a depth of up to 5 m (Beirigo et al. 2010; Nascimento 2012).

In the abandoned lobe, the non-flooded areas are mostly formed by mounds, paleochannels, paleo floodplains and paleolevees, covered with savanna (*Cerrado*) vegetation (Nascimento 2012; Beirigo et al. 2010). The mounds, locally known as *murundus*, are random or aligned small elevations that have been eroded during the Holocene, some of them being probably built with the contribution of successive termite nests (Nascimento et al. 2015; Oliveira Junior et al. 2017a). They are 1–1.5 m above the flooded areas and have rounded and convex shapes with diameters of between 2 and 30 m (Oliveira Junior et al. 2017a). The last three geomorphic features are elongated stripes, 1–6 m higher than the surroundings that are locally called *cordilheiras*. The flooded areas are mainly composed of paleo floodplains and eroded paleolevees and paleochannels, covered by herbaceous vegetation. In general, they are large inundation fields between the elongated non-flooded areas, locally known as *vazantes* (Nascimento 2012; Beirigo et al. 2010). However, some of the low topographical areas are closed or open depressions delimited by higher non-flooded paleochannels and paleo levees (Nascimento 2012).

The distribution of the soils in the RPPN is intimately associated with the paleo and current sedimentary and geomorphological processes. The flooded zones, including the Cuiabá river floodplain and the *vazantes* of the São Lourenço's abandoned lobe, are characterized by gleyic and fluvic properties, with mainly dystric and alumic or eutric Gleysols and eutric Arenosols. Dystric and alumic Plinthosols are also common in the *vazantes* (Beirigo et al. 2010; Nascimento 2012). The non-flooded geomorphic features of the abandoned lobe are mostly occupied by Solonetz, Planosols, Luvisols, and Arenosols. Solonetz and Planosols are generally dystrophic in the near-surface horizons and eutrophic in the B horizons, whereas Luvisols and Arenosols tend to be entirely eutrophic (Beirigo et al. 2010; Nascimento 2012; Oliveira Junior et al. 2017b). Besides the Solonetzes, soils such as Planosols, Fluvisols and Plinthosols can present relatively high ESP (generally $\geq$6%), whereas Luvisols and Arenosols may present high Mg^{2+} saturation (Beirigo et al. 2010; Couto et al. 2017).

Detailed studies performed in the RPPN (Nascimento 2012; Nascimento et al. 2013, 2015; Oliveira Junior 2015; Oliveira Junior et al. 2017a, b, 2019a) demonstrate a close relationship between soils and the geomorphological features identified by Beirigo et al. (2010) and, especially, by Nascimento (2012). Using the classification of salt-affected soils (USSL Staff 1954), adopted from now on in this subitem, Sodic and Solodized Solonetz (both Solonetzes in the WRB classification) generally occur at the highest positions of the landscape, especially in mounds and *cordilheiras* (paleolevees and paleochannels), respectively. Solods (mainly Planosols in WRB) occur in the lower geomorphic units, such as depressions and *vazantes*, mostly in eroded paleochannels and paleoplains. Finally, Solodized Solonetzes (Planosols and Solonetz in WRB) and Solods (mainly Planosols in WRB) occur in intermediate landscape positions. A and E horizons occur in all these salt-affected soils, with Btn being more common in the Sodic soils, Btgn in the Solodized Solonetzes and Btg in the Solods (Nascimento 2012; Oliveira Junior et al. 2017a, b) (Fig. 5).

A horizons tend to be light colored (e.g., light yellowish brown, light gray) and loamy sand to sandy loam in the Sodic soils and darker (e.g., very dark brown, very dark grayish brown) and loam to sandy clay loam in the Solodized Solonetzes and Solods. E is lighter (e.g., white, light gray) and loamy sand, with an abrupt transition to the underlying B natric in all soils. In general, Btn presents a strong brown color, with clay loam or silty clay loam textures, whereas Btgn is grayish (e.g., light brownish gray, dark gray) and sandy clay loam to clay loam. Btg presents grayish colors (e.g. gray, dark gray) and silt clay loam to clay loam textures (Nascimento 2012; Nascimento et al. 2015; Oliveira Junior et al. 2017a, b). Micromorphological descriptions of Sodic and Solodized Solonetz soils show coalescent cavities at the abrupt boundary between E-Bt (Oliveira Junior et al. 2017a; 2019a) and within the Bt horizons (Nascimento 2012). Irregular carbonates nodules with signs of dissolution have also been observed (Oliveira Junior et al. 2017a, 2019a) as well as clay coatings and infillings with sharp extinction under polarized light (Nascimento et al. 2015; Oliveira Junior et al. 2017a, 2019a).

In chemical terms, EC is $\leq$2 dS m^{-1} in the whole profile of all salt-affected soils, not reaching the values to be considered saline ($\geq$4 dS m^{-1}). The variations in pH and ESP in the Sodic soils are, respectively, 6.2–8.0 and 0–26% in the A and E horizons

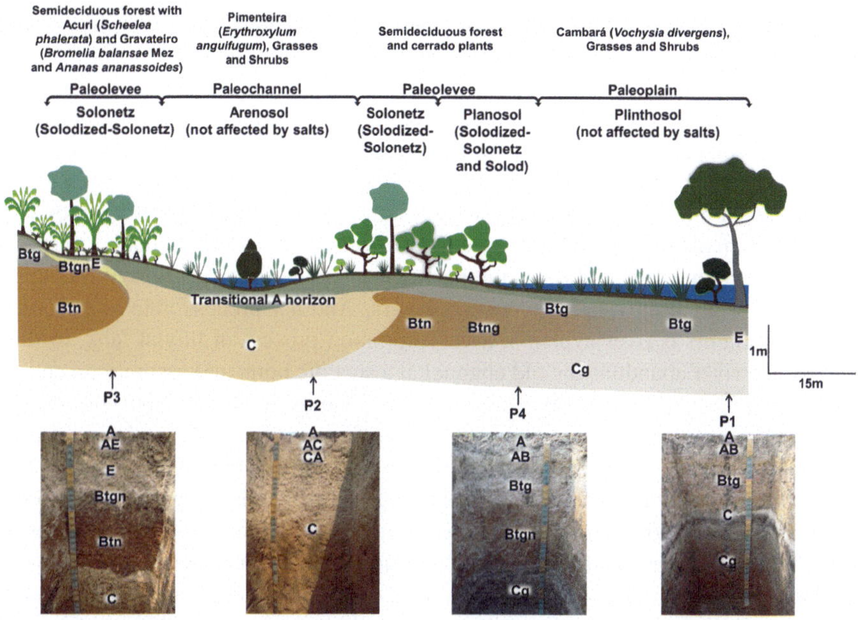

Fig. 5 Morphology of a transect containing salt-affected soils (Sodic, Solodized-Solonetz and Solod) in the RPPN, Barão de Melgaço subregion (adapted from Nascimento 2012)

and 8.8–9.7 and 20–60% in the B natric horizons (Oliveira Junior 2015; Oliveira Junior et al. 2017a, b, 2019a). Solodized Solonetzes present ESP and pH varying from 0.2 to 33.0% and 4.3–6.0 in the surface and/or near-surface horizons and from 2.0 to 35.0% and 5.5–6.6 in the B horizons, respectively. Although the Na^+ saturation is $\geq$15% in some horizons, the pH is mostly $\leq$6, preventing their classification as Sodic soils (pH $\geq$ 8.5). Finally, Solods generally present ESP of 0–3% and 4–10% and pH of 4.5–5.8 and 5.6–6.5 near the surface and in deeper horizons, respectively (Nascimento 2012; Nascimento et al.; 2015; Oliveira Junior et al. 2017a, b, 2019a).

Although Nascimento et al. (2015) partially attributed the genesis of the mounds to termites activity, the mounds containing Sodic soils, studied by Oliveira et al. (2017a, b), are likely to have formed by a sequence of sedimentary and geomorphological events, inasmuch as only low biological activity was found. According to them, these high areas were probably paleolevees surrounding abandoned river channels (e.g., oxbow lakes). The lentic conditions of the lakes facilitated the concentration of water, resulting in Na^+ accumulation and precipitation of clays and carbonates on the bottom, forming a layer with finer texture. The subsequent erosion of the adjacent and higher sandier paleolevees was responsible for the burial of the clayey layer with coarser sediments, giving rise to the textural contrasts. The erosion was probably caused by the increase in the humidity in Pantanal about 3000 years BP, as already pointed out in item 3.3. The continuity of denudation generated a landform inversion, with higher altitudes mainly centered in the ancient lakes, where the clayey horizons prevent major erosion. Also, current denudation is still narrowing and rounding these mounds, but preserving the original alignment of the former paleochannels/paleolevees.

Nascimento (2012) and Nascimento et al. (2013) suggest a different genesis of the textural contrasts in the region. Successive flooding caused by overbank flow can generate horizons with alternate coarser and finer particle sizes (fluvic characteristics). Seasonal inundations generally show variations in energy, depositing finer and coarser sediments during slower and quicker flows (e.g., small channels), respectively. Also, inundation fields also experience erosion, which can carry out finer sediments, originating relatively coarser layers. Another process able to generate textural gradients is river avulsion, a very common process in fluvial fans. In this process, the river abandons the old channel at a specific point and assumes an active one, breaching the adjacent levee and spreading sandy sediments upon the floodplain, which is commonly occupied by finer sediments. The authors observed that soil profiles entirely composed by layers/horizons with variable textures and total C content are probably formed by overbank flow. On the other hand, profiles with only a few coarser layers/horizons, clearly distinguished from under and overlying finer layers/horizons, are probably formed by avulsion, which is a rarer event.

Thus, the genesis of the salt-affected soils of the RPPN seems to have an important influence of drier paleoclimates, influencing salt accumulation, and past processes linked to the dynamics of the São Lourenço megafan, generating textural contrasts. However, current pedological processes are also responsible for soil formation in the region. Solodization seems to be a major process nowadays, provoking many changes in chemical and morphological characteristics. This is corroborated by the fact that all soils present low EC and soils with higher PST and pH are located in the highest landscape positions and in subsurface horizons, when compared to the lower adjacent areas and subsurface horizons. Also, whereas Sodic soils present mainly Btn, Solodized Solonetzes present Btgn and Solods Btg horizons. Thus, the seasonal freshwater flooding that reaches the region every year is responsible for leaching most of the salts from the Solods in the *vazantes*, leaving no more natric horizons in these soils (Btg). Intermediate positions and higher areas are progressively less affected by inundations and leaching, preserving more salts in these positions (Btn, Btng) and triggering redoximorphic processes in the lowest and intermediate areas, where Btng and Btg are present, respectively. However, the degradation of carbonates observed in Sodic soils shows that even the highest areas are reached by the floods, but in a lower frequency and intensity (Oliveira Junior et al. 2017b, 2019a).

The influence of leaching and solodization in these soils is corroborated by the fine clay ($\leq 0.2\ \mu m$) mineralogical analyses performed by Oliveira et al. (2019b) on samples from a long trench that extended from higher to lower zones of the RPPN. In the higher zone of the trench, rarely reached by the floods, the authors found an endmember illite and different K-S MLM, ranging from kaolinite-rich to smectite-rich minerals. The lower zone, more frequently reached by inundations, presented a similar range of K-S, but with lower amounts of the smectite-rich phase. Besides, a K-I MLM was detected, probably originating from the endmember illite. Thus, such as in low Nhecolândia (Andrade et al. 2020), smectite and illite have been transformed into kaolinite in the more diluted conditions of lower geomorphic features, through processes that do not involve the total dissolution of these minerals.

However, it is likely that, in higher areas, the current evapotranspiration rates of Pantanal are contributing, albeit in minor proportions, to the concentration of solutions and Na^+ accumulation, compensating part of the loss in occasional inundations (Nascimento et al. 2015; Oliveira Junior et al. 2019b). The water chemistry of the São Lourenço river, coming from the surrounding highlands, corroborates the current sodification in northern Pantanal, inasmuch as these river presents much higher values of alkalinity than Ca^{2+}, which is a requirement for the genesis of alkaline-sodic waters in evaporative environments (Rezende Filho et al. 2012; Al-Droubi et al. 1980). Other pedological processes currently observed include argilluviation and ferrolysis. Although most of the textural gradients are due to past activity of the São Lourenço megafan, some clay coatings and infillings indicate that argilluviation occurs in the profiles, triggered by the high ESP and pH caused by the sodic and protosodic conditions. The cavities observed in the thin sections and the typical water table oscillations of the region also suggest the action of ferrolysis, a process that provokes clay dissolution by changes in Fe valence, caused by seasonal hydric saturation. Both argilluviation and ferrolysis seem to be responsible for thickening of the E horizon in these soils (Nascimento et al. 2015; Oliveira Junior et al. 2017a, 2019a).

5 Other Examples

Even though most of the research efforts for the understanding of the salt-affected soils of Pantanal were in the low Nhecolândia and Barão de Melgaço subregions, some other studies were developed in different areas, representing more isolated study cases. Rueda et al. (1998), for example, showed that the subregions of Miranda and Aquidauana, in the central to southern part of Pantanal (Fig. 3), are regional depressed zones with a hydrological system of low circulation due to the tilting caused by the Rio Negro fault. Thus, this structural condition coupled with high evapotranspiration rates allowed the concentration of the waters and the consequent salinization, sodification and carbonate accumulation in the soils. The study shows that the highest ESP ($\geq$15%) and amounts of carbonate are found mainly in paleolevees and mounds, described as higher positions of the landscape (cordilheiras and capões). In the former, they found Mollisols[1] and soils with Bt characterized by planosolic and vertic properties. Relatively high Na^+ saturation (ESP 6–15% and $\geq$15%) was also observed in Mollisols[1] of current levees and *vegas*, both landforms with intermediate altitudes, that are occasionally reached by inundations.

In the subregion of Poconé, located in northern Pantanal (Fig. 3), Sousa (2003) studied a profile of a Luvisol in high altitudinal zones (cordilheira), and a profile of

[1]The term Mollisol, a Soil taxonomy class, was kept in the text to respect the classification of the authors.

a Gleysol in the seasonal inundation field (*vazante*). The Luvisol presented A, Bt and C horizons, with an abrupt transition between the first two, pH ≤6.3 and PST of 6.7 in the B and up to 16% in the C. The Gleysol, composed by A, Btg and C horizons, present ESP ≤1.5% and pH ≤5.8. Gleyzation was attributed as the main process responsible for the differences between the soils, inasmuch as the flooding waters reach the lower zones every year, provoking redoximorphic conditions and originating the Gleysols. Lithological discontinuities were identified in both soils and was assigned to the parent material (Pantanal Formation). Although the author did not emphasize the genesis of sodic and solodic conditions in the cordilheiras, the same pattern similar to previous studies was described, with low exchangeable Na^+ in the *vazantes* probably due to a significant influence of solodization.

6 Conclusions

Although the Pantanal is a huge wetland characterized by complex environmental conditions, only a few systematic and general soils surveys, published mostly in the 1980s, have been carried out in this region. Thus, new systematic soil investigations must occur, in more detailed scales, to improve knowledge about the distribution of the soils, including the salt-affected soils. Even though initial mapping efforts were very important to a better understand of the Pantanal's soils, continued research in the region has revealed, for example, an important occurrence of salt-affected soils in the Barão de Melgaço subregion (RPPN) and an overestimation of Podzols coupled with a massive presence of salt-affected soils associated with alkaline–saline lakes and new *vazantes* in low Nhecolândia. Also, the research focused on the salt-affected soils is mainly concentrated in these two subregions, with a huge gap in other parts of the wetland.

Based on the current knowledge, it is possible to notice important similarities and differences between the processes occurring in the Pantanal. Solodization seems to be a dominant process in most of the studies, leaving Saline–Sodic and Sodic soils in higher landforms and generating degraded Sodic soils (Solonetz Solodized and Solods) in lower positions. In low Nhecolândia and in the RPPN, leaching is related to ancient lobes of the Taquari and São Lourenço fluvial megafans, respectively, that have been eroded by seasonal inundation flows, probably due to the more humid conditions over the last 3000 years BP. The mineralogical processes triggered in the acid pedoenvironments caused by solodization are also similar, with the transformation of illitic into smectitic and kaolinitic mixed-layered minerals with progressive leaching.

Many features of the salt-affected soils of the RPPN were derived from past climatic, sedimentary, geomorphological and hydrological dynamics, such as most of the Na^+ accumulation and textural contrasts. In Nhecolândia, however, current evaporative concentration in the confined alkaline–saline lakes is the main responsible for the genesis of Saline Sodic soils, promoting the authigenic formation of clays typically originated in highly evaporative environments, such as Mg-smectite,

Fe-illite and glauconite. The neoformation of Fe-micas in the subsurface is probably provoking the main textural gradients in these soils, increasing the clay contents in the B natric horizons. It is important to highlight that even in low Nhecolândia, where salts are still accumulating, Sodic and degraded Sodic soils have been formed by leaching. More research about the salt-affected soils must be performed in order to understand if this salt balance is positive, with salt accumulation, or negative, with loss of salts, in the Pantanal wetland.

References

Adámoli J (1986) A dinâmica das inundações no Pantanal. Anais do 1° Simpósio sobre Recursos Naturais e Sócio-Econômicos do Pantanal. EMBRAPA, Corumbá, pp 51–61

Al-Droubi A, Fritz B, Gac JY, Tardy Y (1980) Generalized residual alkalinity concept; application to prediction of the chemical evolution of natural waters by evaporation. Am J Sci 280:560–572. https://doi.org/10.2475/ajs.280.6.560

Alfonsi RR, Camargo MBP (1986) Condições climáticas para a região do Pantanal matogrossense. Anais do 1° Simpósio sobre Recursos Naturais e Sócio-Econômicos do Pantanal. EMBRAPA, Corumbá, pp 29–42

Alho C (2011) Concluding remarks: overall impacts on biodiversity and future perspectives for conservation in the Pantanal biome. Braz J Biol 71:337–341. https://doi.org/10.1590/S1519-69842011000200013

Allen BL, Hajek B (1989) Mineral occurrence in soil environments. In: Dixon JB, Weed S (eds) Minerals in soil environments. Soil Science Society of America Inc. (SSSA), Madison, p 199–278

Almeida TIR, Karmann I, Paranhos Filho AC, Sígolo JB, Bertolo RA (2010) Os diferentes graus de isolamento da água subterrânea como origem de sua variabilidade: evidências isotópicas, hidroquímicas e da variação sazonal do nível da água no Pantanal da Nhecolândia. Geol USP Série Científica 10:37–47. https://doi.org/10.5327/Z1519-874X2010000300003

Almeida TIR, Sígolo JB, Fernandes E, Queiroz Neto JP, Barbiero L, Sakamoto AY (2003) Proposta de classificação e gênese das Lagoas da Baixa Nhecolândia-MS com base em sensoriamento remoto e dados de campo. Rev Bras Geoc 33:83–90. https://doi.org/10.25249/0375-7536.200333s28390

Alvarenga SM (1984) Estudo geomorfológico aplicado a Bacia do Alto Paraguai e Pantanais Mato-Grossense. Ministério das Minas e Energia, Projeto RADAMBRASIL, Salvador

Amaral Filho ZP (1986) Solos do Pantanal. In: Anais do 1° Simpósio sobre recursos naturais e sócio-econômicos do Pantanal. EMBRAPA, Corumbá, pp 91–103

Amouric M, Olives J (1998) Transformation mechanisms and interstratification in conversion of smectite to kaolinite: an HRTEM study. Clays Clay Miner 46:521–527. https://doi.org/10.1346/CCMN.1998.0460505

Andrade GRP, Furquim SAC, Vidoca T, Brito AC, Camargo G, Souza GC (2020) Transformation of clay minerals in salt-affected soils, Pantanal wetland, Brazil. Geoderma, 371:114380. https://doi.org/10.1016/j.geoderma.2020.114380

Assine M (2003) Sedimentação na bacia do Pantanal Mato-Grossense, Centro-Oeste do Brasil. Dissertation, Universidade Estadual Paulista (UNESP), Instituto de Geociências e Ciências Exatas, Rio Claro, 106 p

Assine ML, Merino ER, Pupim FN, Macedo HA, Santos MGM (2015) The quaternary alluvial systems tract of the Pantanal Basin, Brazil. Braz J Geol 45:475–489. https://doi.org/10.1590/2317-4889201520150014

Assine ML, Merino ER, Pupim FN, Warren LV, Guerreiro RL, McGlue MM (2016) Geology and geomorphology of the Pantanal Basin. In: Bergier I, Assine ML (eds) Dynamics of the Pantanal Wetland in South America. Springer International Publishing, Cham, pp 23–50

Assine ML, Padovani CR, Zacharias AA, Angulo RJ, Souza MC (2005) Compartimentação geomorfológica, processos de avulsão fluvial e mudanças de curso do rio Taquari, Pantanal Mato-Grossense. Rev Bras Geomorfol 6:97–108. https://doi.org/10.20502/rbg.v6i1.43

Barbiero L, Berger G, Rezende Filho AT, Meunier J, Martins-Silva ER, Furian S (2016) Organic control of dioctahedral and trioctahedral clay formation in an alkaline soil system in the Pantanal Wetland of Nhecolândia, Brazil. PLoS One 1–23. https://doi.org/10.1371/journal.pone.0159972

Barbiero L, Queiroz Neto JP, Ciornei G, Sakamoto A, Capellari B, Fernandes E, Valles V (2002) Geochemistry of water and ground water in the Nhecolândia, Pantanal of Mato Grosso, Brazil: variability and associated processes. Wetlands 22:528–540. https://doi.org/10.1672/0277-5212(2002)022[0528:GOWAGW]2.0.CO;2

Barbiero L, Filho AR, Furquim SAC, Furian S, Sakamoto AY, Valles V, Graham RC, Fort M, Ferreira RPD, Queiroz JPQ (2008) Soil morphological control on saline and freshwater lake hydrogeochemistry in the Pantanal of Nhecolândia, Brazil. Geoderma 148:91–106. https://doi.org/10.1016/j.geoderma.2008.09.010

Barbiero L, Furquim SC, Valles V, Furian S, Sakamoto A, Filho AR, Fort M (2007) Natural arsenic in groundwater and alkaline lakes at the Upper Paraguay Basin, Pantanal, Brazil. In: Bhattacharya P, Mukherjee AB, Bundschuh J, Zevenhoven R, Loeppert RH (eds) Trace metals and other contaminants in the environment, Elsevier, pp 101–126

Barbiero L, Queiroz Neto JP, Sakamoto AY (2000) Características geoquímicas dos solos relacionadas à organização pedológica e à circulação da água (Fazenda Nhumirim: Embrapa CPAP, Nhecolândia, MS). Anais do 3° Simpósio sobre recursos naturais e socioeconômicos do Pantanal. EMBRAPA, Corumbá, pp 90–100

Beirigo RM, Vidal-Torrado P, Stape JL, Couto EG, Andrade GRP (2010) Solos da Reserva Particular do Patrimônio Natural SESC Pantanal. Rio de Janeiro. SESC, 76 p

Berbery EH, Barros VR (2002) The hydrologic cycle of the La Plata Basin in South America. J Hydrometeorol 3:630–645. https://doi.org/10.1175/1525-7541(2002)003%3c0630:THCOTL%3e2.0.CO;2

Bohn HL, McNeal BL, O'Connor GA (1985) Soil chemistry, 2nd, 3rd edn. Wiley, New York

Bohn, HL, McNeal, BL, O'Connor, GA (2001) Soil chemistry. 3rd edn. JohnWiley and Sons, New York, p 322

Cardoso EL, Santos SA, Urbanetz C, Carvalho Filho A, Naime UJ, Silva MLN, Curi N (2016) Relação entre solos e unidades da paisagem no ecossistema Pantanal. Pesq Agrop Bras 51:1231–1240. https://doi.org/10.1590/s0100-204x2016000900023

Collischonn W, Tucci CEM, Clarke RT (2001) Further evidence of changes in the hydrological regime of the River Paraguay: part of a wider phenomenon of climate change? J Hydrol 245:218–238. https://doi.org/10.1016/S0022-1694(01)00348-1

Costa M, Telmer KH, Evans TL, Almeida TI, Diakun MT (2015) The lakes of the Pantanal: inventory, distribution, geochemistry, and surrounding landscape. Wetl Ecol Manag 23:19–39. https://doi.org/10.1007/s11273-014-9401-3

Couto EG, Oliveira VA (2010) The soil diversity of the Pantanal. In: Junk WJ, Silva CJ, Cunha CN, Wantzen KM (eds) The Pantanal of Mato Grosso: ecology, biodiversity and sustainable management of a large neotropical seasonal wetland. Pensoft Publishers, Sofia, pp 71–102

Couto EG, Oliveira VA, Beirigo RM, Oliveira Junior JC, Nascimento AF, Vidal-Torrado P (2017) Solos do Pantanal Matogrossense. In: Curi N, Ker JC, Novais RF, Vidal-Torrado P, Schaefer CEGR (eds) Pedologia. Solos dos biomas brasileiros, Sociedade Brasileira de Ciência do Solo, Viçosa, pp 303–352

Cunha NG (1980) Considerações sobre os solos da sub-região da Nhecolândia, Pantanal Mato-Grossense. Circular Técnica Embrapa 1:1–45

Del'Arco JO, Silva RH, Tarapanoff I, Freire FA, Pereira LGM, Souza SL, Luz DS, Palmeira RCB, Tassinari CCG (1982) Geologia da folha SE.21-Corumbá e parte da folha SE.20. In: RADAMBRASIL- Levantamento dos Recursos Naturais. Rio de Janeiro, pp 25–160

Dudek T, Cuadros J, Fiore S (2006) Interstratified kaolinite-smectite: nature of the layers and mechanism of smectite kaolinization. Am Mineral 91:159–170. https://doi.org/10.2138/am.2006.1897

Evans TL, Costa M (2013) Land cover classification of the lower Nhecolândia subregion of the Brazilian Pantanal Wetlands using ALOS/PALSAR, RADARSAT-2 and ENVISAT/ASAR imagery. Remote Sens Environ 128:118–137. https://doi.org/10.1016/j.rse.2012.09.022

Fanning DS, Fanning MCB (1989) Soil, morphology, genesis, and classification. Wiley, New York

Fernandes E (2000) Caracterização dos elementos do meio físico e da dinâmica da Nhecolândia (Pantanal Sulmatogrossense). Universidade de São Paulo (USP), Departamento de Geografia, São Paulo, Master dissertation, p 132p

Fernandes E (2007) Organização espacial dos componentes da paisagem da baixa Nhecolândia - Pantanal de Mato Grosso do Sul. Universidade de São Paulo (USP), Departamento de Geografia, São Paulo, Doctoral dissertation, 177 p

Fernandes E, Sakamoto AY, Queiroz Neto JP, Lucati HM, Capellari B (1999) Le "Pantanal da Nhecolândia" Mato Grosso: Cadre physique et dynamique hydrologique. Geogr Fis Din Quat 22:13–21

Freitas JG, Furquim SAC, Aravena R, Cardoso EL (2019) Interaction between lakes' surface water and groundwater in the Pantanal wetland Brazil. Environ Earth Sci 78:139. https://doi.org/10.1007/s12665-019-8140-4

Furian S, Martins ERC, Parizotto TM, Rezende Filho AT, Victoria RL, Barbiero L (2013) Chemical diversity and spatial variability in myriad lakes in Nhecolândia in the Pantanal wetlands of Brazil. Limnol Oceanogr 58:2249–2261. https://doi.org/10.4319/lo.2013.58.6.2249

Furquim SAC (2007) Formação de carbonatos e argilominerais em solos sódicos do pantanal sul-mato-grossense. Universidade de São Paulo (USP), Departamento de Geografia, São Paulo, Doctoral dissertation, p 222

Furquim SAC, Graham RC, Barbiero L, Queiroz Neto JP, Vidal-Torrado P (2010) Soil mineral genesis and distribution in a saline lake landscape of the Pantanal wetland, Brazil. Geoderma 154:518–528. https://doi.org/10.1016/j.geoderma.2009.03.014

Furquim SAC, Barbiero L, Graham RC, Queiroz Neto JP, Ferreira RPD, Furian S (2010) Neoformation of micas in soils surrounding an alkaline-saline lake of Pantanal wetland, Brazil. Geoderma 158:331–342. https://doi.org/10.1016/j.geoderma.2010.05.015

Furquim SAC, Graham RC, Barbiero L, Queiroz Neto JP, Vallès V (2008) Mineralogy and genesis of smectites in an alkaline-saline environment of Pantanal wetland, Brazil. Clays Clay Miner 56:579–595. https://doi.org/10.1346/CCMN.2008.0560511

Furquim SAC, Santos MA, Vidoca TT, Balbino MA, Cardoso EL (2017) Salt-affected soils evolution and fluvial dynamics in the Pantanal wetland, Brazil. Geoderma 286:139–152. https://doi.org/10.1016/j.geoderma.2016.10.030

Gardi C, Angelini M, Barceló S, Comerma J, Cruz Gaistardo C, Encina Rojas A, Jones A, Krasilnikov P, Brefin MLMS, Montanarella L, Muniz Ugarte O, Schad P, Vara Rodriguez MI, Vargas R (2014) Atlas de suelos de América Latina y el Caribe. Oficina de Publicaciones de la Unión Europea, Luxembourg, p 176

Gedroiz KK (1912) Colloidal chemistry as related to soil science. I. Colloidal substances in the soil solution. Formation of sodium carbonate in the soil. Alkali soil and saline soils. Zhurnal Opitnoi Agron 13:363–420

Gedroiz KK (1917) Saline soils and their improvement. Zhurnal Opitnoi Agron 18:122–140

Gedroiz KK (1925) Soil absorbing complex and the absorbed soil cations as a basis of genetic soil classification. Nosovka Agric Exp Stn 38

Godoi Filho JD (1986) Aspectos geológicos do Pantanal Mato-Grossense e de sua área de influência. Anais do 1° Simpósio sobre recursos naturais e socioeconômicos do Pantanal. EMBRAPA, Corumbá, pp 63–76

Hamilton S, Sippel SJ, Melack JM (1996) Inundation patterns in the Pantanal wetland of South America determined from passive microwave remote sensing. Arch Hydrobiol 137:1–23

Hamilton SK (1999) Potential effects of a major navigation project (Paraguay-Paraná Hidrovia) on inundation in the Pantanal floodplains. Regul Rivers Res Manage 15:289–299. https://doi.org/10.1002/(SICI)1099-1646(199907/08)15:4%3c289::AID-RRR520%3e3.0.CO;2-I

Hamilton SK (2002) Comparison of inundation patterns among major South American floodplains. J Geophys Res 107:8038. https://doi.org/10.1029/2000JD000306

IUSS Working Group WRB (2015) World reference base for soil resources 2014, update 2015 International soil classification system for naming soils and creating legends for soil maps. World Soil Resources Reports No. 106. FAO, Rome

Junk WJ, da Cunha CN, Wantzen KM, Petermann P, Strüssmann C, Marques MI, Adis J (2006) Biodiversity and its conservation in the Pantanal of Mato Grosso, Brazil. Aquat Sci 68:278–309. https://doi.org/10.1007/s00027-006-0851-4

Karathanasis AD, Hajek BF (1983) Transformation of smectite to kaolinite in naturally acid soil systems: structural and thermodynamic considerations. Soil Sci Soc Am J 47:158–163. https://doi.org/10.2136/sssaj1983.03615995004700010031x

Martins ERC (2012) Tipologias de lagoas salinas no Pantanal da Nhecolândia (MS). Universidade de São Paulo (USP), Departamento de Geografia, São Paulo, Doctoral dissertation, p 210

McBride M (1994) Salt-affected and swelling soils. In: McBride MB (ed) Environmental chemistry of soils. Oxford University Press, 18 p

McGlue MM, Silva A, Zani H, Corradini FA, Parolin M, Abel EJ, Cohen AS, Assine ML, Ellis GS, Trees MA, Kuerten S, Gradella FS, Rasbold GG (2012) Lacustrine records of Holocene flood pulse dynamics in the Upper Paraguay river watershed (Pantanal wetlands, Brazil). Quat Res 78:285–294. https://doi.org/10.1016/j.yqres.2012.05.015

Nascimento AF (2012) Relações pedologia-geomorfologia-sedimentologia no Pantanal Norte. Doctoral dissertation, Universidade de São Paulo (USP), Escola Superior de Agricultura Luiz de Queiroz (Esalq), Piracicaba, 201 p

Nascimento AF, Furquim SAC, Couto EG, Beirigo RM, Oliveira Júnior JC, Camargo PB, Vidal-Torrado P (2013) Genesis of textural contrasts in subsurface soil horizons in the Northern Pantanal-Brazil. Rev Bras Ciênc Sol 37:1113–1127. https://doi.org/10.1590/S0100-06832013000500001

Nascimento AF, Furquim SAC, Graham RC, Beirigo RM, Oliveira Junior JC, Couto EG, Vidal-Torrado P (2015) Pedogenesis in a Pleistocene fluvial system of the Northern Pantanal—Brazil. Geoderma 255–256:58–72. https://doi.org/10.1016/j.geoderma.2015.04.025

Oliveira Junior JC (2015) Processos Pedogenéticos atuais e pretéritos em solos alcalino-sódicos do Pantanal Norte. Doctoral dissertation, Universidade de São Paulo (USP), Escola Superior de Agricultura Luiz de Queiroz (Esalq), Piracicaba, p 185

Oliveira Junior JC, Beirigo RM, Chiapini M, Nascimento AF, Couto EG, Vidal-Torrado P (2017) Origin of mounds in the Pantanal wetlands: an integrated approach between geomorphology, pedogenesis, ecology and soil micromorphology. PLoS ONE 12:e0179197. https://doi.org/10.1371/journal.pone.0179197

Oliveira Junior JC, Chiapini M, Nascimento AF do, Couto EG, Beirigo RM, Vidal-Torrado P (2017b) Genesis and classification of Sodic soils in the Northern Pantanal. Rev Bras Ciênc Sol 41:1—19. https://doi.org/10.1590/18069657rbcs20170015

Oliveira Junior JC, Furquim SAC, Nascimento AF, Beirigo RM, Barbiero L, Valles V, Couto EG, Vidal-Torrado P (2019) Salt-affected soils on elevated landforms of an alluvial megafan, northern Pantanal, Brazil. CATENA 172:819–830. https://doi.org/10.1016/j.catena.2018.09.041

Oliveira Junior JC, Andrade GRP, Barbiero L, Furquim SAC, Vidal-Torrado P (2019b) Flooding effect on mineralogical and geochemical changes in alkaline-sodic soil system of northern Pantanal wetlands, Brazil. Eur J Soil Sci. https://doi.org/10.1111/ejss.12871

Orioli AL, Amaral Filho ZB, Oliveira AB (1982) Pedologia da Folha SE.21 Corumbá e parte da Folha SE.20. In: RADAMBRASIL (ed) Levantamento de Recursos Naturais, v 27. Projeto RADAMBRASIL, Rio de Janeiro, pp 25–321

Padovani CR (2010) Dinâmica espaço-temporal das inundações do Pantanal. Doctoral dissertation, Universidade de São Paulo (USP), Escola Superior de Agricultura Luiz de Queiroz (Esalq), Piracicaba, 175 p

Parizotto TM (2012) Estudo morfológico e hidroquímico de pequenas depressões na Nhecolândia, Pantanal, MS. Master dissertation, Universidade de São Paulo (USP), Departamento de Geografia, São Paulo, 142 p

Por FD (1995) The Pantanal of Mato Grosso (Brazil)—world's largest wetland. Springer, Dordrecht

Pott A, Silva JSV (2016) Terrestrial and aquatic vegetation diversity of the Pantanal wetland. In: Bergier I, Assine ML (eds) Dynamics of the Pantanal wetland in South America. Springer International Publishing, Cham, The Handbook of Environmental Chemistry, pp 111–131

Rengasamy P, Sumner M (1998) Processes involved in sodic behavior. In: Sumner ME, Naida R (eds) Sodic soils, distribution, properties, management, and environmental consequences, Oxford University Press, pp 35–50

Rezende Filho AT, Furian S, Victoria RL, Mascré C, Valles V, Barbiero L (2012) Hydrochemical variability at the Upper Paraguay Basin and Pantanal wetland. Hydrol Earth Syst Sci 16:2723–2737. https://doi.org/10.5194/hess-16-2723-2012

Ribeiro MR, Freire FJ, Montenegro AAA (2003) Solos halomórficos no Brasil: Ocorrência, gênese, classificação, uso e manejo sustentável. In: Curi N, Marques JJ, Guilherme LRG, Lima JM, Lopes AS, Alvarez VH (eds) Tópicos em ciência do solo. SBCS, Viçosa, pp 165–208

Rueda JRJ, Pessoti JES, Mattos JT (1998) Modelo para o estudo da dinâmica evolutiva dos aspectos fisiográficos dos pantanais. Pesq Agrop Bras 33:1763–1773

Ryan PC, Huertas FJ (2009) The temporal evolution of pedogenic Fe–smectite to Fe–kaolin via interstratified kaolin–smectite in a moist tropical soil chronosequence. Geoderma 151:1–15. https://doi.org/10.1016/j.geoderma.2009.03.010

Sakamoto AY (1997) Dinâmica hídrica em uma lagoa salina e seu entorno no pantanal da Nhecolândia: contribuição ao estudo das relações entre o meio físico e a ocupação, Fazenda São Miguel do Firme (MS). Universidade de São Paulo (USP), Departamento de Geografia, São Paulo, Doctoral dissertation, p 183

Schaetzl RJ, Anderson S (2005) Soils, genesis and geomorphology. Cambridge University Press, Cambridge

Schiavo JA, Dias Neto AH, Pereira MG, Rosset JS, Secretti ML, Pessenda LCR (2012) Characterization and classification of soils in the Taquari river basin—Pantanal region, state of Mato Grosso do Sul, Brazil. Rev Bras Ciênc Sol 36:697–708. https://doi.org/10.1590/S0100-068320 12000300002

Seidl AF, Silva JSV, Moraes AS (2001) Cattle ranching and deforestation in the Brazilian Pantanal. Ecol Econ 36:413–425. https://doi.org/10.1016/S0921-8009(00)00238-X

Silva J, Abdon M, Silva S, Moraes J (2011) Evolution of deforestation in the Brazilian Pantanal and surroundings in the timeframe 1976–2008. Geografia 36:35–55

Silva JSV, Abdon M (1998) Delimitação do Pantanal brasileiro e suas sub-regiões. Pesq Agropec Bras 33:1703–1711

Silva MH, Sakamoto A (2003) Perfis pedomorfológicos do Pantanal da Nhecolândia-MS: um estudo comparativo. In: XII Encontro Sul-Matogrossense de Geografia. Três Lagoas, pp 544–552

Silva MH, Sakamoto YA, Barbiero L, Queiroz Neto JP, Furian S (2004) Morfologia do solo de três topossequências na área da lagoa Salina do Meio, Fazenda Nhumirim, Pantanal da Nhecolândia, MS. In: IV Simpósio sobre Recursos Naturais e Sócio- Econômicos do Pantanal. Corumbá

Silva TC (1986) Contribuição da Geomorfologia para o conhecimento e valorização do Pantanal. Anais do 1° Simpósio sobre recursos naturais e socioeconômicos do Pantanal. EMBRAPA, Corumbá, pp 77–90

Sousa JB (2003) Caracterização e gênese de solos em ambientes de cordilheira e campo de inundação periódica da sub-região do pantanal de Poconé. Doctoral dissertation, Universidade Federal de Viçosa (UFV), Viçosa, Mato Grosso, p 84

Sparks DL (2003) Environmental soil chemistry, 2nd edn. Elsevier, San Diego

Sumner ME, Rengasamy P, Naidu R (1998) Sodic soils: a reappraisal. In: Sumner ME, Naidu R (eds) Sodic soils, distribution, properties, management, and environmental consequences. Oxford University Press, 3–17 p

Tardy Y, Bocquier G, Paquet H, Millot G (1973) Formation of clay from granite and its distribution in relation to climate and topography. Geoderma 10:271–284. https://doi.org/10.1016/0016-7061(73)90002-5

Tarifa JR (1986) O sistema climático do Pantanal: da compreensão do sistema à definição de prioridades de pesquisa climatológica. Anais do 1° Simpósio sobre recursos naturais e socioeconômicos do Pantanal. EMBRAPA, Corumbá, pp 9–27

USSL Staff (1954) Diagnosis and improvement of saline and alkali soils. Handbook 60. U. S. Department of Agriculture, United States Salinity Laboratory (USSL), Washington, 160 p

van Breemen N, Buurman P (2003) Soil formation, 2nd edn. Springer, Netherlands, Dordrecht

Velde B (1995) Origin and mineralogy of clays. Springer, Berlin Heidelberg, Berlin, Heidelberg

Westin FC (1953) Solonetz soils of Eastern South Dakota: their properties and genesis. Soil Sci Soc Am J 17:287–293. https://doi.org/10.2136/sssaj1953.03615995001700030026x

Whitney BS, Mayle FE, Punyasena SW, Fitzpatrick KA, Burn MJ, Guillen R, Chavez E, Mann D, Pennington RT, Metcalfe SE (2011) A 45 kyr palaeoclimate record from the lowland interior of tropical South America. Palaeogeogr Palaeoclimatol Palaeoecol 307:177–192. https://doi.org/10.1016/j.palaeo.2011.05.012

Zani H, Assine ML, Silva A, Corradini FA (2009) Redes de drenagem distributária e formas posicionais no megaleque do Taquari, Pantanal: uma análise baseada no MDE-SRTM. Rev Bras Geomorfol 10:21–28

Temperate Coastal Salt Marsh Soils—Effects of Grazing and Management Alternatives

Carla E. Di Bella, Adriana M. Rodríguez, Miguel A. Taboada, and Agustín A. Grimoldi

Abstract Salt marshes are important ecosystems that provide a wide range of ecosystem services. Temperate coastal salt marshes at Samborombón Bay have been subjected to increasing grazing intensity. Continuous grazing on this salt marsh during the cold-humid seasons (autumn, winter, and early spring) generated negative changes in soil and vegetation properties. At medium and low elevation level of the salt marsh, grazing increased soil salinity, associated with a higher proportion of bare soil and groundwater proximity (i.e., upward flow of salts from saline groundwater). At the high elevation level, grazing increased the proportion of creeping species, avoiding the presence of bare soil and soil salinization. Grazing changed vegetation composition and decreased forage quality due to a decrease of forage legumes (*Melilotusofficinalis*) and an increase in non-palatable dicotyledonous forbes (*Sarcocornia perennis*). This chapter proposes a new management alternative to improve forage quality and soil properties of the grassland. It consists in late summer patch burns followed by cattle exclusion during autumn and rotational stocking from winter to late summer. This grazing alternative could promote sustainable use of this salt marsh favoring both, its conservation and productive use.

Keywords Salt marshes · Grazing · Fauna · Samborombón Bay · Wetlands

---

C. E. Di Bella (✉) · A. A. Grimoldi
IFEVA-CONICET, Facultad de Agronomía, Universidad de Buenos Aires, Buenos Aires, Argentina
e-mail: dibella@agro.uba.ar

A. A. Grimoldi
e-mail: grimoldi@agro.uba.ar

C. E. Di Bella · A. M. Rodríguez · A. A. Grimoldi
Cátedra de Forrajicultura, Departamento de Producción Animal, Facultad de Agronomía, Universidad de Buenos Aires, Buenos Aires, Argentina
e-mail: arodrigu@agro.uba.ar

M. A. Taboada
Instituto de Suelos, Instituto Nacional de Tecnología Agropecuaria, Buenos Aires, Argentina
e-mail: mtaboada@agro.uba.ar

CONICET, Buenos Aires, Argentina

E. Taleisnik and R. S. Lavado (eds.), *Saline and Alkaline Soils in Latin America*, https://doi.org/10.1007/978-3-030-52592-7_13

1 General Characteristics of Samborombón Bay Salt Marshes

Wetlands and coastal ecosystems (e.g., salt marshes, mangroves, coral reefs, seagrass beds, etc.) provide a wide range of ecosystem services that contribute to human well-being and ecological functions such as: coastal erosion protection and pollution control, water supply, climate regulation, recreational opportunities, habitat for characteristic fauna and flora, among others (Millennium Ecosystem Assessment 2005; Barbier et al. 2011). Among these environments, salt marshes are intertidal grasslands that are formed in low-energy, wave-protected shorelines along continental margins, and therefore, they are regulated by physical processes such as flooding and salinity (Barbier et al 2011). In relation to other types of wetlands and coastal ecosystems, salt marshes are those most affected by human activity. In fact, almost 50% of salt marshes in the world are lost or degraded worldwide due to the increasing intensity of human activity (Barbier et al. 2011).

Anthropogenic disturbances, such as grazing by domestic ungulates, have contrasting impacts on salt marshes around the world. Grazing by domestic herbivores may be beneficial because this activity increases floristic diversity by decreasing the cover of the dominant species with respect to abandoned or ungrazed salt marshes (Bakker 1985; Bos et al. 2002). Alternatively, in other salt marshes, grazing appears to be detrimental because selective and intense defoliation and trampling are considered responsible for the reduction in rare species and the inability of vegetation to regenerate or reestablish (Laegdsgaard 2006). Besides, when grazing increases bare soil, its effect on soil salinity does not show conclusive results. In some cases, grazing increased bare soil and promoted soil salinity, due to higher evaporation and the consequent rise of salts to the surface (Yu and Chmura 2009). However, in other salt marshes, grazing increased bare soil but not soil salinity (Bakker 1985). Pennings and Bertness (1999) predicted that in US salt marshes, salinity might imply greater importance in those at low-latitude rather than in high-latitude salt marshes. This would be the result of the higher temperatures and increased evaporation of the former, which can be ameliorated by plant shade (or increase basal cover). The effects of cattle and sheep grazing on the structure and functioning of Europe, North America, and Australian salt marshes have been extensively studied. On the contrary, in salt marshes of South America, with a different vegetation composition and climatic conditions, the study of grazing effects is scarce and limited to native small herbivores, without focusing on the impact of grazing on vegetation structure and soil salinity (Jackson and Giulietti 1988; Bortolus and Iribarne 1999; Costa et al. 2003; Cardoni et al. 2007; Vila et al. 2008).

The temperate salt marshes of Samborombón Bay, located in the eastern limit of Argentinean Pampas grasslands, extend along 2440 km^2 from Punta Piedra (35° 27′ S, 56° 45′ W) to Punta Rasa (36° 17′ S, 56° 46′ W) (Fig. 1). The climate is sub-humid to humid, and mesothermal, with scarce to null water deficiency. The annual average temperature and precipitation are 15 °C and 1000 mm, respectively. The relief is plain, with a slope of 0.01% and an average altitude of 1.6 m above

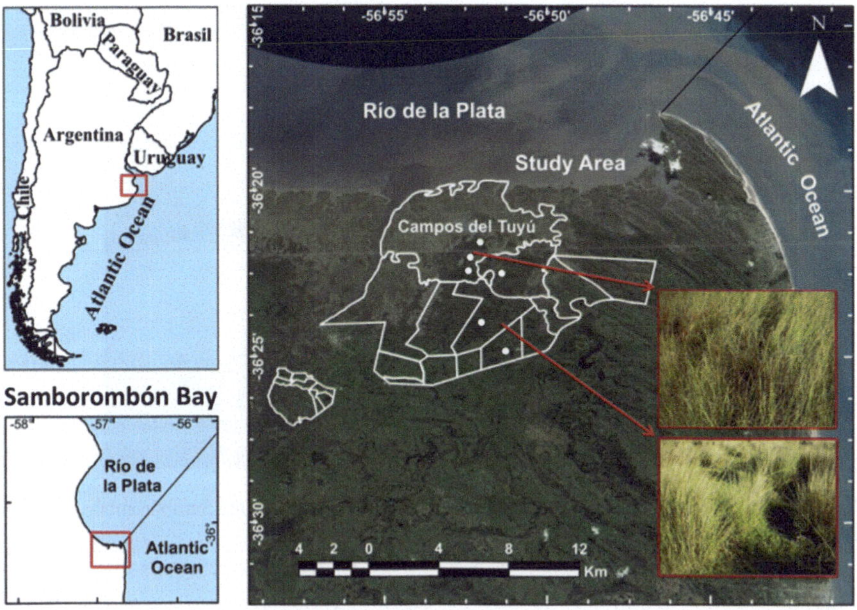

Fig. 1 Location map and characteristics of the study area. The area where the transects were located is represented by dots. Insets show the vegetation structure at the higher elevation level of the ungrazed "Campos del Tuyu" National Park (upper photograph) and at the adjacent livestock farms (lower photograph) (In Di Bella et al. 2015). The detailed area of the Samborombón Bay salt marsh can be observed in Isacch et al. (2006)

mean sea level (Carol et al. 2008). Tides are mixed (two high and two low tides per tidal day), predominantly semi-diurnal, with tidal ranges lower than 2 m. The tidal wave coming from the southern Atlantic enters, and the numerous tidal channels distributed all over the area (Carol et al. 2009). Floods are caused by the periodic movements of tides (Fig. 2), and the frequency of flooding is variable along the elevation gradient. Floods occur daily at the low elevation level, while they are less frequent at the medium and high elevation levels. Soil salinity depends on the depth of the water table and on the groundwater salt concentration, resulting from salt dissolution and accumulation processes (Carol et al. 2009). In this salt marsh, tidal water (surface water) receives the contribution of inland freshwater through artificial channels (see Chap. 10 by Imbellone et al., this book), and therefore, the salt content is lower than that of the groundwater (Carol et al. 2009). In this study, tidal water had an average electrical conductivity of 26 ± 0.3 dS m^{-1}, and groundwater could reach 44 dS m^{-1} (Carol et al. 2009).

Dominant soils are clayey and smetitic Vertisols (US Soil Taxonomy), which have an aquic regime, remaining wet almost all year round (Damiano 2009; see also the chapter by Imbellone et al., in this book: "Origin, Properties and Management of salt-affected soils in the Flooding Pampas, Argentina"). The depth of the surface horizon (horizon A) is approximately 15 cm, followed by a strong clay horizon B.

Fig. 2 Image of the tidal channels under high (**a**) and low tide (**b**) in the Samborombón Bay salt marsh (*Photo* Carla E. Di Bella)

The most extended plant community of these salt marshes is the *Spartina densiflora* community, which covers more than 50% of the area (Vervoorst 1967; Isacch et al. 2006). This community is dominated by *S. densiflora* in association with *Sarcocornia perennis*, *Juncus acutus*, *Cortaderia selloana*, and other species in minor proportions as *Apium sellowianum*, *Limonium brasiliense*, *Distichlis spicata*, and *Malvella leprosa* (Cagnoni and Faggi 1993). *S. densiflora* communities occupy the high, middle, and low elevation levels (Bortolus 2006). *S. densiflora* is a native, rhizomatous, and tussock growing grass (up to 1.5 m high). It has a C_4 photosynthetic regime so its higher growing rates occur during the summer, while in winter, its growth is minimum or null. In those communities where *S. densiflora* cover is higher than 85%, aerial net primary productivity (ANPP) can reach 2300–4000 kg DM ha^{-1} $year^{-1}$ in dry and humid years, respectively. It has a low forage quality, since its digestibility ranges 48–59%, and the crude protein content is 6–8% (Jacobo et al. 2015). Two other communities present in these salt marshes are the mono-specific community of *Spartina alterniflora* and the mono-specific community of *Sarcocornia perennis*, located in the lower intertidal zones exposed to longer flood periods (Cagnoni 1999).

Samborombón Bay has been declared wetland of international importance (Ramsar Convention Secretariat 2011), “WildLife Shelter” (Buenos Aires Provincial Government Decree Law No. 12016/97), “Valuable Grassland Area” and “Relevant Area for Bird Conservation” (Bilenca and Miñarro 2004), due to the high biological diversity and because it is a valuable habitat for migratory birds and for the conservation of the endangered Pampas deer (*Ozotoceros bezoarticus celer*). For this reason, it is of great importance to promote the conservation of this area through the rational and sustainable use of its resources. These salt marshes have been sporadically grazed at a low stocking rate due to the low forage quality or palatability. But, in the last decade, the expansion of agriculture generated a displacement of livestock, increasing grazing pressure on these marginal areas. This chapter describes the current livestock management of these temperate coastal salt marshes, a brief overview of its effects on soil and vegetation properties, and finally, a sustainable grazing management is proposed for this system.

2 Current Grazing Management on Samborombón Bay Salt Marshes

Despite the conservation status of Samborombón Bay, less than 15% of its area corresponds to natural reserves (or National Parks). The rest of the area corresponds to private lands used mainly for cattle breeding. These farmlands lack subdivisions (i.e., paddocks) and artificial water-points (windmills). For this reason, the stocking method is mainly "seasonal" (according to Allen et al. 2011; i.e., a method to restrict use of a land unit(s) to one or more specific seasons of the year) during the cold-humid seasons (autumn, winter, and early spring, or from March to October), when natural water-points are formed with the accumulation of rainwater. During those months, "continuous" stocking method (i.e., a method of grazing on a specific unit of land where livestock have unrestricted and uninterrupted access throughout the time when grazing is allowed) is performed. The installation of artificial water-points by windmills is not a viable option due to the salinity content of the groundwater, making it unsuitable for animal consumption. Therefore, the absence of grazing during the warm season, when *S. densiflora* has the highest growth rates, favors the dominance of this species, decreasing floristic diversity and accentuating the loss of forage quality of the grassland. Consequently, cattle consume during winter the dry biomass generated during the summer, which is of very low quality. To avoid this situation, a common management strategy performed by farmers is the burning of "espartillares" (i.e., grasslands dominated by *S. densiflora*) in late summer, to remove dry biomass and then promote the re-growth of *S. densiflora* (i.e., better quality) and the establishment of winter species. As a consequence of this practice, Pampa's deer migrate from the "Campos del Tuyu" National Park (where no disturbance is made) to neighboring cattle fields, to take advantage of the better forage quality. This situation, however, exposes them to poaching. In the following section, we are presenting more consequences of this stoking method used in this salt marsh on soil and vegetation properties.

3 Grazing Effects on Soil Properties and Vegetation Composition on Samborombón Bay Salt Marshes

Studies were conducted in the south of the Samborombón Bay (General Lavalle, Province of Buenos Aires, Argentina, 36° 40′ S, 56° 96′ W; Fig. 1) to assess the effect of grazing (according to the stocking method described in the previous section, i.e., seasonal-continuous stocking method) on soil and vegetation properties along a topographic gradient (Di Bella et al. 2014, 2015). Ungrazed sites were located within the "Campos del Tuyú" National Park (3000 ha, 36°19′ S, 56°50′ W), where domestic grazing has been excluded since 1979; while the grazed sites were located on an adjacent livestock farm (1650 ha), where continuous stocking method is performed at stocking rates that have been increasing in the last decades, reaching 0.6 animals ha^{-1} at present. In both sites, three different ground elevation levels were identified: (i) high elevation level at 0.80 m above sea level (masl), flooded only twice a year, during the vernal and autumnal equinoxes; (ii) medium elevation level at 0.70 masl, flooded once or twice a month, when there is a new or full moon; and (iii) low elevation level at 0.60 masl, flooded once or twice a day. Groundwater depth is around 0.50, 0.40, and 0.30 m at each elevation level, respectively.

It was found that soil salinity (i.e., superficial electrical conductivity, 0–15 cm) was greater on grazed sites at the medium and low elevation levels in comparison with the ungrazed sites (Elevation x Grazing interaction, $P = 0.002$; Fig. 3). Besides, it was significantly greater in summer than in other seasons at all elevation levels (rmANOVA, $P_{time} = 0.001$; Fig. 3). This increase in soil salinity can be explained by the upward flow of salts from saline groundwater (located at <40 cm depth at the medium and lower elevation level of the study site), driven by the increase of bare soil on grazed sites (Table 1; Lavado and Taboada 1987; Alconada et al. 1993; Yu and Chmura 2009). On the higher elevation, soil salinity did not increase as compared with the other elevation levels, probably due to the deeper groundwater and the lower bare soil proportion (Table 1). Similar results were also found in low-latitude salt marshes on both US coasts, where salinity levels reach a peak in the middle marsh because of evaporation processes, and salinities may reach higher levels than those found in sea water (Pennings and Bertness 1999, 2001). In contrast, in high-latitude salt marshes, salinity tends to decline from the low to the high marsh, and it is rarely greater than levels found in seawater (Pennings and Bertness 1999). In contrast to this, in our temperate salt marsh, the proximity of saline groundwater and the greater proportion of bare soil on grazed sites, in comparison with the ungrazed sites, would be the main factors driving soil salinity increases on the medium and low elevations, showing a high vulnerability to soil salinization. Below 15 cm soil depth, salinity was greater on grazed than on ungrazed sites at all elevation levels, with salinity values also greater in summer than in the other seasons ($P_{grazing} = 0.017$, rmANOVA $P_{time} < 0.001$; Fig. 3).

Grazed sites, regardless of the elevation level, presented negative effects on soil physical properties. Surface compaction (0–15 cm; measured as soil bulk density) in grazed sites increased (Di Bella et al. 2015). Similar results were also found in

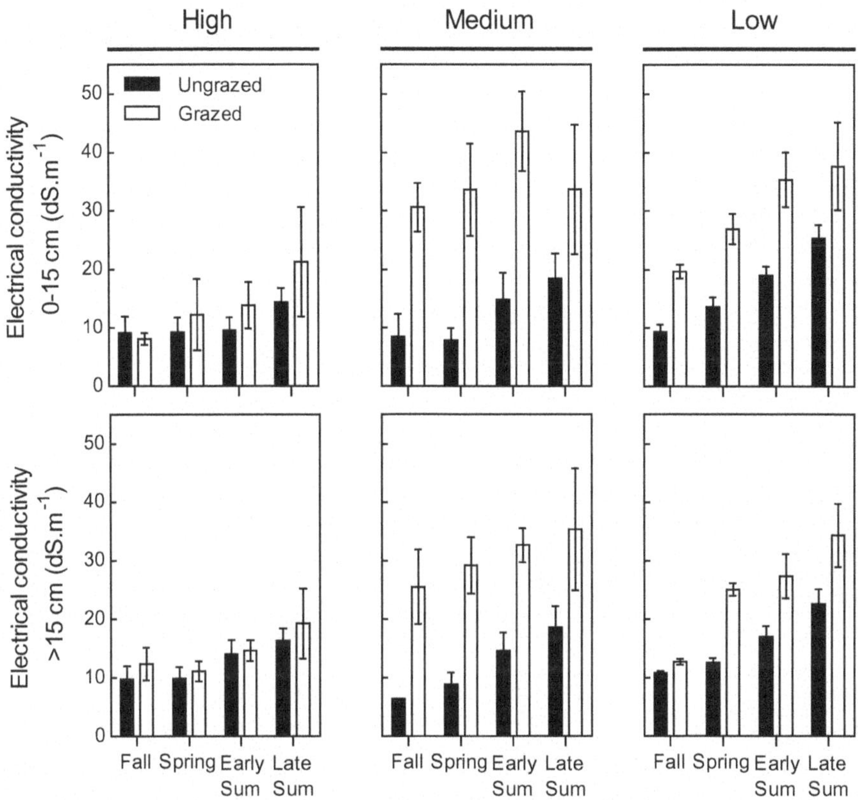

Fig. 3 Soil electrical conductivity in the 0–15 cm (upper panel) and >15 cm (lower panel) soil layers at high, medium, and low elevation levels of grazed (black bars) and ungrazed (white bars) sites of the salt marsh. Samples were taken in fall, spring, early summer (Early Sum), and late summer (Late Sum). Values are means ± SE ($n = 3$). (Adapted from Di Bella et al. 2015)

soils from different geographical areas and different intrinsic properties (Greenwood and McKenzie 2001; Taboada et al. 2011) as well as in other salt marshes (Elschot et al. 2013; Nolte et al. 2013). However, despite the significant increase in soil bulk density in the grazed sites of this study, the values obtained were lower than those considered critical for plant growth in clay soils (>1.3 Mg m^{-3}; Álvarez and Taboada 2008), when root penetration is likely to be severely restricted. Gravimetric water content decreased in grazed sites, especially during fall and spring (rmANOVA, time x grazing, $P < 0.001$; Fig. 4). This result coincides with those found in other grasslands in the world (Greenwood and McKenzie 2001). However, in these salt marshes, the decrease in soil moisture of grazed sites is leading to humidity levels below field capacity (defined from the equivalent moisture value calculated for the soils in the area: 57.7%). This, added to the greater hardness of the soils, could generate problems in the establishment and growth of propagules. In this sense, the observed changes in the values of soil salinity, soil bulk density, and gravimetric water content, in

Table 1 Basal cover of the main species (%), bare soil (%), forage quality index[a], richness, and diversity, for each elevation level (high, medium, and low) of the salt marsh, in ungrazed and grazed sites[b]

Species	High			Medium			Low		
	Ungrazed	Grazed		Ungrazed	Grazed		Ungrazed	Grazed	
Warm season (C_4) tussock grasses									
Spartina densiflora	33.6	20.4	**	38.6	40.4		58.8	48.6	
Warm season (C4) creeping grasses									
Stenotaphrum secundatum	0.0	8.0	**						
Cool season legumes									
Melilotusofficinalis (E)	7.0	2.6		6.6	0.3	**	1.0	0.0	
Warm season dicotyledonous forbs									
Phyla canescens	0.0	8.5	**						
Centaurium pulchellum (E)	0.3	8.0	**	4.3	0.0	**	1.6	0.0	
Sarcocornia perennis (E)	7.0	2.7		9.3	29.4	**	8.9	14.6	*
Bare soil	1.5	5.4	B	1.7	6.9	B	20.7	30.8	A
Forage quality index	1.94a	1.37b	A	1.58a	1.34b	B	1.40a	1.18b	B
Richness	16	28	A	12	9	B	5	3	B
Diversity(Simpson's Index)	0.79	0.88	A	0.75	0.66	A	0.36	0.44	B

Adapted from Di Bella et al. 2014

[a]The forage quality index was calculated as the average of each species-specific quality value (qi) weighted by its relative contribution in the site, according to Cahuepé et al. (1985)

[b]Values are means of each treatment. Statistical significance between grazing treatments is indicated with $^{**}P < 0.01$ and with $^{*}P < 0.05$ according to the indicator species analysis and Monte Carlo test. E: exotic species. For bare soil, forage quality index, richness, and diversity results, different uppercase letters indicate significant differences among elevation levels (elevation level effect $P < 0.05$) and different lowercase letters indicates significant differences between treatments within each elevation level (treatment effect $P < 0.05$) according to HSD Tukey's test

grazed sites, could act as important drivers of the vegetation composition in these salt marshes.

The effect of grazing on vegetation composition depends, on the one hand, on the cattle selection of palatable species, and on the other hand, on the indirect effects of grazing on soil properties, which can modify the establishment and growth of some species. The effect of grazing differs among elevation levels of the salt marsh (Di Bella et al. 2014). At the high elevation level, grazing changed floristic composition and tended to increase floristic richness (i.e., number of species) and diversity. Defoliation by cattle released the competition exerted by the dominant species (i.e.,

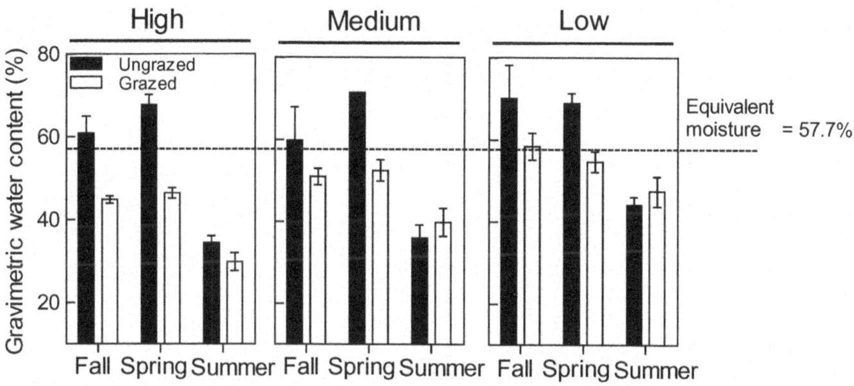

Fig. 4 Soil gravimetric water content at each elevation level (high, medium, and low), of ungrazed (black bars) and grazed (white bars) sites of the salt marsh. Samples were taken in fall, spring, and summer. Values are means ± SE ($n = 3$). Dotted line indicates the equivalent moisture value of "Bahia de Samborombón" soil series

Spartina densiflora) and allowed the replacement by shorter and creeping species that avoid grazing and have lower forage quality (i.e., *Phyla canescens*, *Stenotaphrum secundatum*; Table 1). A similar response was found in high-latitude salt marshes of the Northern Hemisphere, where cattle and sheep grazing shifted floristic composition and increased species richness and diversity due to the reduction of the dominant tall species (Bakker 1985; Bos et al. 2002). The rapid installation of creeping species observed in this salt marsh avoids the increase of bare soil (only 5.4%), which together with the greater depth of the water table prevented salt rise to the surface (Fig. 3). Grazing also reduced forage quality at the high elevation level due to a decrease in the basal cover of *Melilotusofficinalis* (cool season forage legume highly selected by livestock; Table 1). At medium elevation level, grazed sites presented a higher proportion of bare soil which can also promote an increase in soil salinity (Table 1 and Fig. 3). The increase in soil salinity together with cattle selection favored the decrease of high quality forage species, as *M. officinalis*, and an increase of low quality species, although more salt tolerant, such as *Sarcocornia perennis* (Table 1). At the low elevation level, the higher soil salinity in grazed sites did not change the vegetation composition (Table 1), because at that elevation level the species present are already tolerant to high levels of salinity (i.e.,*S. densiflora* and *S. perennis*).

M. officilalis is almost the only cold season forage legume able to grow in this restricted environment. The importance of *M. officinalis* in the salt marsh lies in the ability of legumes to fix atmospheric nitrogen, which can potentially increase soil nitrogen (Bruning et al 2015). Besides, the high tissue protein content improves the forage quality of the grassland (Phelan et al. 2015). Grazing on Samborombón Bay salt marsh eliminated *M. officinalis* from the soil seed bank and hinders its establishment at the high elevation level (Di Bella et al. 2010). In grazed sites, no seeds of *M. officinalis* were found in the soil seed bank (Table 2). This coincides with the lower basal cover of adult plants of this species at the high levels of grazed salt

Table 2 *Melilotusofficinalis* seeds present in the soil seed bank and number of seedlings established after a month of sowing at the high elevation level of the salt marsh, in ungrazed and grazed sites subjected to burned or non-burned treatments (Adapted from Di Bella et al. 2010)

Treatments	*M. officinalis* (seeds m^{-2})	*M. officinalis* (seedlings m^{-2})
Ungrazed—Non-burned	1207	–
Ungrazed—Burned	555	209
Grazed—Non-burned	–	–
Grazed—Burned	–	–

marsh (Table 1). Instead, in ungrazed sites, 500 and >1000 seeds m^{-2} were found in burned and non-burned sites, respectively (Table 2). Besides, one month after *M. officinalis* was sown, established plants were only registered in ungrazed and burned sites (Table 2). These results suggest that grazing stocking method used in this salt marsh would negatively affect *M. officinalis* seed production by repeated grazing of this palatable species, considering that cattle stays in the field from March to October, coinciding with the growth period of this species. Besides, trampling would affect germination and seedling establishment by compacting the soil and increasing its salinity (Fig. 3). In ungrazed sites, with better soil physical and chemical conditions, a disturbance such as fire (or patch burns) is necessary to open the canopy and allow germination and establishment of this important legume.

All this information allowed us to develop a conceptual model summarizing the main results obtained in relation to the effect of grazing on soil and vegetation properties in this temperate coastal salt marsh (Fig. 5). In this conceptual model, grazing increases salinity at the medium and low elevation levels and changes vegetation

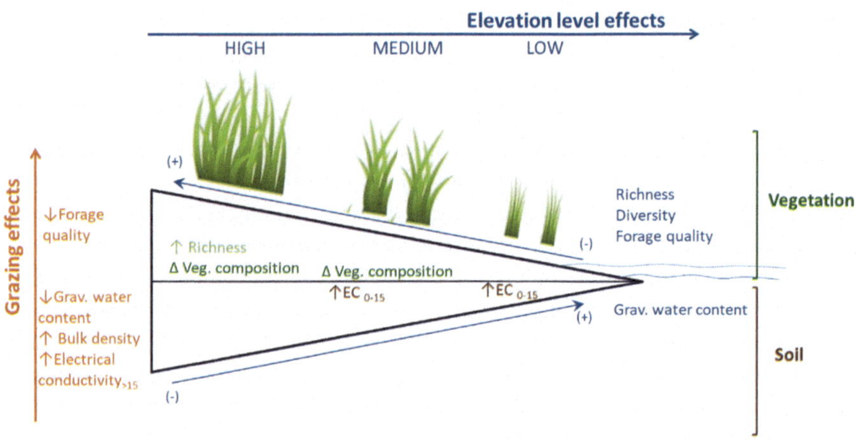

Fig. 5 Grazing (left) and elevation level (right) effects and the interaction (inside inverted triangle) scheme, on soil physical and chemical properties (below) and on vegetation (above). EC_{0-15}: Electrical conductivity of saturation extract for the first 15 cm of soil

composition at the high and medium elevation level (Fig. 5; inside the inverted triangle). Besides, grazing increases soil surface compaction (i.e., soil bulk density), decreases soil gravimetric water content (Fig. 4), increases soil subsurface salinity (>15 cm; Fig. 3), and decreases forage quality, regardless of the elevation level. This decrease in forage quality is related with cattle selectivity and the loss of palatable species as *M. officinalis*, and the increase of less palatable species as *S. secundatum*, *P. canescens* y *C. pulchellum* (at the medium elevation level), and *S. perennis* (at the medium and low elevation level; Table 1. In addition, in general, it is observed that toward the low elevation level, the water content of the soil increases while richness, diversity, and forage quality decrease, both in grazed and ungrazed sites (Fig. 5 and Table 1). This is due to the fact that toward the low part of the salt marsh, the environmental conditions become more stressful (greater frequency of flooding and higher salinity levels), which determine that only a small group of species, tolerant to these conditions, can be established and grown (i.e., species with lower forage quality but more stress-tolerant).

4 Cattle Management Alternatives for Samborombón Bay Salt Marshes

A sustainable management strategy at the high elevation level to avoid salt marsh degradation consists of late summer patch burns, followed by cattle exclusion during autumn and rotational stocking method from winter to late summer,[1] resulting in:

1. Elimination of the low digestibility biomass of *S. densiflora*, accumulated during summer, by late summer patch burns. This disturbance allows the germination of *Melilotus* sp. and other C_3 grasses of high forage value (Rodríguez et al. 2016), when seeds are available in the seed-bank.
2. Rest from grazing during fall ensures the establishment of *Melilotus* sp. and other C_3 grasses seedlings and prevents soil compaction and poaching, which occur under water-saturated condition, a frequent situation during the autumn wet period.
3. Increased forage offer of higher digestibility in winter, due to the contribution of *Melilotus* sp. and other C_3 grasses, and the new green tissues of *S. densiflora* in spring.
4. Application of a grazing frequency equal to or lower than 60 days during spring and summer to maintain *S. densiflora* plants at a low senescence rate (Jacobo et al. 2015; Rodríguez et al. 2016). This grazing frequency reduces the accumulation of senescent biomass and the requirement of burning each late summer.

[1]To perform this management strategy, electric fences to separate high from medium and low elevation level and the access to freshwater (especially during spring and summer) in the paddocks is needed. The access to freshwater was adopted by some farmers through water pumping from a near non-saline source and the transfer to the paddocks by pipes.

5. Patch burning causes a structural vegetation heterogeneity of short and tall patches, with young and green leaves or old, tall, and senescent biomass. This burning practice provides a better fire control, because small areas are burnt. The tall patches provide wind protection and shadow to cattle, in this community without trees, as well as shelter and habitat to wild mammals and birds. Short patches offer high nutritive forage both to domestic and wild herbivores.

Patch burn is not recommended at the medium elevation level because the bare soil promotes soil salinization. Conversely, rotational stocking method would maintain the vegetation cover and avoid soil salinization. Low elevation level areas are a relevant habitat for wild fauna, like crabs and birds. Grazing is not recommended because of the high risk of soil denudation and salinization.

5 Conclusions

In the temperate coastal salt marshes located in the Samborombón Bay, seasonal and continuous stocking method has a negative effect on soil and vegetation traits. Negative effects on soil properties, associated with increased salinity, were observed at the medium and low elevation levels. These effects may have been magnified by the higher temperatures present in these temperate salt marshes with respect to cold salt marshes (located in colder latitudes), where the salt rising process does not usually occur under grazing conditions (Pennings et al. 2005). Besides, grazing promotes changes in the floristic composition in detriment of the forage quality throughout the topographic gradient. These facts demonstrate the great susceptibility of these coastal salt marshes to the current stocking method and the need for new management alternatives, tending to reverse this deterioration situation and improve their use from the productive and conservation point of view.

References

Alconada M, Ansin OE, Lavado RS, Deregibus VA, Rubio G, Gutierrez VH (1993) Effect of retention of run-off water and grazing on soil and on vegetation of a temperate humid grassland. Agric Water Manag 23:233–246

Allen VG, Batello C, Berretta EJ, Hodgson J, Kothmann M, Li H, McIvor J, Milne J, Morris C, Peeters A, Sanderson M (2011) An international terminology for grazing lands and grazing animals. Grass Forage Sci 66:2–28. https://doi.org/10.1111/j.1365-2494.2010.00780.x

Álvarez CR, Taboada MA (2008) Indicadores de la fertilidad física del suelo. In: Taboada MA, Álvarez CR (eds) Fertilidad física de los suelos, 2° edición. Editorial Facultad de Agronomía Universidad de Buenos Aires, pp 180–206

Bakker JP (1985) The impact of grazing on plant communities, plant populations and soil conditions on salt marshes. Plant Ecol 62:391–398

Barbier EB, Hacker SD, Kennedy C, Koch EW, Stier AC, Silliman BR (2011) The value of estuarine and coastal ecosystem services. Ecol Monogr 81:169–193. https://doi.org/10.1890/10-1510.1

Bilenca D, Miñarro F (2004) Identificación de Áreas Valiosas de Pastizal (AVPs) en las Pampas y campos de Argentina. Uruguay y sur de Brasil, Fundación Vida Silvestre, Buenos Aires, Argentina

Bortolus A (2006) The austral cordgrass *Spartina densiflora* Brong.: its taxonomy, biogeography and natural history. J Biogeogr 33:158–168. https://doi.org/10.1111/j.1365-2699.2005.01380.x

Bortolus A, Iribarne O (1999) Effects of the SW Atlantic burrowing crab *Chasmagnathus granulata* on a *Spartina* salt marsh. Mar Ecol Prog Ser 178:79–88. https://doi.org/10.3354/meps178079

Bos D, Bakker JP, de Vries Y, van Lieshout S (2002) Long-term vegetation changes in experimentally grazed and ungrazed back barrier marshes in the Wadden Sea. Appl Veg Sci 5:45–54

Bruning B, van Logtestijn R, Broekman R, de Vos A, Andrés Parra González A, Rozema J (2015) Growth and nitrogen fixation of legumes at increased salinity under field conditions: implications for the use of green manures in saline environments. AoB Plants 7. https://doi.org/10.1093/aobpla/plv010

Cagnoni M (1999) Espartillares de la costa bonaerense de la República Argentina. Un caso de humedales costeros. In: Málvárez A (ed) Tópicos sobre humedales subtropicales y templados de Sudamérica. Programa sobre el Hombre y la Biósfera de la UNESCO. Montevideo, Uruguay, pp 551–567

Cagnoni MA, Faggi A (1993) La Vegetación de la reserva de Vida Silvestre Campos del Tuyú. Parodiana 101–112

Cahuepé MA, Hidalgo MG, Galatoire A (1985) Aplicación de un índice de valoración zootécnica en pastizales de la Depresión del Salado. Rev Argentina Prod Anim 5:681–690

Cardoni DA, Isacch JP, Iribarne OO (2007) Indirect effects of the intertidal burrowing crab *Chasmagnathus granulatus* in the habitat use of Argentina's South West Atlantic salt marsh birds. Estuaries Coasts 30:382–389. https://doi.org/10.1007/BF02819385

Carol ES, Kruse E, Pousa J (2008) Environmental hydrogeology of the southern sector of the Samborombon Bay wetland, Argentina. Environ Geol 54:95–102. https://doi.org/10.1007/s00254-007-0796-5

Carol ES, Kruse E, Mas-Pla J (2009) Hydrochemical and isotopical evidence of ground water salinization processes on the coastal plain of Samborombón Bay, Argentina. J Hydrol 365:335–345. https://doi.org/10.1016/j.jhydrol.2008.11.041

Costa CSB, Marangoni JC, Azevedo MG (2003) Plant zonation in irregularly flooded salt marshes: relative importance of stress tolerance and biological interactions. J Ecol 91:951–965. https://doi.org/10.1046/j.1365-2745.2003.00821.x

Damiano F (2009) Estudio de caso: ordenamiento ambiental de la planicie costera en el partido de General Lavalle. In: Taboada M, Lavado RS (eds) Alteraciones de la Fertilidad de los Suelos: El halomorfismo, la acidez, el hidromorfismo y las inundaciones. Editorial Facultad de Agronomía, Buenos Aires, pp 185–215

Di Bella CE, Borghetti S, Jacobo E, Rodríguez AM (2010) ¿Qué pasó con *Melilotus officinalis* en los humedales costeros del sur de la Bahía de Samborombón (Argentina)? IV Reunión binacional de Ecología, FCEN, UBA, Buenos Aires

Di Bella CE, Jacobo E, Golluscio RA, Rodríguez AM (2014) Effect of cattle grazing on soil salinity and vegetation composition along an elevation gradient in a temperate coastal salt marsh of Samborombón Bay (Argentina). Wetl Ecol Manag 22:1–13. https://doi.org/10.1007/s11273-013-9317-3

Di Bella CE, Rodríguez AM, Jacobo E, Golluscio RA, Taboada MT (2015) Impact of cattle grazing on temperate coastal salt marsh soils. Soil Use Manag 31:299–307. https://doi.org/10.1111/sum.12176

Elschot K, Bouma TJ, Temmerman S, Bakker JP (2013) Effects of long-term grazing on sediment deposition and salt-marsh accretion rates. Estuar Coast Shelf Sci 133:109–115. https://doi.org/10.1016/j.ecss.2013.08.021

Greenwood K, McKenzie B (2001) Grazing effects on soil physical properties and the consequences for pastures: a review. Aust J Exp Agric 41:1231–1250. https://doi.org/10.1046/j.1523-1739.2002.99536.x

Isacch JP, Costa CSB, Rodríguez-Gallego L, Conde D, Escapa M, Gagliardini DA, Iribarne OO (2006) Distribution of saltmarsh plant communities associated with environmental factors along a latitudinal gradient on the south-west Atlantic coast. J Biogeogr 33:888–900. https://doi.org/10.1111/j.1365-2699.2006.01461.x

Jackson JE, Giulietti J (1988) The food habits of Pampas deer *Ozotoceros bezoarticus celer* in relation to its conservation in a relict natural grassland in Argentina. Biol Conserv 45:1–10. https://doi.org/10.1016/0006-3207(88)90048-1

Jacobo EJ, Rodriguez AM, Fariña CM, Paggi Y (2015) Tidal suppression negatively affects soil properties and productivity of *Spartina densiflora* salt marsh. Rangel Ecol Manag 68:276–284. https://doi.org/10.1016/j.rama.2015.03.005

Laegdsgaard P (2006) Ecology, disturbance and restoration of coastal saltmarsh in Australia: a review. Wetl Ecol Manage 14:379–399

Lavado RS, Taboada MA (1987) Soil salinization as an effect of grazing in a native grassland soil in the flooding Pampa of Argentina. Soil Use Manag 3:143–148. https://doi.org/10.1111/j.1475-2743.1987.tb00724.x

Millennium Ecosystem Assessment (2005) Ecosystems and human well-being: wetlands and water. Island Press, Washington, D.C

Nolte S, Müller F, Schuerch M, Wanner A, Esselink P, Bakker JP, Jensen K (2013) Does livestock grazing affect sediment deposition and accretion rates in salt marshes? Estuar Coast Shelf Sci 135:296–305. https://doi.org/10.1016/j.ecss.2013.10.026

Pennings SC, Bertness MD (1999) Using latitudinal variation to examine effects of climate on coastal salt marsh pattern and process. Curr Top Wetl Biogeochem 3:100–111

Pennings SC, Bertness MD (2001) Salt marsh communities. In: Hay ME (ed) Marine community ecology. Sinauer Associates, Sunderland, pp 289–316

Pennings SC, Grant MB, Bertness MD (2005) Plant zonation in low-latitude salt marshes: disentangling the roles of flooding, salinity and competition. J Ecol 93:159–167. https://doi.org/10.1111/j.1365-2745.2004.00959.x

Phelan P, Moloney AP, McGeough EJ et al (2015) Forage legumes for grazing and conserving in ruminant production systems. CRC Crit Rev Plant Sci 34:281–326. https://doi.org/10.1080/07352689.2014.898455

Rodríguez A, Jacobo E, Roitman G, Miñarro F, Preliasco P, Beade M (2016) Management of forage offer in the National Park Campos del Tuyú and neighbouring cattle farms for Pampas's deer conservation. Ecol Austral 26:150–165

Taboada MA, Rubio G, Chaneton EJ (2011) Grazing impacts on soil physical, chemical, and ecological properties in forage production systems. In: Hatfield JL, Sauer TJ (eds) Soil management: building a stable base for agriculture. American Society of Agronomy and Soil Science Society of America, Madison, pp 301–320

Vervoorst FB (1967) La Vegetación de la República Argentina. VII. Las comunidades vegetales de la Depresión del Salado (Provincia de Buenos Aires). INTA, Instituto de Botánica Agrícola. Castelar

Vila AR, Beade MS, Barrios Lamunière D (2008) Home range and habitat selection of pampas deer. J Zool 276:95–102. https://doi.org/10.1111/j.1469-7998.2008.00468.x

Yu OT, Chmura GL (2009) Soil carbon may be maintained under grazing in a St Lawrence Estuary tidal marsh. Environ Conserv 36:312–320. https://doi.org/10.1017/S0376892910000184

Limitations and Sustainable Management of Halohydromorphic Soils of the Santa Fe Province, Argentina

Silvia Imhoff and José Luis Panigatti

Abstract The province of Santa Fe is located between 28° and 34° S and 59° and 63° W. Saline and alkaline soils are found in the center of the province and especially in the region known as *Bajos Submeridionales*. These soils present processes of hydromorphism, salinization, and sodification, which generally occur superimposed or sequentially and are influenced by poor-quality water tables, frequent floods, and droughts. Natural drainage varies from drained to poorly drained. These limitations are a challenge for productive activities and the management of native plant species, both grassland and forest. The introduction of the "sustained multiple use of the territory" approach, i.e., the use of natural resources in a combination best suits the needs of the population without damaging those resources. The technology of management by environments would allow for production increases while maintaining the sustainability of the vulnerable environments of the *Bajos Submeridionales*.

Keywords Bajos submeridionales · Land-use planning · Hydromorphism · Soil management · Sodification · Salinization

1 Introduction

The province of Santa Fe, in Argentina, is located between parallels 28° and 34° S and 59° and 63° W. It occupies an area of 133,007 km^2. The province is a plain with a slight northwest-southeast slope that presents an array of climate, vegetation, and soil environments that determine seven natural regions: Chaco Plain, Flat Pampa, Rolling Pampa, basin with internal drainage, Submeridian lows (Bajos Submeridionales), Wooded Wedge and Relief linked to the Paraná River (Fig. 1, Panigatti 2017).

S. Imhoff (✉)
ICiAgro Litoral, Facultad de Ciencias Agrarias, Universidad Nacional del Litoral, CONICET, Esperanza, Argentina
e-mail: imhoff.silvia@gmail.com

J. L. Panigatti (Deceased)
Instituto Nacional de Tecnología Agropecuaria (INTA), Buenos Aires, Argentina

E. Taleisnik and R. S. Lavado (eds.), *Saline and Alkaline Soils in Latin America*, https://doi.org/10.1007/978-3-030-52592-7_14

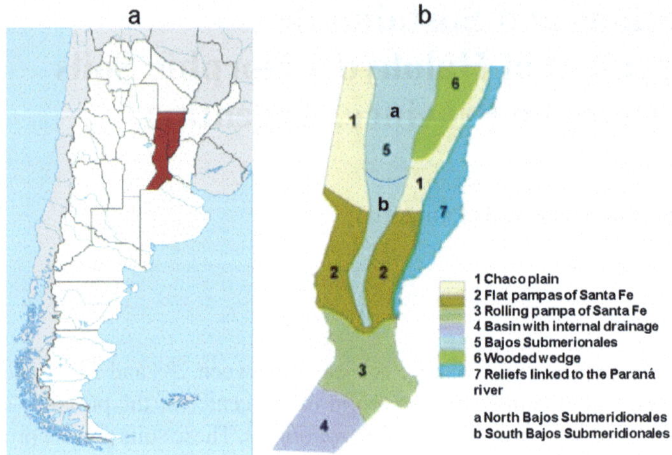

Fig. 1 Map of Argentina (**a**) and geomorphological units of the province of Santa Fe (**b**). (Adapted from Panigatti 2017)

The relief varies from flat to smoothly rolling, with an extensive central depression, called *Bajos Submeridionales* (BS), which occupies about 40,000 km^2 that correspond to a rainfall contribution basin of 67,000 km^2. The climate is variable, alternating periods of floods and droughts. Precipitations (average 1000 mm) decrease from the east to west. The geomorphology, the low permeability of the soils, and the rainfall intensity favor the occurrence of frequent floods (Lagos and Jaeschke 1979). Temperature decreases from north to south (winter average 14–15 °C, summer average 26–27 °C) (Gómez 2005).

The *Bajos Submeridionales* region (BS) extends from 26° N in the province of Chaco to 33° S in the province of Santa Fe. Two subregions, divided by the Salado River, can be distinguished: North BS (NBS) and South BS (SBS) (Panigatti and Mosconi 1978/79). The NBS is divided in four areas from west to east: the Western plain, a transition area with very gentle slopes, the central depression, and the Eastern plain (Fig. 2) (FVSA and FUNDAPAZ 2008).

In the *Bajos Submeridionales*, the main productive activity was livestock production. However, in the last decades, in the Western plain, agriculture has displaced livestock, and the natural grasslands and forests have practically disappeared. In the transition area with very gentle slopes, the savanna is being replaced by agriculture in the best soils. In the central depression, the hydrohalophilic savanna is being degraded by the exploitation of the natural forest and overgrazing. In the Eastern plain, the high-altitude humid forest alternates, as mosaics, with grasslands adapted to salinity, and some sectors with bare soils. Forests are also decreasing by excessive wood exploitation and are being replaced by livestock and agriculture in some sectors. These changes are causing accelerated degradation of soils, vegetation, and water quality. As a result, the risks for productive activities and the entropy of these ecosystems are markedly increasing (Soto et al. 2018).

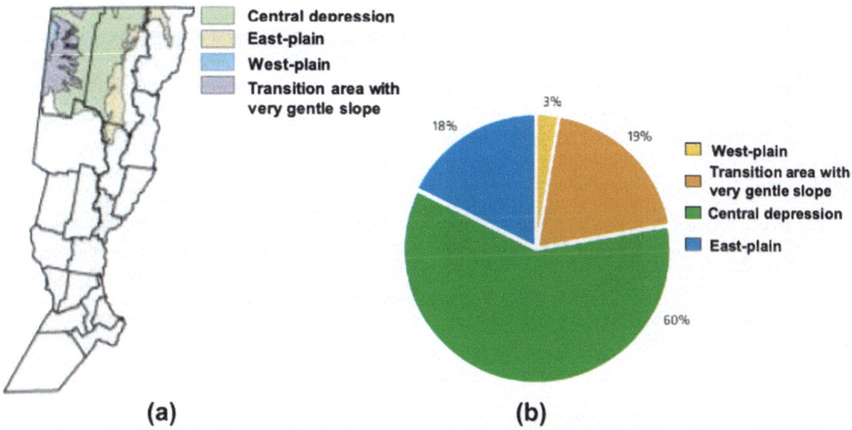

Fig. 2 Subareas of the *North—Bajos Submeridionales* (**a**) and percent distribution of each one (**b**) (Adapted from Soto et al. 2018)

In general, the NBS is characterized by a concave flat landscape with <0.1% slopes, poorly defined runoff paths, a large number and types of estuaries, semi-permanent lagoons, and the absence of permanent rivers. The water excesses drain in numerous lagoons and streams in the east of the NBS that gradually bring the waters to the Salado River and from there to the Paraná River.

The SBS is a smaller, narrower strip. Several features distinguish the SBS from the NBS. The SBS is an extended flat landscape of subnormal or subnormal-concave relief, with numerous runoff paths that are poorly defined, composed of a succession of depressions of slow drainage, and often water-flooded. Unlike the NBS, drainage is favored by a series of permanent streams and runoff paths, which lead the waters to the main collectors and finally to the Paraná River.

A differential feature of BS is the monotony of the landscape. The natural vegetation is composed of communities that vary in their composition according to their position in the relief. In the highest and best drained soils, a savanna with *Elionurus viridulus* predominates, with some tree species as *Geoffroea decorticans*, *Prosopis nigra*, and *Locust prosopis* in small groups or isolated. In the saline-alkaline lower parts, *Spartina argentinensis* predominates in association with *Chloris virgata, Polypogon chilense, Leptochloa chloridiformis, Distichlis spicata, Vicia graminea*, and others. In the estuaries, *Echinochloa helodes, E. polystachya, Paspalum distichum, Paspalidium paludivagum*, and other species can be found (Morello et al. 1975; Bissio 1979). For a detailed list, please see also the chapter by Pensiero et al. in this book: "Native and Naturalized Forage Plant Genetic Resources for Saline Environments of the Southernmost Portion of the American Chaco." Figure 3 shows examples of typical vegetation of the BS.

Fig. 3 Typical vegetation of the *Bajos Sumeridionales*. Typical forest (**a**) and savanna (**b**) found in the highest and best drained soils. Savanna with degraded grassland (**c**) and savanna with improved pastures inter-seeded in the natural grasslands (**d**). Photos S Imhoff

Three very important processes, generally superimposed, limit crop choice and management in the BS. These processes are hydromorphism, salinization, and sodification. As a result, an important characteristic of the BS is the spatial heterogeneity, composition and dynamics of the soils that integrate cartographic units (CU) called "complexes" and "undifferentiated complexes", which can differ to a level of subgroup, series, or phases in a few meters. The distribution and complexity of the CU vary with the topography in the W-E direction and also, as the altitude decreases in relation to the sea level. The intricate complexity of soils, which in a few meters may differ at the level of subgroups, series, or phases, makes their separation very difficult. In these cases, two formation factors of soils coexist, which respond to the general topography and within this (in each cartographic unit) to the micro-topography (Espino et al. 1983; National Agricultural Technology Institute-INTA 1990a). The minerals that predominate in the clay fraction are illite, montmorillonite, interstratified clay, and kaolinite (Morrás 2017; Panigatti and Mosconi 1978/79). The proportion of each one varies depending on the sedimentary material and relief. West of the BS illite predominates, with low proportion of other types of clay. Interstratified minerals, such as montmorillonite-illite and smectite, predominate in the east of the BS. Soils of the central part of the BS have a greater quantity of expandable clay types than those of the western part. These characteristics reflect the diversity of parent materials. In the west of the BS, the parent materials come from the Pampeanas and Andes Mountains, whereas at the east, they mainly come from the Paraná River basin. In the center of the BS, parent materials are a mixture of both, the west and east materials.

The presence of various types of clay produces marked differences in the physical properties of the soils, especially in the water dynamics, soil resistance to root penetration when dry, and in the capacity to support machinery traffic and animal trampling when wet. Other limitations are: reduced depth of A horizon, presence of a saline water table near the surface, high sodium content, slow soil permeability, alternation of droughts and floods (Panigatti 1990).

In the north and center of Santa Fe, the typical toposequence of soils from the well-drained areas to the center of the BS generally includes following soils: Typic Argiudoll, Aquic Argiudoll, Typic Argialboll, Natralboll, Natraquoll, Natraqualf (Imbellone et al. 2010) (Fig. 4). The distribution and proportion of each soil in the complexes and undifferentiated complexes vary with the topography. Generally, they are integrated by three or more soil series that are very difficult to separate.

The degree of hydromorphism increases from the Typic Argiudoll to the Natraqualf. The first three soils (Typic Argiudoll, Aquic Argiudoll, Argialboll) are found in the higher areas of the landscape.

In general, their land capability use is IIw or IIIws according to the land capability classification (Klingebiel and Montgomery 1961). The land capability classification places all the soils in eight capability classes and several subclasses. Soils in each class have about the same responses to management systems of commonly cultivated crops

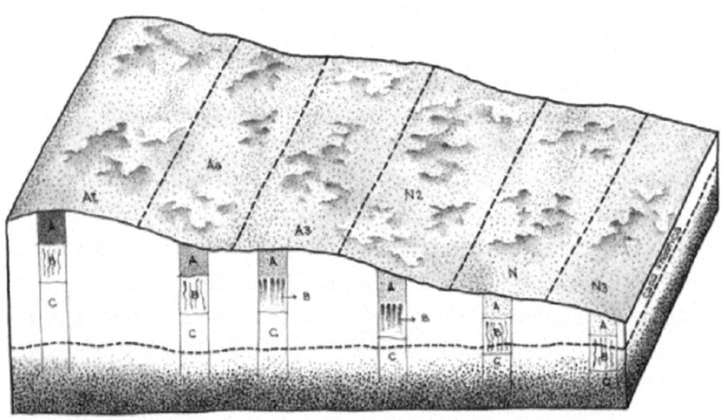

Fig. 4 Soil toposequence according to the location in the relief from the highest to the lowest part (center) of the BS. At: Typic Argiudoll, Aa: Aquic Argiudoll, A3: Argialboll, N2: Natralboll, N: Natraquoll, N3: Natraqualf. (Adapted from Imbellone et al. 2010)

and pasture. The subclasses indicate that soils have similar kinds of limitations and hazards. Four general kinds of limitations or hazards are recognized: erosion hazard (e), wetness (w), rooting zone limitations (s), and climate (c). The risks of soil damage or limitations in use progressively increase from class I to class VIII. In general, soils in class I and II are suitable for crops; soils in class III and IV are suitable for cultivated pastures and some crops, but require special management practices to avoid soil deterioration. Soils in classes V and VI are suitable for multiannual grasslands, while soils in class VII are only suitable for natural grasslands that are tolerant to severe environmental restrictions. Soils in class VIII have severe limitations for commercial crop production, thus they are mainly suitable for recreation.

In the BS, soils in class IIw can be used for agriculture and those in class IIIws can be used for some cultivated pastures and some crops. Soils have limitations that restrict the choice of species and management practices due to water excess (subclass w) and low-moderate salinity or sodium in the B horizon (subclass s). Soils present differences because of their genesis, influenced in turn by the general flat relief and the very frequent presence of micro-reliefs.

The other three soils (Natralboll, Natraquoll, Natraqualf) are located in lower areas of the landscape. Natracualfs present lower content of organic matter, small thickness of the massive E horizon, abrupt E/B limit, B horizon with high clay content, sodium, and salts, presence of more superficial carbonates and water table than Natralbolls. The physical environment of these two soils, and especially the water economy, is very different, even where they are only a few meters apart.

In general, the land capability use of Natralbolls, Natraquolls, and Natraqualfs varies from IVws to VIIws. Soils in class IV are suitable for occasional pasture cultivation. Soils in classes V, VI, and VII have severe limitations that make them unsuitable for cultivation and limit their use largely to grasslands. They have deficient drainage because water tables frequently reach the surface, maintaining the soil horizons saturated. High water table supplies sodium salts, generating sodic horizons. Thus, they are practically unproductive.

An integrative approach is required to understand the productive capability of these soils. Spatial and temporal heterogeneity, linked to their genesis, the processes of natural hydromorphism, salinization and sodification, and those introduced by soil management and infrastructure, must be considered together. Therefore, a productivity index (PI) was developed in order to establish homogeneous areas in terms of their agricultural productivity, in a semi-detailed scale (1: 50,000). Through this system, originally developed by FAO and adapted to the conditions of the Pampas region by the National Agricultural Technology Institute (INTA 1990b), all soil series and cartographic units of the soil maps of Argentina were assessed.

The PI is a parametric land evaluation system, which considers ten selected properties. Critical ranges are established for each property, and a coefficient is assigned to each one, depending on the level of limitation that the property imposes to the

Table 1 Agricultural productive capacity classes (APCC) and productivity index ranges (PI ranges)

APCC	PI range
Very high	100–85
High	84–70
Average	69–55
Low	54–40
Very low (non-agricultural areas)	<39

land use. The properties considered by the method are: climate condition, drainage, surface texture, subsurface texture, cation exchange capacity, organic matter, effective rooting depth, salinity and sodicity, current erosion, and potential erosion. The PI of each soil taxonomic unit (series and its phases) is calculated after introducing each property in the equation. Finally, the PI of the cartographic units is established from the PI of the constituent series and their percentage participation.

Sets of intervals of the weighed PI of the cartographic units are expressed in agricultural productive capacity classes (APCC). According to the procedure adopted to establish the interpretative grouping of the soil units PI greater than 40 define agricultural areas (INTA 1990b), whereas non-agricultural areas are defined by grouping cartographic units with values of PI lower than 39. Productivity index ranges and APCC are shown in Table 1.

In the BS, the PI varies from 81 to 58 in class IIw and IIIws, and from 22 to 3 in classes IVws to VIIws (Table 2). The PI of three toposequences of the *Bajos Submeridionales* is shown in Table 2, and characteristics and limitations of the best and the worst soil for agricultural production are shown in Table 3 (Panigatti 2017).

The classic toposequence found in the BS may have another component, the Albaqualf, when the lowest part of the landscape is a lagoon or the floodplain of streams. Because of the water table dynamics, rainfall, and surface drainage characteristics, this soil has lower content of exchangeable sodium and salts than soils with natric horizons. This toposequence is found from 28° S to 33° S. However, as the width of the BS decreases from north to south, the proportion of more suitable soils increases in the same direction. The differences in productivity are mainly determined, from west to east, by the drainage characteristics and, from north to south by the climate because the annual average temperature decreases in this direction, whereas rainfalls decrease from east to west.

The region of the BS has frequent floods due to several factors: wet multi-year cycles, irregular distribution of rainfall (more frequently in autumn and spring), high

Table 2 Soils series and productivity index of three toposequences of the *Bajos Submeridionales*

Latitude	Typic Argiudoll	Aquic Argiudoll	Argialboll	Natralboll	Natracuoll	Natracualf
30°10′ S	72	65	49	22	14	3
31°30′ S	81	65	58	28	26	8
32°30′ S	90	75	65	31	–	16

Table 3 Properties of Argiudolls and Natraqualfs, and their environments (Panigatti 2017)

Properties/item	Typic Argiudoll	Typic Natraqualf
A Horizon Depth	Deep	Absent
E Horizon	Absent	Present
A (E) structure	Blocks	Massive
Superior limit of B horizon	Gradual	Abrupt
Columnar structure	Absent	Present sometimes
Organic matter content	High	Low
Total nitrogen	High	Medium
Cation exchangeable capacity	High	High
Salinity	Absent	Medium/high
Lixiviation and salinity	Absent	Dynamic
Alkalinity	Absent	High (15%)
Drainage	Well drained	Poorly drained
Runoff	Moderate	Null to slow
Permeability	Moderate	Null to slow
Suspended water table	Absent	Present, discontinuous
Water table	Deep	Shallow, fluctuating
Depth to CO_3Ca	>100 cm	<50 cm
Floods	Absent	Present, irregular
Waterlogging	Scarce and short	Common and durable
Anoxia	Absent	Present
Vegetation	Crops	Native grassland
Landscape	Normal/convex	Subnormal/concave
Resilience	Moderate/high	Low
Natural fertility	High	Moderate/low
Productivity	Very high	Low
Aptitude	Intensive agriculture	Extensive livestock
Compaction	By machinery	By animal trampling

intensity of rainfall, level of the water table variable as a function of the previously mentioned factors, low capacity of drainage, and human activity that modifies the natural conditions.

2 Soil Management Criteria for the Bajos Submeridionales

To manage the soils of the BS, it is necessary to define the depth of soil sampling in order to determine the fertility of a field. The problem is mainly caused by the great variability of horizon thicknesses. As an example, the profiles of the typical soils are shown in Table 4.

A sampling depth from 0 to 15 cm includes a mixture of A horizons when Typic Argiudolls, Aquic Argiudolls, and Typic Argialbolls integrate complex CU. Thus, it may be considered that the homogeneity and representativeness are adequate. On the other hand, a sampling depth from 15 to 30 cm includes a mixture of three horizons, increasing the magnitude of errors. Even greater errors may be induced when sampling undifferentiated soil complexes. In these cases, soil heterogeneity is natural, however, it can be intensified by animal trampling or tillage. As a result, soil analysis interpretation becomes very difficult without information on sample precedence and represents a great challenge for professionals.

Well-drained soils located in the high areas of the west and east of the BS may have phases of poor drainage. These soil phases present minor limitations for the evolution of the main crops (*Medicago sativa* sp., *Triticum aestivum* sp., *Zea maiz* sp., *Glycine max* sp., *Helianthus annuus* sp.) and have a similar behavior to the main soils they are associated with. Thus, from a practical point of view, they are considered similar. In addition, their separation is difficult based on the available cartography (1:50,000 scale). Consequently, the same management practices are generally applied. One alternative for maximizing crop productivity in these soils is precision agriculture. Crop yield maps of detailed scales, generated by precision agriculture technology, can reveal some soil differences that help to identify and spatially separate the causes for productive limitations. The challenge is to interpret if limitations are due to genetic or anthropic reasons, and whether they are reversible or not, in order to apply correct management practices.

Maps of crop yield have been made mainly in areas with good agricultural aptitude soils and crops yields, where the high investments in knowledge, technology, and machinery required for their preparation are justified. In the best soils, agriculture

Table 4 Thickness (cm) of the upper horizons of the typical soils of the *Bajos Submeridionales*

Typic Argiudoll	Aquic Argiudoll	Typic Argialboll	Natralboll	Natraquoll	Natraqualf	Albaqualf
A 0–27	A 0–23	A 0–18	A1 0–17	A1 0–17	E 0–13	E1 0–31
BA –37	BA –34	E –25	E –26	BA –25	Bt –43	E2 –58
Bt1 –82	Bt1 –73	Bt1 –63	Bt1 –53	Bt1 –53		

can be combined with livestock production, employing rotations that include pastures that help to recover soil fertility and, simultaneously, provide forage of good quality.

A more complex situation occurs when dealing with soils that are located in the central depression of the BS. A cartographic base of greater detail than that of the 1: 50,000 scale is needed to identify the causes of low productivity of the poorly drained soils, such as the hydromorphic soils located in the lower areas of the BS, which have great pedological and productive heterogeneity. The numerous components of the cartographic units are difficult to inventory because they require careful observation and field and laboratory work to identify and separate them. Due to their low productive potential (low soil use capacity), studies with greater detail to distinguish the components are not justified from the economic point of view. In these cases, it is convenient to focus on the complete system, because an integrated management of natural vegetation, soils, and water will have the greatest impact on its potential productivity.

It is important to keep in mind that these soils constitute a fragile resource because they are susceptible to salinization, hydromorphism, and, also, erosion although the landscape is flat. Even though the slope of the main cartographic units is minimal, they are very long, exceeding 2 km, particularly in the NBS. The loss of soil can promptly begin in the banks of streams without vegetation, even in sectors with little erosion. Surface runoff is increased by torrential rainfalls and can generate laminar erosion or in furrows. Animal trails from higher areas of the relief to lower areas with freshwater favor water runoff and erosion in furrows. The formation of furrows in the distal part of the floodplain of streams illustrates the fragility of the environments in the central depression of the BS. Moreover, crop production in these environments has caused the loss of natural grasslands, the decrease in the soil support capacity, and the salinization of surface horizons because of the increase of soil temperature, soil evaporation, and the rise of the water table.

Therefore, focusing on the management of the complete system should be an essential procedure since it allows to maintain the functionality of the environment, even when soils are different, mainly in complex cartographic units.

3 Sustainable Land-Use Planning for the Bajos Submeridionales

Hydromorphic soils and environments with water excesses, such as those in the BS, require land-use planning and the creation of a multi-dimensional and interconnected network of water reservoirs, based on the basic concepts of environmental agro-ecology. Land-use planning is important to: (1) assign environment use according to the soil use capacity for optimizing productivity, (2) conserve natural resources, soils, water, vegetation, fauna, and (3) implement actions toward sustainable development (Imbellone et al. 2010). The use of technologies, such as capture, processing and interpretation of sensor and satellite images and the elaboration of digital elevation models of the topography, could contribute to elaborate an updated mapping of environments with cultivated pastures, grasslands, and forests to guide the land-use planning (Pérez 2008).

The extensive areas with the worst hydromorphic soils (Natraqualf, Albaqualf) in the NBS, which produce very little biomass for livestock, could integrate an interconnected network of places with natural vegetation that act as true wildlife refuges, extending the benefits of a greater biodiversity to the surrounding areas.

In the SBS, soils with lower productivity that occupy small areas could integrate environmental conservation and production networks, enhancing their ecosystem services and controlling the degradation of fragile environments.

Soils with intense processes of hydromorphism, salinization, and alkalization could integrate natural conservation networks at various magnitudes of basin scale. These networks could be interconnected on a large scale to define public resource conservation environments. In addition, these ecological networks could be interconnected with existing reserves, roads, railways, and field boundaries. Hydrographic networks may be used as a regional instrumental basis of information for designing conservation networks, as they clearly indicate differences in climatic, geomorphological, and edaphic conditions. The superposition of specific and more detailed information to the hydrographic networks would allow conservation and productive areas to be defined more precisely.

Beef production is one of the most important productive activities in the BS. This area can be an optimal environment for sustainable production with a protocol of good practices, especially in relation to a differentiated livestock of high value for a market that requires meat from animals that grow in natural environments (Soto et al. 2018). The more important native species that can support beef production are *Spartina argentinensis* sp. (sweet pastures), *Paspalum* sp., *Echinochloa helodes, Diplachne uninervia*, and others. Legumes of the genus *Prosopis* are the dominant species in the best soils while in the worse soils predominates *Cyclolepis genistoides* (Morello et al. 1975; Biasatti et al. 2016).

Natural grasslands can be combined with cultivated pastures, such as *Lotus* sp., *Melilotusalbus, Agropyron elongatum*, but only when more detailed information of the soil characteristics is available and special management practices are applied (Fossatti et al. 1979; Panigatti et al. 1983). Otherwise, soil degradation will be increased.

Productive activities in fields with hydromorphic, alkaline, and saline soils, as those of the BS, have a promising future because factors related to ethical principles are becoming important for the new generations. These principles, e.g., animal welfare, environmental pollution, and soil sustainability, will affect the beef consumption in the future (García 2002). Numerous studies carried out at INTA have demonstrated that beef from natural systems may be a true differentiated product, more adapted to the nutritional requirements of the new generations than that from intensive systems (Aiello and Marino 2015).

In recent years, a notable increase in beef production has been achieved through the application of simple practices, such as field subdivision, improvement of natural grasslands, and rotational grazing according to the vegetation condition (Feldman

2017). Therefore, to assign environments with serious limitations for crop production to direct grazing of natural grasslands could be one of the best alternatives. Genetic engineering opened paths to produce stress-tolerant crops, e.g., to drought, salinity, and other adverse conditions, which constitute useful tools to develop genetically improved pastures that could be inter-seeded in the natural grasslands to improve productivity and quality using guided fertilization at special times. Also, the use of prescribed fire as a management tool should be considered (Kunst et al. 2003). The productive value of these areas can be increased by using them simultaneously as providers of environmental services, for honey production, sport hunting, ecotourism, and recreation. However, the incorporation of both input and process technology is still scarce, among others because of the lack of infrastructure.

Natural landscapes of heterogeneous characteristics, when connected, better preserve intrinsic soils conditions for long-term sustainable production and biological control. Areas with limited agricultural capability can be used as reserves of standing forage for grazing with high loads for very short periods, reducing the impacts of animal trampling that causes soil compaction. This management should maintain maximum cover and allow for natural reseeding.

In the BS, one of the most recommended practices to manage saline and alkaline soils is the design and construction of hydrological circuits inside the fields (Barbagallo 1983; Aiello and Marino 2015). In this system, the field is divided into plots, some of which remain under dry-land regime and others remain flooded. In the flooded plots, the water sheet decreases the negative effects of salts and sodium, which allows a better development of the natural vegetation. A rotational system of grazing between flooded and dry plots is applied, pastures are grazed in each plot when they have the highest volume and quality and, producing, as a result, an increase in animal load per hectare. Improved pastures, such as *Chloris gayana, Thinopyrum ponticum, Melilotusalba*, and *Lotustenuis*, can be grown in the best parts of the dry lots. According to Aiello and Marino (2015), the use of planned water retention allows the efficiency of harvesting and use of forage to increase by 35% and, at least, to double the production of beef per hectare. Additionally, it is important to build dams in strategic locations in the dry plots to supply animals with water and also to store some potential water excess.

Another important aspect is the native forest conservation that generally forms small islets in slightly higher areas. These open forests of native species (*Prosopis* sp.) accompanied by grasslands offer food and shade for animals. Moreover, reforestation with native or foreign species (*Tamarix halophile, Eucaliptus* sp.) should be included in the design to increase the productivity and services of the system (Feldman 2017). The inclusion of forest species would allow the recovery of native forest and associated grasslands for increasing the infiltration of rainwater, and the wood production for multiple uses, avoiding the exploitation of native species, which are of slow growth.

The major challenge for the development of the region is the sustainable management of water resources. The infrastructure to be built for providing water of good quality for human consumption, animals, and irrigation should be complemented with an adequate system for collection and management of rainwater. Also, road networks should be built to facilitate the entry and exit of regional products.

It is essential to strengthen the ecological coherence and resilience of productive systems for the sustained development of natural environments. The goal should be to achieve synergy between the areas under protection and those under production, seeking to achieve maximum sustainable production with an acceptable level of conservation of the biodiversity, so that these systems provide their services to agriculture, livestock, forestry, and other productive systems, improving the quality of life. In this sense, the application of the multiple-use sustained of the territory (MUST) approach would allow the inhabitants of the BS to achieve that goal. This approach had its origin in the Law of Multiple-Use Sustained-Yield legislated by the Congress of the USA in June 1960.

The MUST approach implies the management of all renewable resources in a territory so that they are used in the combination that best suits the needs of the population without damaging those resources (Gastó et al. 1997). This vision would also contribute to resizing the importance, destination, and use of areas with hydromorphic soils as the BS, including areas of reduced size, since one of the ways to reverse land degradation is to consider and integrate the contributions and functions of biodiversity to the productivity of the agro-ecosystems (Pilatti et al. 2002).

4 Concluding Remarks

Some basic points must be taken into account for a correct management of the agro-ecosystem in the BS are:

- Land use according to its capacity.
- Infrastructure improvement: field subdivision, areas of closure, road construction.
- Knowledge of the native vegetation and evaluation of its forage value before replacing it.
- To reduce machinery traffic and use the no-till system when possible.
- To manage animal load, to reduce animal trampling and overgrazing.
- To improve natural grassland by inter-seeding genetically improved pastures, such as *Chloris gayana, Panicumcoloratum, Digitaria decumbens*, and others.
- When necessary, to induce grasslands regrowth through the use of controlled fire.
- To graze grasslands and pastures at the optimal moment and use deferred grazing when soil conditions are inadequate.
- To develop hydrological circuits to improve water, soil, and pasture management.
- To build dams to store rainwater since the water in the aquifer is of poor quality.
- To build elevated areas for animal rest during periods of slight flooding.
- To reforest areas with natural or foreign species of trees.
- To promote the diversification of activities (goat and sheep farming, beekeeping, horticulture, and the production of native fruits).
- To promote ecotourism.

According to the MUST, each territory has a potential for various uses that must be assessed taking into account all the stakeholders in that territory. In a vulnerable environment, such as the BS, the use of natural resources should be less than local renewal capability and the emission of pollutants should be lower than territorial capability to purify the environment (Pilatti et al. 2002). To achieve a sustainable development of the *Bajos Submeridionales*, territory, land-use planning, and technology application should be harmonized with the livelihood of the regional population.

References

Aiello F, Marino G (2015) Manejo del agua superficial: implementación de retenciones hídricas en beneficio de la ganadería y la biodiversidad. Aves Argentinas, AOP, Buenos Aires

Barbagallo JF (1983) Los Bajos Submeridionales del norte de Santa Fe y Chaco. Un planteo agrohidrológico para el ordenamiento del recurso hídrico, INTA, Unidad de Agrohidrología, Departamento de Suelos, Buenos Aires

Biasatti N, Rozzatti JC, Fandiño B, Pautaso A, Mosso E, Marteleur G, Algarañaz N, Giraudo A, Chiarulli C, Romano M, Llorenz P, Vallejos L (2016) Las ecoregiones: su conservación y las áreas protegidas de la provincia de Santa Fe. Ministerio de Medio Ambiente, Santa Fe

Bissio JC (1979) Clasificación de los pastizales naturales de los Bajos Submeridionales Santafesinos. Fundación José María de Aragón, Publicación N°12. Buenos Aires

Espino LM, Seveso MA, Sabatier MA (1983) Mapa de suelos de la provincia de Santa Fe. Tomo II, MAG-INTA, Santa Fe

Feldman I (2017) Alternativas productivas sustentables en suelos halo-hidromórficos de la provincia de Santa Fe. In: Taleisnik E, Lavado R (eds) Ambientes salinos y alcalinos de la Argentina: recursos y aprovechamiento productivo. Universidad Católica de Córdoba, Buenos Aires, Orientación Gráfica Editora, pp 135–151

Fossati JL, Bruno OA, Panigatti JL, Gambaudo SP (1979) Comportamiento de forrajeras estivales en los Bajos Submeridionales. INTA, EEA Rafaela, Informe Técnico N° 1

FVSA (Fundación Vida Silvestre Argentina), FUNDAPAZ (Fundación para el Desarrollo en Justicia y Paz) (2008) Zonificación de los Bajos Submeridionales del Norte Santafesino. Una herramienta para la planificación del desarrollo productivo y la conservación de la biodiversidad del humedal. Vida Silvestre Argentina, Buenos Aires

García PT (2002) Limitaciones al consumo de carne bovina. Factores de riesgo. IDIA XXI 2:195–199

Gastó J, Vélez LD, D´Angelo C (1997) Gestión de recursos vulnerables y degradados. In: Viglizzo E (Ed) Elementos para una política agroambiental en el Cono Sur. PROCISUR, Montevideo, pp 75–116

Gómez, S (2005) Base de datos agroclimática, ICyA, CIRN, INTA, Serie 1981–2005. https://geointa.inta.gov.ar/webfm/Descargas/L%C3%A1minas/Laminas_de_Suelos

Imbellone PA, Giménez JE, Panigatti JL (2010) Suelos de la región pampeana. Procesos de formación, INTA, Buenos Aires

INTA (Instituto Nacional de Tecnología Agropecuaria) (1990a) Atlas de suelos de la República Argentina. Escala 1:500000 y 1:1000000. Tomo I y II. CIRN, Instituto de Evaluación de Tierras, Buenos Aires

INTA (Instituto Nacional de Tecnología Agropecuaria) (1990b) Carta de suelos de la República Argentina. Hoja 3163–5 y 6. EEA Rafaela

Klingebiel AA, Montgomery PH (1961) Land-capability classification (No. 210). Soil Conservation Service, US Department of Agriculture
Kunst CR, Bravo S, Panigatti JL (eds) (2003) Fuego en los ecosistemas argentinos. INTA, Santiago del Estero
Lagos F, Jaeschke J (1979) Las precipitaciones en los años de inundación en los Bajos Submerionales de Santa Fe. Fundación José María de Aragón, Publicación N°10. Buenos Aires
Morello J, Sancholuz L, Alfonso D, Sulzberger S (1975) Informe de la vegetación de los Bajos Submeridionales. Programa Desarrollo Agropecuario para la Región Bajos Submeridionales. Convenio Bajos Submeridionales, CFI- INCYTH, Santa Fe, Chaco y Santiago del Estero
Morrás H (2017) Propiedades químicas y físicas de suelos halomórficos de la fracción norte de los Bajos Submeridionales. In: Taleisnik E, Lavado R (eds) Ambientes salinos y alcalinos de la Argentina: recursos y aprovechamiento productivo. Universidad Católica de Córdoba, Buenos Aires, Orientación Gráfica Editora, pp 29–53
Panigatti JL (2017) Conocimientos y manejo de suelos halohidromórficos. Caso de los Bajos Submeridionales de Santa Fe. In: Taleisnik E, Lavado R (eds) Ambientes salinos y alcalinos de la Argentina: recursos y aprovechamiento productivo. Orientación Gráfica Editora, Universidad Católica de Córdoba, Buenos Aires, pp 89–115
Panigatti JL (1990) Bajos Submeridionales. In: FECIC (Ed) Manejo de Tierras anegadizas. Prosa, Buenos Aires, pp 279–296
Panigatti JL, Mosconi FP (1978/79) Arcillas de suelos de centro de Santa Fe y centro-oeste de Córdoba. RIA Serie XIV (3):137–148
Panigatti JL, Fossati JL, Gambaudo S, Bruno O (1983) Siembra aérea y convencional de trébol de olor de flor blanca en los Blancos Submeridionales. Informe Técnico. 17, INTA EERA Rafaela, Rafaela
Pérez M (2008) Cartografía actualizada de montes en el área del Consejo Regional de los Bajos Submeridionales. Convenio Consejo Federal de Inversiones - provincia de Santa Fe., Buenos Aires, p 12
Pilatti MA, D´Angelo CH, Marano RP, Pensiero JF, Potente HJ, López Calderón A (2002) Ordenamiento territorial de la cuenca de Los Saladillos (Santa Fe). Estudio de prefactibilidad del uso actual y potencial de los recursos naturales. Ediciones UNL, Santa Fe
Soto V, Basan M, Castro G, Biasatti R (2018) Plan de manejo integral de los Bajos Submeridionales santafesinos. Gobierno de la provincia de Santa Fe, Santa Fe

Effects of Supplementary Irrigation on Soils and Crops in Humid and Sub-humid Areas in the Pampas Region of Argentina

Carina Rosa Alvarez, Helena Rimski Korsakov, and Martín Torres Duggan

Abstract Drought events are frequent in the Pampas region of Argentina where most field crops are being managed under rainfed farming conditions and hence supplementary irrigation allows farmers to increase and stabilize yield crops. Supplementary irrigation for grain production supplies between 10 and 30% of the total water received annually by the cropping system. Most irrigation water across the Pampas is characterized by high sodium bicarbonate concentrations. According to a synthetic analysis from representative field experiments carried out across different farming systems, supplementary irrigation increased pH by 9%, electric conductivity by 34% and exchangeable sodium percentage by 177%. There is conflicting information on the impact of increased exchangeable sodium percentage on soil physical characteristics. Some field experiments show higher values of bulk density, mechanical resistance, or lower water infiltration rates in supplementary-irrigated fields, but available data is not sufficient to build a general conceptual model for the whole Pampas region. More research should be conducted to assess the long-term impact and sustainability of supplementary irrigation systems across different farming and irrigation conditions.

Keywords Irrigation · Exchangeable sodium percentage · Soil physical properties · Crop yield · Rainfed agriculture

C. R. Alvarez (✉) · H. Rimski Korsakov
Soil Fertility and Fertilizer, School of Agronomy, University of Buenos Aires, Buenos Aires, Argentina
e-mail: alvarezc@agro.uba.ar

H. Rimski Korsakov
e-mail: rimski@agro.uba.ar

M. Torres Duggan
Tecnoagro, Buenos Aires, Argentina
e-mail: mjjtorresduggan@gmail.com

E. Taleisnik and R. S. Lavado (eds.), *Saline and Alkaline Soils in Latin America*, https://doi.org/10.1007/978-3-030-52592-7_15

1 Supplementary Irrigation in Field Crops

The Pampas region is a vast plain with an area of more than 60,000 km^2 that extends between 28° and 38° S and 68° and 57° W in Argentina. The predominant climate is humid to sub-humid and temperate without a marked dry season but there is a great variation in total rainfall, across and within years. Thus, water stress is frequent in the region and, consequently, droughts could occur any time (Maddonni 2012; Rubio et al. 2018). Although most field crops (soybean, maize, wheat, sunflower, barley) are being managed under rainfed conditions, supplementary irrigation is applied from the 1970s and its use is slowly but steadily increasing. Supplementary irrigation is a technology that supplies water to crops during periods of water deficiency, even in humid areas prone to random droughts. While other systems of irrigation usually make cropping possible, supplementary irrigation has two main objectives: to improve crop yields and stability over time (Torres Duggan et al. 2016).

The effects of supplementary irrigation on the increase and stabilization of crop yields over time are shown in Table 1.

The productive and derived economic benefits have played a key role in expanding supplementary irrigation in the Pampas region, mainly for maize and other field crops. The irrigated area in the region is currently 551,223 ha (Sánchez et al. 2016), of which 54% is irrigated by sprinkling (e.g., center pivot equipment, linear irrigation systems, canyon, etc.), the preferred irrigation method used for grain crops. Typically, supplementary irrigation for grain production consists in a water supply of 100–250 mm $year^{-1}$, which represents between 10 and 30% of the total water received annually by the cropping system (Sánchez et al. 2016). However, higher water irrigation rates are frequently applied in maize seed production systems (inbred lines production and seed multiplication). The reasons for this are related to a very particular production system aimed at maintaining soil water contents close to field capacity over the whole cropping season.

The effects of supplementary irrigation on winter and summer crops are shown in Table 2, indicating significant increase and stabilization of maize and soybean yield, and sunflower to a lesser extent. Wheat shows a significant response to supplementary irrigation in the NW of the Pampas region where rainfalls are more concentrated in the summer, with low rainfall over winter.

Table 1 Average performance and coefficient of variation (CV) in the NW of the Pampas region in the period 1996–2016 (Salinas 2018)

	Rainfed		Irrigated	
Crop	Yield (kg ha^{-1})	CV (%)	Yield (kg ha^{-1})	CV (%)
Wheat	2058	47	4564	32
Soybean II*	2398	28	3001	17
Maize	7951	28	12,016	17

*Soybean can be seeded as the only annual crop, called soybean I, or in the annual sequence wheat/soybean, called soybean II

Table 2 Yield increases due to supplementary irrigation for different field crops in the Pampas region. N° years: number of years evaluated. Soybean I, full season soybean; maize II (and Soybean II), crops sowed after wheat harvest. NA: data not available. DM: Dry matter. Figures in parenthesis indicate percentage yield increase due to irrigation

Author	Location/soil	Crop	N° years	Yield increases due to irrigation compared with rainfed situations
Marano and Pilatti (2017)	NE	Silage maize	6	From 2000 to 11,600 kg DM ha^{-1}
	Argiudoll	Soybean I	1	From 700 to 850 kg grain ha^{-1}
		Wheat	2	From 1200 to 1518 kg grain ha^{-1}
		Sunflower	1	1064 kg grain ha^{-1}
Andriulo et al. (2017)	NE	Maize	NA	2000 kg grain ha^{-1} (22%)
	Argiudoll	Wheat	NA	750 kg grain ha^{-1} (20%)
		Soybean II	NA	850 grain ha^{-1} (35%)
Salinas et al. (2016)	NW	Wheat	20	2595 kg grain ha^{-1} (116%)
	Haplustoll	Soybean I	20	1218 kg grain ha^{-1} (42%)
		Soybean II	20	822 kg grain ha^{-1} (36%)
		Maize I	20	2931 kg grain ha^{-1} (33%)
		Maize II	16	1442 kg grain ha^{-1} (22%)
Severina et al. (2016)	NW	Soybean	4	441 kg grain ha^{-1} (17%)
	Haplustoll	Maize	4	3673 kg grain ha^{-1} (45%)
		Wheat	3	2057 kg grain ha^{-1} (81%)

2 Assessing the Impact of Supplementary Irrigation on Soil Quality Features

2.1 Salinity and Sodicity Constraints in Supplemented-Irrigated Crops

Although supplementary irrigation brings about remarkable productive and economic benefits, negative impacts on soil properties raise concerns among farmers, professionals, and researchers. Among factors impacting soil salinization and/or sodification, the quality and the amount of the irrigation water applied (as a proportion of the total water received by the crops) are key factors to consider for their impact on soil quality. The main variables to characterize water quality are: electrical conductivity (EC), sodium adsorption ratio (SAR), residual sodium carbonate (RCS) as well as the concentration of specific ions that can produce direct toxicity effects. Most irrigation water applied across Pampas fields comes from the so-called Puelche aquifer and, to a lesser extent, from the Pampeano aquifer. The water from both aquifers contains high sodium bicarbonate concentrations (Table 3).

According to local classification systems, most irrigation waters have low soil salinization risk and higher risk of sodification (SAR and RSC) for field crops supplementarily irrigated (INTA 1999; Rampoldi et al. 2010). Considering that irrigation impact on soil physical properties depends also on other soil properties (texture, organic matter, among others), an integrated approach is recommended for a proper irrigated field management evaluation.

The salts added by the irrigation water may produce soil salinization and/or sodification processes in the soil in arid or semiarid regions. This is because crop water requirements are covered mostly by irrigation and salt leaching by rainfall is scarce. Supplementary irrigation impacts on soil properties in humid or sub-humid regions may be different, because rainfall can leach salts from the root zone and thus keep low soil electrical conductivity (EC) there (Genova 2011). When irrigation water contains significant amounts of sodium as bicarbonates, soil sodification may occur. The increase in the exchangeable sodium percentage (ESP) is directly associated with the quality of the irrigation water. The application of high SAR irrigation water, keeping other factors unchanged, rises the soil ESP at the expense of the exchangeable calcium or magnesium.

The standard ESP threshold considered to define a soil as "sodic" (ESP > 15%) was proposed for arid and/or semi-arid regions where soils are characterized by high

Table 3 Average, minimum, and maximum pH, electrical conductivity (EC), sodium adsorption ratio (SAR), residual sodium carbonate (RSC), cations and bicarbonates figures in irrigation water samples obtained in the Pampas region. n = number of samples considered

	Mean	Min	Max
pH	7.84 ($n = 783$)	6.5 ($n = 1323$)	9 ($n = 1323$)
EC (dS m^{-1})	1.25 ($n = 1163$)	0.15 ($n = 1703$)	4.7 ($n = 1703$)
SAR	12 ($n = 1163$)	0.38 ($n = 1703$)	57.7 ($n = 1703$)
RSC	7.11 ($n = 597$)	−0.2 ($n = 1137$)	36.5 ($n = 1137$)
Na^{+} (mmolc L^{-1})	12.41 ($n = 456$)	1 ($n = 996$)	68 ($n = 996$)
Ca^{+2} (mmolc L^{-1})	1.14 ($n = 456$)	0.1 ($n = 996$)	9.7 ($n = 996$)
Mg^{+2} (mmolc L^{-1})	0.9 ($n = 456$)	0 ($n = 996$)	22.4 ($n = 996$)
HCO_3^{-} (mmolc L^{-1})	8.81 ($n = 455$)	1 ($n = 995$)	24 ($n = 995$)

Source Peineman et al. (1998); Costa (1999); Pilatti et al. (2006); Andreani (2009); Romanelli et al. (2012); Costa and Aparicio (2015); Alvarez et al. (2016); Andriulo et al. (2017); Wilson and Cerana (2017)

EC values. On the contrary, in humid or sub-humid regions, where supplementary irrigation is applied, the ESP critical limits may be significantly lower since there is no presence of sufficient salts concentration for the soil colloids to flocculate. Therefore, in this situation, there is a higher probability to affect soil quality features (e.g., structure stability, water infiltration, mechanical resistance, etc.).

According to a synthetic analysis from representative field experiments carried out across different farming systems of the Pampas region, supplementary irrigation rose pH by 9%, EC by 34%, and ESP by 177% (Fig. 1).

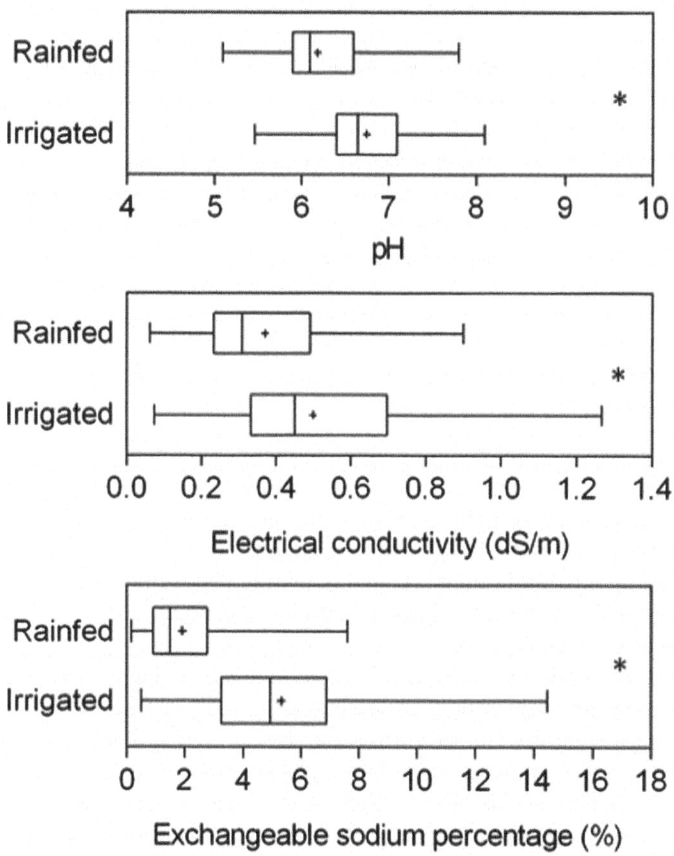

Fig. 1 Boxplot for pH, salinity (electrical conductivity) and sodicity (exchangeable sodium percentage) in the upper 0–20 cm of rainfed and supplementary-irrigated soils. The external lines of the boxes represent quartiles 0.25 and 0.75, and the internal lines, the median (0.50). The ends of the horizontal lines represent the minimum and maximum values. The + signs correspond to the mean figures of treatments. Significant differences between treatments: * (t-couplet; $p < 0.05$). Andriulo et al. (1998), Caviglia and Paparotti (2000), Pilatti et al. (2006), Andriani (2009), Torres Duggan et al. (2012), Costa and Aparicio (2015), Etcheverry and Génova (2015), Giubergia and Rampoldi (2017), Alvarez et al. (2019) and Peralta et al. 2019

In spite of the significant increase in EC under supplementary-irrigated fields, salinity levels were low and far from critical salinity thresholds for sensitive crops. Similarly, the rise in soil pH was slight, without further implications on soil chemical dynamic (e.g., nutrient bioavailability). As can be seen in Fig. 1, the greatest effect of supplementary irrigation was on ESP values. ESP rose three-fold due to irrigation. Similar processes have been described in other parts of the world, like the Great Plains of North America (i.e., Wienhold and Trooien 1995) and Europe (Nunes et al. 2007).

Different patterns regarding the dynamics of salinity and sodicity over time, on supplementary-irrigated soils, have been observed. Thus, according to Genova (2011), ESP levels rise during the irrigation season with a subsequent reduction caused by rainfall and salt leaching while EC remains low with few variations over time. On the other hand, Torres Duggan et al. (2017a, b) observed situations where ESP rose during the irrigation season without a subsequent drop. The reasons behind these findings should be related with the shallow water-table levels that prevent sodium from leaching out of the root zone. Moreover, because most groundwater in the humid farming areas of the Pampas region tends to be poor (high SAR values), a subsurface sodification might occur in fields with shallow water tables.

2.2 *Effects of Supplementary Irrigation on Soil Physical Properties*

The sodification process can affect soil physical quality. Information related to the impact of irrigation on soil physical properties in the Pampas region is scarce and has shown different trends. These studies evaluated variables associated with water dynamics in the soil (infiltration rate and hydraulic conductivity), soil compaction (bulk density and mechanical resistance), and soil features related to aggregation (size and aggregate stability).

Processes related to water dynamics, infiltration rate, and hydraulic conductivity can be negatively affected by the increase in ESP as a consequence of colloid dispersion and the degradation of the porous system. However, if ESP increase is followed by an increase in EC, the effect can be mitigated or reversed by colloid flocculation (Irurtia and Peinemann 1986). According to field experiments carried out in the Pampas region aimed at evaluating the changes in water infiltration rates due to supplementary irrigation (Torres Duggan et al. 2012; Giubergia and Rampoldi 2017; Alvarez et al. 2019; Peralta et al. 2019) water infiltration rates decreased by 11–48%, but differences were not statistically significant ($p > 0.05$; Fig. 2).

It should be pointed out that the evaluated physical properties showed high variability within the plots, thus statistical differences were not detected. In this regard, Torres Duggan et al. (2012) observed no statistically significant differences between supplementary-irrigated and rainfed plots, although infiltration mean values were

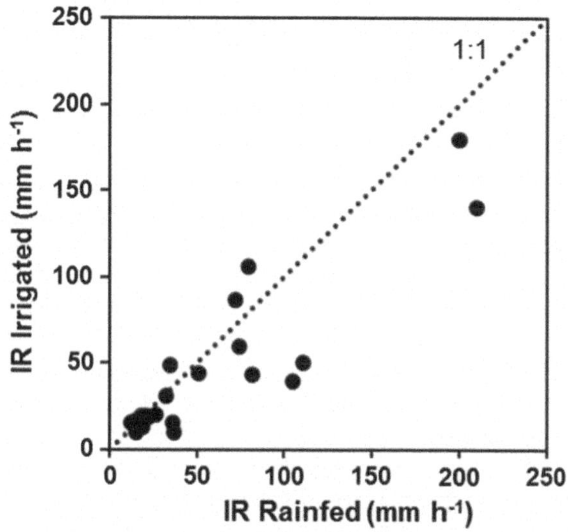

Fig. 2 Infiltration rate (IR) in rainfed and irrigated situation. Curve = 1:1 (Torres Duggan et al. 2012; Giubergia and Rampoldi 2017; Alvarez et al. 2019; Peralta et al. 2019)

moderate for irrigated plots and moderately rapid for rainfed fields. The only registered study in the Pampas where statistically significant decreases in the infiltration rate were reported is by Pilatti et al. (2006). According to these authors, in a regional field experiment network, 4 of the 11 evaluated sites where reductions in infiltration rates were measured with irrigation, showed increased ESP values. ESP increased between 6 and 20% on the average.

On the other hand, as far as we know, there are only two studies where the effect of irrigation water on hydraulic conductivity was measured in the Pampas. One (Ghiberto et al. 2007) was carried out in the North of the Pampas region, it reports decreases in hydraulic conductivity from 10 to 79%. In this situation, ESP reached values of 8 and 9% and the EC was always less than 1 dS m^{-1}. The other study (Costa and Aparicio 2015) was developed in the Southeast of the Pampas region where decreases were detected in hydraulic conductivity, when irrigation increased the RAS of the soil above 3.5.

In supplementary-irrigated environments, compaction may also impact soil physical condition, because soils maintain higher soil water contents, hence increasing susceptibility to compaction. Several studies carried out in the Pampas to evaluate soil bulk density did not observe significant differences between irrigated and rainfed situations, regardless of the increase in ESP figures (Torres Duggan et al. 2012; Peralta et al. 2019). In the case of field experiments in which differences between treatments were statistically significant, the increases in bulk density under supplemented-irrigated soils were low. These cases were associated with greater machinery traffic in those plots. In the Southeast of the Pampas region, soil bulk density rose by 1.6% (from 1.20 to 1.22 g cm^{-3}) under supplementary-irrigated treatments (Costa and Aparicio 2015). In addition, Giubergia and Rampoldi (2017) observed a 6% increment in bulk density under irrigated situations in the Northwest of the Pampas region.

This figure was related to a higher soil water content under irrigation and hence a greater susceptibility to compaction. Furthermore, Alvarez et al. (2019) reported a 5% increase soil bulk density in the 10–15 cm layer (1.32 versus 1.39 g cm^{-3}) and no impact at the upper 5 cm of top soil in experiments carried out in the Southwest of the Pampas region.

Although mechanical resistance often represents a sensitive variable to assess the impact of agronomic practices on soil properties, research findings were variable depending of the cropping region. Thus, different authors were not able to detect effects of irrigation on this physical property either in the Southwest (Alvarez et al. 2019) or North (Peralta et al. 2019) of the Pampas region, even in plots with high ESP. By contrast, in the Northern´s Pampas, Torres Duggan et al. (2012) reported mechanical resistance to rise by 15% in the supplementary-irrigated fields at 40 cm soil upper layer. However, statistically difference between treatments was only detected for the 35–40 cm.

There is consensus that the increase in soil bulk density and/or mechanical resistance on supplementary-irrigated fields is not associated with the irrigation per se but on the lower soil support capacity caused by wet irrigated soils, compared to soils of non-irrigated fields.

Torres Duggan et al. (2012) found higher aggregate average diameter under irrigation compared to rainfed situations. This increase was associated with the greater soil mechanical resistance measured in supplementary-irrigated treatments. However, the structural instability indices measured in this study were very low but significantly lower in irrigated plots. The structural instability values were: irrigated = 0.52 mm; rainfed = 0.60 mm. These low values are common in soil subjected to no tillage systems in the region (Alvarez et al. 2009). On the other hand, Giubergia and Rampoldi (2017) reported greater aggregate stability in the irrigated treatment related with higher organic matter, while other authors like Pilatti et al. (2006) showed no statistically significant differences between irrigated and rainfed situations.

3 Conclusions

Research information regarding the effects of supplementary irrigation on field crops of the Pampas region indicates the impact on soil sodicity (i.e., ESP) is greater than on salinity (EC). Under these conditions, ESP increased sharply reaching levels that might constraint root growth and/or crop yields. However, information is not consistent on whether such increase in ESP may impact on soil physical features. Some field experiments have shown higher figures of bulk density, mechanical resistance, or lower water infiltration rates in supplementary-irrigated fields, but there is not enough data to build a general conceptual model for the whole Pampas region. In contrast, the positive influence of supplementary irrigation on crop yield is generalized.

References

Alvarez CR, Rimski Korsakov H, Torres Duggan M (2016) Calidad de agua e impacto del riego complementario en la Región Pampeana. Actas del XXV Congreso Argentino de la Ciencia del Suelo, Río Cuarto, Córdoba (27/06/2016)

Alvarez CR, Rimski-Korsakov H, Steinbach HS, González A, Mayol F, Rosso Alba J, Sabatté E, Peralta G (2019) Efecto del riego complementario en el sudoeste bonaerense sobre las propiedades físico químicas del suelo. Actas VI Congreso RAS, Buenos Aires (22/07/2019)

Alvarez CR, Taboada MA, Gutiérrez Boem FH, Bono A, Fernández PL, Prystupa P (2009) Topsoil properties as affected by tillage systems in the Rolling Pampa region of Argentina. Soil Sci Soc Am J 73:1242–1250

Andriani J (2009) Impacto del Agua de riego sobre las propiedades químicas del suelo. Para mejorar la producción, INTA Oliveros 41:55–59

Andriulo A, Galetto ML, Ferreyra C, Cordone G, Sasal C, Abrego F, Galina J, Rimatori F (1998) Efecto de once años de riego complementario sobre un Argiudol Típico Pampeano. C Suelo 16:125–127

Andriulo A, Lopresti M, Milesi L (2017) Norte de Buenos Aires y Sur de Santa Fe. Riego suplementario con aguas de mediana a baja calidad en la ecorregión pampeana y del espinal (Argentina). Efectos sobre suelos y cultivos. In: Taleisnik E, Lavado RS (eds) Ambientes salinos y alcalinos de la Argentina: recursos y aprovechamiento productivo. Orientación Gráfica Editora, Universidad Católica de Córdoba, Buenos Aires, pp 199–207

Caviglia OP, Paparotti OF (2000) Efecto del uso de aguas de riego de calidad dudosa sobre algunas propiedades químicas del suelo en el centro oeste de Entre Ríos. Actas XVII Congreso Argentino de la Ciencia del Suelo, Mar del Plata, Buenos Aires, 11/04/2000

Costa JL (1999) Effect of irrigation water quality under supplementary irrigation on soil chemical and physical properties in ten Southern Humid Pampas of Argentina. J Crop Prod 2:85–99

Costa JL, Aparicio VC (2015) Quality assessment of irrigation water under a combination of rain and irrigation. Agric Water Manag 159:299–306

Etcheverry M, Génova L (2015) Uso sustentable de los recursos hídricos y edáficos para riego complementario de maíz y soja en la Cuenca del Río Arrecifes, Provincia de Buenos Aires. Rev Fac Agron, La Plata 114:125–141

Génova LJ (2011) Calidad del agua subterránea para riego complementario en la Pampa Húmeda argentina. Rev Fac Agron, La Plata 119:63–81

Ghiberto PJ, Pilatti MA, Imhoff S, de Orellana JA (2007) Hydraulic conductivity of Molisolls irrigated with sodic-bicarbonated waters in Santa Fe (Argentina). Agric Water Manag 88:192–200

Giubergia JP, Rampoldi EA (2017) Córdoba. Riego complementario en la provincia de Córdoba. Riego suplementario con aguas de mediana a baja calidad en la ecorregión pampeana y del espinal (Argentina). Efectos sobre suelos y cultivos. In: Taleisnik E, Lavado RS (eds) Ambientes salinos y alcalinos de la Argentina: recursos y aprovechamiento productivo. Orientación Gráfica Editora, Universidad Católica de Córdoba, Buenos Aires, pp 211–221

INTA (1999) Recomendaciones para la utilización de aguas para riego en función de su calidad, síntesis de los trabajos realizados en las unidades del INTA-IPG. Seminario de capacitación. Impacto ambiental del riego complementario. Pergamino, Argentina

Irurtia CB, Peinemann N (1986) Efecto de la relación de adsorción de sodio y la concentración de sales sobre la conductividad hidráulica de diferentes suelos. C Suelo 2:165–177

Maddonni GA (2012) Analysis of the climatic constraints to maize production in the current agricultural region of Argentina, a probabilistic approach. Theor Appl Climatol 107:325–345

Marano R, Pilatti MA (2017) Santa Fe. Riego suplementario con aguas de mediana a baja calidad en la ecorregión pampeana y del espinal (Argentina). Efectos sobre suelos y cultivos. In: Taleisnik E, Lavado RS (eds) Ambientes salinos y alcalinos de la Argentina: recursos y aprovechamiento productivo. Orientación Gráfica Editora, Universidad Católica de Córdoba, Buenos Aires, pp 183–193

Nunes JM, López-Piñeiro A, Albarrán A, Muñoz A, Coelho J (2007) Changes in selected soil properties caused by 30 years of continuous irrigation under Mediterranean conditions. Geoderma 139:321–328

Peinemann N, Diaz-Zorita M, Villamil MB, Lusarreta H, Grunewald D (1998) Consecuencias del riego complementario sobre propiedades edáficas en la llanura pampeana. C Suelo 16:39–42

Peralta G, Agosti B, Gil RC, Rimski Korsakov H, Alvarez CR (2019) Impacto del riego complementario sobre las propiedades físico químicas del suelo en el norte de la Provincia de Buenos Aires. Actas VI Congreso RAS, Buenos Aires, 22/07/2019

Pilatti MA, Imhoff S, Ghiberto P, Marano RP (2006) Changes in some physical properties of Mollisols induced by supplemental irrigation. Geoderma 133:431–443

Rampoldi EA, Boccardo M, Álvarez C, Martelloto E, Salinas AI, Lovera EF, Giubergia JP, Bustos V (2010) Calidad de las aguas subterráneas utilizados para riego suplementario en la provincia de Córdoba. INTA Manfredi

Romanelli A, Lima ML, Quiroz Londoño OM, Martínez DE, Massone HE (2012) A GIS-based assessment of groundwater suitability for irrigation purposes in flat areas of the Wet Pampa Plain, Argentina. Environ Manag 50:490–503

Rubio G, Lavado RS, Pereyra FX (eds) (2018) Soils of Argentina. In: Hartemink AE (ed) World soils book series. Springer International Publishing AG, New York

Salinas A (2018) Con riego suplementario se puede hasta duplicar el rendimiento en trigo https://inta.gob.ar/documentos/con-riego-suplementario-se-puede-hasta-duplicar-el-rendimiento-en-trigo

Salinas A, Saverina I, Bocardo M, Giubergia J (2016) Actualización de los resultados productivos con riego suplementario: sistema de producción en siembra directa continua, módulo demostrativo y experimental INTA Manfredi. En: 5° Reunión Internacional de Riego. Uso eficiente del agua para riego. Ediciones INTA, pp 114–120

Sánchez RM, Dunel Guerra L, Scherger M (2016) Evaluación de las áreas bajo riego afectadas por salinidad y/os sodicidad en Argentina. INTA Ediciones, Buenos Aires, Argentina

Severina I, Boccardo M, Aimar, F, Giubergia JP, Salinas, A (2016) Resultados productivos en trigo, soja y maíz en riego por goteo subterráneo. En: 5° Reunión Internacional de Riego. Uso eficiente del agua para riego. Ediciones INTA, pp 141–144

Torres Duggan M, Alvarez CR, Rimski-Korsakov H (2016) Evaluación y monitoreo de suelos bajo riego complementario en la Región Pampeana. In: 5° Reunión Internacional de Riego. Uso eficiente del agua para riego, Ediciones INTA, Buenos Aires, pp 13–39

Torres Duggan M, Álvarez CR, Rimski Korsakov H (2017a) Relevamiento de la calidad de agua y sodicidad edáfica de suelos regados en forma complementaria en norte de Buenos Aires y sur de Santa Fe. In: Taleisnik E, Lavado RS (eds) Ambientes salinos y alcalinos de la Argentina: recursos y aprovechamiento productivo. Universidad Católica de Córdoba, Buenos Aires, Orientación Gráfica Editora, pp 208–210

Torres Duggan M, Alvarez CR, Rimski-Korsakov H (2017b) Evaluación de la calidad del agua y del suelo regado en forma complementaria en la región pampeana argentina. Informaciones Agronómicas de Hispanoamérica 25(14):23

Torres Duggan M, Álvarez CR, Taboada MA, Celesti T, Vignarolli F, D'ambrosio D (2012) Riego complementario en un Argiudol típico de la pampa ondulada Argentina bajo siembra directa: Efectos sobre algunas propiedades químicas y físicas del suelo. C Suelo 30: 201–207

Wienhold BJ, Trooien TP (1995) Salinity and sodicity changes under irrigated alfalfa in the Northern Great Plains. Soil Sci Soc Am J 59:1709–1714

Wilson MG, Cerana JA (2017) Entre Ríos. Riego suplementario con aguas de mediana a baja calidad en la ecorregión pampeana y del espinal (Argentina). Efectos sobre suelos y cultivos. In Taleisnik E, Lavado RS (ed). Ambientes salinos y alcalinos de la Argentina: recursos y aprovechamiento productivo.UCC-OGE. Buenos Aires. 194–198 pp

Conceptual and Practical Framework to Address Gypsum Management in Salt-Affected Soils

Martín Torres Duggan and Mónica B. Rodríguez

Abstract This chapter gives theoretical and practical information for proper gypsum management in salt-affected soils. Although gypsum is well known as amendment an to ameliorate soil physical conditions, key factors should be carefully considered to optimize gypsum agronomic efficiency and effectiveness at farm level. Some of the main factors are the gypsum quality for agriculture use and physiographical and soil features related with salinity and sodicity constraints. Under typically non-irrigated sodic-soils, where cattle production is the most frequent land use, surface cover and water table dynamics play a significant role in defining the gypsum technological management. This is valid, particularly, in terms of application methods. In high-yielding crops receiving supplementary irrigation with poor quality water, gypsum management is quite different. Some of the key differences arise from surface sodification while subsoil sodicity is not a constraint in these agricultural soils. However, other issues like surface compaction and exchangeable sodium percentage (ESP) are relevant diagnosis factors to address gypsum management in irrigated land. Moreover, combined application of gypsum and organic amendments like animal manure are considered effective technologies to improve the physical quality of salt-affected soils alongside prevention practices.

Keywords Amendments · Sodic soils · Calcium sulfate · Soil cover · Minerals

1 Introduction

Minerals represent resources of great economic importance since they are widely used in different applications such as the chemical industry, construction, as well as fertilizers and amendments to improve plant nutrition and soil fertility features (Van Straaten 2014; Herrmann and Torres Duggan 2017). The main minerals used

M. Torres Duggan (✉)
Tecnoagro, Buenos Aires, Argentina
e-mail: mjjtorresduggan@gmail.com

M. B. Rodríguez
Facultad de Agronomía, Universidad de Buenos Aires, Buenos Aires, Argentina

E. Taleisnik and R. S. Lavado (eds.), *Saline and Alkaline Soils in Latin America*, https://doi.org/10.1007/978-3-030-52592-7_16

in agriculture are carbonates (either calcium or calcium-magnesium carbonates), phosphate rock, gypsum, and more recently zeolites minerals.

Gypsum is the most common mineral used for sodic soils amelioration at global level. In addition, this mineral is applied for improving the physical condition of agriculture soils being affected by secondary alkalinization because of waterlogging, flooding, irrigation with alkaline waters, among other processes (Lavado and Taboada 2009; Torres Duggan et al. 2017). Likewise, gypsum is a key resource to complement and improve the agronomic effectiveness of limestone in acid and/or acidified soils (Vázquez 2011; Vázquez et al. 2013).

It is important to highlight that gypsum is a widespread source to provide plant-available sulfur for crops. Moreover, this source is the most frequent solid sulfur fertilizer type being used by farmers in the fertilization of field crops in the non-saline/sodic soils of the Pampas Region of Argentina (Torres Duggan et al. 2012a; Rodríguez and Torres Duggan 2015).

The main reasons why gypsum is nowadays the most popular mineral adopted by farmers to ameliorate sodic soils at worldwide level is linked with some factors like its great abundance in nature, its chemical composition, and reaction in soil. Furthermore, the wide supply of gypsum in different countries and regions worldwide explains its lower cost compared with other soluble calcium materials available in the fertilizer marketplace.

Most of the published scientific and technological literature related with gypsum management in sodic soils has been focused on soils under conventional tillage, where the mineral is incorporated into topsoil using different tillage machinery. However, this situation is not applicable for most salt-affected soils across Pampas soil environments, where alkalinization issues can arise from natural processes like flooding or waterlogging. Soil cover is also maintained in saline soils to prevent upward salt movement from water table and/or B sodic horizons (Torres Duggan and Malfitano 2019).

Considering the context described above, this chapter aims to provide insights and conceptual models regarding gypsum management for salt-affected soils of Argentina with the following specific objectives:

(1) Characterize gypsum agronomic properties and aptitude as conditioner/amendment to ameliorate or improve salt-affected soils, mainly under no tillage farming environments.
(2) Provide conceptual and practical frameworks for a proper gypsum management either in typical sodic soils environments or fields with secondary alkalinization, mainly under supplementary irrigated soils in humid and sub-humid areas of the Pampas region.

2 Gypsum Geological Resources in Argentina and Key Mineralogical and Chemical Features

Gypsum is an abundant mineral with a profuse regional distribution in Argentina. Mesozoic deposits are found in the Neuquén Basin, in the provinces of Mendoza, Río Negro, and Neuquén, while the tertiary deposits are located in the extra-Andean strip (e.g., Chubut, Salta, Catamarca, Tucumán, Santiago del Estero, San Luis, La Rioja, and San Juan). In Entre Ríos, La Pampa, and Buenos Aires, gypsum deposits are quaternary (SEGEMAR-UNSAM 2008). In nature, gypsum minerals are the first to crystallize after the evaporation of saline waters. Gypsum deposits can be from marine or lacustrine origin, and based on this, two genetic models of mineral deposits are defined: marine evaporites and lacustrine evaporites (Herrmann and Torres Duggan 2016). The main deposits are of the first type, they occur in marginal marine basins whose source of calcium sulfate is sea water, such deposits are extensive and considerable thick, with an average tonnage and grade (purity) of 280 Mt and 90.7% of purity, considering the main deposits in the world. Lacustrine deposits are smaller, although they can be up to 30 m thick with average tonnages of 14 Mt and 85% purity.

In Argentina, various surveys of the available gypsum resources have been carried out in the different provinces (Ponce and Torres Duggan 2005; SEGEMAR-UNSAM 2008). Although gypsum deposits of the Neuquén basin and in particular in Mendoza province are considered the best in terms of quality (i.e., high mineralization and purity), gypsum resources available in other geological basins and provinces are also adequate for the agriculture sector. Further details on gypsum geological deposits and their geographical distribution, and characteristics, which are beyond the scope of this chapter, can be obtained in SEGEMAR-UNSAM (2008).

Regarding mineralogical standpoint, gypsum is a calcium sulfate mineral phase in nature and corresponds to calcium sulfate dihydrate ($CaSO_4 . 2H_2O$). Its greater solubility and molecular water content differentiate it from the basanite and the anhydrite minerals (Table 1).

Gypsum is a soluble-sulfate salt (~2.4 g L^{-1} at 0 °C), with a low salt index and neutral reaction in the soil (Gowariker et al. 2008). These attributes, alongside its wide geographic availability, and its lower relative cost to other types of calcium fertilizers promote its use for both sodic soil amelioration and sulfur crop fertilization (Torres Duggan et al. 2012a; Herrmann and Torres Duggan 2016). Although other minerals like elemental sulfur are mentioned in scientific literature as a conditioner or amendment for improving alkaline soils, it is important to point out that its main supply comes from the petrochemical industry, while the contribution of natural deposits is minor. Therefore, the relatively uncommon availability of elemental sulfur brings about a reduced use of this mineral to ameliorate alkaline environments. On the other hand, taking into consideration that most salt-affected soils in the humid and sub-humid areas of the Pampas region are sodic soils (i.e., high exchangeable sodium contents), elemental sulfur can be used for reducing pH levels. However, it is not adequate for improving physical features, since this requires the exchange of

Table 1 Main calcium sulfate minerals available in nature

Calcium sulfate specie	Chemical composition	Nutrient content	Key agronomic features
Gypsum	$CaSO_4 . 2H_2O$	S = 18.6% Ca = 23%	– Plant-available sulfate source – Main solid sulfur fertilizer used by farmers in Argentina – Sulfur range between 15 and 18% – Neutral reaction in soil – Proper to ameliorate sodic or alkalinized soils – Useful for improving limestone. Agronomic effectiveness under no tillage systems
Basanite	$CaSO_4 . 0.5\ H_2O$	S = 22% Ca = 27%	– Rapid conversion to gypsum during reaction in soil (hydration) – Similar agronomic effectiveness and efficiency to gypsum – Generally, marketed in the construction sector. Companies obtain it from high purity gypsum raw materials – Greater S and Ca concentration because of chemical composition
Anhydrite	$CaSO_4$	S = 23.5 Ca = 29.4%	– Low value as sulfur fertilizer or soil conditioner/amendment due to low water solubility and reactivity in soils

sodium for calcium which is a typical gypsum effect. However, in alkaline soils with high calcium saturation or presence of free carbonates, the addition of elemental sulfur reacts producing gypsum, a compound that is ultimately responsible for the effects on the physical condition of the soil (Richards 1973). Thus, when gypsum is available at the production site, it is more reasonable to directly use this mineral since research information regarding elemental sulfur oxidation and its agronomic effectiveness under Pampas soils environments is not available, existing a significant scientific information gap.

3 Gypsum Quality for Agricultural Use

To address the use of minerals as fertilizers, soil conditioners or amendments both geological and agronomic issues should be taken into account, like geological-mining issues, processing as well as the distribution throughout the supplying system (i.e., from mine or fertilizer plant to field). In this regard, significant progress has been made in Argentina in relation to the quality of the gypsum offered in the marketplace as both sulfur fertilizer and mineral amendment. Local norms developed by the National Normalization and Certification Institute of Argentina (IRAM) provide the specific physical and chemical requirements of the gypsum for agriculture use. This information framework was incorporated into the current local legal regulations for the registration and use of gypsum-based fertilizers or amendments (Rodríguez and Torres Duggan 2015; Torres Duggan and Malfitano 2019). Those local norms were based on the American Society for Testing and Materials (ASTM) normative for agricultural gypsum.

The gypsum quality for agricultural use depends on its chemical and physical properties. The main feature related with chemical quality is the purity of the product (percentage of gypsum mineral in the commercial product) that defines the nutrient content of the product. The physical quality is addressed by the analysis of the particle size distribution both in terms of range and dust content (Rodríguez and Torres Duggan 2015). Furthermore, the concentration of specific elements like sodium, iron, and aluminum are considered when high gypsum rates are applied, typically for sodic soil amelioration or correction.

According to current Argentine normative, a minimum of 85% calcium sulfate dihydrate is necessary to market gypsum in the agriculture sector. Table 2 provides a detail on the quality requirements to market agricultural gypsum locally.

The granulated gypsum is the most frequent physical presentation used to market this product as sulfur fertilizer, while powder gypsum or the so-called "pelletized" gypsum is mainly applied to ameliorate sodic soil conditions. The granulated gypsum is the one that after extraction from geological deposits is processed through primary

Table 2 Chemical and physical quality requirements for the commercialization of agricultural gypsum at the argentine marketplace

Properties	Quality requirement
Purity (% of $CaSO_4 . 2H_2O$)	Min 85% (w/w)
Sodium chloride	Max. 0.5% (w/w)
Iron oxide	Max. 1% (w/w)
Aluminum oxide	Max. 1.2% (w/w)
Free water	Max. 1% (w/w)
Particle size of granulated gypsum	90% between 1 y 4 mm Max. 10% below 1 mm
Particle size of pelletized gypsum	97% entre 1 y 4 mm Max. 3% below 1 mm

Source SENASA (2011)

and second crushing to reach a granulometry range between 1 and 5 mm (Torres Duggan and Malfitano 2019).

On the other hand, pelletized gypsum is obtained by agglomerating fine gypsum using specific equipment such as pelletizing disks or agglomerating drums (Rodríguez and Torres Duggan 2015). It is important to point out that, regardless of the improved physical properties of the pelletized material (e.g., particle size uniformity), there was not observed differences in agronomic effectiveness or agronomic efficiency when comparing pelletized and un-pelletized gypsum products (Torres Duggan et al. 2012b). This is reasonable since pelletizing process does not modify the solubility of the mineral or the reaction of the mineral in the topsoil. The improved disaggregation speed characteristics of the pelletized gypsum, while an interesting feature for sales purposes, it does not impact the dynamic of the gypsum in the soil–plant system. In other words, both gypsum types, pelletized, and un-pelletized, have the same capability to provide plant-available nutrients. However, pelletized gypsum might have an advantage for amending sodic or alkalized soils due to the capability to provide a very fine material, which is necessary for a rapid reaction of the product in top soil under no tillage systems, and at the same time, it can be applied by using typical fertilizer spreaders without dust generation. On the contrary, when a very fine gypsum is broadcast over the soil surface, using gypsum without pelletizing, a poor distribution of the product can be observed under windy application conditions.

4 General Framework to Address Soil Sodicity Amelioration Though Gypsum Application

It is important to point out that improvement, and rehabilitation of sodic soils and/or agricultural soils which have suffered sodification processes due to water table rise requires an integral diagnosis (Cisneros et al. 2008; Torres Duggan et al. 2017). In general terms, the main objective of a proper management of saline or saline-sodic environments is to maintain or recover the vegetation cover in order to avoid the capillary upward movement of salts from the water table. Among management practices aimed at improving forage production at farm level (e.g., rotational grazing, tillage amelioration, direct seeding, subsoiling, foresting, fertilization, among others), gypsum application in bands localized in the sowing zone is an effective strategy to improve soil water infiltration rate and the soil physical quality that, in turn, improves the production capacity of the site (Cisneros et al. 2008; see also Chap. 11 in this book).

The first effect of the gypsum addition to the soil is to produce clay flocculation, ameliorating the physical condition, particularly the water infiltration rate (Lavado 2010). Later on, as the mineral interacts with the soil solid phase, a progressive ionic exchange takes place (Eqs. 1 and 2).

$$CaSO_4 \cdot 2H_2O\ (\text{solid}) \rightarrow CaSO_4\ (\text{solution}) + 2H_2O \qquad (1)$$

$$CaSO_4 \text{ (dissolved)} \rightarrow Ca_2X + Na_2SO_4 \text{ (solution)} \quad (2)$$

The main effect of calcium is to improve the soil structural condition and reduce adsorbed sodium, replaced by calcium, and sodium ions react with sulfate forming sodium sulfate in solution. This chemical compound must be eliminated through soil drainage in order to complete the soil chemical correction. Therefore, adequate soil permeability is necessary and the water table must be located deep enough to prevent the salt upward movement from water table (i.e., capillary salt movement). As the sodium is removed from the soil, alkalinity constraints are ameliorated (Richards 1973; Lavado and Taboada 2009).

The evaluation of the water table dynamics, the amount of vegetation cover, and the profile distribution of alkalinity are major decision variables to take into account for a proper evaluation of the viability and expected agronomic effectiveness of gypsum application in sodicity-affected environments (Lavado 2010). The depth at which the water table brings about an upward salt moment toward upper soil layers is so-called "critical depth". It depends mainly on the soil texture and structure. For the central Pampas region, this critical depth is located roughly in the range of 100 and 130 cm in sandy loam and loamy soils, respectively, being a little higher in clayey soils (Cisneros et al. 2008).

The above-mentioned theoretical or idealized dynamic of groundwater is quite different under soils having subsurface horizons with high swelling clay contents. These soil types are widespread in the flooding Pampas (Lavado and Taboada 2009). The impermeable subsurface horizons of Natraquolls and Natraqualfs keep the groundwater confined below the Bt horizon and hence limit the upward movement of salts and the subsequent salinization and/or sodification constraints. The empirical evidence for these phenomena is the formation of calcium carbonate at the base of Bt horizons, which are named "tosquillas" locally. Nevertheless, under situations of low vegetation cover due to overgrazing and high atmospheric water demand (e.g., under spring or summer), there might be situations of surface salinization/sodification even under lowland soils with high clay subsurface horizons due to capillary continuity across A and B horizons that allow the upward salt movement.

When surface salinization and/or sodification from groundwater takes place either because the water table is located over the critical depth or due to salt upward movement from high clay Bt subsurface horizons, the massive application of gypsum is an ineffective practice because of the high probability of re-sodification over time (Cisneros et al. 2008). Therefore, under these conditions, the direct seeding of adapted forage crops alongside the application of low gypsum rates in bands over the sowing row are recommended practices to improve soil fertility and forage productivity. The banded-applied gypsum at lowlands, by improving physical and chemical topsoil features, optimizes seedlings germination as well as crop emergency and establishment. However, where vegetation cover is very low due to a strong salinization, it is recommended in the first place to restore the cover by different practices like grazing closures, mulches, re-vegetation techniques, among others. Once the vegetation is

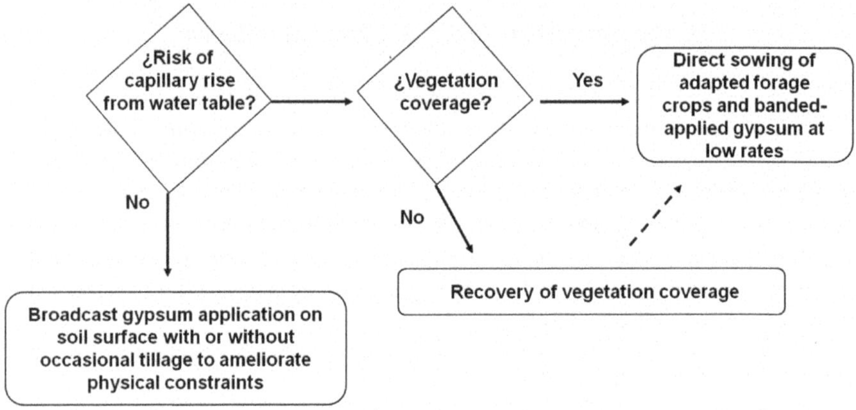

Fig. 1 Decision model to define gypsum application technology under sodic or sodified soils of the Pampas region. *Source* Based on Cisneros et al. (2008) and Taboada and Lavado (2009)

established, banded-applied gypsum can be used to improve the sowing microenvironment as mentioned above (Cisneros 2008; see also Chap. 11 by Taboada et al. “Origin, Management and Reclamation Technologies of Salt-affected and Flooded Soils in the Inland Pampas of Argentina,” in this book), (Fig. 1).

When the water table is located below the critical depth, the salts are not able to reach the topsoil by capillarity. Thus, if the soil presents a proper internal drainage, the possibilities of applying gypsum in greater rates and broadcasted over the soil surface are higher. Likewise, when the presence of hard crust on the surface is observed, the combination of mechanical amelioration (with low or no cover removal) alongside gypsum application is effective to overcome compaction constraints and improve soil fertility features (Fig. 2).

The definition of the accurate gypsum rate to be applied at the site or farm level should be based on regional and subregional calibrations obtained by local experimentation. Nowadays, this information is not available for most Pampean environments with the exception of some experiments carried out at field or laboratory scale. Hence, in this context of scarce experimental information, a theoretical estimation of “gypsum requirements” can be a first approximation. Gypsum rates are calculated considering the amount of sodium (meq) to be replaced from the soil CEC by the gypsum application, assuming a specific depth of reaction. However, this method has limitations and requires adjustments in the calculation taking into account other factors like gypsum purity, soil characteristics, and application methods (Richards 1973; Lavado 2010). Likewise, in sodic soils with very high ESP levels, gypsum rates calculated by using the above theoretical calculation are quite high and hence frequently are not viable from the economic and/or logistic standpoint.

Because of those limitations, the use of low rates of granulated or powder gypsum sources, tailoring the management according to the available machinery and application conditions, represents an effective technology to improve sodic environments growing adapted forages. Due to the lack of experimental information regarding

Fig. 2 Direct drilling planters with adapted decompaction attachments used by farmers to improve forage production under salt-affected soils in the flooded Pampas of Argentina (A y B) and decompaction equipment developed by INTA San Luis (C y D). Photos: M. Torres Duggan

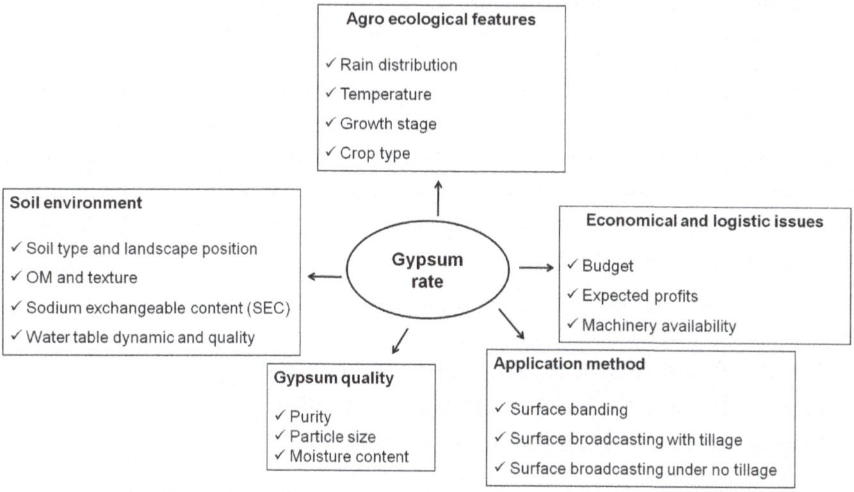

Fig. 3 Necessary information for an integrating evaluation of gypsum rates at farm level. *Source* Own elaboration

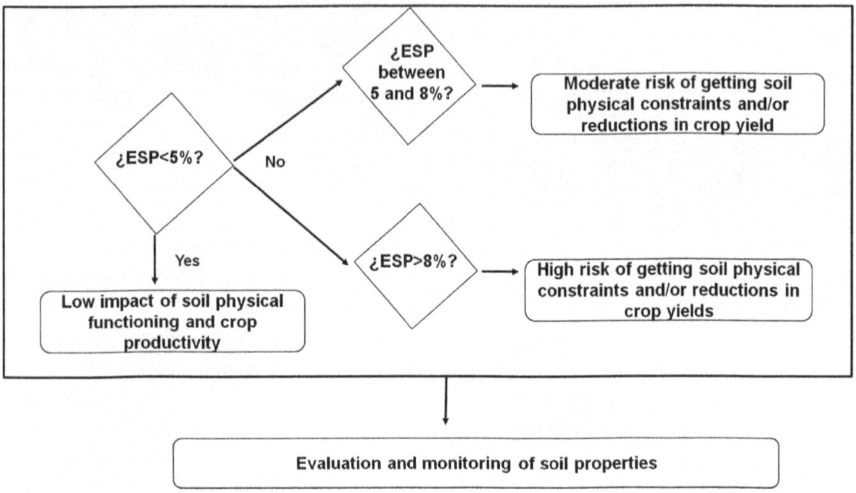

Fig. 4 Conceptual model to assess the sodification risk and its impact on crop yields under supplementary-irrigates environments of the Pampas region of Argentina. *Source* Torres Duggan et al. (2016)

gypsum use in sodic soils, agronomic advisors play a key role in addressing proper gypsum management, integrating the different factors that affect gypsum agronomic effectiveness under salt-affected soils (Fig. 3).

As pointed out, the expected efficiency and effectiveness of the gypsum application in soils being affected by groundwater movements near soil surface is quite low. The better the soil physical condition and the lower the influence of the salt upward movement from the water table, the greater the efficiency and the perdurability of gypsum applications. However, such effectiveness is associated with several factors as gypsum rates, purity, and particle size; gypsum placement and application time; initial soil ESP; soil texture, and organic matter. In addition, the environmental conditions during and after the application, mainly soil temperature and humidity are key factors to be considered.

5 Strategies to Manage Agricultural Gypsum to Ameliorate Sodicity Constraints Under Supplementary Irrigated Environments

The use of agricultural gypsum in production systems where complementary irrigation is applied must be considered as a part of an integral diagnosis. This approach should be focused on analyzing key aspects like the rate and quality of irrigation water, soil properties, and agro-ecological conditions (Torres Duggan et al. 2017). Hence, the first step to evaluate the need to apply gypsum is to carry out a proper soil fertility diagnosis based on the consideration of indices like the ESP as a proxy for soil sodicity (Fig. 4 and Table 3).

From the amelioration of physical constraints standpoint, soluble calcium provided by the gypsum improves surface topsoil structure and reduces the sodium accumulation (ESP) in irrigated fields where poor irrigation water is applied (e.g., high sodium bicarbonate content). Nowadays, there is no experimental available information regarding gypsum amending under complementary irrigated environments, particularly under no tillage systems which are the most frequent in the Pampas region of Argentina.

The combination of mechanic amelioration without removal of the stubble cover and the addition of gypsum in low rates (e.g., 2–3 Mg ha^{-1}) feasible to distribute with farm available machinery is an effective technology to improve the topsoil physical fertility in supplementary irrigated soils showing moderate sodification constraints (Ponce and Torres Duggan 2005; Torres Duggan et al. 2017). Nevertheless, the gypsum application at low rates may require more than one application, thus increasing the cost of the practice. As a consequence, the advantages and disadvantages of each technology must be evaluated for each production system taking into account not only the mentioned agronomic aspects but also their economic viability and logistic. Likewise, re-sodification due to capillary salt rise toward topsoil should be considered because it might reduce sharply the agronomic effectiveness of gypsum amending due to re-sodification issues.

The above-mentioned theoretical approach used for defining gypsum application rates for sodic soils can also be considered for non-sodic soils which are the typical conditions in supplementary irrigated environments. However, the ESP levels of the latter soils and the subsequent gypsum requirements are much lower than sodic

Table 3 Management practices for diagnosing, preventing, and correcting sodification constraints under supplementary-irrigated environments of the Pampas region of Argentina

Context	Recommended practices
Diagnosis	✓ Soil analysis and selected variables for monitoring purposes (e.g., pH, EC, OM, CEC, and cations) and ESP calculation ✓ Analysis of irrigation water (pH, EC, anions, and cations) and calculation of SAR or adjusted SAR ✓ Observation of soil profile (structure; presence of crusts, roots depth; etc.) ✓ Evaluation of soil physical condition (e.g., water infiltration rate, mechanical resistance, bulk density, etc.)
Prevention	✓ Irrigation rates tailored to crop water requirements ✓ Control of machinery transit at harvest time ✓ Rotation with cereal and cover crops ✓ Balanced fertilization
Correction	✓ Reduction of irrigation rates ✓ Organic and mineral amendment application ✓ Tillage amelioration by removing compacted zones either at soil surface or subsoil (e.g., paratill/paraplow)

Source Adapted from Torres Duggan et al. (2017)

soils. For this calculations, traditional literature like Richards (1973) and more recent like Lavado and Taboada (2009) and Lavado (2010) can be consulted by readers. Tables 4 and 5 show reference gypsum application rates for moderately sodified supplementary-irrigated soils.

Another topic that has not been studied locally is the combined application of organic animal residues and gypsum, to improve the soil physical condition though organic carbon and calcium addition (e.g., higher water infiltration rates and structural stability figures). In this regard, a recent field research study conducted by Gabioud et al. (2019) showed an improved surface structure at silty soils receiving combined application of poultry manure and gypsum. Furthermore, the organo-mineral amending strategy not only improves the soil physical condition but also improves plant-available N. enhancing crop yield and quality. The eventual losses of nitrogen due to ammonia volatilization processes can be mitigated either by incorporating the organic residue into subsoil or by using nitrogen stabilizers (e.g., urease inhibitors, zeolites, among others) that overall increase the nitrogen use efficiency (Rasoulzadeh and Yaghoubi 2010; Larney and Angers 2012; Torres Duggan 2017).

Table 4 Theoretical gypsum requirements (Mg ha^{-1}) to reach 2% ESP at supplementary-irrigated soils of the Pampas region

	CEC (meq 100 g^{-1})				
Initial ESP (%)	10	15	20	25	Efficiency (%)
5	0.9	1.3	1.8	2.3	50
8	1.8	2.7	3.7	4.6	50
11	2.5	3.9	5.2	6.9	60
14	3.5	5.2	6.9	9.2	60

Source Own elaboration
Assumptions: water table under the critical depth, depth = 20 cm; bulk density = 1.2 g cm^{-3}; gypsum purity = 100%; equivalent mass of gypsum = 86
The exchange efficiency varies according to initial ESP level and was considered the efficiency figure proposed by Richards et al. (1973)

Table 5 Theoretical gypsum requirements (Mg ha^{-1}) to reach 5% ESP at supplementary-irrigated soils of the Pampas region

	CEC (meq 100 g^{-1})				
Initial ESP (%)	10	15	20	25	Efficiency (%)
8	0.9	1.3	1.8	2.3	50
11	1.7	2.6	3.4	4.3	60
14	2.6	3.9	5.2	6.5	60

Source Own elaboration
Assumptions: water table under the critical depth, depth = 20 cm; bulk density = 1.2 g cm^{-3}; gypsum purity = 100%; equivalent mass of gypsum = 86
The exchange efficiency varies according to initial ESP level and was considered the efficiency figure proposed by Richards et al.(1973)

The adoption of organo-mineral materials to supply plant-available nutrients (e.g., nitrogen, phosphorus, etc.) and improve soil fertility features (e.g., structure, C stocks, water infiltration, microbiological activity, among many others) is particularly relevant, considering that nowadays, most of the animal wastes are not being re-used at farm level for crop and pasture fertilization/amending. On the contrary, in general, these animal organic residues are concentrated and accumulated at farming operations, leading to soil, air, or groundwater contamination (Herrero and Gil 2008). Thus, it is important to highlight the value of these materials as strategic resources for crops fertilization, mainly as nitrogen and phosphorus sources (Herrero 2014; Torres Duggan and Rodriguez 2015).

6 Concluding Remarks

The information discussed in this chapter highlights the extensive amount of available information in Argentina regarding gypsum mineralogical and geology issues as well as about gypsum quality for agriculture use. In contrast, very scarce experimental information is available at regional and/or subregional level regarding gypsum and organo-mineral amending of salt-affected soils either at sodic and saline-sodic environments or at agriculture soils, subjected to supplementary irrigation and showing moderate surface sodification constraints.

Because of the great complexity of sodic soils management, an integral approach is necessary to evaluate and define amelioration strategies including gypsum management. Several factors should be taken into account like physiographic characteristics at regional and farm levels, soil properties, hydrological, and groundwater dynamics, among the main ones. In addition, the machinery available at farm level to apply gypsum plays a central role in determining the expected efficiency and effectiveness of the gypsum as soil conditioner or amendment. The main objectives of its utilization should be to improve the overall soil fertility conditions by different mechanisms and processes like increase of soil water infiltration rates, maintenance and/or restoration of the soil cover (key factor to avoid the upward movement of salts from groundwater), ameliorate surface crust that tend to be very hard due to sodification and compaction, among others. In this context and specially under high surface ESP levels, gypsum application at low rates is a cheap and effective technology to improve forage production for both sown adapted forage crops or under natural grassland of the flooding Pampas.

On the other hand, the application of gypsum and other mineral and organo-mineral amendments are also key technological tools to overcome moderate sodicity issues in supplementary irrigated environments of the Pampas region, where an overlap and interaction of sodification and compaction processes are quite frequent under no tillage farming systems. There is still a huge research gap in relation to sodicity diagnosis at regional, subregional, and farm levels; gypsum application technology by soil type and crop; integrated assessment of sodicity, and compaction, among other research needs.

References

Cisneros JM, Degioanni A, Cantero JJ, Cantero A (2008) Caracterización y manejo de suelos salinos en el área Pampeana. In: Taleisnik E, Grunberg K, Santa María G (eds) La salinización de suelos en la Argentina, su impacto en la producción agropecuaria. Editorial Universidad Católica de Córdoba, Córdoba, pp 17–46

Gabioud EA, Sasal MC, Wilson MG, Seehaus MS, Van Opstal NA, Beghetto WAB (2019) Addition of organic and inorganic amendments to regenerate the surface structure of silty soils. Soil Use Manag. https://doi.org/10.1111/sum.12567

Gowariker V, Krishnamurthy VN, Gowariker S, Dhanorkkar M, Paranjape K (2009) In: The fertilizer encyclopaedia. Wiley, NJ, USA, p 861

Herrero MA, Gil SB (2008) Consideraciones ambientales de la intensificación en producción animal. Ecología Austral 18:273–289

Herrero MA (2014) Manejo y tratamiento de estiércol y purines. In: Gil SB, Rebuelto M, Sardi GMI (eds) MA Herrero MA. La producción animal y el ambiente. BMPRESS, Buenos Aires, pp 97–126

Herrmann C, Torres Duggan M (2016) Fertilizantes y enmiendas de origen mineral: caracterización y uso en la Argentina. In: Pereyra FX, Torres Duggan M (eds) Suelos y Geología Argentina. Asociación Argentina de la Ciencia del Suelo y Asociación Geológica Argentina, Buenos Aires, pp 329–355

Larney FJ, Angers DA (2012) The role of organic amendments in soil reclamation: a review. Can J SoilSci 92:19–38

Lavado RS, Taboada MA (2009). Alteraciones de la fertilidad del suelo causado por halomorfismo. In: Taboada MA and Lavado RS (eds). Alteraciones de la fertilidad de los suelos. El halomorfismo, la acidez, el hidromorfismo y las inundaciones. Editorial Facultad de Agronomía (Universidad de Buenos Aires), Buenos Aires, pp 1–44

Lavado RS (2010) Salinidad y alcalinidad: propiedades, efectos sobre los cultivos y manejo. In: Alvarez R, Rubio G, Alvarez C, Lavado RS (eds) Fertilidad de suelos, caracterización y manejo en la Región Pampeana. Universidad de Buenos Aires, Buenos Aires, Editorial Facultad de Agronomía, pp 21–44

Ponce B, Torres Duggan M (2005)Yeso. In: Nielson H, Sarudiansky R (eds) Minerales para la agricultura en Latinoamérica. CEPS. Buenos Aires

Rasoulzadeh A, Yaghoubi A (2010) Effect of cattle manure on soil physical properties on a sandy clay loam soil in North-West Iran. J Food Agric Environ 8:976–979

Richards LA (1973) Diagnóstico y rehabilitación de suelos salinos y sódicos. Editorial Limusa, México

Rodríguez MB, Torres DM (2015) Caracterización de los fertilizantes, enmiendas, abonos y su calidad agronómica. In: Alvarez R, Rubio G, Alvarez C, Lavado RS (eds) Fertilidad de suelos, caracterización y manejo en la Región Pampeana. Universidad de Buenos Aires, Buenos Aires, Editorial Facultad de Agronomía, pp 369–400

SEGEMAR-UNSAM (2008) Yacimientos portadores de sulfato de calcio. En: Carbonato y sulfato de calcio. Publicación técnica N°15, pp 81–98

SENASA (2011) https://www.senasa.gov.ar/normativas/resolucion-264-2011-senasa-servicio-nacional-de-sanidad-y-calidad-agroalimentaria

Torres Duggan M, Melgar R, Rodriguez MB, Lavado RS, Ciampitti IA(2012 a). Sulfur fertilization in the argentine Pampas region: a review. Agronomia & Ambiente, 32: 61–73

Torres Duggan M, Alvarez CR, Taboada MA, Celesti T, Vignarolli F, D'ambrosio D (2012b) Riego complementario en un Argiudol típico de la Pampa Ondulada argentina bajo siembra directa: efectos sobre algunas propiedades químicas y físicas. C Suelo, 30:201–207
Torres Duggan M, Rodríguez MB (2015) Biomasa y aprovechamiento de agua y nitrógeno de raigrás en un Hapludol éntico tratado con residuos de feedlot y zeolitas. In: García F, Correndo A (eds) Simposio Fertilidad 2015. Nutriendo los suelos para las generaciones del futuro, Rosario, pp 217–225
Torres Duggan M, Álvarez CR and Rimski Korsakov H (2016) Evaluación y monitoreo de suelos bajo riego complementario en la Región Pampeana. In: 5º Reunión Internacional de Riego. INTA Manfredi. Córdoba, pp 13–39
Torres Duggan M (2017) Productividad forrajera en ambientes ganaderos mejorados con la aplicación de residuos de feedlot, roca fosfática y zeolita natural. Doctoral dissertation. Facultad de Ciencias Veterinarias. Universidad de Buenos Aires
Torres Duggan M, Alvarez CR, Rimski-Korsakov H (2017) Evaluación de la calidad de agua y del suelo regado en forma complementaria en la región Pampeana. Informaciones Agronómicas de Hispanoamerica 25:17–23
Torres Duggan M and M Malfitano (2019) Panorama de mercado de agrominerales. Yeso agrícola. Secretaría de Política Minera. Ministerio de Producción de la Nación. Dirección de Gestión de Servicios y Prestaciones Complementarias de la Minería. p 68
Van Straaten P (2014) Rocks for crops: the use of locally available minerals and rocks to enhance soil productivity. Proceedings of the 16th World Fertilizer Congress of CIEC. Technological innovations for a sustainable tropical agriculture. October 20–24. Río de Janeiro, Brazil, pp 55–58
Vázquez M (2011) Causas de la acidificación en el ámbito templado argentino, consecuencias y avances para su diagnóstico. In: García F, Correndo A (eds) Simposio Fertilidad 2011. La nutrición de cultivos integrada al sistema de producción, Rosario, pp 13–29
Vázquez M, Terminiello A, Millán G, Daverede I, Baridon E (2013) Dynamics of soil liming materials brodcast on a thaptoargic Hapludoll soil in Argentina. C Suelo 31:23–32

Plant Resources from Saline Soils and Their Contribution to Ecological Sustainability

Ecological Restoration and Productive Recovery of Saline Environments from the Argentine Monte Desert Using Native Plants

Pablo E. Villagra, Carlos B. Passera, Silvina Greco, Carmen E. Sartor, Pablo A. Meglioli, Juan A. Alvarez, Sofía Dágata, Cecilia Vega Riveros, Liliana I. Allegretti, María Emilia Fernández, Bárbara Guida-Johnson, Nerina B. Lana, and Mariano A. Cony

Abstract The accumulation of soluble salts in soils is one of the main environmental factors that contribute to the productive capacity limitations of Argentina's arid ecosystems. Salinization processes lead to critical states of degradation and desertification. The challenge to recover and improve the productivity of such degraded areas is complex because it should consider restoration strategies that will be integrated with local economic, cultural, and social activities. The integrative use and

P. E. Villagra (✉) · P. A. Meglioli · J. A. Alvarez · C. Vega Riveros · B. Guida-Johnson · N. B. Lana
Instituto Argentino de Nivología, Glaciología y Ciencias Ambientales—CONICET Mendoza, Mendoza, Argentina
e-mail: villagra@mendoza-conicet.gob.ar

P. A. Meglioli
e-mail: pmeglioli@mendoza-conicet.gob.ar

J. A. Alvarez
e-mail: jalvarez@mendoza-conicet.gov.ar

C. Vega Riveros
e-mail: cvegariveros@mendoza-conicet.gob.ar

B. Guida-Johnson
e-mail: bguidaj@mendoza-conicet.gob.ar

N. B. Lana
e-mail: blana@mendoza-conicet.gob.ar

P. E. Villagra · C. B. Passera · S. Greco · C. E. Sartor · P. A. Meglioli · J. A. Alvarez · S. Dágata · L. I. Allegretti
Facultad de Ciencias Agrarias, Universidad Nacional de Cuyo, Mendoza, Argentina
e-mail: cpassera@uncu.edu.ar

S. Greco
e-mail: sgreco@fca.uncu.edu.ar

C. E. Sartor
e-mail: csartor@fca.uncu.edu.ar

S. Dágata
e-mail: sdagata@fca.uncu.edu.ar

E. Taleisnik and R. S. Lavado (eds.), *Saline and Alkaline Soils in Latin America*, https://doi.org/10.1007/978-3-030-52592-7_17

management of native species in remediation programs are an attractive restoration tool that could improve the productivity capacity of degraded areas. Native species have developed numerous strategies and adaptations that could ensure their survival in saline environments. Nevertheless, species selection, management, and appropriate technologies to be used in afforestation programs may be limited because of partial information on the potential and requirements of each native species and the environmental characteristics of each site. In this chapter, we analyze not only the problem of soil salinization and the challenge of restoring saline environments in Argentina, but also the characteristics of native species of trees, shrubs, and grasses of the Monte region, considering their salt tolerance and the provision of goods and services to local populations, which can be useful in restoration programs.

Keywords Desertification · Remediation programs · Afforestation · Environmental services

1 The Problem of Salinity in Arid Lands of Argentina

Drylands represent 41% of the Earth's land surface and 69% of Argentina's territory (ONDTyD 2010). They are characterized by scarce, infrequent, and irregular rainfall, a large daily and seasonal thermal amplitude, and soils with low organic matter and water content. These features make arid ecosystems inherently fragile. Thus, intense anthropogenic pressures such as deforestation, overgrazing, or unsustainable agriculture (UNCCD 1994) can undermine the resilience of the ecosystems triggering degradation processes. Land degradation is defined as the long-term loss of the functioning and productivity of an ecosystem, caused by a disturbance from which it cannot recover to the original state without assistance (Bai et al. 2008). As a consequence, the productivity, biodiversity, and the production of ecosystem services and nature's contribution to people are reduced or lost, affecting the life quality of the local population (Díaz et al. 2018). Therefore, one of the main challenges is to determine which restoration or management strategies result in reversing these degradation trends in different environments.

L. I. Allegretti
e-mail: lia@mendoza-conicet.gob.ar

L. I. Allegretti · M. E. Fernández · M. A. Cony
Instituto Argentino de Investigaciones de las Zonas Áridas (IADIZA)—CONICET Mendoza, Mendoza, Argentina
e-mail: mefernandez@mendoza-conicet.gob.ar

M. A. Cony
e-mail: mcony@mendoza-conicet.gob.ar

B. Guida-Johnson · N. B. Lana
Facultad de Ciencias Exactas y Naturales, Universidad Nacional de Cuyo, Mendoza, Argentina

Environmental heterogeneity at the landscape scale in arid zones is associated with geomorphological and edaphic processes, and their effects on water distribution and availability (Rundel et al. 2007; Bisigato et al. 2009). In the Monte biogeographic province (known as Monte desert), located in the arid West of Argentina, these factors determine typical landscape units, such as rocky slopes, alluvial cones, valley bottoms, and sedimentary plains. Each unit receives different effective precipitation depending on the incident precipitation, and the rates of runoff, infiltration and evaporation, sediment accumulation, and irradiation. Particularly in low-lying areas, with the accumulation of fine sediments, a shallow phreatic level, and high evaporation rates, salts tend to accumulate on the surface, becoming the limiting factor for biological communities. In this case, the vegetation is distributed in concentric rings depending on the salinity gradient and the tolerance of different species to salinity.

Agricultural production in drylands depends on systematized irrigation, which generates foci of productivity within the arid ecosystems. In the Monte region, irrigation is performed with water from the snowmelt of the mountain, complemented with groundwater. In this territory, unsustainable agricultural practices determine nutrient loss, salinization, acidification, desiccation, compaction, sealing, or accumulation of toxic substances in the soil (Abraham 2002). Particularly, salt accumulation in the soil profile reduces the productivity of large cultivated areas, decreasing the value of the land. As a consequence, these areas are abandoned due to low or no productivity (Abraham et al. 2014).

Salinization is a complex problem that leads to critical states of degradation and desertification in drylands. This has negative impacts over natural ecosystems or agroecosystems, from environmental, economic, and social viewpoints. Given this situation, it is necessary to quantify the origin and evolution of soil salinization processes and to evaluate management decisions to revert salinization processes and to restore ecosystem functions and increase land productivity. Ecological restoration is defined as the process of assisting the recovery of an ecosystem that has been degraded, damaged, or destroyed (SER 2004). It has been recognized as a priority by many global agreements, such as the Aichi Biodiversity Targets 14 and 15 of the Convention on Biological Diversity or the Land Degradation Neutrality Target of the UN Convention to Combat Desertification or the Bonn Challenge on Forest Landscape Restoration (Aronson and Alexander 2013). More recently, in March 2019, the United Nations General Assembly declared 2021–2030 the UN Decade on Ecosystem Restoration, a global call for action that is expected to scale up restoration initiatives. Related to these initiatives, the so-called restorative activities are inspired by the values and principles underpinning ecological restoration. They are aimed at increasing ecosystem services and reducing environmental impacts while improving the ecological sustainability and production systems (McDonald et al. 2016). These activities become especially relevant in the search for global sustainability, considering the need to increase the food production rate for a growing population, and to maintain biodiversity (Foley et al. 2011).

2 Spatial Characterization of Natural and Anthropogenic Salinity in the Monte

Salinization is the ionic accumulation process in the soil profile, with two main types being differentiated according to their origin as primary and secondary salinization (Amezketa 2006; Zhou et al. 2013). The first is the result of natural processes that take place in areas where the parental material is rich in salts, and the evapotranspiration rate is higher than the precipitation rate. Other factors that may induce primary salinization are natural drainage patterns or topographic features, the geological structure, or the distance to the sea.

Secondary salinization occurs when salt accumulation is a consequence of unsustainable anthropogenic activities. There are two different triggers, according to whether it occurs in irrigated or non-irrigated drylands (Thomas 1993; Rengasamy 2006; Zhou et al. 2013). On the one hand, the salinization of non-irrigated drylands occurs when the perennial native vegetation (with deep roots) is replaced by annual crops or grasslands with shallow roots systems. Reduced evapotranspiration alters the natural water balance and causes the phreatic level to rise. The accumulation of salts from the groundwater will depend on both climatic conditions and soil hydraulic properties. On the other hand, the salinization of irrigated drylands occurs as a consequence of excessive irrigation and lack of adequate drainage. This process can be accelerated by poor irrigation water quality, low hydraulic conductivity, and high evaporation conditions (Abraham et al. 2002; Jobbágy et al. 2008).

Argentina ranks third in the world regarding its extent of salt-affected soils (Marinoni et al. 2019), with approximately 35 million hectares (12.6% of the total area) being affected by primary salinity (INTA 1990). The Monte Biogeographical Province, which comprises approximately 460,000 km^2 from the province of Salta to the Atlantic coast in Chubut, displays regional differences related to predominant types of soil (Fig. 1). The Northern Monte is occupied mostly by rock outcrops (49%) and Entisols (42%), and the Central Monte is dominated by Entisols (77%), whereas in the Southern Monte, Aridisols (60%) are the prevalent soil type. According to data produced by INTA (1990), Central and Southern Monte are affected by primary salinity in extensions that cover 28% and 29% of each region, respectively. In general, a quarter of the Monte soils (25%) are affected by the presence of natural salinity in the first 50 cm of the soil profile (Fig. 1).

Only 5% of the total productive area in Argentina corresponds to irrigated areas (2.1 million ha), and it is estimated that 24% of this area is affected by salinization or sodicity processes (Sánchez and Dunel Guerra 2017). In the Monte region, 25% of the irrigated land is affected by secondary salinization, with a wide range of variation that goes from 8% in Salta to 68% in Chubut (Table 1).

Within the great territorial extension of the Monte, different natural conditions and processes give rise to high spatial heterogeneity in terms of quality and quantity of ions present in the soil. This variability is observed even at fine scales, so it is necessary to know the superficial and deep distribution of salts when implementing projects for the recovery of degraded areas. In semiarid degraded ecosystems, besides

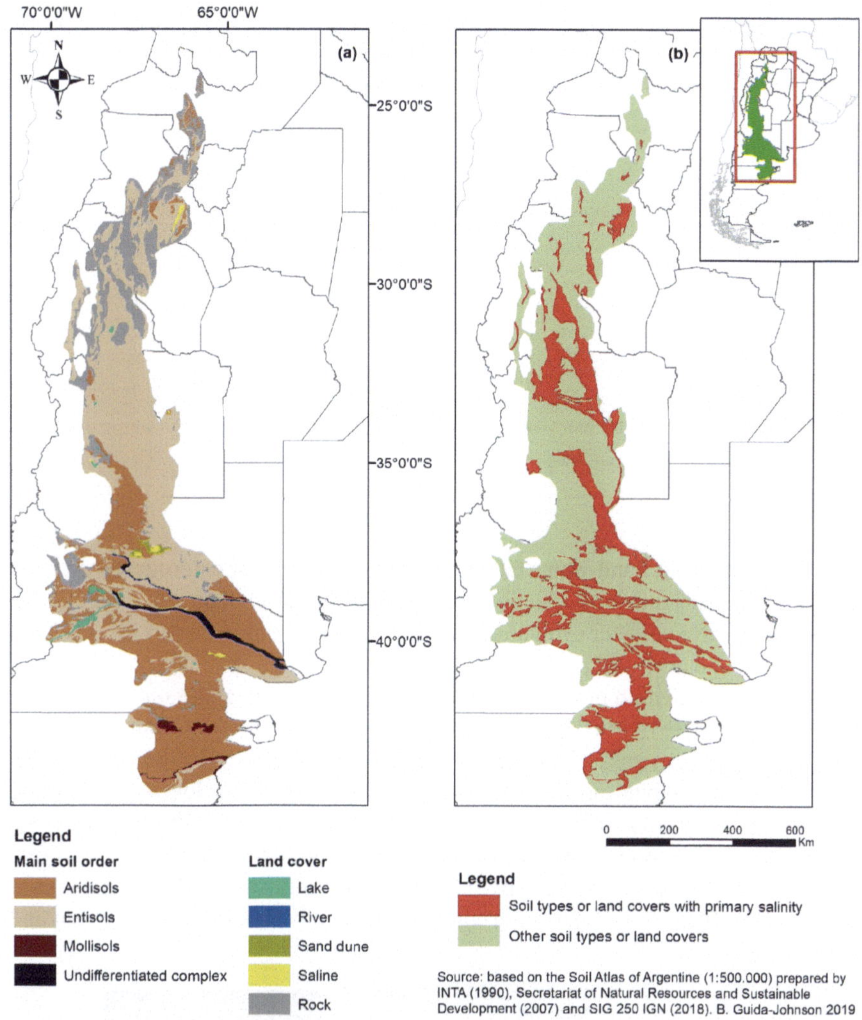

Fig. 1 Main soil orders (**a**) and zones affected by primary salinity (**b**) in the Monte region

the effect of post-planting drought, the small-scale heterogeneity of abiotic factors plays an important role in the survival of transplanted seedlings (Maestre et al. 2003). For example, microtopography, soil texture, vegetation cover, and water table can generate microsites that favor salt accumulation. The combination of these environmental variables and the biological characteristics (salinity tolerance, growth rate, root development) of the plant species used can generate synergistic effects that threaten the viability of restoration projects. The quali/quantitative study of these variables allows the development of conceptual tools tending to recover soils affected by salinization processes.

Table 1 Irrigated areas and areas affected by salinity or sodicity in Argentine provinces located in the Monte region (Sánchez and Guerra 2017)

Province	Irrigated areas (ha)	Salinity affected areas (ha)	Proportion (%)
Salta	118,898	8937	8
Tucuman	108,484	10,438	10
Catamarca	61,847	13,134	21
La Rioja	51,738	5300	10
San Juan	95,704	53,830	56
Mendoza	276,324	73,213	26
La Pampa	4600	460	10
Neuquén	14,600	4380	30
Río Negro	79,320	22,500	28
Chubut	26,050	17,749	68
Total	837,565	209.941	25

In this context, soil and water factors that affect the spatial heterogeneity of soil salinity were analyzed in the locality of Media Agua (San Juan province). In 16 plots, soil and groundwater evaluations were carried out, based on topsoil (<0.25 m deep) sample and deep soil profiles (up to 4.5 m deep). High spatial heterogeneity of salinity was found within plots and with increasing variability at the superficial layer (Meglioli et al. 2018). Consequently, this fine-scale variability in salinity should be considered and represents a challenge for the planning of restoration projects (Fig. 2).

The different edaphic profiles showed similar patterns, with levels of electrical conductivity decreasing with increasing depth. However, high salt concentrations (>20 mS cm^{-1}) were also observed even in deep soil samples, especially in fine texture soils (silty clay loam and silty loam). These results highlight that soil salinization is a problem that affects both surface and deep layers. In turn, the depth and

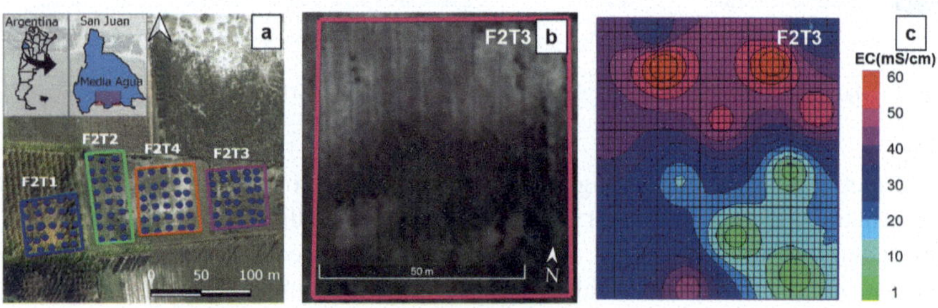

Fig. 2 Evaluation of spatial variability of saline condition in an irrigated farmland (poplar farm) of Media Agua (San Juan). **a** Examples of plot establishment in a site of the study area. The four plots represent different salinity conditions within farmland. Each blue point represents a superficial soil sample. **b** Detailed image of the plot F2T3. **c** Analysis of spatial interpolation of salinity levels for F2T3 (GMS 6.5 software) (Meglioli et al. 2018)

quality of the water table are relevant to the development of revegetation programs. In this study case, the phreatic level is relatively shallow (between 1 and 4 m deep), which may increase the chances of survival if the implanted species reach this water reservoir. However, measurements of the electrical conductivity of the phreatic for the different analyzed plots indicate that the quality of this water resource is variable spatially and temporally (Meglioli et al. 2018).

3 Restoration of Salinized Environments Using Native Species

The ecosystem restoration approach as a practice to reverse salinization processes has allowed diminishing the economic and logistical difficulties associated with some management strategies related to a more engineering approach (Chhabra and Thakur 1998; Hamidov et al. 2007; Ram et al. 2007). Actions that can be used to restore soils affected by secondary salinization should result in the reduction of groundwater table level, the promotion of infiltration and the improvement of the physicochemical conditions of the soil (Taleisnik and López Launestein 2011). In this sense, different strategies have been tested in arid zones: reforestation to promote a decrease in the water table, an effect known as "biodrainage" (Hbirkou et al. 2011); or revegetation to improve the soil structure, that favors drainage and therefore salt leaching. Revegetation could be performed either with halophyte species (Basavaraja et al. 2007) or with non-halophyte species inoculated with microorganisms. In this case, arbuscular mycorrhizal fungi (Zhang et al. 2011) or rhizosphere bacteria (Nabti et al. 2010) are commonly used for this purpose. Another alternative is phytoremediation, which, in general terms, involves the removal of contaminants from the environment. In this case, the intention is to remove salts from the soil using halophyte plant species due to their bioaccumulatory capacity (Ravindran et al. 2007; Hasanuzzaman et al. 2014). This ability can be confirmed by detecting a decrease in the concentration of salts in the soil, along with its increase in plant tissues. This could be corroborated for different species, both native and exotic, either under controlled greenhouse conditions (Rabhi et al. 2009) or in field experiments (Hamidov et al. 2007).

The selection of suitable species in restoration projects is a key factor that can greatly determine the efficiency and success of the process. The species to be introduced in degraded ecosystems should preferably be selected from those existing in nearby non-degraded environments, in order to guarantee their adaptation to the adverse conditions in which they will be established (Cortina et al. 2004). In addition, they must also be able to endure in the context of climate change according to the concept of restoration (Butterfield et al. 2017). The use of native species in restoration programs requires knowledge of propagules collection sites and times, seed quality and germination requirements, as well as preconditioning techniques to overcome transplant shock and improve seedling survival (Fernández et al. 2016, 2019).

Besides, degraded arid lands are associated with significantly altered soil conditions, therefore knowing the different physiological adaptations of native species also allowed to evaluate their suitability to restore a particular site or microsite. The availability of information and appropriate techniques to achieve germination and subsequent production of seedlings of some native species are a limiting factor in their selection.

4 Physiological Adaptations of Native Species to Be Used in Saline Environments from the Monte Desert

The use of native plants is an especially appropriate tool in restoration projects as they can ensure their adaptation to stressful environmental conditions. High salinity affects plants both by salt toxicity and by dehydration caused by low water potential. Native species exhibit physiological, morphological, and anatomical mechanisms that allow them to live in saline environments. These include the elimination of salts from specialized organs, ion compartmentation, ion selectivity, osmotic adjustment, partitioning of assimilates, specialized conduction systems, reduced leaves, modification of stomatal density, presence of aquifer parenchyma, thickened cuticle, and deposition of waxes (Villagra et al. 2010, 2017; Taleisnik and López Launestein 2011; Hasanuzzaman et al. 2014).

Germination is a critical stage in the plant life cycle, particularly in fluctuating and stressful environments such as those affected by salinity. The success of germination and seedling establishment is conditioned by the existence of adaptive mechanisms at these stages. The presence of dormant states in seeds is widespread in most of the native halophytes shrubs (Flowers and Colmer 2015; Tug and Yaptrak 2019). The production of heteromorphic seeds is another adaptive strategy, in which the response of each type of seed depends on different environmental factors and dormancy mechanisms. Several species of the family Asteraceae and Amarantaceae show such adaptation (Liu et al. 2018). However, the production of heteromorphic seeds for woody shrub species native from the Monte has not been described.

The propagules of many halophytes may remain viable in the soil seed bank at high salinity levels and germinate when stress decreases. Field germination for many species of saline environments usually occurs after the rainy season, when salinity levels decrease, improving the chances of seedlings surviving (Piovan 2016). On the other hand, it has been found that some species present higher germination percentages after preconditioning of the seeds with moderate salinity concentrations (<250 Mm NaCl). Rapid germination (<4 days), when environmental conditions are favorable, is another adaptive strategy found in species from saline environments (Tug and Yaptrak 2019).

In the Monte region, there are several species adapted to saline environments or with potential for use in the ecological restoration of degraded environments (Villagra et al. 2017; Villagra and Alvarez 2019). They can be used by diverse human populations because they provide forage, firewood, charcoal, medicinal and aromatic substances, and elements for the construction of housing in these extreme areas with low vegetation cover (Alvarez and Villagra 2009; Montani et al. 2010; Taleisnik and López Launestein 2011).

In the following sections, we review eco-physiological and morphological aspects and possible uses of species that present themselves as potential tools in the restoration of saline environments in the Monte desert, whereas they can be added to future eco-technological packages to be implemented.

4.1 Woody Species

Native woody species with tolerance to salinity are a group of species belonging to several taxonomic families, which have different mechanisms of adaptation to environments that impose conditions of high salt toxicity and strong water stress (Taleisnik and López Launestein 2011; Villagra et al. 2011, 2017). The studies of physiological aspects of native species of the Monte have focused particularly on the species of the genus *Prosopis* and *Atriplex*. For some species of these genera, there are already technological packages that allow their multiplication and establishment in different field situations such as for *P. flexuosa, P. chilensis, A. lampa*, and *A. cordobensis* (Passera and Borsetto 1989; Aiazzi et al. 2005; Villagra et al. 2010; Passera 2017). While for many species of saline environments, endemic and native, key aspects for their use in restoration tasks are still unknown, such as *Ehretia cortesia, Allenrolfea vaginata, Lycium tenuispinosum* and *Suaeda divaricata*, among others. Currently, research is being carried out by our working group in search of the productive recovery of degraded environments and/or the current demand for eco-technological packages to restore mining environmental liabilities.

4.1.1 Trees and Shrubs of Fabaceae Family

Afforestation with species of the genus *Prosopis* has been suggested as having a high potential for the productive restoration of saline environments in the Monte, taking into account their tolerance to salinity and their potential uses (Guida-Johnson et al. 2017; Villagra et al. 2017).

Prosopis shows a high inter and intraspecific variability in tolerance to salt and other environmental stresses. Moreover, salt tolerance varies during ontogeny and is associated with different physiological processes. The tolerance is lower during the initial stages of the life cycle, which seems to be critical for their persistence in saline environments (Cony and Trione 1998; Villagra et al. 2010). Salinity tolerance mechanisms observed in the genus include both the ability to regulate the absorption and transport of ions and thus to avoid the toxic effect of salinity, as well as the ability to counteract the effects of low osmotic potentials through osmotic adjustment, regulation of stomatal opening and induction of anatomical changes in roots, stems, and leaves (Zhao and Harris 1992; Villagra and Cavagnaro 2005; Villagra et al. 2011).

The inter- and intraspecific variability in *Prosopis* species from the Monte has been studied in different trials. From these studies, it is possible to identify a gradient of salinity tolerance among the species of the genus, from glycophytes to euhalophytes. This gradient appears to be related to the distribution of the species. Among trees, the germination capacity of *P. flexuosa* seeds is greater than that of *P. chilensis* in −0.3 M NaCl solutions (Cony and Trione 1998) and the survival of *P. flexuosa* seedlings is greater than that of *P. chilensis* in 0.5 M NaCl solutions (Rhodes and Felker 1987). In shrubs, *P. alpataco,* a dominant species in clayey and saline soils, shows a higher germination capacity and a lower reduction on its growth in saline conditions than *P. argentina*, which comes from sandy soils with very low saline content (Villagra 1997; Villagra and Cavagnaro 2005). In contrast, *P. alpataco* is more affected by water stress than *P. argentina*, which has shown to be very tolerant of low water potential but not of the toxic effect of salinity. Similarly, *P. chilensis* is more tolerant to water stress than to saline toxicity (Zhao and Harris 1992). These evidences suggest that the tolerance to salinity in some species of the genus is the result of their ability to avoid ion toxicity. In this sense, *P. alpataco* is capable of regulating the absorption and transport of ions, since it maintains low Na^+ contents and high K^+ and Ca^{++} contents in its leaves when it grows in saline conditions (Villagra and Cavagnaro 2005; Vega Riveros et al. 2011). In *P. strombulifera*, a species found in highly saline areas, it has been found that low NaCl concentrations stimulate growth indicating that this species is an euhalophyte with respect to NaCl; however, SO_4Na_2 concentrations determine growth inhibition and severe toxicity (Sosa et al. 2005). This species has been studied as a model of salinity tolerant species and presents numerous anatomical and physiological adaptations (Luna et al. 2017).

For the study of intraspecific variability of different species of *Prosopis*, seeds from randomly selected mother plants have been collected in different areas within the Monte. In total, seeds of approximately 240 origin plants of different native species of the Monte are conserved in a germplasm bank in Mendoza (Cony 1993). Besides, these seeds have been used to establish 12 ha of progeny-procedure trials for *P. flexuosa* and *P. chilensis* in the town of Bermejo (Mendoza). The comparison of the growth characters, bioform, and physiology among different origin enabled the identification of rapid growth provenances in both species (e.g., Bolsón de Fiambalá, Catamarca) (Cony 1996; Cony and Trione 1998; Mantován 2002, 2005). The observed differences among provenances are consistent with the structural differences among woodlands in their natural distribution (Villagra et al. 2005; Alvarez et al. 2015). Genetic studies in *P. flexuosa* suggest that these morphological and physiological differences have a significant genetic basis and may be explained by local adaptation (Bessega et al. 2019). However, the potential of these provenances for afforestation of areas with water or salt stress has not been extensively evaluated for Monte species, while initial studies have been conducted on species in Chaco and other sites (Velarde et al. 2003; Felker et al. 2008). In this sense, differences have been observed among *P. flexuosa* provenances at germination stages in saline conditions, with some

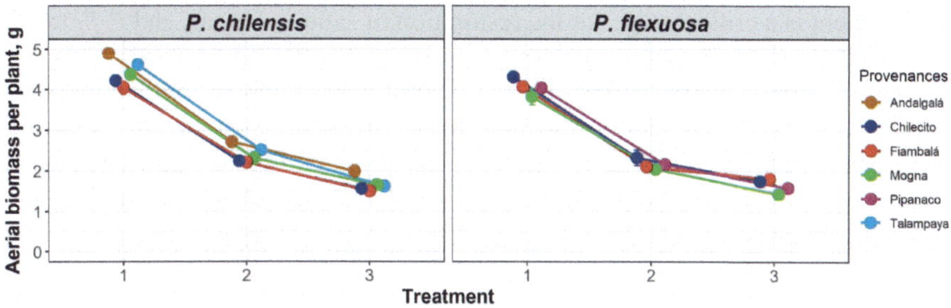

Fig. 3 Aerial biomass of different provenances of *P. chilensis* and *P. flexuosa* growing under three saline treatments (T1 = 0.0; T2 = 0.10; T3 = 0.25 M NaCl) in a greenhouse trial

of them emerging as potential provenances to select for salinity tolerant (Cony and Trione 1998).

The high genetic variability observed (Bessega et al. 2019) and the differences between the provenances in salinity tolerance at germination stage in *P. flexuosa* and *P. chilensis* (Westphal et al. 2015; Cony and Trione 1998) lead us to think about the possibility of selecting suitable provenances of these species for the implantation of areas with environmental stress. To test this idea, we performed greenhouse and field experiments to evaluate the effects of salinity on growth performances of four provenances of *P. flexuosa* and five provenances of *P. chilensis*.

In the greenhouse trial, a split-plot design was arranged, containing the combination of species and provenances, and three treatments of sodium chloride solutions (T1 = 0.02 M; T2 = 0.10 M; T3 = 0.25 M) were applied to blocks, during the summer 2015. Results showed that all species and provenances decreased their growth and eco-physiological parameters (such as water potential) under increased salinity stress. Plant growth indicators of salinity tolerance showed that aerial biomass per plant was reduced between 41 and 48% for T2 and between 55 and 65% for T3 in relation to T1 (control) (Fig. 3).

The field trial was performed in salinized soils of irrigated areas of Media Agua (San Juan Province). These sites were chosen because there is a large proportion of land with salinized soils and/or irrigation problems. In areas with non-saline soils, poplar, olive, grapevine, and fruit plantations are found, whereas salinized areas are excluded from the local productive system. In this trial, a factorial experimental design was used according to the following detail: (a) two species: *P. flexuosa* and *P. chilensis*; (b) three provenances of each species: local origin, from highly saline environments, and origin with high growth rate according to previous trials (Cony and Trione 1998); and (c) four site qualities: site with high soil salinity, site with medium salinity, site with low salinity but not suitable for poplar production and site suitable for poplar production (as a control to evaluate potentiality of *Prosopis* species in high-quality sites). The first results suggest that both species present some tolerance to salinity, *P. flexuosa* appears to be less affected by salinity than *P. chilensis*. Both species show a 50% of survival under medium to high salinity conditions (about 40 mS cm^{-1}) and a decrease in growth was observed as salinity increased. The establishment is a critical stage for the restoration of salinized areas, and *P. flexuosa* from Fiambalá showed advantages during this period with respect to other origins. The spatial heterogeneity of the soil is a determinant of the restoration. In very saline treatments (>50 mS cm^{-1}), the mortality of the seedlings occurs in a spatially heterogeneous way in response to the spatial distribution of soil salinity (Villagra et al. 2019). The detailed analysis of this data will be interesting since it would allow the establishment of a tolerance limit for each species and to plan with more precision the implantation strategy.

4.1.2 Shrubs of Chenopodiaceae Family

Many native species of the genus *Atriplex* are appreciated for their nutritional quality and fodder potential. In addition, some species of this genus have been used extensively in revegetation tasks of degraded environments (Passera 1990; Fernández et al. 2016; Piovan 2019). Under laboratory-controlled trials, seeds of some native species can germinate in high proportions under water potentials of −1.5 MPa. Some species such as *A. undulata* are sensitive to saline stress, but seeds that do not germinate at high salt concentrations maintain their viability and are able to germinate when transferred to distilled water (Piovan et al. 2014). On the other hand, seeds of *A. lampa* have the capacity to germinate in oil drill cuttings which contain alkaline salts and heavy metals and petroleum hydrocarbons (Pentreath et al. 2015; Dalmasso et al. 2018).

Seedlings of *A. lampa* and *A. cordobensis* perform an osmotic adjustment, allowing them to maintain turgidity in leaves with low water potential and to continue absorbing water from the soil both in saline and water stress conditions (Passera and Allegretti 1996; Aiazzi et al. 2005; Fernández et al. 2018b). As in species of the genus *Prosopis*, there is great intra- and interspecific variability in response to different types of stress in the genus *Atriplex* (Ungar 1987; Piovan et al. 2014). For instance, Aiazzi et al. (2005) found that seedlings of *A. cordobensis* respond differently to salinity according to their provenance.

Atriplex lampa shows low water potentials and reduced growth as a response to increasing soil salt concentrations (Caraciolo et al. 2002). The ability to achieve such negative water potentials may be due to the high protein contents associated with compatible osmolites, such as proline and betaine found after plants have been subjected to water deficit (Trione and Passera 1993). *A. lampa* can reach water potentials of −5.47 MPa under conditions of severe water stress (Fernández et al. 2018b). In addition, a lower shoot/root ratio has been observed for this species under salt and water stress (Passera and Allegretti 1996; Caraciolo et al. 2002; Villagra et al. 2011). This mechanism can be advantageous since the maintenance of root growth in saline soils allows plants to explore a larger volume of soil in search for water or less saline soil strata, and thus mitigate the effect of salinity. In field trials, *A. lampa* has shown high levels of survival (>90%) in different environments and soil stress conditions (Dalmasso et al. 2018; Fernández et al. 2018a; CB Passera, personal communication).

Another adaptive character to saline environments of some species of the genus *Atriplex* is the presence of epidermal salt-secreting glandular trichomes, which avoid ions accumulating in other tissues (Pérez Cuadra 2012).

Two endemic shrub species of the genus *Allenrolfea, A. vaginata* and *A. patagonica,* are present in Argentina and both are found in saline and alkaline soils. They present adaptations in the morphology of their leaves and the anatomy of their vascular system. Both species have reduced and stem welded leaves, and aquifer parenchyma that gives succulence and allows osmotic regulation. The stem of *A. patagonica* presents anomalous secondary growth that probably gives the phloem greater resistance to cavitation (Pérez Cuadra and Cambi 2010). *A. vaginata* has a highly specialized water conduction system. This species presents very numerous, small, short, multiple, and grouped vessels. The fibrotracheids and tracheids collaborate in the conduction of water (Giménez et al. 2008). In *A. vaginata,* the utricles are dispersed with perianth remains. Both structures restrict but do not inhibit seed imbibition, constituting an adaptive mechanism to prevent germination under unfavorable conditions (Dágata, unpublished data). In *A. patagonica,* the seeds must be scarified to eliminate physical dormancy (Piovan et al. 2014). The germination of *A. patagonica* is more affected under conditions of alkaline than saline stress and can germinate in concentrations of 0.4 M of NaCl (Piovan et al. 2019).

4.1.3 Shrubs of Other Families

Atamisquea emarginata (Capparaceae) is a foraging, melliferous, and medicinal shrub. This species is a facultative phreatophyte (Jobbágy et al. 2011) moderately tolerant to salinity. It can be easily propagated by seeds or asexually (Eynard et al. 2017). Seeds may have endogenous dormancy. However, without pre-germination treatments, 50% of seeds can germinate at 25 °C (Fernández et al. 2016). Seedlings of this species reduce foliar area and stomatal conductance but maintain good growth rates under moderate water stress. Besides, under severe water stress (4–5% soil water content), it can reach water potentials of −8.2 MPa and have a high survival rate (>80%). The manipulation of irrigation, as a preconditioning treatment to drought, could favor a greater survival of the seedlings transplanted to the field, becoming an interesting technique for revegetation purposes (Fernández et al. 2016).

Lycium tenuispinosum (Solanaceae) is a xero-halophyte shrub widely distributed in the Monte. This species has been used for the restoration of degraded slopes (Dalmassò and Ciano 2015). *L. tenuispinosum* is sensitive to hydric stress during its germinative stage, decreasing significantly the percentage of germinated seeds at water potential lower than −0.28 MPa and being inhibited the germination at −2. 24 MPa. It can germinate at pH of 5–9 (Dágata 2018). However, adult specimens have been found in saline-sodic soils. Branches that touch the ground can originate new plants by asexual multiplication, and propagation by layering is feasible. In conditions of water stress, it loses the foliage and it can recover quickly after new precipitations.

Baccharis spartioides (Asteraceae) occurs in open environments with high salinity (20 dS/m). It is an endemic perennial, aromatic, and medicinal shrub. In addition, it is used by local people for the manufacture of brooms. It has rhizomes and gemiferous roots that allow it to sprout (Dalmasso et al. 2016). It presents small leaves, a common characteristic in species of xeric environments. Its stomata are not protected, being elevated or at the level of other epidermal cells. It tolerates salinity but does not tolerate severe water stress (Pérez Cuadra 2012).

Cyclolepis genistiodes (Asteraceae) inhabits saline and alkaline soils. It is forage and medicinal shrub. In the leaves and stems of this species, the stomata are pseudo-sunken. The vascular bundles in the leaves are immersed in aquifer parenchyma (Pérez Cuadra 2012). The seeds of this species are dormant, but the rupture mechanism has not been clarified. There are problems due to fungal infection of the seeds under laboratory conditions (Peter et al. 2014). Piovan et al. (2014) found that the threshold at which germination does not occur is −1.5 MPa. *C. genistiodes* is sensitive to saline stress at the germination stage. However, seeds remain viable under such conditions and can germinate when transferred to distilled water; this is an important adaptive feature for species in saline environments (Piovan et al. 2019).

4.2 Grasses

Native forage grasses of the Monte are able to resist conditions of water stress and saline toxicity (Céccoli et al. 2015) and allow livestock production in marginal areas. Besides, they enhance other associated environmental benefits such as soil fixation, the addition of organic matter and benefits to soil microorganisms.

Pappophorum phillippianum and *Leptochloa crinita* are important forage species of the Monte owing to their presence and coverage, they even grow in saltpeter beds (Candia and Guevara 1973; Ragonese and Piccinini 1978; Pérez Cuadra and Cambi 2010). *L. crinita* was indicated as tolerant to salinity by Zabala et al. (2011). The leaves of *L. crinita* and leaves and stems of *P. phillippianum* have salt glands that excrete salts from metabolically active tissues (Taleisnik and Anton 1988; Pérez Cuadra and Cambi 2010). According to that fact it may be probable that *Pappophorum caespitosum, Chloris castilloniana, Cottea pappophoroides, Sporobolus cryptandrus*, and *Panicum urvilleanum* are resistant to saline soils, since they also have cogeneric species that have glands or salt hairs (Céccoli et al. 2015).

Moreover, intraspecific variability has been found within these species, and for this reason, the selection of specific genotypes more adapted to saline soils could allow reducing the risks of mortality during the plant establishment. In this context, *L. crinita* is a promising species because it grows in a wide variety of environments with fast establishment and colonization. The fact that it is an autogamous species allows us to maintain relatively pure lines of genotypes throughout time (Kozub et al. 2017). Cavagnaro et al. (2006) classified 20 accessions of *L. crinita* from different environments of the Monte into three groups, based on the forage production per plant: high productivity (>110 g dry matter plant^{-1} year^{-1}), medium productivity

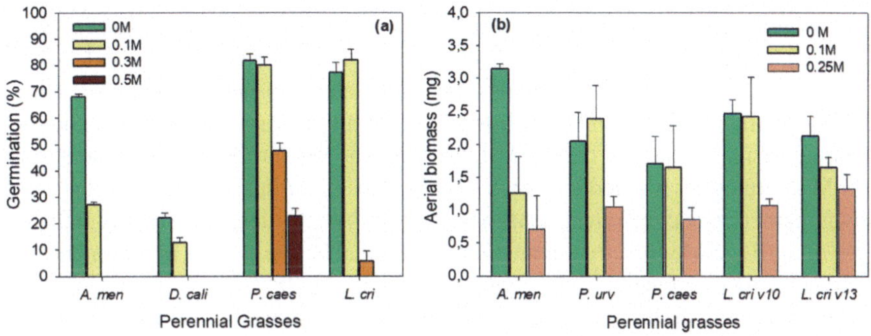

Fig. 4 Germination (**a**) and biomass (**b**) of perennial grasses under different NaCl solutions. References: ***A. men****: Aristida mendocina,* ***D. cali****: Digitaria californica,* ***P. caes****: Pappophorum caespitosum;* ***L. cri****: Leptochloa crinita,* ***P. urv****: Panicum urvilleanum,* ***L. cri v10****: Leptochloa crinita v10,* ***L. cri v13****: Leptochloa crinita v13*

(between 110 and 75 g dry matter plant^{-1} year^{-1}), and low productivity (<75 g dry matter plant^{-1} year^{-1}). Seeds of high productivity genotypes were collected from plants that grow in saline soils (apparent electrical conductivity >2 dS m^{-1}); thus, they could be used in restoration programs for saline soils.

In germination trials, the proportion of *Digitaria californica* and *A. mendocina* seeds that germinated under a saline solution of NaCl 0.1 M at 30 °C was reduced by half than those incubated under distilled water. On the other hand, *Leptochloa crinita* and *Pappophorum caespitosum* maintained the same proportion of germinated seeds as the control treatment. When the seeds were placed in more concentrated saline solutions, 0.3 M and 0.5 M, only *P. caespitosum* seeds germinated (20% and 50% of the control, respectively) while the other species did not germinate or did so in very low proportion (Greco and Cavagnaro 2004). Therefore, *L. crinita* and *P. caespitosum* have shown to be promising species for revegetation programs of saline soils by the incorporation of seeds to the soil (Fig. 4a).

These perennial forage grasses also show different performance when growing with saline watering. *A. mendocina, P. urvilleanum, P. caespitosum,* and *L. crinita* were submitted to a trial in pots. Irrigation was applied with two saline solutions of NaCl: low salinity (0.1 M) and high salinity (0.25 M), and a control irrigated with running water (Próspero et al. 2018). Most of the *A. mendocina* plants died for saline treatment at 0.25 M. With the 0.1 M treatment, plants decreased in crown diameter, number of tillers and total aerial dry matter compared to the control treatment plants (Fig. 4b). *P. urvilleanum* plants survived the 0.1 and 0.25 M watering treatments, with very little reduction in crown diameter in the 0.1 M treatment compared to the control and a significant decrease for the 0.25 M treatment. Salinity did not affect the number of tillers. Total aerial biomass was not affected by an irrigation of 0.1 M, and it was reduced by half with the treatment of 0.25 M (Fig. 4). In the case of *P. caespitosum* plants, both crown diameter and number of tillers, were not affected by the 0.1 M treatment but were significantly reduced with the 0.25 M treatment compared to the control. Aerial biomass was not affected by 0.1 M treatment compared to control and was half reduced with 0.25 M watering (Fig. 4). Two accessions of *L. crinita* were evaluated, one of high productivity (*L. crinita* v13) and the other of low productivity (*L. crinita* v10). *L. crinita* v10 plant crown diameter was not affected by 0.1 and 0.25 M treatments. The number of tillers was slightly lower with saline treatments. The total aerial biomass of the 0.1 M treatment was similar to the control, and with 0.25 M treatment, it was reduced by half (Fig. 4b). *L. crinita* v13 plants had similar crown diameters between treatments and the number of tillers was slightly reduced with 0.1 and 0.25 M saline treatments. The total aerial biomass in 0.1 M treatment was reduced by 30% compared to the control and by 40% in 0.25 M treatment (Fig. 4b).

Summarizing, all the four species of native grasses showed tolerance to watering with saline water, although differences between species were observed. In the case of *A. mendocina*, it was the only species that did not survive with watering of 0.25 M, while it did so at 0.1 M. The rest of the species grew in both 0.1 and 0.25 M treatments. When salinity is low (0.1 M), *P. urvilleanum, P. caespitosum*, and *L. crinita* v10 show little variation in their total aerial biomass compared to control. With higher salinity, the aerial biomass of all species was affected, decreasing by half. *L. crinita* v13 is a promising variety to produce fodder biomass under saline conditions because it has high productivity and it only showed a 40% decrease in the aerial biomass with salinity.

Another promising species could be *Distichlis spicata*, named "saltgrass," which is common in drainages, salty and humid soils, mainly in the floody valleys of southern Mendoza. It is considered a species of regular preference for livestock (Passera and Borsetto 1989) being consumed by goats and to a minor degree by horses (CB Passera, personal communication). In addition, it has been indicated as a component of the livestock diet in 8–24% in the summer season in humid areas (Brizuela et al. 1990).

5 Objectives of the Restoration of Saline Environments

5.1 Livestock Management in Saline Areas

The most important livestock activities in the central Monte desert, Mendoza, are goat husbandry in the northeast, and cow–calf operations in the southeast, which are managed using extensive production systems (Guevara et al. 2009). Shrublands and open woodlands play an important role because they provide forage for grazing animals throughout the year. Continuous grazing is the dominant strategy employed. In the other hand, semi-intensive or intensive livestock production, mainly cattle production, takes place in irrigated oases. In both areas, we can find areas with soils with high salinity or salinized by anthropic action, a characteristic that constitutes one of the main restrictions for agricultural and livestock production (Taleisnik et al. 2008).

The possibility of establishing pastoral systems in saline or salinized areas should be evaluated according to the environmental conditions, the carrying capacity of agrosystems, the objectives of livestock production, and the possibilities of capital investment. In some cases, grazing management can be a tool to improve these environments. The rotation of grazing sectors, depending on the availability of fodder and times of the year, together with the use of a moderate animal load adjusted to carrying capacity, would allow the maintenance of plant cover and diversity and thus the improvement of these environments. In other cases, it is necessary to resort to techniques such as the introduction of forage species, especially native species tolerant or moderately tolerant to salinity. In Mendoza, Martínez Carretero (2001) indicates for saline, warm, and cold temperate deserts, the presence of species such as *Suaeda divaricata, Lycium chilense*, and *Atriplex lampa* mentioned in this work as possible species to be used for restoration. Other species also described here, both woody and herbaceous, that could be used in this type of soil are *P. flexuosa, A. lampa, T. usillo, L. crinita*, and *P. urvilleanum*, which are native fodder species selected by goats and cattle in different proportions, according to their diet (Allegretti et al. 2012a, b; Egea et al. 2014).

The establishment of forage species in irrigated areas affected by salinity was studied in Mendoza at the INTA (National Institute for Agricultural Technology) with promising results. Besides, Ochoa (2011) evaluated the implementation and production of dry matter of introduced fodder species, such as tall wheatgrass (*Thinopyrum elongatum*), white clover (*Melilotusalbus*), and yellow clover (*Melilotus officinalis*). The results obtained in these studies indicate that it is feasible to implant these species in saline areas, becoming potential material to revegetate these areas.

5.2 Forestry

In the salinized fields of the irrigated oases from the Monte, traditional forest crops are not successful. The occupation of these fields with other forest species that are less demanding in terms of soil quality and resistance to various stress conditions is a productive alternative, as has been suggested for other irrigated areas of the world (Ridley and Pannell 2005). According with the results shown in the previous section of this chapter, native trees such as *P. flexuosa* and *P. chilensis* appear to be promising species given their multiple uses (production of poles, charcoal and firewood and in some cases, wood for timber) and environmental benefits, such as nitrogen fixation, nutrient cycling, and generation of microhabitats for other species, among others (Fig. 5a, b). Although these are perhaps the species in which studies for their use are most advanced, the lack of proven technologies for their establishment and management in these extreme environments still makes it difficult for producers to consider this alternative species and their use for productive purposes. In plantations with both *Prosopis* species under saline soils, we observed great variability in plant growth, even within the same provenances, which is an aspect to be improved in order to achieve a homogeneous plantation for productive purposes. Further study is still needed on the selection of adequate germplasm, transplanting methods, irrigation requirements, the control of weed competition, the use of fertilisers and manures,

Fig. 5 **a** Afforestation with *Prosopis* spp. in moderately saline irrigated lands in Media Agua, San Juan (Photograph: P.E. Villagra). **b** Restoration of highly salinized irrigated lands in Media Agua, San Juan (Photograph: P.E. Villagra). **c**) Ecological restoration in an oil well location in Tupungato, Mendoza (Photograph: M.E. Fernández). **d** Restoration of areas degraded by mining activities, Catamarca (Photograph: C.B. Passera)

and pruning management. Additionally, it is important to evaluate the costs related to the different stages of production and planting of the seedlings, as well as the management and monitoring of the forested plots.

Both *Prosopis* species are also proposed for the afforestation of non-irrigated saline environments of the Monte desert. *P. flexuosa* grows in saline areas of the Monte. However, the distribution of this species woodlands in saline environments gives us an idea of the importance of considering the spatial heterogeneity of salinity when setting the implementation strategy in afforestation tasks.

5.3 Restoration of Areas Degraded by Mining Activities

Mining can severely affect natural ecosystems in arid lands. The observed environmental changes by mining activities depend on the type of techniques used and the exploited field. The main types of environmental modifications are loss of the vegetation cover, complete loss of surface soil, erosive processes, environmental liabilities, soil compaction, changes in the landform, salinity, and sodicity problems, and drainage changes (Ciano et al. 2005; Arce et al. 2015). To restore these environments, it is first necessary to improve soil conditions through modifications of the topography, soil scarification, application of new soil covers, soil amendments, and water-retention polymers (Fig. 5c, d) (Ciano et al. 2005; Cony et al. 2013; Cony 2016; Busso and Pérez 2018).

Due to the harsh environmental conditions, the main technique to restore the native vegetation of these areas is the transplanting of nursery-grown seedlings. This allows higher survival rates than seed-based restoration (Cony et al. 2013; Pérez et al. 2019). Many of the species mentioned in this chapter have been successfully used to restore areas degraded by the mining industry. Some of them are *P. flexuosa, A. lampa, A. sagittifolia,* and *P. caespitosum.* Survival percentages of transplanted seedlings are highly variable, around 40–90% (Ciano et al. 2005; Cony et al. 2013; Dalmasso et al. 2015; Cony 2016), and the main causes of mortality are soil heterogeneity, water availability, and herbivory.

5.4 Biomass Production for Energy

An alternative to be evaluated is the implantation of species with energy potential, whose objective is the carbon fixation and the subsequent transformation into energy from biomass. This is an underdeveloped line of study in the Argentine Monte region, but it is beginning to develop in other arid areas of the world (Paneque 2013). Possible products include firewood, thermal power generation, biofuel and biogas production. Of these alternatives, firewood production is traditional for local people, where there are native species of excellent quality and calorific value, such as *P. flexuosa* and *P. chilensis* (Alvarez and Villagra 2009). There are no major advances in the rest of the products although it is interesting to start exploring the heating power of species from highly saline environments, such as *Allenrolfea vaginata, Suaeda divaricata,* and *Atriplex crenatifolia.* Some observations in these species suggest that they have relatively high productivity in these environments. The productivity of other *Atriplex* species has been evaluated in the coastal desert of Chile, finding under-irrigation yields of 5200 kg ha^{-1} $year^{-1}$ and a calorific value between 3500 and 4500 kcal kg^{-1} (Paneque 2013).

6 Final Considerations and General Conclusions

The restoration of soils affected by salinization is, therefore, a challenge that requires interdisciplinary studies that broaden the knowledge not only of native species and origins to be used, the appropriate implantation practices and technologies but also of environmental factors and management practices that generate the saline conditions (Meglioli et al. 2018). The diversity of ecosystems and agroecosystems in the Monte region, with their different states of degradation of their environments, and the range

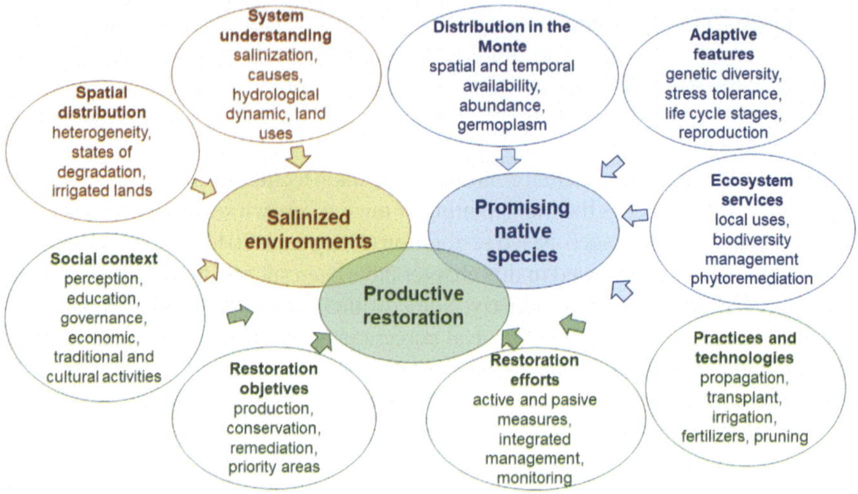

Fig. 6 Some aspects of native plant species, salinized environments and productive restoration that would be relevant to consider for improving the success of ecological restoration programs

of economic and cultural activities, imprints on us particular dynamics of resource use and management, which should be integrated into any restoration project (Fig. 6).

The selection of native species mixtures and plant functional groups to be applied in restoration programs, still requires detailed information about the biology and ecophysiology of the species, their interactions with biotic (included herbivory) and abiotic stress, which coexist with those imposed by salinization (Fig. 6). The restoration and productive recovery of saline environments in the Argentine Monte region present different development according to the objective and the species to be considered for restoration. While a few species have been studied for a long time and are in stages of evaluation and selection of germplasm to take in cultivation, others are practically unknown on their physiology and behavior in different environmental conditions. The species of the genera *Prosopis* and *Atriplex*, and the grasses *L. crinita* and *P. caespitosum* appear initially as the most promising. It is important to advance with the studies of other species, in order to determine productive potential and degree of tolerance to saline stress. The high variability on their adaptations to salinity found in native species implies a high potentiality for germplasm selection.

The restoration and productive recovery of saline environment can improve the provision of ecosystem services and benefits, including traditional activities in the region as livestock management and forestry, and non-traditional as energy production.

Acknowledgements The studies mentioned in this chapter received funding from CONICET, Universidad Nacional de Cuyo, UCAR- Ministerio de Agroindustria, Agencia Nacional de Promoción Científica y Tecnológica (PICT 2017-762) y PIO SECITI (San Juan). We thank the company Frutos del Sur S.A. and foundation Tikún for their contribution to the land for plantations and the logistics of the restoration activities. We are grateful to the local families who have helped in the plantations and maintenance of plots. We would like to thank Leandro Alvarez, Patricia Baldaccini, Gualberto Zalazar, Marina Morsucci, Aldana Cantón, Gabrielle Le Gall, Erica Cesca, Gisela Rábida, Diego Zeverini, Hugo Debandi, Dimitri Lesik, Georgina Marianetti, Sofia Hinrich, Vanesa García, Ricardo Elias, and Vladimir Matskovsky for their contribution to the development of the greenhouse and field trials that provided data to this chapter.

References

Abraham EM (2002) Lucha contra la desertificación en las tierras secas de Argentina. El caso de Mendoza. In: Fernández Cirelli A, Abraham EM (eds) El agua en Iberoamérica. De la escasez a la desertificación. Cooperación Iberoamericana CYTED Programa Iberoamericano de Ciencia y Tecnología para el desarrollo. Buenos Aires, pp 27–44

Abraham E, Rubio C, Salomón M, Soria D (2014) Desertificación: problema ambiental complejo de las tierras secas. In: Torres L, Abraham E, Pastor G (eds) Una ventana sobre el territorio. Herramientas teóricas para comprender las tierras secas. EDIUNC, Mendoza, pp 187–264

Aiazzi MT, Carpane P, Deza C (2005) Efecto de la salinidad sobre el crecimiento de plantas de *Atriplex cordobensis*. Multequina 14:39–46

Alvarez JA, Villagra PE (2009) *Prosopis flexuosa* DC. (Fabaceae, Mimosoideae) Kurtziana 35:49–63

Alvarez JA, Villagra PE, Cesca EM, Rojas F, Delgado S (2015) Estructura, distribución y estado de conservación de los bosques de *Prosopis flexuosa* del Bolsón de Fiambalá (Catamarca). B Soc Argent Bot 50:193–208

Allegretti L, Sartor C, Paez S, Egea V, Fucili M, Passera C (2012a) Effect of physiological state of criollo goats on the botanical composition of their diet in NE Mendoza, Argentina. Small Ruminant Res 103(2–3):152–157

Allegretti L, Savietto M, Tacchini F, Iglesias G, Peralta I (2012b) Principales especies de interés forrajero en la zona árida de General Alvear. In: Tacchini F (ed) Ganadería bovina de General Alvear. Mendoza, pp 38–73

Amezketa E (2006) An integrated methodology for assessing soil salinization, a pre-condition for land desertification. J Arid Environ 67(4):594–606

Arce ME, Walicki V, Castro I, Valenzuela, MF, de Sosa Tomas A, Rack MG, Ferro L (2015) Evaluación de la revegetación natural en canteras de áridos en dos sitios de la provincia de Chubut. In: Martínez Carretero E, Dalmasso AD (eds) Restauración ecológica en la Diagonal Árida de la Argentina, pp 355–383

Aronson J, Alexander S (2013) Ecosystem restoration is now a global priority: time to roll up our sleeves. Restor Ecol 21:293–296

Bai ZG, Dent DL, Olsson L, Schaepman ME (2008) Proxy global assessment of land degradation. Soil Use Manage 24:223–234

Basavaraja PK, Sharma SD, Badrinath MS, Sridhara S, Hareesh GR (2007) *Prosopis juliflora*—an efficient tree species for reclamation of salt affected soils. Karnataka J Agric Sci 20(4):727–731

Bessega C, Cony M, Saidman BO, Aguiló R, Villagra PE, Alvarez JA, Pometti C (2019) Genetic diversity and differentiation among provenances of *Prosopis flexuosa* DC (Leguminosae) in a progeny trial: implications for arid land restoration. For Ecol Manage 443:59–68

Bisigato AJ, Laphitz RML (2009) Ecohydrological effects of grazing-induced degradation in the Patagonian Monte, Argentina. Austral Ecol 34:545–557

Brizuela MA, Cid MS, Miñon DP, Fernández Grecco R (1990) Seasonal utilization of saltgrass (*Distichlis* spp.) by cattle. Anim Feed Sci Technol 30:321–325

Busso CA, Pérez DR (2018) Opportunities, limitations and gaps in the ecological restoration of drylands in Argentina. Ann Arid Zone 57(3–4):191–200

Butterfield BJ, Copeland SM, Munson SM, Roybal CM, Wood TE (2017) Prestoration: using species in restoration that will persist now and into the future. Restor Ecol 25:S155–S163

Candia RJ, Guevara JC (1973) Las comunidades vegetales del campo fiscal de La Llave y su valor forrajero. Deserta 4:93–108

Caraciolo LM, Passera CB, Allegretti LI, Robles Cruz AB (2002) Efecto de diferentes niveles de salinidad sobre el crecimiento de *Atriplex lampa*. Rev FCA, UNCUYO 36:101–106

Cavagnaro PF, Cavagnaro JB, Lemes JL, Masuelli RW, Passera CB (2006) Genetic diversity among varieties of the native forage grass *Trichloris crinita* based on AFLP markers, morphological characters, and quantitative agronomic traits. Genome 507:906–918

Céccoli G, Ramos J, Pilatti V, Dellaferrera I, Tivano JC, Taleisnik E, Vegetti AC (2015) Salt glands in the Poaceae family and their relationship to salinity tolerance. Bot Rev 81:162–178. https://doi.org/10.1007/s12229-015-9153-7

Chhabra R, Thakur NP (1998) Lysimeter study on the use of biodrainage to control waterlogging and secondary salinization in (canal) irrigated arid/semi-arid environment. Irrig Drain Syst 12:265–288

Ciano NF, Luque JL, Nakamatsu V, Vicente CV, Lisoni C (2005) Plan de abandono de canteras y picadas en la cuenca del Golfo San Jorge-Patagonia Argentina. EEA Chubut. INTA. https://inta.gob.ar/documentos/plan-de-abandono-de-canteras-y-picadas-en-la-cuenca-del-golfo-san-jorge-patagonia-argentina

Cony MA (1993) Programa de Conservación y Mejoramiento de Especies del Género *Prosopis* en la Provincia Fitogeográfica del Monte, Argentina. In: IADIZA (ed) Contribuciones mendocina a la quinta reunión regional para América Latina y el Caribe de la Red de Forestación del CIID. Conservación y Mejoramiento de Especies del Género *Prosopis*. IADIZA-CRICYT-CIID, Mendoza, pp 37–72

Cony MA (1996) Genetic variability in *Prosopis flexuosa* D.C., a native tree of the Monte phytogeographic province, Argentina. For Ecol Manag 87:41–49

Cony MA (2016) Requerimientos y avances sobre restauración ecológica en la Ecoregión del Monte. In: Libro de resúmenes de la VI Reunión Binacional de Ecología organizada por la Asociación Argentina de Ecología, Puerto Iguazú, Misiones, del 18 al 22 de septiembre de 2016

Cony MA, Trione SO (1998) Inter- and intraspecific variability in *Prosopis flexuosa* and *P. chilensis*: seed germination under salt and moisture stress. J Arid Environ 40:307–317

Cony MA, Paez J, Fernández ME (2013) Revegetation of mining fields in Argentinean arid regions. In: Proceedings of the society of ecological restoration world conference: reflections on the past, directions for the future, pp 6–11

Cortina J, Bellot J, Vilagrosa A, Maestre FT, Rubio E, Ortíz de Urbina JM, Bonet A (2004) Restauración en semiárido. In: Vallejo VR, Alloza JA (eds) Avances en el estudio de la gestión del monte mediterráneo. Fundación Centro de Estudios Ambientales del Mediterráneo, Valencia, pp 345–406

Dágata SL (2018) Arbustos nativos de la provincia de Mendoza, con potencial de uso en tareas de restauración: factores que afectan la germinación. Universidad Nacional de Cuyo. https://bdigital.uncu.edu.ar/10870

Dalmasso AD, Ciano N (2015) Restauración de taludes con especies nativas para zonas áridas y semiáridas. Experimentia 5:1–80

Dalmasso AD, Márquez J, Scaglia J, Hadad M, Cáceres JP, Brizuela R, Carnino J, Gómez A (2015) Rehabilitación de un área disturbada en la Sierra Chica de Zonda, San Juan. In: Martínez Carretero E, Dalmasso AD (eds) Restauración ecológica en la Diagonal Árida de la Argentina, pp 401–412

Dalmasso AD, Cruces E, Caprio S (2016) Respuesta al corte de *Baccharis spartioides* (Hook. & Arn. ex DC). J Rémy Multequina 25:23–28

Dalmasso AD, Parera V, Martínez Carretero E, Elmida C, Moreno D, Oneto ME (2018) Utilización de recortes de perforación petrolera para la restauración ecológica de zonas áridas y semiáridas. Experimentia 6:19–36

Díaz S, Pascual U, Stenseke M, Martín-López B, Watson RT, Molnár Z, Hill R, Chan KM, Baste IA, Brauman KA, Polasky S (2018) Assessing nature's contributions to people. Science 359:270–272

Egea AV, Allegretti L, Paez Lama S, Grilli D, Sartor C, Fucili M, Guevara JC, Passera C (2014) Selective behavior of creole goats in response to the functional heterogeneity of native forage species in the central Monte desert, Argentina. Small Ruminant Res 120:90–99

Eynard C, Calviño A, Ashworth L (2017) Cultivo de plantas nativas: propagación y viverismo de especies de Argentina central. Editorial Universidad Nacional de Córdoba

Felker P, Ewens M, Velarde M, Medina D (2008) Initial evaluation of *Prosopis alba* Griseb clones selected for growth at seawater salinities. Arid Land Res Manag 22:334–345

Fernández ME, Passera CB, Cony MA (2016) Sapling growth, water status and survival of two native shrubs from the Monte Desert, Mendoza, Argentina, under different preconditioning treatments. Rev FCA, UNCUYO 48:110–125

Fernández ME, Passera CB, Cony MA, Pissolito CI, Cortina J (2018a) Influencia del preacondicionamiento a la sequía y del agregado de compost en la sobrevivencia y crecimiento inicial de plantines de *Atriplex lampa*, *Larrea cuneifolia* y *Senecio subulatus* en un área degradada del monte. In: Libro de resúmenes del I Taller Internacional de Restauración Ecológica y IV Taller de Rehabilitación y Restauración en la Diagonal Árida de la Argentina, Salta 2018

Fernández ME, Pissolito CI, Passera CB (2018b) Water and nitrogen supply effects on four desert shrubs with potential use for rehabilitation activities. Plant Ecol 219(7):789–802

Fernández ME, Cony MA, Passera CB (2019) Germination temperatures and seed dormancy of two *Larrea* species (Zygophyllaceae) from the Monte Desert, Argentina. Rev FCA, UNCUYO 51:235–247

Flowers TJ, Colmer TD (2015) Plant salt tolerance: adaptations in halophytes. Ann Bot 115:327–331

Foley JA, Ramankutty N, Brauman KA, Cassidy ES, Gerber JS, Johnston M, Mueller ND, O'Connell C, Ray DK, West PC, Balzer C, Bennett EM, Carpenter SR, Hill J, Monfreda C, Polasky S, Rockström J, Sheehan J, Siebert S, Tilman D, Zaks DPM (2011) Solutions for a cultivated planet. Nature 478:337–342

Giménez AM, Figueroa ME, Hernández P, Cejas M (2008) Jume negro (*Allenrolfea vaginata* Kuntze, Chenopodiaceae): potencialidades de usos en la rigurosidad del ambiente salino. Quebracho 15:32–36

Greco SA, Cavagnaro JB (2004) Efecto de la salinidad en la germinación de 4 gramíneas forrajeras del Monte. In: Libro de resúmenes de la II Reunión Binacional de Ecología, Mendoza, Argentina

Guevara JC, Grünwaldt EG, Estevez OR, Bisigato AJ, Blanco LJ, Biurrun FN, Ferrando CA, Chirino CC, Morici E, Fernández B, Degiorgis A, Allegretti LI Passera CB (2009) Range and livestock production in the Monte Desert, Argentina. J Arid Environ 73:228–237

Guida-Johnson B, Abraham EM, Cony MA (2017) Salinización del suelo en tierras secas irrigadas: Perspectivas de restauración en Cuyo, Argentina. Rev FCA, UNCUYO 49:205–215

Hamidov A, Beltrao J, Neves A, Khaydarova V, Khamidov M (2007) *Apocynum lancifolium* and *Chenopodium album*—Potential species to remediate saline soils. WSEAS Trans Environ Dev 3:123–128

Hasanuzzaman M, Nahar K, Alam MM, Bhowmik PC, Hossain MA, Rahman MM, Prasad MNV, Öztürk M, Fujita M (2014) Potential use of halophytes to remediate saline soils. BioMed Res Int. https://doi.org/10.1155/2014/589341

Hbirkou C, Martius C, Khamzina A, Lamers JPA, Welp G, Amelung W (2011) Reducing topsoil salinity and raising carbon stocks through afforestation in Khorezm, Uzbekistan. J Arid Environ 75:146–155

INTA (1990) Atlas de Suelos de la República Argentina. Available in https://www.geointa.inta.gob.ar/2013/05/26/suelos-de-la-republica-argentina/

Jobbágy EG, Nosetto M, Santoni C, Baldi G (2008) El desafío ecohidrológico de las transiciones entre sistemas leñosos y herbáceos en la llanura Chaco-Pampeana. Ecol Austral 18:305–322

Jobbágy EG, Nosetto MD, Villagra PE, Jackson RB (2011) Water subsidies from mountains to deserts: their role sustaining groundwater-fed oases in a sandy landscape. Ecol Appl 21:678–694

Kozub PC, Barboza K, Galdeano F, Quarin CL, Cavagnaro JB, Cavagnaro PF (2017) Reproductive biology of the native forage grass *Trichloris crinita* (Poaceae, Chloridoideae). Plant Biol 19:1–9

Liu R, Wang L, Tanveer M, Song J (2018) Seed heteromorphism: an important adaptation of halophytes for habitat heterogeneity. Front Plant Sci 871:1–10

Luna V, Llanes A, Reginato M (2017) Respuestas ecofisiológicas a la salinidad por sulfato de sodio en una halófita nativa de Sudamérica. In: Taleisnik E, Lavado RS (eds) Ambientes salinos y alcalinos de la Argentina. Universidad Católica de Córdoba - Orientación Gráfica Editora, Buenos Aires, pp 517–540

Maestre FT, Cortina J, Bautista S, Bellot J, Vallejo R (2003) Small-scale environmental heterogeneity and spatiotemporal dynamics of seedling establishment in a semiarid degraded ecosystem. Ecosystems 6(7):630–643

Mantován NG (2002) Early growth differentiation among *Prosopis flexuosa* D.C provenances from the Monte phytogeographic province, Argentina. New For 23:19–30

Mantován NG (2005) Variabilidad intra-específica de *Prosopis flexuosa* DC. var. *flexuosa* en el Monte. Su estudio morfo-fisiológico. Tesis Doctoral. Universidad Nacional de Cuyo

Marinoni LR, Zabala JM, Taleisnik EL, Schrauf GE, Richard GA, Tomas PA, Giavedoni JA, Pensiero JF (2019) Wild halophytic species as forage sources: key aspects for plant breeding. Grass Forage Sci 1–24. https://doi.org/10.1111/gfs.12410

Martínez Carretero E (2001) Esquema sintaxonómico de la vegetación de regiones salinas de Argentina. Multequina 10:67–74

McDonald T, Gann GD, Jonson J, Dixon KW (2016) International standards for the practice of ecological restoration—including principles and key concepts. Soc Ecol Restor, Washington, D.C.

Meglioli PA, Villagra PE, Lana NB, Alvarez L, Alvarez JA, Cony MA (2018) Forestación de suelos salinizados y factores ambientales incidentes en la potencialidad de restauración en Cuyo. In: Actas de trabajos completos Encuentro International Center for Earth Sciences 13. Malargüe, Mendoza. ISBN 978-987-1323-63-0

Montani M, Vega Riveros C, Ribas Y, Karlin U, Gaviorno M, Hadad M, Meglioli, P, Inojosa M, Ordóñez C, Damiani O (2010) Raíces Huarpes: Uso medicinal de plantas en la comunidad de Lagunas del Rosario, Mendoza, Argentina. Editorial Universidad Nacional de San Juan. ISBN 978-950-605-619-3

Nabti E, Sahnoune M, Ghoul M, Fischer D, Hofmann A, Rothballer M, Schmid M, Hartmann A (2010) Restoration of growth of durum wheat (*Triticum durum* var. waha) under saline conditions due to inoculation with the rhizosphere bacterium *Azospirillum brasilense* NH and extracts of the marine alga *Ulva lactuca*. J Plant Growth Regul 29:6–22

Ochoa M (2011) Producción de forraje en suelos salinos. EEA Rama Caída. Mendoza. INTA. https://inta.gob.ar/documentos/produccion-de-forraje-en-suelos-salinos

ONDTYD (2010) *Índice de aridez climático para Argentina*. https://www.desertificacion.gob.ar/indicadores/indice-de-aridez-segun-unep/?ref=h

Paneque M (2013) Biomasa en el desierto de Chile. Facultad de Ciencias - Universidad de Chile, Santiago de Chile

Passera CB (1990) Efecto de la temperatura y del estrés hídrico en la germinación de *Atriplex lampa* Moq. (Chenopodiaceae). Stvdia Oecológica VII:77–84

Passera CB (2017) Guía para la propagación e implantación de Zampa (*Atriplex lampa*). Boletín de Extensión Científica, Nº 9; 20 Pág. Instituto Argentino de Investigaciones de las Zonas Áridas, IADIZA-CONICET Mendoza. Inca Editorial, Mendoza

Passera CB, Allegretti LI (1996) Relaciones hídricas, productividad y partición de asimilados en plántulas de *Atriplex lampa* sometidas a estrés hídrico. In: Libro de resúmenes de la XXI Reunión Argentina de Fisiología Vegetal, pp 230–231

Passera CB, Borsetto O (1989) Aspectos ecológicos de *Atriplex lampa*. Investigación Agraria: Producción y Protección Vegetales 4(2):179–198 (ISSN: 0213-5000)

Pentreath V, González E, Barquín M, Ríos SM, Perales S (2015) Bioensayo de toxicidad aguda con plantas nativas para evaluar un derrame de petróleo Rev. Salud Ambient 15:4–12

Pérez Cuadra V (2012) Anatomía ecológica de la vegetación del Salitral de la Vidriera. Tesis doctoral. Universidad Nacional del Sur

Pérez Cuadra V, Cambi V (2010) Anatomía vegetativa comparativa entre Chloridoideae (Poaceae) halófilas de importancia forrajera. Phyton 79:69–76
Pérez DR, González F, Ceballos C, Oneto ME, Aronson J (2019) Direct seeding and outplantings in drylands of Argentinean Patagonia: estimated costs, and prospects for large-scale restoration and rehabilitation. Restor Ecol 27(5):1105–1116
Peter G, Leder C, Calvo D, Cruz A, Torres Robles S (2014) Evaluación de tratamientos pregerminativos en seis especies nativas del ecotono Monte-Espinal del noreste de la Patagonia. Bioscriba 7:1–10
Piovan MJ (2016) Controles geomorfológicos sobre la presencia y estructura de humedales costeros en el estuario de Bahía Blanca. Tesis doctoral. Universidad Nacional del Sur
Piovan MJ, Zapperi G, Pratolongo P (2014) Seed germination of *Atriplex undulata* under saline and alkaline conditions. Seed Sci Technol 42:286–292
Piovan MJ, Pratolongo P, Donath TW, Loidy A, Eckstein RL (2019) Germination response to osmotic potential, osmotic agents, and temperature of five halophytes occurring along a salinity gradient. Int J Plant Sci 4:345–355
Próspero J, Greco S, Sartor C, Kozub C, Vega Riveros C (2018) Selección de gramíneas perennes nativas del Monte para tareas de revegetación en la provincia de Mendoza. In: Libro de resúmenes XXVIII Reunión Argentina de Ecología. Mar del Plata, 29 de octubre al 2 de noviembre de 2018
Rabhi M, Hafsi C, Lakhdar A, Hajji S, Barhoumi Z, Hamrouni MH, Abdelly C, Smaoui A (2009) Evaluation of the capacity of three halophytes to desalinize their rhizosphere as grown on saline soils under nonleaching conditions. Afr J Ecol 47:463–468
Ragonese AE, Piccinini BG (1978) Consideraciones sobre la vegetación de las Salinas de Mascasín (La Rioja - San Juan, República Argentina). Darwiniana 21:49–60
Ram J, Garg VK, Toky OP, Minhas PS, Tomar OS, Dagar JC, Kamra SK (2007) Biodrainage potential of *Eucalyptus tereticornis* for reclamation of shallow water table areas in north-west India. Agrofor Syst 69:147–165
Ravindran KC, Venkatesan K, Balakrishnan V, Chellappan KP, Balasubramanian YT (2007) Restoration of saline land by halophytes for Indian soils. Soil Biol Biochem 39:2661–2664
Rengasamy P (2006) World salinization with emphasis on Australia. J Exp Bot 57(5):1017–1023
Ridley AM, Pannell DJ (2005) The role of plants and plant-based research and development in managing dryland salinity in Australia. Aust J Exp Agric 45:1341–1355
Rhodes D, Felker P (1987) Mass screening *Prosopis* (mesquite) seedlings for growth at sea water salinity. For Ecol Manag 24:169–176
Rundel PW, Villagra PE, Dillon MO, Roig-Juñent S, Debandi G (2007) Arid and semi-arid regions and ecosystems. In: Veblen TT, Young K, Orme A (eds) The physical geography of South America, pp 158–183
Sánchez MR, Dunel Guerra L (2017) Salinidad y sodicidad de suelos bajo riego en zonas áridas y semiáridas y su efecto sobre los cultivos. In: Taleisnik E, Lavado RS (eds). Ambientes salinos y alcalinos de la Argentina. Universidad Católica de Córdoba - Orientación Gráfica Editora, Buenos Aires, pp 227–250
SER (2004) The SER international primer on ecological restoration. Available in https://cdn.ymaws.com/www.ser.org/resource/resmgr/custompages/publications/SER_Primer/ser_primer.pdf
Sosa L, Llanes A, Reinoso H, Reginato M, Luna V (2005) Osmotic and specific ion effects on the germination of *Prosopis strombulifera*. Ann Bot 96:261–267
Taleisnik EL, Anton AM (1988) Salt glands in *Pappophorum* (Poaceae). Ann Bot 62:383–388
Taleisnik E, López Launestein D (2011) Leñosas perennes para ambientes afectados por salinidad. Una sinopsis de la contribución argentina a este tema. Ecol Austral 21:3–14
Taleisnik E, Grunberg K, Santa María G (eds) (2008) La salinización de suelos en la Argentina: su impacto en la producción agropecuaria. Editorial Universidad Católica de Córdoba, Córdoba
Thomas DS, Middleton NJ (1993) Salinization: new perspectives on a major desertification issue. J Arid Environ 24(1):95–105
Trione SO, Passera CB (1993) Growth and nitrogen status of *Atriplex lampa* seedlings under different water regimes. J Arid Environ 25:331–341

Tug GM, Yaptrack AE (2019) An overview of the germination behavior of halophytes and their role in food security. In: Hasanuzzaman M, Nahar K, Öztürk M (eds) Ecophysiology, abiotic stress responses and utilization of halophytes. https://doi.org/10.1007/978-981-13-3762-83

UNCCD (1994) Convention to combat desertification in those countries experiencing serious drought and/or desertification, particularly in Africa. Final text of the convention

Ungar IA (1987) Population ecology of halophyte seeds. Bot Rev 53:301–334

Vega Riveros C, Meglioli PA, Villagra PE (2011) *Prosopis alpataco* Phil. (Fabaceae, Mimosoideae) Kurtziana 36:53–64

Velarde M, Felker P, Degano C (2003) Evaluation of Argentine and Peruvian *Prosopis* germplasm for growth at seawater salinities. J Arid Environ 55:515–531

Villagra PE (1997) Germination of *Prosopis argentina* and *P. alpataco* seeds under saline conditions. J Arid Environ 37:261–267

Villagra PE, Alvarez JA (2019) Determinantes ambientales y desafíos para el ordenamiento forestal sustentable en los algarrobales del Monte, Argentina. Ecol Austral 29:146–155

Villagra PE, Cavagnaro JB (2005) Effects of salinity on the establishment and early growth of *Prosopis argentina* and *Prosopis alpataco* seedlings in two contrasting soils: implications for their ecological success. Austral Ecol 30(3):325–335

Villagra PE, Boninsegna JA, Alvarez JA, Cony M, Cesca E, Villalba R (2005) Dendroecology of *Prosopis flexuosa* woodlands in the Monte desert: Implications for their management. Dendrochronologia 22:209–213

Villagra PE, Vilela A, Giordano CV, Alvarez JA (2010) Ecophysiology of *Prosopis* species from the arid lands of Argentina: what do we know about adaptation to stressful environments? In: Ramawat KG (ed) Desert plants. Biology and biotechnology, pp 322–354

Villagra PE, Giordano C, Alvarez JA, Cavagnaro JB, Guevara A, Sartor C, Passera CB, Greco S (2011) Ser planta en el desierto: Estrategias de uso de agua y resistencia al estrés hídrico en el Monte Central de Argentina. Ecol Austral 21:29–42

Villagra PE, Passera CB, Greco S, Sartor C, Aranibar JN, Meglioli PA, Alvarez JA, Allegretti L, Fernández ME, Cony M, Kozub PC, Vega Riveros C (2017) Uso de plantas nativas en la restauración y recuperación productiva de ambientes salinos de las zonas áridas de la región del Monte, Argentina. In: Taleisnik E, Lavado RS (eds) Ambientes salinos y alcalinos de la Argentina. Universidad Católica de Córdoba - Orientación Gráfica Editora, Buenos Aires, pp 419–444

Villagra PE, Alvarez JA, Meglioli PA, Lana NB, Cony MA (2019) The challenge of restoring salinized environments in irrigated arid lands of Western Argentina: salt tolerance thresholds in *Prosopis* species. In: Abstracts of the 8th world conference on ecological restoration

Westphal C, Gachón P, Bravo J, Navarrete CA, Salas C, Ibáñez C (2015) The potential of Algarrobo (*Prosopis chilensis* (Mol.) Stuntz) for regeneration of desertified soils: assessing seed germination under saline conditions. Environ Manage 59:209–220

Zabala JM, Widenhorn P, Pensiero JF (2011) Germination patterns of species of the genus *Trichloris* in arid and semiarid environment. Seed Sci Technol 39:338–353

Zhang YF, Wang P, Yang YF, Bi Q, Tian SY, Shi XW (2011) Arbuscular mycorrhizal fungi improve reestablishment of *Leymus chinensis* in bare saline-alkaline soil: implication on vegetation restoration of extremely degraded land. J Arid Environ 75:773–778

Zhao KF, Harris PJC (1992) The effects of iso-osmotic salt and water stresses on the growth of halophytes and non-halophytes. J Plant Physiol 139:761–763

Zhou D, Lin Z, Liu L, Zimmermann D (2013) Assessing secondary soil salinization risk based on the PSR sustainability framework. J Environ Manage 128(15):642–654

Native and Naturalized Forage Plant Genetic Resources for Saline Environments of the Southernmost Portion of the American Chaco

José F. Pensiero, Juan M. Zabala, Lorena del R. Marinoni, and Geraldina A. Richard

To date, little attention has been paid to the optimum qualities of our several indigenous species present in natural rangelands in Argentina... agronomists and heads of experimental stations typically perform trials only with exotic plants... completely neglecting the ones that we own.
(Parodi 1919)
It is shocking to know that Paspalum dilatatum, P. notatum, P. urvillei, Axonopus compressus, Bromus unioloides, Cortaderia selloana , and other forage species from Uruguay are cultivated in other countries and not here. We find it very shocking that our seeds are requested from different regions of Africa, Australia, and the USA.
(Rosengurtt et al. 1970)

Abstract The forage value of several of the species occurring in saline soils of the southernmost portion of the American Chaco and particularly their genetic variability has been recognized by breeders from many countries. Such recognition is confirmed by the fact that accessions of these species are conserved in international germplasm banks specialized in exploration, collection, and conservation of forage species. However, these species have received scant attention in Argentina and scarce germplasm is conserved in local banks. The study of these species for the purposes

J. F. Pensiero (✉) · J. M. Zabala · L. R. Marinoni · G. A. Richard
Programa de Documentación, Conservación y Valoración de La Flora Nativa (PRODOCOVA), Facultad de Ciencias Agrarias, Universidad Nacional del Litoral (FCA-UNL), Esperanza, Santa Fe, Argentina
e-mail: jfpensi@fca.unl.edu.ar

J. M. Zabala
e-mail: jmzabala@fca.unl.edu.ar

L. R. Marinoni
e-mail: lmarinoni@fca.unl.edu.ar

G. A. Richard
e-mail: geralrichard@gmail.com

E. Taleisnik and R. S. Lavado (eds.), *Saline and Alkaline Soils in Latin America*, https://doi.org/10.1007/978-3-030-52592-7_18

of breeding and incorporation into agricultural systems is a hard and often discouraging task. Fundamental basic knowledge necessary for breeding programs, such as flowering phenology, mode of reproduction, quantity and quality of produced seeds, presence or absence of seed dormancy, and seedling establishment, is lacking for most of the species. This chapter provides a list of native and naturalized forage species that should be given priority for incorporation into agricultural systems and discusses essential aspects to be considered for such aim. It then reports on experiences that have clearly demonstrated the feasibility of successfully incorporating wild forage species to cultivation.

Keywords American chaco · Breeding programs · Germplasm collection · Germplasm conservation · Legal framework · Wild forage species

1 Introduction

Livestock production on rangelands are the main economic activity in saline areas of the Great Chaco Region. However, vast rangeland areas have been replaced with monocultures and degraded (Paruelo et al. 2004; Grau et al. 2005; Díaz 2007a, b). The sustainable forage offer of these rangelands is based on the adequate management of grazing and rest periods. Rangelands on marginal areas, such as those characteristics of saline soils, are the most fragile and susceptible to degradation due to improper management (Deregibus 1987).

Although there are abundant references about the advantages of rangelands as the base for livestock feeding as well as for environmental conservation, strong economic incentives and the lack of sustainable livestock production policies have been the drivers for deforestation and rangeland replacement with intensive agricultural systems (Murray et al. 2016). Besides providing diverse environmental services, rangelands have an array of productive and economic advantages over cultivated pastures (Lemaire et al. 2005; Hector and Loreau 2005; Batello et al. 2008; Muir et al. 2011; Squires and Glenn 2016). Several valuable native species comprise the forage offer of Argentine rangelands, particularly gramineous species that have evolved for thousands of years and that are adapted to the region. Adaptation sustains the resilience of rangeland species and communities toward drought, floods, fires, and improper management.

Incorporating improved species would contribute to increase rangeland productivity and forage quality (Muir et al. 2011, 2014). The need to increase the offer of forage species and improved cultivars that can adapt to intercropping in rangelands (Batello et al. 2008) has been highlighted. However, forage species for those marginal environments are scarce. Moreover, "Rhodes grass" (*Chloris gayana*), a widely cultivated salt-tolerant forage species, does not adapt to intercropping in rangelands and requires monoculture. In this context, the inclusion of native or naturalized forage species into cropping systems should be considered as a possible strategy (Díaz Maynard 2005). This incorporation would also provide greater production stability, increase forage species diversity, improve nutrient mineralization, and increase activity of microorganisms that promote plant growth, such as mycorrhizae (Schellberg et al. 2013). In this context, native and naturalized species breeding programs can contribute to a new paradigm of sustainable livestock production and reduced environmental footprint (Brummer et al. 2011). This new paradigm includes aspects that have not been addressed in traditional breeding programs, such as ecosystem services, and polyculture. This paradigm does not exclude monoculture of subtropical forage species; rather, their use should be restricted to surface soils that can be properly managed, since the production of these species, which is concentrated in a few months, is very often not accompanied with a suitable stocking density, or with alternative use strategies, such as stockpiling or haymaking.

The incorporation of native forage resources into cropping systems should be framed, from the methodological perspective, within long-term programs that include the following stages: (a) characterization of the target area/s and survey of the forage plant genetic resources, (b) prioritization of species for further breeding program, (c) collection and conservation of germplasm, (d) pre-breeding and breeding strategies, and (e) registration, release, and adoption of cultivars.

No intensive or long-term breeding program has been implemented with Argentine native forage species for the Great Chaco Region; therefore, the offer of cultivars of this type of species is scarce or nonexistent. This issue entails deficiencies in three key aspects: (a) knowledge about native forage plants genetic resources, (b) a national strategy for germplasm collection and characterization, and (c) multidisciplinary research in forage breeding. Such projects should include botanists, biologists, geneticists, microbiologists, physiologists, biotechnologists, ecologists, and bioinformatics experts (Shelton et al. 2005; Nichols et al. 2007; Parra-Quijano et al. 2012; Marinoni et al. 2015). It is also necessary to incorporate the private sector (producers, seed companies) and extension technicians from the start of these programs to collaborate with the process of selection and adoption of new forage species (Nichols et al. 2007). An example that can illustrate these issues is the case of Argentine wheatgrass (*Elymus scabrifolius*), a native species of great value for saline environments, with three registered cultivars in Argentina (Covas 1978; Zabala et al. 2011a, b). Two of these cultivars were registered in the 1980s and the other in 2004 (INaSe 2019). None of them was broadly adopted and their seeds are not regularly obtained in the market. The greatest difficulty involved in the use of those cultivars has been the lack of a pasture management package developed along with seed companies and extension agents.

Another aspect that has discouraged the study of our native forage plant genetic resources is the lack of incentives for producers to make the necessary investments (fencing, water troughs, intercropping equipment, etc.) to adopt technologies that would increase sustainable meat production on rangelands. Accordingly, in Argentina there is a promising policy that consists of subsidies granted by the National Fund for Enrichment and Conservation of Native Forests through the National Act N° 26331 to livestock producers whose farms bear native forests and who can present management plans including sowing native pastures to restore deforested areas.

In the following sections, we analyze these aspects focusing on saline soils from the Great Chaco Region.

2 Characterization of Saline Environments and Their Forage Species in the Southernmost Portion of the Great American Chaco

The Great American Chaco is a vast region extending over an area of 1,066,000 km^2, from tropical to subtropical latitudes (18°–31° S) that includes southeast of Bolivia, west of Paraguay, and center and north of Argentina (Hueck 1972).

In Argentina, Cabrera (1994) defined the southernmost portion of the Great American Chaco as the Chaco phytogeographic province and described four districts within this province: Eastern Chaco District, Western Chaco District, Mountain/Sierra Chaco District, and the Savannas. According to the ecoregion scheme proposed for Argentina (Burkart et al. 1999), two ecoregions are recognized for the Chaco: Humid Chaco and Dry Chaco. We will use the latter classification for the characterization of saline soils.

2.1 Humid Chaco

Briefly, based on the works of Lewis and Pire (1981, 1996), and Lewis et al. (1990), the Humid Chaco saline environments include: uplands with moderately saline soils, midlands with moderately saline non-flooded soils, midlands with saline-flooded soils, and floodable saline lowlands.

2.1.1 Uplands with Moderately Saline Soils

These environments include pure or mixed forests dominated by *Schinopsis balansae*, *Astronium balansae*, *Prosopis nigra*, *P. alba*, *P. hassleri*, *P. algarrobilla*, and *Geoffroea decorticans*. These soils hold a great diversity of species, many of them of very high forage potential. Livestock production activities in the region are mainly cow–calf operations, which are carried out with very scarce technological resources and continuous grazing management. In Argentina, since most of the mentioned forests are within the frame of the National Act 26331 (Minimum Budget for Environmental Protection of Native Forests), their exploitation must be coupled with sustainable management plans. In the case of livestock production activities, most plans include silvopastoral management schemes, usually incorporating exotic tropical forage species that eventually cause the disappearance of native forage species. An option for rangelands associated with these environments, which has been successfully tested in the north of the Santa Fe province, is rational intensive grazing management (high stocking rates in small areas and over short periods). This scheme not only keeps native species but also increases their frequency and results in remarkable rangeland productivity and persistence improvement (Fernández Ridano 2018). Electric fencing is an essential tool for this management system.

2.1.2 Midlands with Moderately Saline Non-flooded Soils

These environments bear pure forests of *S. balansae* and *Prosopis* spp., palm groves of *Copernicia alba* are frequent, as well as herbaceous communities dominated by *Elionurus muticus* "aibe", a bunchgrass forming dense medium-height tussocks. These rangelands, in which fire is frequent and somewhat shapes the community, have a high species richness, with several species of forage potential. The most important leguminous species present include the genera *Desmanthus*, *Desmodium*, *Dolichopsis*, *Galactia*, *Macroptilium*, *Rhynchosia*, and *Vicia*. Likewise, there are numerous grasses belonging to the genera *Bromus*, *Bothriochloa*, *Deyeuxia*, *Digitaria*, *Elymus*, *Eragrostis*, *Nassella*, *Pappophorum*, *Paspalum*, *Piptochaetium*, *Poa*, *Setaria*, and *Tridens*. Continuous grazing is the most frequent management practice in these rangelands, the use of fire in open spaces removes dry residues and favors resprouting of "aibe", which is highly consumed by livestock. Accordingly, communities dominated by *E. muticus* are one of the most important grasslands to consider when collecting plant genetic resources with forage potential for saline environments.

A current practice has been to replace these natural rangelands with monocultures of exotic tropical/subtropical forage species, such as "Rhodes grass". Such practice should be avoided, given these rangelands species richness, their resilience and the forage potential of many of them; rather, technologies aiming at an increase of forage productivity should be considered. Such increase might be achieved by intercropping rangeland with leguminous species, such as *Macroptilium erythroloma* and *M. lathyroides*, which have shown good establishment in rangelands in the northeast of the Santa Fe province. Also, though possibly with slower establishment rates, the incorporation of *Elymus scabrifolius* may be considered. An important aspect to take into account for the incorporation and successful establishment of these species is an adequate grazing management, avoiding the presence of livestock during the establishment period (first year after seeding).

2.1.3 Midlands with Saline-Flooded Soils

In these environments, the most important communities, in terms of surface area cover, are the tall-grasslands of *Sporobolus spartinus* (Syn. *Spartina spartinae*, the so-called espartillares), which are usually flooded during the rainy season but dry in winter.

These communities, apparently homogeneous (monospecific) due to the dominance of *S. spartinus*, have nevertheless several species growing along with it. Mature *S. spartinus* plants are not palatable because of their hard, pointed leaves. Burning is usually a common practice to promote resprouting, which is well consumed by livestock. Both burning and frequent cutting favor the development and growth of numerous species in the open area between "espartillos," many of which have forage potential. Among them, gramineous species of the genera *Echinochloa*, *Elymus*, *Eriochloa*, *Hemarthria*, *Hymenachne*, *Leersia*, *Luziola*, and *Paspalum*, and legume genera *Aeschynomene*, *Desmanthus*, *Dolichopsis*, and *Vicia*.

Other communities that may occur in these environments are the tall-grasslands of *Paspalum intermedium*, *P. quadrifarium*, *Sorghastrum setosum* and, to a lower extent, *Cenchrus pilcomayensis;* and shrublands dominated by *Tessaria dodoneifolia*, *Cyclolepis genistoides*, and *Maytenus vitis-idaea.*

In these tall-grasslands, particularly in sites less prone to flooding, yet not exempt from it (i.e., up to two months), the replacement of those grasses with forage crops has been attempted, unsuccessfully (Bissio 2014a). Forage species used so far, such as "white sweet clover" (*Melilotusalbus)* and "Rhodes grass," do not tolerate flooding. Therefore, forage species tolerant to both salinity and flooding should be considered for these sites.

The following native grasses, which have forage potential and can tolerate both stresses, could be considered for those environments: *Echinochloa helodes*, *Hemarthria altissima*, *Leersia hexandra*, *Luziola peruviana*, *Paspalum acuminatum*, *P. buckleyanum*, *P. denticulatum*, *P. distichum*, *P. vaginatum*, and *Setaria geminata* (Table 1). Most of these species have problems in terms of seed production and/or harvest (null or scarce seed production, or mature seed shedding); important issues that should be investigated in any of them.

Species of the genus *Aeschynomene* stand out among the legumes that can tolerate moderate salinity and flooding in these environments (Table 2). Some species, particularly *Aeschynomene americana*, *A. denticulata*, and *A. rudis*, are highly consumed by livestock. Although these species exhibit uneven flowering and maturity, with a single plant frequently bearing flowers and seeds simultaneously, they have a high production of seeds that can be harvested. Another important characteristic of this genus is nitrogen fixation, and these species show good nodulation even in flooded soils (Skerman et al. 1991). Therefore, they are valuable resources that should be given priority for incorporation into cultivation.

2.1.4 Floodable Saline Lowlands

Communities of hydrophytes or halophytes may occur in these wetlands, depending on the degree of salinity, erosion, and flooding frequency. Hydrophyte communities are dominated by *Schoenoplectus californicus* and *Typha domingensis* in the lowest sectors, which remain flooded for longer periods. In higher sectors, communities of *Thalia geniculata*, *T. multiflora*, and *Ludwigia peploides* are present, as well as flooding grasslands, with a high diversity of grasses and graminoid species,

Table 1 Native and naturalized gramineous forage species prioritized for saline environments of the Humid and Dry Chaco ecoregions in Argentina

Species	Habit	PC	TF	TSa	TDr	Status	Seed	Ecoregion	Remarks
Chloris halophila Parodi	PH	SSA	1	2	3	Nv	**O**	HC, DC	Although it tolerates intermediate salinity and drought levels, its forage potential is low because of its low forage production. However, it is important to maintain soil cover with a high number of plants to reduce erosion by water (Pensiero and Zabala 2017)
Cottea pappophoroides Kunth	PH	SSA	0	2	2	Nv	**O**	DC	Widely distributed species that has good tolerance to grazing (Pensiero 1986, 2012a); exhibits a cleistogamous mechanism (Tivano and Vegetti 2010) that allows its survival under different disturbances
Distichlis acerosa (Griseb.) H. L. Bell & Columbus	PH	SSA	2	3	3	E	**O**	DC	Stoloniferous species; forage value increases in winter, providing forage resource when the remaining herbaceous species are scarce or non-existent (Parodi 1954; Cavanna et al. 2010; Coirini et al. 2010)
Distichlis scoparia (Kunth) Arechav	PH	SSA	3	3	3	E	**O**	HC, DC	Given its tolerance to extreme conditions, it is one of the few herbaceous species capable of providing forage to livestock and soil cover (Ragonese 1951)
Distichlis spicata (L.) Greene	PH	SSA	3	3	3	Nv	**O**	HC, DC	Given its tolerance to extreme conditions, it is one of the few herbaceous species, along with *D. scoparia*, capable of providing forage to livestock and soil cover (Pensiero and Zabala 2017)
Echinocloa helodes (Hack.) Parodi	PH	SSA	3	2	0	Nv	**No data**	HC	Forage species with good quality and high productivity (Fernández et al. 1993; Bissio et al. 1994). It has the disadvantage that its seeds fall as they reach maturity, hindering their harvest (Pensiero and Zabala 2017)

(continued)

Table 1 (continued)

Species	Habit	PC	TF	TSa	TDr	Status	Seed	Ecoregion	Remarks
Elionurus muticus (Spreng.) Kuntze	PH	SSA	0	1	3	Nv	**O**[b]	HC, DC	While the cattle eat only the leaf tips from adult plants, regrowth after burning or cutting is well consumed, becoming a valuable forage resource, particularly in semiarid environments (Schulz 1962; Fernández et al. 1993; Bissio et al. 1994; Pensiero and Zabala 2017)
Elymus scabrifolius (Döll) J. H. Hunz[a]	PH	AWS	2	2	1	Nv	**O**	HC	Due to its forage productivity and quality, it is one of the most valuable native C3 gramineous species for germplasm conservation and use in plans for incorporation into cultivation (Hawkins and Donald 1963; Covas 1978, 1982; Zabala et al. 2011a, b; Pensiero and Zabala 2017)
Eragrostis orthoclada Hack	PH	SSA	1	2	2	Nv	**O**[b]	HC, DC	Species affected by continuous grazing; frequently present in protected sites (Santa Cruz and Quiroga 1998; Quiroga and Correa 2011)
Eragrostis spicata Vasey	PH	SSA	1	2	2	Nv	**O**	HC	Species that has been confused with *Sporobolus phleoides*; it differs mainly by its spikelets with several flowers (www.floraargentina.edu.ar)
Hemarthria altissima (Poir.) Stapf and C. E. Hubb	PH	SSA	3	2	0	Nd	**No data**	HC	There are commercial cultivars, with low availability in Argentina. It has vegetative propagation. Given the forage productivity and quality of some naturalized populations, they should be collected and evaluated (Schulz 1962; Batello et al. 2008; Bissio 2014a, b; Pensiero and Zabala 2017; https://www.tropicalforages.info/)

(continued)

Table 1 (continued)

Species	Habit	PC	TF	TSa	TDr	Status	Seed	Ecoregion	Remarks
Hordeum stenostachys Godr	PH	AWS	1	2	1	Nv	**O**	HC, DC	Of good quality and medium to low productivity. It has the disadvantage that its spikelets fall at maturity, hindering their harvest (Pensiero and Zabala 2017)
Leersia hexandra Sw	PH	SSA	3	3	1	Nv	**O**	HC, DC	Very good forage productivity and quality. It has the disadvantage that its spikelets fall at maturity (Fernández et al. 1993; D'Angelo et al. 1987; Bissio et al. 1994; Batello et al. 2008; Pensiero and Zabala 2017)
Leptochloa fusca (L.) Kunth ssp. *uninervia* (J. Presl) N. W. Snow	PH	SSA	2	3	3	Nv	**O**	HC, DC	It exhibits a noticeable intraspecific variability, both in growth habit and morphological traits. It retains the seeds, which enables harvest. It should be prioritized for plans for collection and incorporation into cultivation (Parodi 1927; Nicora 1995; Batello et al. 2008; Pensiero and Zabala 2017; https://www.floraargentina.edu.ar)
Luziola peruviana Juss. ex J. F. Gmel	PH	SSA	3	3	1	Nv	**No data**	HC, DC	Monoicous species, of excellent forage productivity and quality, although it does not tolerate drought. Its spikelets fall off easily at maturity, hindering harvest (Schulz 1962; Fernández et al. 1993; D'Angelo et al. 1987; Pensiero and Zabala 2017)
Pappophorum caespitosum R. E. Fr[a]	PH	SSA	0	2–3	3	Nv	**O**	HC, DC	One of the few gramineous species with forage potential growing in saline soils of the Dry Chaco. It produces a good amount of seeds that can be easily harvested. Collection and germplasm conservation should be prioritized, as well as its inclusion in breeding plans for its cultivation in saline soils of central and northern Argentina (Pensiero 1986, 2012c; Pensiero and Zabala 2017)

(continued)

Table 1 (continued)

Species	Habit	PC	TF	TSa	TDr	Status	Seed	Ecoregion	Remarks
Paspalum acuminatum Raddi	PH	SSA	3	2	1	Nv	**O**[b]	HC	Forage species of very good productivity and quality; whether local populations produce seeds is unknown (Schulz 1962; Pensiero and Zabala 2017)
Paspalum buckleyanum Vasey	PH	SSA	3	2–3	1	Nv	**O**	HC, DC	Valuable forage species with diverse cytotypes. It should be prioritized for germplasm collection and incorporation into cultivation (Schulz 1962; Burson 1997; Sartor et al. 2011; Pensiero and Zabala 2017)
Paspalum denticulatum Trin	PH	SSA	3	2–3	1	Nv	**O**	HC, DC	Tetraploid apomictic facultative species ($2n = 4x = 40$). It should be prioritized for germplasm collection and incorporation into cultivation (Schulz 1962; Covas 1978; D'Angelo et al. 1987; Bissio et al. 1994; Sartor et al. 2011; Pensiero and Zabala 2017)
Paspalum distichum L.	PH	SSA	2–3	2–3	1	Nv	**O**	HC, DC	Valuable forage species with great potential for incorporation into cultivation in saline and flooded soils. It should be prioritized for germplasm collection and introduction into cultivation (Schulz 1962; D'Angelo et al. 1987; Rogers et al. 2005; Pensiero and Zabala 2017; Halophytes database https://www.sussex.ac.uk/affiliates/halophytes/)
Paspalum vaginatum Sw	PH	SSA	2–3	3	2	Nv	**O**	HC, DC	The most salinity-tolerant species of the genus; it occurs in very saline and flooded soils (D'Angelo et al. 1987; Bissio et al. 1994; Feldman et al. 2008). It is used as turfgrass due to its rusticity. Its collection and introduction into cultivation should be prioritized (Schulz 1962; Rogers et al. 2005; Pensiero and Zabala 2017)

(continued)

Table 1 (continued)

Species	Habit	PC	TF	TSa	TDr	Status	Seed	Ecoregion	Remarks
Setaria geminata (Forssk.) Veldkamp	PH	SSA	3	2	1	Nv	**O**[b]	HC	Excellent forage species that propagates vegetatively. Its seed production and the possibility of conserving germplasm should be analyzed (Bissio et al. 1994; Schinini et al. 2004; Pensiero 2012c; Pensiero and Zabala 2017; Halophytes database https://www.sussex.ac.uk/affiliates/halophytes/)
Sporobolus indicus (L.) R. Br	PH	SSA	1	2	2	Nv	**O**	HC, DC	Good forage species that tolerates intensive grazing and trampling. It should be prioritized for germplasm collection and introduction into cultivation (Hidalgo et al. 1998; Pensiero and Zabala 2017)
Sporobolus phleoides Hack[a]	PH	SSA	2	3	3	E	**O**	HC, DC	Important plant genetic resource for these environments due to its high tolerance to salinity. Although it is endemic to Argentina, its germplasm is conserved in Australian germplasm banks (Rogers et al. 2005; Richard et al. 2015; Pensiero and Zabala 2017; Halophytes database https://www.sussex.ac.uk/affiliates/halophytes/))
Sporobolus pyramidatus (Lam.) Hitchc	PH	SSA	2	2	2	Nv	**O**	HC, DC	Despite its importance as plant genetic resource, it has low forage productivity (Hilgert et al. 2003; Feldman et al. 2008)

(continued)

Table 1 (continued)

Species	Habit	PC	TF	TSa	TDr	Status	Seed	Ecoregion	Remarks
Trichloris crinita (Lag.) Parodi[a]	PH	SSA	0	3	3	Nv	**O**	HC, DC	One of the species with greatest forage plant genetic resource potential for arid and semiarid regions from central-northern Argentina. See text (Parodi 1919; Cavanna et al. 2010; Marinoni et al. 2015; Pensiero and Zabala, 2017; Halophytes database https://www.sussex.ac.uk/affiliates/halophytes/)

Habit: *PH* Perennial herb
PC Phenological cycle: *AWS* Autumn–winter–spring, *SSA* Spring–summer–autumn
TF Tolerance to flood: 0—None, 1—Low, 2—Medium, 3—High
TSa Tolerance to salinity: 0—None. 1—Low, 2—Medium, 3—High
TDr Tolerance to drought: 0—None, 1—Low, 2—Medium, 3—High
Status: *Nv* Native, *Nd* Naturalized, *E* Endemic
Seeds: *O* Orthodox
Ecoregion: *HC* Humid Chaco, *DC* Dry Chaco
[a]Germplasm conserved at the "Ing. Agr. José Mario Alonso" Germplasm Bank, Universidad Nacional del Litoral
[b]No data have been published, but the registration of species of the genus indicates that 100% of species for which there is available information are orthodox (https://data.kew.org)

Table 2 Native and naturalized leguminous forage species prioritized for saline environments of the Humid and Dry Chaco ecoregions in Argentina

Species	Habit	PC	TF	TSa	TDr	Status	Seed	Ecoregion	Main characteristics
Aeschynomene americana L.[a]	PS	SSA	2	0–1	0	Nv	**O**	HC	Forage species with cultivars (Skerman et al. 1991; Tropical Forage 2016). Seed not available in Argentina. In the Chaco region, it occurs on margins of wetlands and low fields (Schulz 1962; Hacker et al. 1996). It has not been recorded in environments with saline soils in this region. There are no published works on its salinity tolerance. It has been recorded in areas affected by secondary salinization in Cuba (Oquendo et al. 2006). Availability of germplasm in other countries, its flood tolerance and high nitrogen fixing capacity make it necessary to evaluate the existing variability in salinity tolerance. It is also necessary to start collection of germplasm of this region
Aeschynomene denticulata Rudd[a]	AH or BH	SSA	2	3	0	Nv	**O**	HC	Forage species recorded in humid areas of Central Chaco in Paraguay (Hacker et al. 1996) and mainly in the humid Chaco in Argentina. No works have been published about its incorporation to cultivation. Accessions collected from the flooded and saline submeridional lowlands are conserved at the Germplasm Bank of the UNL. It is necessary to increase the conserved germplasm, and to develop works related to the species biology and variability for salinity tolerance
Aeschynomene rudis Benth[a]	PS	SSA	2	2	0	Nv	**O**	HC	Forage species that occurs on margins of wetlands and lowlands in the Humid Chaco (Schulz 1962). According to Schulz (1962), it is more preferred by animals than *A. americana*. Collections performed by the Germplasm Bank of the UNL indicate that it occurs mainly in low but not saline fields. It is necessary to increase the conserved germplasm and conduct works related to the species biology and variability for salinity tolerance

(continued)

Table 2 (continued)

Species	Habit	PC	TF	TSa	TDr	Status	Seed	Ecoregion	Main characteristics
Desmanthus tatuyensis Hoehne[a]	PH or PS	SSA	1	2	1–2	Nv	**O**	HC, DC	The only species of the genus occurring in the submeridional lowlands. Although it has a low forage production volume, its salinity tolerance and the scarcity of legumes in the submeridional lowlands make it necessary to collect germplasm and perform agronomic tests
Galactia texana (Scheele) A. Gray var. *texana*	PH or PS	SSA	0	0–1?	2–3	Nv	**O**	HC, DC	Species highly appreciated as forage resource in the Dry Chaco. There are no records about its salinity tolerance. One of the few native legumes with potential for cultivation in those environments (Diaz 2007; Quiroga and Esnarriaga 2014). It is more frequently found in the Dry Chaco in Paraguay (Hacker et al. 1996) and Argentina (Burkart 1971), and less frequently in the Humid Chaco (Burkart 1971; Luchetti 2008). Given its forage potential and variability (Burkart 1971), it is necessary to start a germplasm collection program and evaluation of variability of salinity tolerance for dry areas. Some works have been published on its biology and agronomic assessment (Kraus et al. 2003; Quiroga et al. 2007; Quiroga et al. 2009; Galíndez et al. 2016)
Lotus tenuis Waldst. & Kit. ex Willd[a]	PH	AWS	2–3	2–3		Nd	**O**	HC	Species of high forage potential for saline and flooded environments (Cambareri et al. 2011; Nichols et al. 2013). It is widely cultivated in the Rio Salado Depression in Buenos Aires province, but attempts to introduce it in the Chaco region have failed. The work group of the Germplasm Bank of the UNL found a naturalized population in a saline-flooded lowland of Santa Fe province, and it is being subjected to several agronomic tests
Macroptilium lathyroides (L.) Urb.[a]	AH or BH	SSA	2 (3?)	2	1–2	Nv	**O**	HC, DC	See main text

(continued)

Table 2 (continued)

Species	Habit	PC	TF	TSa	TDr	Status	Seed	Ecoregion	Main characteristics
Medicago polimorpha L.	AH	SSA	1	1 (2?)	0–1?	Nd	**O**	HC	Species occurring in the Submeridional lowlands. In grasslands, it occurs in low frequency and its forage volume is non-significant, but there are records of cultivars used as forage resources in saline environments in Australia (Howie et al. 2007; Nichols et al. 2007). Whether it tolerates salinity or escapes by flowering at the beginning of spring, before salinity increases significantly, is a matter of debate (Craig 2006). Given the records and the scarce presence of herbaceous leguminous species in saline environments, collection of germplasm should be started as well as studies related to introduction to cultivation. It is considered a weed in wheat, but of low importance
Melilotus albus Desr.[a]	AH or BH	AWS	1	2–3	1–2	Nd	**O**	HC, DC	See main text
Melilotus officinalis (L.) Lam	AH or BH	AWS	1	2–3	1–2	Nd	**O**	HC, DC	Forage species cultivated in saline environments of the Pampas region (Ferrari and Maddaloni 2001; Maddaloni 1986). This species is little cultivated in the Chaco region. No cultivars have been developed in Argentina. Seed of identified category is sold without name of cultivar. It has lower coumarin content (Nair et al. 2010) than *M. albus*. It is necessary to start germplasm collection and assessment of salinity and flood tolerance. There is germplasm available at germplasm banks abroad
Melilotus indicus (L.) All	HA	AWS	1-(2?)	2–3	1–2	Nd	**O**	HC, DC	Forage species of saline environments (Al Sheriff 2009). It grows adventitiously in several provinces of Argentina and it is not cultivated at present. Given the sites where it occurs adventitiously and the published records, greater tolerance to floods than *M. albus* is likely (Nichols et al. 2008; Rogers et al. 2008). It has lower levels of coumarin (Nair et al. 2010). It is necessary to start germplasm collection and assessment of salinity and flood tolerance

(continued)

Table 2 (continued)

Species	Habit	PC	TF	TSa	TDr	Status	Seed	Ecoregion	Main characteristics
Neptunia pubescens Benth	PH or PS	SSA	1	2	0	Nv	**O**	HC	Forage species native to saline and flooded environments of Argentina and Paraguay (Bissio et al. 1994; Luchetti 2014; Vogt 2015). It is not cultivated at present. It is necessary to start germplasm collection and assessment of salinity and flood tolerance

Habit: *AH* Annual herb, *BH* Biennial herb, *PH* Perennial herb, *PS* Perennial subshrub, *S* Shrub
PC Phenological cycle: *AWS* Autumn–winter–spring, *SSA* Spring–summer–autumn
TF Tolerance to flood: 0—None, 1—Low, 2—Medium, 3—High
TSa Tolerance to salinity: 0—None, 1—Low, 2—Medium, 3—High
TDr Tolerance to drought: 0—None, 1—Low, 2—Medium, 3—High
Status: *Nv* Native, *Nd* Naturalized
Seeds: *O* Orthodox
Ecoregion: *HC* Humid Chaco, *DC* Dry Chaco
[a]Germplasm conserved at the "Ing. Agr. José Mario Alonso" Germplasm Bank, Universidad Nacional del Litoral

such as *Echinochloa helodes*, *Hymenachne amplexicaulis*, *Leersia hexandra*, *Luziola peruviana*, *Oplismenopsis najada*, *Paspalum denticulatum*, *P. vaginatum*, *Setaria geminata*, and *Eleocharis elegans*.

Regarding forage, flooding grasslands are the most important communities in terms of quality and productivity of species. In general, these species are perennial, of spring–summer–autumn cycle, have high protein and low fiber content, and exhibit rapid growth under high temperatures and surface water availability. Grazing in these communities is usually continuous (unmanaged), and livestock trampling is the most important disturbance factor.

Halophyte communities occur in very saline-flooded soils, where vegetation cover usually does not exceed 30% of the soil surface. Halophytes, such as "saltgrass" *Distichlis spicata* and *Sarcocornia perennis*, are frequent, and although their forage potential is low compared with other communities, they occur under extreme conditions, which make them a valuable resource. One of the most important characteristics of these communities is that the dominant species, *D. spicata*, is a perennial halophytic grass, aggressive, with rhizomes of indefinite growth. Because of these characteristics, if the community is properly managed, the soil becomes covered with a dense carpet of plants that prevent water erosion and the increase of salinity over time, allowing the incorporation of other species with better forage potential.

2.2 Dry Chaco

Following the works of Ragonese (1951), Ruiz Posse et al. (2007), and Karlin et al. (2011, 2012) focusing on the Salinas Grandes basin, the following saline environments can be recognized:

2.2.1 Lowland Environments

Saline Lowlands. The presence of low and sparse shrubs of *Heterostachys ritteriana*, *Allenrolfea patagonica*, *Allenrolfea vaginata* and *Suaeda divaricata* are observed. In these extreme environments of the Dry Chaco, forage offer is almost null; only some grasses may occasionally grow protected by the shrubland, or in the mounds present beneath some shrub clumps.

Flood Plains. These are environments dominated by low and sparse shrubs, such as *Atriplex argentina*, *A. cordobensis*, *A. lampa*, *Cyclolepis genistoides*, and *Maytenus vitis-idaea*. A few forage grasses of low productivity also occur, protected as described above. *Atriplex* species are a valuable forage resource and should be considered for incorporation into cropping systems.

Woodlands with Salinity Influence. Low and sparse woody species grow in these environments (i.e., trees, shrubs, and cacti) such as *Mimozyganthus carinatus*, *Maytenus vitis-idaea*, *Geoffroea decorticans*, and, to a lower extent, *Stetsonia coryne*, and *Atriplex argentina*.

2.2.2 Highland Environments

Dunes. The most frequent species in these environments are the following woody species: *Mimozyganthus carinatus*, *Larrea divaricata*, *Aspidosperma quebracho-blanco*, and *Stetsonia coryne*.

Woodlands with Low Salinity Influence. These environments are characterized by sparse forests dominated by *Aspidosperma quebracho-blanco* in the tree layer, and the shrubs *Capparis atamisquea*, *L. divaricata*, *L. cuneifolia*, *Cercidium praecox* and *Mimozyganthus carinatus*. Grasses with forage potential may be present, the most important being *Trichloris crinita* and *Pappophorum caespitosum*.

In the Santiago del Estero province, and with lower frequency in the west of the Chaco province, the presence of pure forests of the salt-tolerant *Prosopis ruscifolia* in degraded, relatively low sectors with saline soils, is common. The origin of many of these "forests" has been attributed to excessive extraction of other valuable species of trees for wood, such as *Schinopsis lorentzii* and *Prosopis kuntzei*.

After the rainy season, and given the scarce grass forage species availability, some trees (e.g., *Prosopis* sp., *Geoffroea decorticans*, *Cercidium praecox*) and several shrubs (e.g., *Atripex* sp., *Castela coccinea*, *Cyclolepis genistoides*, *Heterostachys ritteriana*, *Lycium boerhaviaefolium*, *Maytenus vitis-idaea*, *Tricomaria usillo*, *Ximenia americana*) become important feed sources. Livestock (cattle, goats, and sheep) browse the soft sprouts, leaves, fruits, and even bark. In critical periods of extreme droughts, succulent species, such as cacti (e.g., *Opuntia sulphurea*, *Stetsonia coryne*), previously devoid of the thorns, are used as forage and, mainly, water sources. The potential of *Opuntia* species in Argentina has been discussed by Grünwaldt et al. (2015).

In those high-altitude saline environments, where sparse forests or shrubby steppes occur, *Trichloris crinita* and *Pappophorum caespitosum* are the most important salt-tolerant grasses with forage potential. Both species are well adapted to grazing and are tolerant to drought. They are widely distributed from sea level to 2300 (*T. crinita*) and 3400 m above sea level (*P. caespitosum*), under an annual rainfall regime ranging between 150 and 1200 mm, corresponding to the Humid Chaco, Dry Chaco, Espinal, Monte, and Puna ecoregions. Although there is abundant information about the forage quality of these species (Cavagnaro and Dalmasso 1983; Dalmasso 1994; Trione and Cavagnaro 1998; Greco and Cavagnaro 2003, 2005; Cavagnaro et al. 2006; Quiroga et al. 2013; Gil Báez et al. 2015), and some *T. crinita* cultivars are registered at the INaSe (2019), there are still no seeds available in the market. Given the variability of some characters, especially in the populations of *T. crinita* (Zabala et al. 2011a; Marinoni 2017), obtaining materials that are well adapted to these environments is feasible.

3 Native and Naturalized Forage Plant Genetic Resources for Saline Environments of the Humid and Dry Chaco Ecoregions

The Convention on Biological Diversity (CBD 1992) defined a genetic resource as any genetic material (of plant, animal, microbial or other origin containing functional units of heredity) with actual or potential value for sustainable human development. Sustainability is based on the use of biodiversity without producing its significant reduction with the aim of its preservation for present or future human needs.

Native plant species, also termed autochthonous or indigenous, are natural components of the vegetation. Introduced species that behave as native (because they are well adapted to environmental conditions) are termed naturalized and were introduced mainly by human activities. In both groups of species, generically known as wild, there may be species of current or potential forage, food, medicinal, ornamental, forestry, and industry value, among others. There can also be species that are sources of genes for breeding other plant species.

Native and naturalized forage plant genetic resources (NNFPGRs) include mainly diverse legumes and grasses. Besides the representatives of these two families, the southernmost portion of the Great Chaco is home to species of other botanical families of forage potential for saline environments of arid and semiarid regions. Specific examples are species of the genera *Atriplex* (Chenopodiaceae), *Ehretia* (Boraginaceae) and *Cyclolepis* (Asteraceae), among others (Cavanna et al. 2010). For instance, the genus *Atriplex* comprises some 300 species (Kadereit et al. 2010), of which 45 are native to South America and 16 are endemic to Argentina (Brignone et al. 2016). Thus, there is a great potential for the development of diverse cultivars from its NNFPGRs, which have been underexploited to date.

An important aspect that should be considered is that the forage value of a species depends on the environmental context where it develops. In this sense, a species like "saltgrass" (*Distichlis spicata*), which is considered as having no forage value, becomes an important forage resource in some environments, since it is one of the few species that tolerate saline and flooded soils (Yensen et al. 1985). Likewise, an intermediate forage production is very often offset with high abundance, persistence, natural reseeding, low nutrient demand or higher tolerance to abiotic stresses. Besides the forage quality of a species, other benefits should be considered, such as their capacity to reduce soil erosion via soil cover and biological nitrogen fixation.

The records provided by botanists, agronomists, producers, and/or qualified informants are a source of relevant information for the prioritization of NNFPGRs. They not only know the species, they can indicate which are most frequently preferred or sought by animals in the rangelands, the possible environments for finding them or any other characteristics that may reveal aptitudes for their incorporation into cultivation. NNFPGRs of Great Chaco have been reported for almost 100 years by different authors, including Parodi (1919), Reichert et al. (1923), Reichert and Parodi (1926), Meyer (1940), Rosengurtt (1946), Hawkins and Donald (1963), Schulz (1962), Covas (1978, 1982), Bordón (1981), Fernández et al. (1993), Díaz (2007a) and Marinoni et al. (2019), among others. Yet, sadly, when a few breeding programs were initiated, it was often difficult to access bibliographic records for conducting the work, and most of the germplasm used have either been lost or is not properly conserved.

Tables 1, 2 and 3 include a list of NNFPGRs from the Humid and Dry Chaco ecoregions that, according to our criteria, should be prioritized for saline areas. Although several forage tree species are present in the saline environments of these ecoregions (mainly species of the genera *Acacia*, *Geoffroea*, *Prosopis*, family Fabaceae), the list includes only the most important herbaceous, shrub or succulent species.

The prioritized species belonging to the family of grasses (Poaceae) are presented in Table 1, and legumes (Fabaceae) are included in Table 2. Species of other botanical families are included in Table 3. Some of them, particularly *Atriplex* species, are commonly consumed by livestock in arid and semiarid regions or in extreme saline environments.

For each species, the following information is listed: accepted scientific name (according to www.floraargentina.edu.ar), botanical family, phenological cycle, tolerance to flooding, salinity and drought (in a subjective manner, see table footnote), status (indicating native, naturalized or endemic species), seed type (for its conservation in germplasm bank), ecoregion where it occurs (Humid Chaco and/or Dry Chaco), and additional relevant remarks. The tables also indicate (with a small letter besides the scientific name) whether seeds are conserved in the "Ing. Agr. José Mario Alonso" germplasm bank of the Universidad Nacional del Litoral.

Regarding grasses, 26 species are listed, of which 22 are native of the Great Chaco, 3 are endemic to Argentina, and one is naturalized. Six of these species are exclusive to the Humid Chaco ecoregion, 2 to the Dry Chaco, and the remaining 18 are common to both ecoregions (Table 1). The list of legumes includes 13 species (Table 2), of which 8 are native and 5 are naturalized. There are no species exclusive to the Dry Chaco, 6 species are exclusive to the Humid Chaco, and 7 grow in both ecoregions. Representatives of the other families include 16 species (Table 3), of which 8 are native, 7 are endemic to Argentina and one is naturalized. None of these species is exclusive to the Humid Chaco; by contrast, 9 of them only occur in the Dry Chaco, whereas the remaining 7 species are present in both ecoregions.

4 Collection and *Ex Situ* Conservation of NNFPGRs for Saline Environments of the Humid and Dry Chaco Ecoregions

One of the consequences of the improper management of rangelands is the loss of valuable forage species and the reduction of their genetic variability (Deregibus 1987; Fernández et al. 2009). Although there is no official information, it is evident that the loss of biodiversity and genetic variability in our rangelands has been of great magnitude. This genetic erosion process has also occurred in diverse parts of the world (Batello et al. 2008). In the Chaco region, this issue is even of greater concern,

Table 3 Non-traditional native and naturalized forage species of other botanical families from saline environments of the Humid and Dry Chaco ecoregions in Argentina

Species	Family	Habit	PC	TF	TSa	TDr	Status	Seed	Ecoregion	Remarks
Allenrolfea patagonica (Moq.) Kuntze	Chenopodiaceae	S	SSA	2	3	3	E	**No data**	CH, DC	Livestock consumes young buds throughout the year (Cavanna et al. 2010; Coirini et al. 2010)
Atriplex cordobensis Grand. & Stuck.[a]	Chenopodiaceae	PS	SSA	1	2	3	Nv	**O**[b]	DC	It is browsed by livestock (Passera et al. 2010)
Atriplex argentina Speg	Chenopodiaceae	PS	SSA	1	2	3	E	**O**[b]	DC	Good forage species throughout the year; good tolerance to browsing and trampling. Polymorphic species of wide distribution in Argentina (Ragonese 1951; Karlin et al. 2012; Cavanna et al. 2010; Coirini et al. 2010). It is one of the species that should be prioritized for plans for collection and incorporation into cultivation for arid and semiarid zones of the Dry Chaco
Atriplex lampa (Moq.) D. Dietr[a]	Chenopodiaceae	PS	SSA	1	2	3	E	**O**[b]	DC	Salinity tolerant species (Caraciolo et al. 2002), of very good forage value (Colomer and Passera 1990; Brignone et al. 2016), with potential for use in revegetation of degraded areas (Dalmasso 2010). Since it is an endemic species with forage potential for these saline environments, its germplasm should be conserved and it should be included in plans for incorporation to cultivation

(continued)

Table 3 (continued)

Species	Family	Habit	PC	TF	TSa	TDr	Status	Seed	Ecoregion	Remarks
Atriplex semmibaccata R. Br	Chenopodiaceae	PH	SSA	1	2	3	Nd	**O**	CH, DC	Species from Australia, cultivated in different countries through seeds (Rogers et al. 2005; Nichols et al. 2014; Halophytes database https://www.sussex.ac.uk/affiliates/halophytes/)
Atriplex undulata (Moq.) D. Dietr	Chenopodiaceae	S	SSA	1	2	3	E	**O**[b]	DC	Species of good forage value; livestock browses the young buds. Since it is an endemic species with forage potential for these saline environments, its germplasm should be conserved and it should be included in plans for incorporation to cultivation (Rogers et al. 2005; Nichols et al. 2014; Halophytes database https://www.sussex.ac.uk/affiliates/halophytes/)
Castela coccínea Griseb	Simaroubaceae	S	SSA	2	2	2	Nv	**O**[b]	HC, DC	Of forage importance for goats, although the fruit gives milk and meat a bitter flavor. It is considered an emerging forage species, since as other woody species, they become important in arid and semiarid environments when herbaceous forage supply is scarce (Cavanna et al. 2010; Coirini et al. 2010)

(continued)

Table 3 (continued)

Species	Family	Habit	PC	TF	TSa	TDr	Status	Seed	Ecoregion	Remarks
Cyclolepis genistoides Gilles ex D. Don	Asteraceae	S	SSA	2	3	3	E	**O**^{c}	HC, DC	Forage potential increases with increasing soil salinity and extreme conditions, since it is browsed by livestock when forage supply is null or scarce (Karlin et al. 2012; Cavanna et al. 2010; Coirini et al. 2010)
Ehretia cortesia Gottschling	Boraginaceae	S	SSA	1	3	3	Nv	**O**^{b}	DC	Good tolerance to browsing and trampling; consumed by livestock almost year round, especially at the fruiting stage (Cavanna et al. 2010; Coirini et al. 2010)
Grahamia bracteata Hook. & Arn	Anacampserotaceae	PS	SSA	1	2	2	E	**No data**	DC	Livestock consume its fleshy leaves, especially during autumn (Cavanna et al. 2010; Coirini et al. 2010)
Heterostachys ritteriana (Moq.) Ung.-Stemb	Chenopodiaceae	S	SSA	1	3	3	Nv	**O**	DC	Considered an emerging forage species; given its high salt content, the animals that eat it need to consume more water (Cavanna et al. 2010; Coirini et al. 2010)
Lycium boerhaviaefolium L. f	Solanaceae	S	SSA	1	2	2	Nv	**O**^{b}	HC, DC	Emerging forest species, browsed by livestock year-round (Coirini et al. 2010)
Maytenus vitis-idaea Griseb	Celastraceae	S	SSA	2	3	3	Nv	**O**^{b}	HC, DC	Emerging forage species, livestock browses young buds. Indicated as a good forage resource for goat (Cavanna et al. 2010; Coirini et al. 2010)

(continued)

Table 3 (continued)

Species	Family	Habit	PC	TF	TSa	TDr	Status	Seed	Ecoregion	Remarks
Opuntia sulphurea Gillies ex Salm-Dyck	Cactaceae	SS	SSA	0	2	3	Nv	**O**[b]	DC	Emerging forage species, especially as water source for animals; for this purpose, the thorns are burnt; when thirsty, cows consume this plant even with thorns (Cavanna et al. 2010; Coirini et al. 2010)
Stetsonia coryne (Salm-Dyck) Britton & Rose	Cactaceae	SS	SSA	0	2	3	Nv	**No data**	HC, DC	Emerging forage species, flowers and fruits are consumed by livestock and are an important source of water during drought periods and in winter; for this purpose, spines are removed (Cavanna et al. 2010)
Tricomaria usillo Hook. & Arn	Malpighiaceae	S	SSA	1	2	3	E	**No data**	DC	Emerging forage species, livestock browses young buds (Cavanna et al. 2010)

Habit: *PH* perennial herb, *PS* perennial subshrub, *S* shrub, *SS* succulent shrub
PC Phenological cycle: *AWS* Autumn–winter–spring, *SSA* spring–summer–autumn
TF Tolerance to flood: 0—None, 1—Low, 2—Medium, 3—High
TSa Tolerance to salinity: 0—None, 1—Low, 2—Medium, 3—High
TDr Tolerance to drought: 0—None, 1—Low, 2—Medium, 3—High
Status: *Nv* native, *Nd* naturalized, *E* endemic
Seeds: *O* orthodox
Ecoregion: *HC* Humid Chaco, *DC* Dry Chaco
[a]Germplasm conserved at the "Ing. Agr. José Mario Alonso" Germplasm Bank, Universidad Nacional del Litoral
[b]No data have been published, but the registration of species of the genus indicates that 100% of species for which there is available information are orthodox (https://data.kew.org)
[c]No data have been published, but the registration of species of the genus indicates that 97.3% of the species for which there is available information are orthodox and the remaining ones are of unknown type (https://data.kew.org)

since there is not sufficient conserved germplasm of most of its NNFPGRs. However, our NNFPGRs have been historically collected by and conserved at research centers from other countries. The same phenomenon has occurred in Uruguay (Rosengurtt et al. 1970). Some of these species have been even cultivated earlier in other countries, as is the case of *Bromus catharticus* (Donald 1939), *Desmanthus virgatus* (Jones and Clem 1997), and *Macroptilium* sp. (Nichols et al. 2007) in Australia.

A successful strategy for the collection and ex situ conservation of germplasm should address four fundamental aspects: the legal framework for access to plant genetic resources, investment in germplasm collecting missions, periodical regeneration of conserved accessions, and the introduction of plant genetic resources into breeding programs.

4.1 Legal Framework for Access to Plant Genetic Resources

Collection and ex situ conservation of plant genetic resources should be framed within a long-term strategy. The international legislation concerning issues related to the use of genetic resources in general and distribution of benefits from their use is addressed by the Convention on Biological Diversity (Argentina agreed through the Act 24375/1994), the Nagoya Protocol (approved by Act 27246/2015 in Argentina), and the FAO International Treaty on Plant Genetic Resources for Food and Agriculture (approved by Act 27182/2015 in Argentina).

The Constitution of Argentina, which was reformed in 1994, in its Section 124 states that "the provinces have the original dominion over the natural resources existing in their territory"; however, only 8 of the 23 Argentine provinces have legislation related to the use of genetic resources (Silvestri 2015). At the same time, there is no national legislation indicating the minimum protection standards about the access to plant genetic resources. In 2019, the Argentine Secretariat of Environment and Sustainable Development published a compilation of laws and regulations on permits for research and access to genetic resources per jurisdiction (https://www.argentina.gob.ar/ambiente/biodiversidad/genetica/nagoya).

4.2 Investment in Germplasm Collection

From the methodological point of view, a germplasm collection must represent the distribution range of the species so that it provides an effective source of genetic variability, which is a basic requirement for the success of any breeding effort. Over the past 40 years, various authors have published indications for designing optimized collection strategies (Guarino et al. 2011). The development of GIS-based tools is particularly interesting (Marinoni et al. 2015). All these actions require economic resources that should be provided by governments (Franco et al. 2014); ensuring that germplasm conserved by public institutions may be made available to anyone requesting it, through material transfer agreements. Future legislation should consider these aspects.

In Argentina, most of the germplasm banks belong to the National Institute of Agricultural Technology (INTA) (nine active banks and one base bank). The last report presented by Argentina about the state of plant genetic resources indicates that at INTA funding of conservation activities depends on national three-year projects. The system does not guarantee conservation of collections in the mid and long term (FAO 2008).

There are also a few collections in other banks mostly located in universities (FAO 2008). These collections contain unique germplasm of many of our plant genetic resources, such as the *Prosopis* Germplasm Bank of Universidad Nacional de Córdoba. However, many of these collections do not have back-up reserves and the lack of budget hinders the deposit of duplicates at base banks. Furthermore, collections performed by universities are financially supported by research projects granted by national and/or international organizations, which overall have a relatively short duration (up to four years).

Specifically, for forage species, the report of FAO (2008) mentions the germplasm banks that conserve seeds of forage species, but fails to list the accession numbers corresponding to wild and exotic species:

- Forage species of temperate climate, INTA (EEA Pergamino, Buenos Aires), 1121 accessions.
- Forage species of temperate climate, INTA (EEA Balcarce, Buenos Aires), 805 accessions.
- Forage species of semiarid temperate climate, INTA (EEA Anguil, La Pampa), 949 accessions.
- Patagonian Shrubby forage species, INTA (EEA Chubut, Chubut), 156 accessions.
- Forage species, Facultad de Agronomía, Universidad Nacional de La Pampa (collection in the custody of INTA Base Bank), 647 accessions.
- Forage species of temperate and subtropical climate, Facultad de Agronomía, Universidad Nacional del Litoral, 400 accessions.
- Native forage species for semiarid temperate environments and trees of *Prosopis* spp. of Monte Ecoregion, Instituto Argentino de Investigaciones de Zonas Aridas, Consejo Nacional de Investigaciones Científicas y Técnicas (IADIZA, CONICET), Mendoza, no data on accessions.

There is scarce information about available accessions, and no exchange among banks, which would allow to have backups of collections, avoid duplicates, lower collection mission costs, widen the genetic base of the studied species and increase the number of characterized environments, among other issues.

Even in countries such as Australia, with a long history of germplasm conservation programs, there are problems related to investments in conservation of forage plant genetic resources (Cox et al. 2009). Until 2013, there were 5 germplasm banks in Australia, 2 for grain crops (Australian Winter Cereals Collection, Tamworth, New South Wales and Australian Temperate Field Crops Collection, Horsham,

Victoria), and 3 for forage species (Australian Tropical Crops and Forages Collection, Biloela, Queensland, Australian *Medicago* Genetic Resource Centre, Adelaide, South Australia, and Australian *Trifolium* Genetic Resource Centre, Perth, Western Australia) (Stoutjesdijk 2013). Currently, with the aim of increasing their efficiency, they have been restructured, and there are now only two banks, one for grain crops (Australian Grains Genebank, Horsham) and another for forage species (Australian Pastures Genebank, Adelaide). To have an idea of the necessary investment for germplasm conservation, the cost of restructuring both centers was approximately US$ 450,000 and the annual budget for both centers is estimated at US$ 2.6 million (Guarino 2015).

4.3 Regeneration of Germplasm

Periodical regeneration of seed collections in a germplasm bank is necessary due to the loss of seed viability after some years. Regeneration and collection are the costliest tasks in germplasm conservation. However, there is an alternative conservation scheme that improves viability of the conserved seeds, reducing the need for periodical regeneration. It is based on the replacement of traditional packaging (trilaminate material) with diverse glass containers, along with the use of a desiccant with moisture indicator (self-indicating silica gel) (Pérez-García et al. 2007). Thus, the moisture condition of seeds can be easily monitored, which is a key aspect for their correct conservation. Conservation in glass containers is applied in one of the world's most important wild species seed banks in Kew, England (Manger et al. 2003). In Argentina, this strategy is used in the horticultural species germplasm bank of INTA (EEA La Consulta, Mendoza) and in the native species germplasm bank "Ing. Agr. José Mario Alonso" of the Universidad Nacional del Litoral (Esperanza, Santa Fe).

An important aspect in the conservation and regeneration of wild species seeds is knowledge of the optimum germination conditions, which is very often lacking. For this reason, in most cases, basic studies to determine those conditions are necessary (Zabala et al. 2009a, b, 2011a; Farley et al. 2013). Knowing those aspects is crucial, especially in species of arid or saline environments, whose seeds usually have dormancy and require the development of strategies to overcome that condition (genetic variation or technological solutions, such as scarifying). At the same time, if information is not available, seed drying and cold storage trials should be conducted to determine their conservation capacity, i.e., to determine if seeds are orthodox, recalcitrant or of intermediate behavior (Hong and Ellis 1996). There is information available on wild species at the Royal Botanical Garden, Kew (https://data.kew.org/sid/) or bibliography of the International Seed Testing Association (ISTA 2018) that guides the search of adequate protocols for conservation and germination.

4.4 Introduction of Plant Genetic Resources in Breeding Programs

Germplasm banks should have an organized work system that considers prioritization of species to be collected and studied, and their effective use; otherwise, they may simply become museums and neglect their fundamental role in biodiversity conservation. In the specific case of NNFPGRs, this use is related to restoration plans and/or development of forage cultivars. Of all conserved germplasm, very few species will certainly reach the final stage of being incorporated into cropping systems; however, a successful scheme implies starting with a broad base of species and germplasm to reach the final stage of the process where some of the species are incorporated into cultivation.

5 Characterization, Selection, and Agronomic Evaluation

Before the materials are commercialized, it is necessary to evaluate them agronomically in different environments/conditions so that adequate cultivation recommendations can be made.

If collection has been successful, a high amount of genetic variability will be available. In the case of the NNPFGRs, since it is impossible to start a breeding program using all the available germplasm, it is necessary to conduct pre-breeding activities, i.e., characterization and preselection of germplasm before applying a selection method.

5.1 Pre-breeding Activities

Pre-breeding activities involve identification of traits, genotypes and/or genes of potential importance for breeding programs. Characterization of the available genetic variability in germplasm banks is one of the main limiting factors for the effective use of plant genetic resources in breeding activities (Marshall 1989). For example, the characterization of the available germplasm of *Desmanthus virgatus* (Zabala et al. 2008) showed that the materials with better forage traits were from a specific region of Argentina; therefore, only that germplasm was included in a breeding program, which was resulted in the registered cultivar Kakan of that species (INaSe 2019).

Characterization involves the classification of the available germplasm through the traits of interest (Zabala et al. 2008; Exner et al. 2010), known in germplasm banks as "descriptors." Reaching consensus of species descriptors among germplasm banks is a priority activity (Gotor et al. 2008). There are descriptors for forage legumes (IPGRI 1984) and forage grasses (IPGRI 1985), which can be used as models in the characterization of NNFPGRs.

5.2 *Relevant Traits to Be Considered for the Incorporation of NNFPGRs into Cropping Systems*

The descriptors to be used depend both on the objective of the characterization and the species. For example, phenology is one of the most important agronomic characteristics of any forage species that should be analyzed (e.g., vegetative period length). For forage legumes, promiscuity (ability to effectively nodulate with native rhizobial strains) can also be included. For summer perennial species, tolerance to cold weather (ability to regrow rapidly after winter) should be considered in the subtropics.

When NNFPGRs are prioritized, it is very important to consider the presence of traits associated with the "domestication syndrome" (Gepts 2002). These traits are different from those associated with the traditional genetic breeding program (forage production and quality, abiotic and biotic stress tolerance, etc.). One of these traits is the suitable production of harvestable seed (Cole and Johnston 2006; Muir et al. 2014). Thus, when prioritizing wild forage species for cultivation schemes, traits such as seed production, fruit dehiscence and fruiting uniformity should be considered (Mitchell et al. 2015). These authors provide an example of a trait associated with the production of harvestable seed, the selection of wild species with exposed reproductive organs in the upper tip of the plant (e.g., which in grasses would be an adequate panicle exertion), a trait that would be desirable for mechanical harvesting. The higher the number of traits associated with the domestication syndrome of a species, the easier its incorporation into cultivation.

On the other hand, within a species, selection could be based on traits that facilitate seed management. In forage species, intraspecific variability has been found for seed production (Davies et al. 2005; Pensiero et al. 2005; Exner et al. 2010), although this strategy has not been widely used in breeding programs. In some cases, selection for higher seed production implies that genotypes redefine their photosynthesis sinks to reproductive organs at the expense of forage production (West and Pitman 2001). Another important aspect for successful establishment is seed size. This trait is moderately to highly heritable and is particularly important in grasses (Casler and Van Santen 2010). In the halophytic species, *Trichloris crinita* an important variation in seed weight was found, with genotypes with greater seed weight having a greater capacity to germinate under suboptimal conditions (Zabala et al. 2011a, b; Marinoni et al. 2018).

Regarding seed production, for *Macroptilium lathyroides,* a native forage legume with indeterminate growth habit and explosive dehiscence (Beyra Matos and Reyes Artiles 2005), genotypes with increased seed yield potential without forage production penalties have been selected; however, problems associated with indeterminate growth and fruit dehiscence still persist (Zabala et al. 2015). For this species, an adequate agronomic management might yield 300 kg/ha of potentially harvestable seeds, with an additional loss of at least 15–50% due to the lack of uniformity in fruiting and fruit dehiscence (Zabala et al. 2015).

It should be noted that occasionally the pre-breeding and breeding processes should be coupled with the development of additional technologies that allow increased harvesting, threshing, and seeding efficiency. The development of harvest technology and the overcoming of dormancy have been widely used strategies for the cultivation of wild forage species (Loch 1980). Although harvesters adapted to work with these species have been developed (NSW 2016), they are not easily available in the Great Chaco.

Finally, in some species, propagation by seeds is limited or based on vegetative mechanisms. *Atriplex* is an example establishment is achieved by transplanting 700–2000 plants per hectare (Guevara et al. 2003; Mitchell et al. 2015), which requires a high initial investment in seedlings, planting, and fencing (Guevara et al. 2003). However, to facilitate the establishment of species of this genus, seed sowing alternatives continue to be evaluated (Nichols et al. 2014). Other species have vegetative propagation by stolons, such as *Hemarthria altissima* and *Hymenachne amplexicaulis*. The former propagates only vegetatively, whereas the latter can also spread by seeds. The availability of equipment adapted to sowing vegetative organs has been a limiting factor to the adoption of this type of species; at present, manual planting is the only mode of cultivation. For the specific case of *Hemarthria altissima*, Bissio (2014a) proposed the adaptation of equipment and a guide to the establishment of this species in low fields of northern Santa Fe province of Argentina.

An important trait of leguminous species is the capacity to establish symbiotic relationships with nodulating rhizobia. Legumes may or may not be specific in terms of the range of rhizobium species that can nodulate them. In this sense, promiscuous species can effectively nodulate with several rhizobia species present in the soil (Peoples and Herridge 1990). Prioritizing promiscuous species can be useful to improve rangeland fertility, with no need to inoculate seeds (Shelton et al. 2005). In the case of prioritizing species of high forage potential, but with records of specificity for nodulation, collection, and characterization of the plant germplasm should be accompanied with collection and evaluation of its nodulating rhizobia (Fornasero et al. 2014) and, if possible, with tests of inoculation with strains available in the market. An example of this is *Melilotussiculus*, an annual forage species from the European Mediterranean introduced in Australia, with high tolerance to salinity and flooding. Despite its potential, commercial use was restricted because it cannot nodulate efficiently with native Australian strains and because of the lack of strains to inoculate seeds. For this reason, the search of rhizobia compatible with this species has become a priority for its incorporation into cropping systems (Nichols et al. 2013). Recently, the *M. siculus* cultivar "Neptuno" was released, together with a rhizobium inoculum.[1]

[1] https://www.seednet.com.au/sites/seednet/files/2018-01/documents/Neptune-messina-factsheet.pdf.

6 Case Studies

Below we present examples of work related to the development of various aspects of a program for incorporating NNFPGRs into cropping systems in general, and one specific example for saline environments.

6.1 Case Studies in Argentina

In Argentina, the earliest prospective surveys of forage genetic resources were conducted in 1947/1948 by the Australian researchers Willian Hartley and John Sthephens (Whyte 1958). In 1958, Dr. Thomas Moir, a FAO field officer, recommended the creation of a collection of forage species at the INTA Experimental Station of Roque Sáenz Peña, Chaco (Whyte 1958). Then, germplasm collections were started until 1988, when the Programa Nacional de Recursos Genéticos (Genetic Resources National Program) was created within INTA. Thus, a germplasm bank was generated in Roque Sáenz Peña, specialized in forage species for the Chaco region. A report published by INTA in 1995 indicates that unique native and introduced forage species accessions were stored (Clausen et al. 1995). A total of 1100 accessions were registered as a result of the work led by Olegario Royo, one of the most important researchers in the introduction of forage species for the Chaco region. The most important species that were conserved in that bank included native legumes of the genera *Aeschynomene, Desmanthus, Macroptilium, Rhynchosia*, and *Tephrosia*. In a report published later (FAO 2008), this collection was not included. Unfortunately, no studies were conducted to characterize this germplasm and further use it in plans for the domestication of those species and their incorporation into cultivation. Part of that collection, already lost in Argentina, is preserved in the Australian Pastures Genebank, Adelaide (former Australian Tropical Crops and Forages Collection).

A pioneer and successful work was developed by the Universidad Nacional del Nordeste, which incorporated into cultivation species of the genus *Paspalum*, which grows in non-saline environments. Basic works on the biology of promising forage species (e.g., Quarín et al. 1997; Espinoza et al. 2001; Quarin et al. 2001; Marcón et al. 2015) have led to the development of new cultivars of *Paspalum guenoarum* (Tropical Forage 2016), *Paspalum atratum* (Tropical Forage 2016), and *Paspalum notatum* (Dr. Mario Urbani, personal communication). This genus includes species that should be prioritized for their introduction in saline environments, such as *Paspalum vaginatum, Paspalum distichum, Paspalum denticulatum*, and *Paspalum buckleyanum*. There is germplasm of these species only in a work collection of the Universidad Nacional del Nordeste (Dr. Mario Urbani, personal communication); therefore, collection and conservation of this germplasm are essential.

During the 1970s and 1980s, several investigations were conducted as part of an agreement between the foundation José María Aragón, INTA, and the government of Santa Fe province; aimed at agricultural development of the Santa Fe Submeridional

Lowlands (the "Bajos Submeridionales," a wetland of almost 5 million hectares that covers part of the provinces of Chaco, Santiago del Estero and Santa Fe, characterized saline soils, minimal slopes and slow drainage). Within this project, important work was conducted during two years, evaluating summer forage species (Fossati et al. 1979). A total of 146 cultivars were evaluated (96 grasses, 44 legumes, 1 Asteraceae species and 2 Chenopodiaceae species). Unfortunately, after those trials, the project was interrupted and no studies for incorporation of the most promising species into cultivation were conducted. On the other hand, the only native species that were included in that study was *Macroptilium lathyroides*, which stood out for its good establishment. Despite the broad distribution of this species in Argentina (Perez et al. 1999), the material used in those trials was from Brazil. This is one of several examples of the lack of a long-term strategy for the use of NNFPGRs in Argentina.

Collection trips conducted by IADIZA personnel in the 1980s and 1990s were pioneer for arid and saline environments, with priority being given to species of the genera *Trichloris, Pappophorum, Digitaria, Setaria* , and *Atriplex* (Lemes 1992). These collections were carried out with a multidisciplinary team, including geneticists, botanists, physiologists, and agronomists. As a result, a total of 1566 accessions were preserved in the germplasm bank of IADIZA. Between 1981 and 1992, 33 exploratory surveys were made, with up to 435 bulk seed samples and 1380 family samples (individual plant harvest) having been collected (Lemes 1992). This collection provided the basis for generating knowledge on the biology and agronomic behavior of several of these species (Cavagnaro and Dalmaso 1983; Cavagnaro et al. 1983, 1989, 2006; Cabeza et al. 1999; Greco and Cavagnaro 2003; Greco and Cavagnaro 2005). Only in *Trichloris crinita* has there been some progress in introducing it to cultivation. Five cultivars of this species were registered at INaSe (Instituto Nacional de Semillas); one cultivar was developed by INTA researchers using germplasm of that institution in 2012; the remaining four were developed in 2014 and 2017 by the Universidad Nacional de Cuyo (INaSe 2019), based on germplasm from IADIZA.

Other works are being conducted at the Facultad de Ciencias Agrarias of the Universidad Nacional del Litoral within the framework of Programa de Documentación, Conservación y Valoración de la Flora Nativa (PRODOCOVA) [Program for the Conservation and Assessment of Native Flora] (https://www.fca.unl.edu.ar/prodocova/index). The earliest steps involved the documentation of the native flora at the "Arturo E. Ragonese" Herbarium (SF), and since 2003, germplasm collection at the "Ing. Agr. José Mario Alonso" germplasm bank. These activities have provided valuable baseline information for many NNFPGRs which favored the development of breeding programs and registration of cultivars at the INaSe. Thus, selected material of *Elymus scabrifolius, Trichloris crinita,* and *Lotustenuis* for saline environments are available, which will be evaluated in the field in the next years. Two cultivars of *Melilotus albus* for saline environments were registered (INaSe 2019); at present, seed multiplication is being performed through a Technology Transfer Agreement with a seed company for further commercialization. Although this species is the only widely cultivated forage legume in saline environments of the Chaco region, breeding programs are scarce (Zabala et al. 2012). Most of the commercialized seed is identified as *M. albus,* but with no variety identification. This species is naturalized in Argentina, and the germplasm that has been collected in this country has been appreciated for its forage potential in other countries (Evans and Kearney 2003; Trigg 2004). INTA has an important germplasm collection of this species, which is in the process of regeneration (Dr. María A. Tomas, pers. comm.). Given the wide distribution of this species in Argentina, the different environments where it occurs, and its forage potential, it is necessary to increase the germplasm collection and include it in breeding programs. Likewise, it is necessary to conduct studies on two other species of the genus that are present in saline environments (Table 2).

6.2 International Case Studies

A work conducted in Australia (Rogers et al. 2005) provides an example of survey and prioritization of plant genetic resources with tolerance to drought and salinity. Seventeen Australian researchers participated in the listing and priority assignment of a group of wild species of saline environments of different regions worldwide, with potential interest as plant genetic resources. Some of these species are native or endemic to Argentina or are naturalized in this country. These authors indicated as of high priority 33 grass species belonging to 19 genera, 6 of which are native to or naturalized in Argentina. On the other hand, many of the mentioned genera include species that are native to Argentina and have forage potential, although they are not listed in that work. Regarding legumes, Rogers et al. (2005) listed 52 species belonging to 11 genera as of priority, with *M. albus* being one of the most important. Moreover, 24 species belonging to 8 genera of other botanical families (Acanthaceae, Asteraceae, Chenopodiaceae, Plantaginaceae) were listed. Three of the listed genera include species native to Argentina.

The Saltland pastures for South Australia guide (Liddicoat and McFarlane 2007) is an example of the implementation of public policies for the diffusion of cultivars of species adapted to saline environments in Australia. The guide includes information about the management of pastures for saline environments of the west of Australia, including species of *Atriplex*, which are native to Argentina. Bennett et al. (2009) state that in Australia, plans for prioritization and collection of plant genetic resources of promising forage species in Australia started in the 1950s with the purpose of revegetating saline environments of the west of Australia, including different Argentine species of the genus *Atriplex*. In Argentina, despite their forage potential for saline environments, their wide distribution (Passera et al. 2010) and their nutritional value (Castellanos et al. 2011), these species have not been included in germplasm collection or genetic breeding plans. Curiously, in Argentina an exotic *Atriplex* species (*A. nummularia*) has been used as a model in species establishment trials (Guevara et al. 2003; Falasca et al. 2014). As with other poorly studied species, the lack of availability of information or germplasm hinders the proposal of production schemes using our NNFPGRs (Nichols et al. 2014).

The Australian Pastures Genebank, located in Adelaide (former Australian Tropical Crops and Forages Collection), has nearly 40,000 accessions; of these, about 11,000 correspond to collections of tropical and subtropical forage grasses and legumes from Africa, Asia, and the Americas. That bank holds a great collection of forage legumes native to South America (e.g., 559 accessions from Argentina corresponding to 109 species). Studies on the characterization and agronomic evaluation of these collections conducted between 1990 and 2005 were the basis for the development of plant breeding programs, resulting in the release of 58 cultivars of 28 forage legume species adapted to different agroecological regions of the south and west of Australia, as well as in other regions of the world. According to Nichols et al. (2007), it is still necessary to increase the offer of cultivars and their adoption. Among the prioritized species, *Melilotus albus*, and *Lotus tenuis* stand out for saline environments (Nichols et al. 2013).

The "South Texas Natives" program[2] is another example involving species of forage interest. Although it does not include species for saline environments, this program provides an interesting example of a public–private initiative from the USA for the collection, characterization, multiplication, and supply of seeds of native species for plans for the restoration and recovery of degraded environments, both public and private ones. According to Muir et al. (2014), the success of this initiative lies in the pressure exerted by the interested livestock producers to conserve wildlife, in the obligation (imposed by the government to private companies) to revegetate with native species on road edge and disturbed natural areas, and in the interest of local seed companies.

7 Conclusions

In South America, the Great American Chaco is the second largest forest region after Amazonia. It comprises a wide diversity of environments, including over 5 million hectares of saline rangelands. This region has suffered a great degradation due to unsustainable agricultural practices. A paradigm that involves agro-silvo-pastoral systems is the most viable alternative to ecosystem restoration and sustainable use of rangeland in the Great American Chaco. In this sense, the reintroduction of wild forage species into saline rangelands is envisioned as a powerful tool. It requires designing strategies for collecting, evaluating and introducing wild forage species into cropping systems, and engaging long-term support and broad stakeholder involvement.

[2]https://ckwri.tamuk.edu/research-programs/south-texas-natives.

References

Al Sherif EA (2009) *Melilotus indicus* (L.) All., a salt-tolerant wild leguminous herb with high potential for use as a forage crop in salt-affected soils. Flora 204:737–746

Batello C, Mannetje L, Martinez A, Suttie J (2008) Plant genetic resources of forage crops, pasture and rangelands. Thematic Background Study, FAO report 5–7, 63 p

Bennett SJ, Barrett-Lennard EG, Colmer TD (2009) Salinity and waterlogging as constraints to saltland pasture production: a review. Agr Ecosyst Environ 129:349–360

Beyra Matos A, Reyes Artiles YG (2005) El género *Macroptilium* (Benth.) Urb. (Leguminosae) en Cuba. Anales Jard Bot Madrid 62:181–190

Bissio JC (2014) Guía para la implantación de pasto clavel en el norte de Santa Fe. Ediciones INTA, EEA Reconquista, pp 1–38

Bissio JC (2014) Los ambientes de pajonal del norte de Santa Fe; técnicas utilizadas para el manejo. EEA Reconquista, Voces y Ecos 32:31–38

Bissio JC, Luisoni LO, Batista W (1994) Fenología de las principales forrajeras nativas de los bajos submeridionales santafesinos. INTA, EEA Reconquista, 24 p

Bordón A (1981) Recursos naturales del chaco árido y semiárido: aprovechamiento forrajero de especies leñosas y herbáceas. Departamento de Producción Animal Doc. 1218. INTA EEA Sáenz Peña (mimeographic material)

Brignone NF, Denham SS, Pozner R (2016) Synopsis of the genus *Atriplex* (Amaranthaceae, Chenopodioideae) from South America. Austral Syst Bot 29:324–357

Brummer EC, Barber WT, Collier SM, Cox TS, Johnson R, Murray SC, Olsen R, Pratt R, Thro AM (2011) Plant breeding for harmony between agriculture and the environment. Front Ecol Environ 9:561–568

Burkart A (1971) El género *Galactia* (Legum.-Phaseoleae) en Sudamérica con especial referencia a la Argentina y países vecinos. Darwiniana 16:663–796

Burkart R, Bárbaro NO, Sánchez RO, Gómez DA (1999) Ecorregiones de la Argentina, Buenos Aires, Administración de Parques Nacionales

Burson BL (1997) Apomixis and sexuality in some *Paspalum* species. Crop Sci 34:1347–1351

Cabeza CE, Brevedan RE, Fioretti MN, Laborde HE (1999) The influence of dehulling and seed storage on the germination of the range grasses of the Caldenal (Argentina). Seed Sci Technol 27:359–363

Cabrera AL (1994) Regiones fitogeográficas de Argentina. Enciclopedia argentina de agricultura y jardinería. Tomo II. Editorial ACME S.A.C.I., Buenos Aires, Argentina

Cambareri G, Castaño J, Fernandez R, Maceira N, Vignolio O (2012) *Lotus tenuis*: un recurso forrajero estratégico para la ganadería de la Pampa Deprimida. Ediciones INTA, 48 p

Caraciolo M, Allegretti LI, Passera CB, Robles C (2002) Efecto de diferentes niveles de salinidad sobre el crecimiento de *Atriplex lampa*. Revista de la Facultad de Ciencias Agrarias (Argentina) 34:101–106

Casler MD, Van Santen E (2010) Breeding objectives in forages. In: Boller B, Posselt UK, Veronesi F (eds) Handbook of plant breeding. Fodder crops and amenity grasses. New York, Springer, pp 115–136

Castellanos EC, Colomer JS, López JB, Passera C (2011) Valor nutritivo de cuatro arbustos forrajeros del género *Atriplex* (*A. nummularia*, *A. cynerea*, *A. undulata* y *A. lampa*). Pastos 16:177–189

Cavagnaro JB, Dalmasso AD (1983) Respuesta a la intensidad y frecuencia de corte en gramíneas nativas de Mendoza. I. *Pappophorum caespitosum* y *Trichloris crinita*. Deserta 7:203–208

Cavagnaro JB, Dalmasso AD, Candia RJ (1983) Distribución vertical de materia seca en gramíneas nativas del este de Mendoza. Deserta 7:271–289

Cavagnaro JB, Lemes J, Ventura JL, Passera CB (1989) Variabilidad ecotípica y producción forrajera de *Trichloris crinita*. Revista Asociación Argentina de Producción Animal 9 (Supl. 1):70

Cavagnaro PF, Cavagnaro JB, Lemes JL, Masuelli RW, Passera CB (2006) Genetic diversity among varieties of the native forage grass *Trichloris crinita* based on AFLP markers, morphological characters, and quantitative agronomic traits. Genome 49:906–918

Cavanna J, Castro G, Karlin U, Karlin M (2010) Ciclo ganadero y especies forrajeras en Salinas Grandes, Catamarca, Argentina. Zonas Áridas 14:170–180

CDB (Convenio sobre la Diversidad Biológica) (1992) Available at https://www.cbd.int/doc/legal/cbd-es.pdf. Accessed 21 Feb 2017

Clausen AM, Ferrer ME, Gómez S, Tillería J (1995) Informe nacional para la conferencia tecnica Internacional de la FAO sobre los recursos fitogenéticos (Leipzig, 1996). INTA, Buenos Aires, Argentina, p 80

Coirini RO, Karlin M, Reati GJ (2010) Manejo Sustentable del Ecosistema Salinas Grandes, Chaco Árido. Encuentro Grupo Editor, Córdoba

Cole IA, Johnston WH (2006) Seed production of Australian native grass cultivars: an overview of current information and future research needs. Anim Prod Sci 46:361–373

Colomer JS, Passera CB (1990) The nutritional value of *Atriplex* spp. as fodder for arid regions. J Arid Environ 19:289–295

Covas G (1978) Forrajeras indígenas: Especies que requieren un plan de conservación de germoplasma. Ciencia e Investigación 34:209–213

Covas G (1982) Región pampeana. In: Grassi M (ed) Conservación de la vegetación natural en la República Argentina. Fundación Miguel Lillo, pp 53–61

Cox K, Scrivener N, Kilpatrick F (2009) Reversing the declining quality of the tropical forages collection. Trop Grassl 43:259–262

Craig A (2006) Salt tolerance of balansa clover-a myth busted? CRC Salinity Factsheet 3. www.saltlandgenie.org.au/LiteratureRetrieve.aspx?ID=23842. Accessed 2 Mar 2017

D´Angelo CD, Prado DE, Stofella SL, Lewis JP (1987) The Subchaquenian vegetation of the province of Santa Fe (Argentina). Phytocoenologia 15: 329–352

Dalmasso AD (1994) Fenología de cinco gramíneas nativas de interés forrajero. Multequina 3:9–24

Dalmasso AD (2010) Revegetación de áreas degradadas con especies nativas. B Soc Argent Bot 45:149–171

Davies CL, Waugh DL, Lefroy EC (2005) Variation in seed yield and its components in the Australian native grass *Microlaena stipoides* as a guide to its potential as a perennial grain crop. Crop Pasture Sci 56:309–316

Deregibus VA (1987) Importancia de los pastizales naturales en la República Argentina: situación presente y futura. In: V Simposio Argentino de Producción Animal. Sección: Producción y Utilización de Pasturas; Paraná. Entre Ríos, AR

Díaz Maynard A (2005) América Latina y su riqueza fitogenética: conservación, domesticación y sistemas productivos: un desafío técnico-político. Agrociencia 9:19–28

Díaz RO (2007a) Utilización de pastizales naturales, 1st edn. Editorial Brujas, Córdoba, Argentina

Díaz RO (2007b) Características de los recursos forrajeros. In: Díaz, RO Utilización de pastizales naturales, 1st edn. Editorial Brujas, Córdoba, pp 31–57

Donald CM (1939) Strain variation in *Bromus unioloides*. J Counc Sc Indust Res 12:212–226

Espinoza F, Urbani MH, Martínez EJ, Quarin C (2001) The breeding system of three *Paspalum* species with forage potential. Trop Grassl 35:211–217

Evans PM, Kearney GA (2003) *Melilotus albus* Medic is productive and regenerates well on saline soils neutral to alkaline reaction in the high rainfall zone of south-western Victoria. Aust J Exp Agr 43:349–355

Exner E, Zabala JM, Pensiero JF (2010) Variación en la fenología de la floración y en el éxito reproductivo en *Setaria lachnea*. Agrociencia 44:779–789

Falasca SL, Pizarro MJ, Mezher RN (2014) The agro-ecological suitability of *Atriplex nummularia* and *A. halimus* for biomass production in Argentine saline drylands. Int J Biometeorol 58:1433–1441

FAO (2008) Informe Nacional sobre el estado de los recursos fitogenéticos para la agricultura y la alimentación. Argentina. Mecanismo nacional de intercambio de información sobre la aplicación del plan de acción mundial para la conservación y la utilización sostenible de los recursos fitogenéticos para la agricultura y la alimentación, 45 p

Farley GJ, Bellairs SM, Adkins SW (2013) Germination of selected Australian native grass species, with potential for minesite rehabilitation. Aust J Bot 61:283–290

Feldman SR, Bisaro V, Biani NB, Prado DE (2008) Soil salinity determines the relative abundance of C3/C4 species in Argentinean grasslands . Global Ecol Biogeogr 17:708–714

Fernández G, Benítez CA, Royo Pallarés YO (1993) Principales forrajeras nativas del medio este de la provincia de Corrientes. INTA, EEA Mercedes, Corrientes, Argentina. Serie Técnica 23, 2nd edn, 80 p

Fernández OA, Gil ME, Distel RA (2009) The challenge of rangeland degradation in a temperate semiarid region of Argentina: the Caldenal. Land Degrad Dev 20:431–440

Fernández Ridano CE (2018) Sistema de pastoreo racional. Conocimiento y práctica. Orientación Gráfica Editora, Buenos Aires, 182 p

Ferrari L, Maddaloni J (2001) Trebol de olor blanco y Trebol de olor amarillo. In: Forrajeras y Pasturas del Ecosistema Templado Húmedo Argentino. UNZ (eds), pp 303–315

Fornasero LV, Del Papa MF, López JL, Albicoro FJ, Zabala JM, Toniutti MA, Pensiero JF, Lagares A (2014) Phenotypic, molecular and symbiotic characterization of the rhizobial symbionts of *Desmanthus paspalaceus* (Lindm.) Burkart that grow in the province of Santa Fe, Argentina. PloS one 9:e104636

Fossati J, Bruno O, Panigatti J, Gambaudo S (1979) Comportamiento de forrajeras estivales en los Bajos Submeridionales. INTA EEA Rafaela. Informe Tec. N°1, 36 p

Franco R, Condón F, Jaurena M, Tiscornia G, Reyno R, Beyhaut E (2014) Colecta de especies forrajeras nativas: un proyecto INIA para el futuro. INIA 38:29–32

Galíndez G, Ceccato D, Malagrina G, Pidal B, Chilo G, Bach H, Fortunato R, Ortega-Baes P (2016) Physical seed dormancy in native legume species of Argentina. Boletín de la Sociedad Argentina de Botánica 51:73–78

Gepts P (2002) A comparison between crop domestication, classical plant breeding, and genetic engineering. Crop Sci 42:1780–1790

Gil Báez C, Ordinola Agüero R, Ernst RD, Ruiz MA (2015) Caracterización morfológica, biomasa aérea y calidad en distintas poblaciones de *Trichloris crinita*. Archivos de Zootecnia 64:49–56

Gotor E, Alercia A, Rao VR, Watts J, Caracciolo F (2008) The scientific information activity of Bioversity International: the descriptor lists. Genet Resour Crop Evol 55:757–772

Grau HR, Aide TM, Gasparri NI (2005) Globalization and soybean expansion into semiarid ecosystems of Argentina. Ambio 34:267–268

Greco SA, Cavagnaro JB (2003) Effects of drought in biomass production and allocation in three varieties of *Trichloris crinita* (Poaceae) a forage grass from the arid Monte region of Argentina. Plant Ecol 164:125–135

Greco SA, Cavagnaro JB (2005) Growth characteristics associated with biomass production in three varieties of *Trichloris crinita* (Poaceae), a forage grass native to the arid regions of Argentina. Rangeland J 27:135–142

Grünwaldt JM, Guevara JC, Grünwaldt EG, Martínez Carretero E (2015) Cacti (*Opuntia* sps.) as forage in Argentina dry lands. Revista de la Facultad de Ciencias Agrarias, Universidad Nacional de Cuyo 47:263–282

Guarino L (2015) The cost-benefit of Australian genebanks. Agricultural Biodiversity Weblog. https://agro.biodiver.se/2015/03/the-cost-benefit-of-australian-genebanks/. Accessed 5 July 2016

Guarino L, Ramanatha Rao V, Goldberg E (2011) Collecting plant genetic diversity: technical guidelines—2011 Update. Bioversity International, Rome. Available online https://cropgeneb ank.sgrp.cgiar.org

Guevara JC, Colomer JS, Estevez OR, Paez JA (2003) Simulation of the economic feasibility of fodder shrub plantations as a supplement for goat production in the north-eastern plain of Mendoza, Argentina. J Arid Environ 53:85–98

Hacker JB, Glatzle A, Vanni R (1996) Paraguay-a potential source of new pasture legumes for the subtropics. Trop Grassl 30:273–281

Hawkins HS, Donald CM (1963) Pasture development in the beef cattle regions of Argentina. Part II. Grass Forage Sci 18:56–61

Hector A, Loreau M (2005) Relationships between biodiversity and production in grasslands at local and regional scales. In: Grassland: a global resource. Wageningen Academic, Wageningen, pp 295–304

Hidalgo LG, Cauhepe MA, Erni AN (1998) Digestibilidad de materia seca y contenido de proteína bruta en especies de pastizal de la Pampa Deprimida, Argentina. Invest Agr Prod Sanid Anim 13:165–177

Hilgert NI, Pensiero JF, Marino G, Lewis JP, D'Angelo C (2003) Vegetation of the Saladillo area (Province of Santa Fe) in the south of the Chaco Region, Argentina. Interciencia 28:512–520

Hong TD, Ellis EH (1996) A protocol to determine seed storage behaviour

IPGRI Technical Bulletin n. 1, 64 p. ISBN: 978-92-9043-279-1. ISBN: 92-9043-279-9

Howie J, Lloyd D, Revell C (2007). Spineless burr medic. Pastures Australia Fact Sheets. https://keys.lucidcentral.org/keys/v3/pastures/Html/index.htm#M. Accessed 2 Mar 2017

Hueck K (1972) As florestas da América do Sul. Ecologia, composição e importancia econômica. Trad. H. Reichardt. Universidade de Brasília y Editora Polígono, São Paulo

INaSe (2019) Catálogo Nacional de Cultivares https://www.inase.gov.ar/consultaGestion/gestiones. Accessed 10 July 2019

IPGRI (1984) Forage Legume Descriptors. Roma, Italia, 24 p. https://www.bioversityinternational.org/index.php?id=244ytx_news_pi1%5Bnews%5D=382ycHash=1966306fb581318def0709c9db8640dd. Accessed 7 July 2016

IPGRI (1985) Forage grass descriptors. Roma, Italia, 30 p. https://www.bioversityinternational.org/e-library/publications/detail/forage-grass-descriptors/. Accessed 7 July 2016

ISTA (2018) International Rules for Seed Testing, International Seed Testing Association, Bassersdorf, Switzerland

Jones RM, Clem RL (1997) The role of genetic resourses in developing improved pastures in semiarid and subhumid Northern Australia. Trop Grassl 31:315–319

Kadereit G, Mavrodiev EV, Zacharias EH, Sukhorukov AP (2010) Molecular phylogeny of Atripliceae (Chenopodioideae, Chenopodiaceae): implications for systematics, biogeography, flower and fruit evolution, and the origin of C4 photosynthesis. Am J Bot 97:1664–1687

Karlin MS, Bachmeier O, Dalmasso A, Sayago JM, Sereno R (2011) Environmental dynamics in Salinas Grandes, Catamarca (Argentina). Arid Land Res Manag 25(328):350

Karlin MS, Buffa EV, Karlin UO, Contreras AM, Coirini RO, Ruiz Posse EJ (2012) Relaciones entre propiedades de suelo, comunidades vegetales y receptividad ganadera en ambientes salinos (Salinas Grandes, Catamarca, Argentina). Revista Latinoamericana de Recursos Naturales 8:30–45

Kraus TA, Bianco CA, Weberling F (2003) Root system morphology of Fabaceae species from central Argentina. Wulfenia 10:61–72

Lemaire G, Wilkins R, Hodgson J (2005) Challenges for grassland science: managing research priorities. Agr Ecosyst Environ 108:99–108

Lemes JL (1992) Metodología de la exploración del recurso fitogenético de forrajeras nativas. Xerophytia 3:1–21

Lewis JP, Pire EF (1981) Reseña sobre la vegetación del Chaco Santafesino. La vegetación de la República Argentina. Serie Fitogeográfica 18, 42 p

Lewis JP, Pire EF (1996) Los mogotales de palo azul (*Cyclolepis genistoides* Don) de áreas deprimidas de la región chaqueña. Rev Asoc Cienc Nat Litoral 27:109–117

Lewis JP, Pire EF, Prado DE, Stofella SL, Franceschi EA, Carnevale NJ (1990) Plant communities and phytogeographical position of a large depression in the Great Chaco, Argentina. Vegetatio 86:25–38

Liddicoat C, McFarlane J (2007) Saltland pastures for South Australia. Department of Water, Land and Biodiversity Conservation (South Australia) on behalf of the Sustainable Grazing on Saline Lands (SGSL) sub-program, 144 p

Loch D (1980) Selection of environment and cropping system for tropical grass seed production. Trop Grassl 14:159–168

Luchetti AM (2008) Claves para el Reconocimiento de las Especies Santafesinas de Leguminosas Papilionoideas. Nat Neotrop 39:47–81

Luchetti AM (2014) Las Leguminosas Mimosoideas de la provincia de Santa Fe. Nat Neotrop 45:21–43

Maddaloni J (1986) Forage production on saline and alkaline soils in the humid region of Argentina. In: Barrett-Lennard EG, Malcolm CV, Stern WR, Wilkins SM (eds), Reclamation and revegetation research: Amsterdam, Elsevier, Netherlands, pp 11–16

Manger KR, Adams J, Probert RJ (2003) Selecting seed containers for the Millennium Seed Bank Project: a technical review and survey. In: Smith RD, Dickie JB, Linington SH, Pritchard HW, Probert RJ (eds) Seed Conservation: turning science into practice. Royal Botanic Gardens, Kew, UK, pp 637–652

Marcón F, Urbani MH, Quarin CI, Acuña CA (2015) Nutritional characteristics and cattle preference in *Paspalum atratum* and *P. lenticulare*. In: 5th International symposium of forage breeding, Buenos Aires

Marinoni L, Bortoluzzi A, Parra-Quijano M, Zabala JM, Pensiero JF (2015) Evaluation and improvement of the ecogeographical representativeness of a collection of the genus *Trichloris* in Argentina. Genet Resour Crop Evol 62:593–604

Marinoni L (2017) Variabilidad en el peso de semillas del género *Trichloris* (Poaceae) en Argentina y su efecto en la respuesta al estrés hídrico y salino. Doctoral thesis, Facultad de Ciencias Agrarias,Universidad Nacional del Litoral, Argentina, 163 p

Marinoni L, Zabala JM, Parra-Quijano M, Fernández RJ, Pensiero JF (2018) Genetic and environmental variation of seed weight in *Trichloris* species (Chloridoideae, Poaceae) and its association with seedling stress tolerance. Plant Ecol Divers 11(2):173–184

Marinoni L., Zabala JM, Taleisnik EL, Schrauf GE, Richard GA, Tomas PA, Giavedoni JA, Pensiero JF (2019) Wild halophytic species as forage sources: key aspects for plant breeding. Grass Forage Sci 74:321–344

Marshall DR (1989) Limitations to the use of germplasm collections. In: Brown AHD, Frankel OH, Marshall DR, Williams JT (eds) The use of plant genetic resources. Cambridge University Press, Cambridge, pp 105–120

Meyer T (1940) Preliminary note on forage grasses of the Chaco. Revista Argentina de Agronomía 7:95–104

Mitchell ML, Norman HC, Whalley RDB (2015) Use of functional traits to identify Australian forage grasses, legumes and shrubs for domestication and use in pastoral areas under a changing climate. Crop Pasture Sci 66:71–89

Muir JP, Pitman WD, Foster JL (2011) Sustainable, low-input, warm-season, grass–legume grassland mixtures: mission (nearly) impossible? Grass Forage Sci 66:301–315

Muir JP, Dubeux JC, Dos Santos MV, Maposse IC, Pitman WD, Butler TJ (2014) Challenges to domesticating native forage legumes. Trop Grassl 2:94–96

Murray F, Baldi G, Von Bernard T, Viglizzo EF, Jobbágy EG (2016) Productive performance of alternative land covers along aridity gradients: ecological, agronomic and economic perspectives. Agr Syst 149:20–29

Nair RM, Whittall A, Hughes SJ, Craig AD, Revell DK, Miller SM, Powell T, Auricht GC (2010) Variation in coumarin content of *Melilotus* species grown in South Australia. New Zeal J Agr Res 53:201–213

Nichols P, Loi A, Nutt BJ, Evans PM, Craig AD, Pengelly BC, Dear BS, Lloyd DL, Revella CK, Nair RM, Ewing MA, Howieson JG, Auricht GA, Howie JH, Sandral GA, Carr SJ, de Koning CT, Hackney BF, Crocker GJ, Snowball R, Hughes SJ, Hall EJ, Foster KJ, Skinner PW, Barbetti MJ, You MP (2007) New annual and short-lived perennial pasture legumes for Australian agriculture—15 years of revolution. Field Crops Res 104:10–23

Nichols PGH, Craig AD, Rogers ME, Albertsen TO, Miller SM, McClements DR, Hughes SJ, D'Antuono MDF, Dear BS (2008) Production and persistence of annual pasture legumes at five saline sites in southern Australia. Anim Prod Sci 48:518–535

Nichols PG, Revell CK, Humphries AW, Howie JH, Hall EJ, Sandral GA, Ghamkhar K, Harris CA (2013) Temperate pasture legumes in Australia—their history, current use, and future prospects. Crop Pasture Sci 63:691–725

Nichols PG, Yates RJ, Loo C, Wintle BJ, Stevens JC, Titterington W, Moore GA, Dixon KW, Barrett-Lennard EG (2014) Direct seeding of chenopod shrubs for saltland and rangeland environments. Future Farm Industries CRC Technical Report 10, 73 p

Nicora EG (1995) Los géneros *Diplachne* y *Leptochloa* (Gramineae, Eragrosteae) de la Argentina y países limítrofes. Darwiniana 33:233–256

NSW (2016) Department of Primary Industries. Grassed up—methods of seed harvesting. https://www.dpi.nsw.gov.au/content/agriculture/pastures/pastures-andrangelands/rangelands/publications/grassedup/seed-harvesting/methods. Accessed 10 Feb 2016

Oquendo G, Machado R, Acosta M, Bernal MA, Cisneros M (2006) Identificación y colecta de plantas forrajeras en suelos de un agroecosistema ganadero afectados por la salinidad. Pastos y Forrajes 29(2):145–154

Parodi LR (1927) Revisión de las gramíneas argentinas del género *Diplachne*. Revista Fac Agron Veterin 6:21–43

Parodi LR (1954) Nota preliminar sobre el género *Monanthochloe* (Gramineae) en Argentina. Physis (Buenos Aires) 20:1–3

Parodi LR (1919) Las Chlorideas de la República Argentina. Revista Fac Agron Veterin 2:233–339

Parra-Quijano M, Iriondo JM, Lamas ET (2012) Strategies for the collecting of wild species. In: Guarino L (ed) Collecting plant genetic diversity. Technical guidelines, pp 1–23

Paruelo JM, Oesterheld M, Del Pino F, Guerschmann JP, Verón SR, Piñeiro Guerra G, Vasallo M (2004) Patrones espaciales y temporales de la expansión de soja en Argentina: Relación con factores socio-económicos y ambientales. Informe final LART/FAUBA al Banco Mundial. https://www.agro.uba.ar/users/lart/bancomundial/INFORME_final.pdf. Accessed 2 Jan 2017

Passera CB, Cavagnaro JB, Sartor C (2010) Plantas C3; C4 y CAM nativas del Monte árido argentino, adaptaciones y potencial biológico. Libro homenaje al Dr. J. Lopez Gorgé, Madrid, pp 165–176

Pensiero JF (1986) Revisión de las especies argentinas del género *Pappophorum* (Gramineae-Eragrostoideae-Pappophoreae). Darwiniana 27:65–87

Pensiero JF (2012a) *Cottea* Kunth. In: Zuloaga FO, Rúgolo ZE Anton (eds) Flora vascular de la República Argentina. Monocotyledoneae, Poaceae: Aristidoideae a Pharoideae. AM, vol 3, Tomo 1, Graficamente Ediciones, Córdoba, Argentina, p 176

Pensiero JF (2012b) *Pappophorum* Schreb. In: Zuloaga FO, Rúgolo ZE, Anton AM (eds) Flora vascular de la República Argentina. Monocotyledoneae, Poaceae: Aristidoideae a Pharoideae, vol 3, Tomo 1, Graficamente Ediciones, Córdoba, Argentina, pp 158–163

Pensiero JF (2012c) *Setaria* P. Beauv. In: Zuloaga FO, Rúgolo ZE, Anton AM (eds) Flora vascular de la República Argentina. Monocotyledoneae, Poaceae: Aristidoideae a Pharoideae, vol 3, Tomo 1, Graficamente Ediciones, Córdoba, Argentina, pp 446–472

Pensiero JF, Gutiérrez HF, Exner E (2005) Sistema de polinización y su efecto sobre la producción y el peso de semillas en nueve especies sudamericanas del género *Setaria*. Interciencia 30:495–500

Pensiero J, Zabala JM (2017) Recursos fitogenéticos forrajeros nativos y naturalizados para los Bajos Submeridionales: prospección y priorización de especies para planes de introducción a cultivo. Revista FAVE - Secc Cs Agrarias 16(1):67–98

Peoples MB, Herridge DF (1990) Nitrogen fixation by legumes in tropical and subtropical agriculture. Adv Agron 44:155–223

Perez SM, Camardelli MC, Juarez F, Bianchi AR, Newman R (1999) Geographic distribution of *Macroptilium* species in Argentina. Trop Grassl 33:22–33

Pérez-García F, González-Benito ME, Gomez-Campo C (2007) High viability recorded in ultra-dry seeds of 37 species of Brassicaceae after almost 40 years of storage. Seed Sci Technol 35:143–153

Quarín CL, Valls JFM, Urbani MH (1997) Cytological and reproductive behavior of *Paspalum atratum*, a promising forage grass for the tropics. Trop Grassl 31:114–116

Quarin CL, Espinoza F, Martinez EJ, Pessino SC, Bovo AO (2001) A rise of ploidy level induces the expression of apomixis in *Paspalum notatum*. Sex Plant Reprod 13:243–249

Quiroga A, Correa RJ (2011) Gramíneas forrajeras presentes en el Chaco Árido de Catamarca. Revista de Divulgación Técnica Agrícola y Agroindustrial. Facultad de Ciencias Agrarias, Universidad Nacional de Catamarca 16:1–12

Quiroga A, Esnarriaga DN (2014) Diversidad de forrajeras nativas consumidas por el ganado caprino en el área pedemontana del chaco árido, Catamarca. Biología en Agronomía 4:121–147

Quiroga RE, Blanco LJ, Ferrando CA, Leal KV (2007) Caracterización forrajera de una leguminosa nativa del Chaco Árido sembrada en consociación con pastos nativos. In: 4to. Congreso Argentino y 1ro. del Mercosur para el Manejo de Pastizales Naturales, p 23

Quiroga RE, Namur P, Blanco LJ, Orionte EL (2009) Velocidad de crecimiento en progenies de la leguminosa *Galactia texana*. In: Congreso Argentino de Producción Animal, Mendoza, Argentina

Quiroga RE, Fernández RJ, Galluscio RA, Blanco LJ (2013) Differential water-use strategies and drought resistance in *Trichloris crinita* plants from contrasting aridity origins. Plant Ecol 214:1027–1035

Ragonese AE (1951) La vegetación de la República Argentina. II. Estudio fitosociológico de las Salinas Grandes. Revista de Investigaciones Agrícolas 5:1–233

Reichert F, Parodi LR (1926) Las plantas forrajeras indígenas y cultivados en la República Argentina (Segunda contribución). Revista Fac Agron Veterin 5:272–309

Reichert F, Trelles RA, Parodi LR, Hauman L (1923) Las plantas forrajeras indígenas y cultivadas de la República Argentina. Talleres S.A. Casa Jacobo Peuser. Primera contribución, Buenos Aires, p 575

Richard G, Pensiero JF, Cerino CM, Galati BG, Gutiérrez JF (2015) Reproductive biology of *Sporobolus phleoides* Hack. (Poaceae), an endemic halophyte grass of Argentina. Plant Syst Evol 301:1937–1945

Rogers ME, Colmer TD, Frost K, Henry D, Cornwall D, Hulm E, Deretic J, Hughes SR, Craig AD (2008) Diversity in the genus *Melilotus* for tolerance to salinity and waterlogging. Plant Soil 304:89–101

Rogers ME, Craig AD, Munns R, Colmer TD, Nichols PGH, Malcolm CV, Barrett-Lennard EG, Brown AJ, Semple WS, Evans PM, Cowley K, Hughes SJ, Snowball R, Bennett SJ, Sweeney GC, Dear BS, Ewing M (2005) The potential for developing fodder plants for the salt-affected areas of southern and eastern Australia: an overview. Aust J Exp Agr 45:301–329

Rosengurtt B (1946) Gramíneas y leguminosas de Juan Jackson. Comportamiento en el campo y en cultivo. Estudios sobre praderas naturales del Uruguay, 5ª Contribución, Montevideo, pp 216–346

Rosengurtt B, Arrillaga de Maffei B, Izaguirre de Artuccio YP (1970) Gramíneas Uruguayas. Universidad de la República, Montevideo, p 489

Ruiz Posse E, Karlin UO, Buffa E, Karlin M, Giai Levra C, Castro G (2007) Ambientes de las Salinas Grandes de Catamarca, Argentina. Multequina 16:123–137

Santa Cruz RH, Quiroga A (1998) Efectos de una clausura tradicional en la recuperación de un área degradada en el campo comunero Las Peñas, Dpto. La Paz. In: Congreso Regional de Ciencia y Tecnología, Secretaría de Ciencia y Tecnología, Univ. Nac. de Catamarca, Tomo II, pp 1–11

Sartor ME, Quarin CL, Urbani MH, Espinoza F (2011) Ploidy levels and reproductive behaviour in natural populations of five *Paspalum* species. Plant Syst Evol 293:31–41

Schellberg J, Verbruggen E, Michalk DL, Millar GD, Badgery WB, Broadfoot KM (2013) New frontiers and perspectives in grassland technology. Revitalising grasslands to sustain our communities: Proceedings, 22nd International Grassland Congress. Australia, Sydney, pp 44–55

Schinini A, Ciotti EM, Tomei CE, Castelán ME, Hack CM (2004) Especies nativas de campos bajos con potencial valor forrajero. Agrotecnia 12:18–22

Schulz AG (1962) Plantas forrajeras indígenas del chaco. INTA, EEA Colonia Benítez. Folleto 4, 24 p

Shelton HM, Franzel S, Peters M (2005) Adoption of tropical legume technology around the world: analysis of success. Trop Grassl 39:198–209

Silvestri LC (2015) La conservación de la diversidad genética argentina: tres desafíos para implementar el régimen de acceso a los recursos genéticos y la distribución de los beneficios. Ecol Austral 25:273–278

Skerman PJ, Cameron DG, Riveros YF (1991) *Aeschynomene*. In: FAO (ed) Leguminosas forrajeras tropicales. Organización de las Naciones Unidas para la Agricultura y la Alimentación, Roma, pp 233–241

Squires VR, Glenn EP (2016) Creating an economic linkage between fossil fuel burning, climate change, and rangeland restoration. In: West NE (ed) Rangelands in a sustainable biosphere. Proceedings of the Fifth International Rangeland Congress, Salt Lake City, Utah, USA, pp 531–532

Stoutjesdijk P (2013) Plant genetic resources for food and agriculture: second national report—Australia, ABARES, Technical Report 13.11, Canberra, Australia

Tivano JC, Vegetti AC (2010) Growth forms in Pappophoreae (Poaceae). Flora 205:291–301

Trigg P (2004) *Melilotus albus* (Sweet clover) 'Jota.' Plant Varietal J 17:127–128

Trione SO, Cavagnaro JB (1998) Water shortage and associated changes in organic nitrogen between *Popphorum caespitosum* (Gramineae) provenances. J Arid Environ 38:519–528

Tropical Forage (2016) List of forages species. https://www.tropicalforages.info/key/Forages/Media/Html/index.htm. Accessed 23 Sept 2016

Vogt (2015) Clasificación de las comunidades halófilas de las estepas salinas en la cuenca del riacho Yakaré Sur, chaco boreal, Paraguay. Bol Mus Nac Hist Nat Parag 19:41–49

West SH, Pitman WD (2001) Seed production technology of tropical forages. In: Sotomayor-Rios SA, Pitman WD (ed) Tropical forage plants: development and use. CRC Press, pp 143–166

Whyte RO (1958) Prospección, recogida e introducción de especies vegetales. Estudios Agropecuarios de la FAO N° 41, Roma, 123 p

Yensen NP, Yensen SB, Weber CW (1985) A review of *Distichlis* spp. for production and nutritional values. In: Whitehead EE, Hutchinson CF, Timmermann BN, Varady YRG (eds) Arid lands today and tomorrow. Westview Press, Boulder, pp 809–822

Zabala JM, Pensiero JF, Tomas P, Giavedoni JA (2008) Morphological characterisation of populations of *Desmanthus virgatus* complex from Argentina. Trop Grassl 42:229–236

Zabala JM, Tomas PA, Schrauf GE, Giavedoni JA (2009a) Seed dormancy in *Elymus scabrifolius* (Döll) J. H. Hunz. Seed Sci Technol 37:241–244

Zabala JM, Tomas PA, Schrauf GE, Giavedoni JA (2009b) Effect of temperature and storage on seed germination in *Elymus scabrifolius* (Döll) J.H. Hunz. Seed Sci Technol 37:245–250

Zabala JM, Taleisnik E, Giavedoni JA, Pensiero JF, Schrauf GE (2011) Variability in salt tolerance of native populations of *Elymus scabrifolius* (Döll) J. H. Hunz from Argentina. Grass Forage Sci 66:109–122

Zabala JM, Widenhorn P, Pensiero JF (2011) Germination patterns of species of the genus *Trichloris* in arid and semiarid environments of Argentina. Seed Sci Technol 39:338–353

Zabala JM, Schrauf G, Baudracco J, Giavedoni J, Quaino O, Rush P (2012) Selection for late-flowering and greater number of basal branches increases the leaf dry matter yield in *Melilotus albus* Desr. Crop Pasture Sci 63:370–376

Zabala JM, Pensiero JF, Forni M, Sosa N, Testa M, Giavedoni J, Aiello F, Yost A, Quarin P (2015) Valorización de los recursos fitogenéticos a través de pequeñas empresas productoras de semillas de forrajeras nativas: evaluación de algunos factores que afectan la producción de semillas en leguminosas forrajeras. In: Lallana VH (ed) Red de Cultivos no Tradicionales de Agricultura Familiar. XVII Foro de Decanos de Facultades de Agronomía del Mercosur, Bolivia y Chile. UNER, Paraná, pp 65–73

Plant Tolerance Mechanisms to Soil Salinity Contribute to the Expansion of Agriculture and Livestock Production in Argentina

Edith Taleisnik, Andrés Alberto Rodríguez, Dolores A. Bustos, and Darío Fernando Luna

Abstract This chapter addresses salt tolerance mechanisms in crops and woody species cultivated in Argentina, highlighting the contribution of local research to these topics. Work on forages and woody species represents approximately half of this research that has been published by Argentine authors in international journals. Basic research on plant salinity mounts to only 12% of the total, indicating that it still does not attract sufficient consideration among researchers. Among forage plants, attention in this chapter is focused on Rhodes grass (*Chloris gayana* Kunth), while in woody perennials, salt tolerance mechanisms in *Prosopis,* which have been extensively investigated locally, are reported. Despite the importance of soybean in Argentine economy, as well as that of other crops such as maize, wheat, sunflower, relatively little research attention has been paid to salinity issues in these major field crops. This situation may reflect the fact that they are mostly cultivated in non-saline soils.

E. Taleisnik (✉) · A. A. Rodríguez
Consejo Nacional de Investigaciones Científicas y Técnicas (CONICET), Buenos Aires, Argentina
e-mail: etaleisnik@conicet.gov.ar

A. A. Rodríguez
e-mail: andresrodriguez@conicet.gov.ar

E. Taleisnik · D. A. Bustos · D. F. Luna
Instituto de Fisiología y Recursos Genéticos Vegetales (IFRGV), Centro de Investigaciones Agropecuarias (CIAP), Instituto Nacional de Tecnología Agropecuaria (INTA), Córdoba, Argentina
e-mail: lolibustos@hotmail.com

D. F. Luna
e-mail: luna.fernando@inta.gob.ar

E. Taleisnik
Facultad de Ciencias Agropecuarias, Universidad Católica de Córdoba, Córdoba, Argentina

A. A. Rodríguez
Laboratorio de Estrés Abiótico y Biótico en Plantas, Unidad de Biotecnología 1, Instituto Tecnológico de Chascomús (INTECH), Chascomús, Argentina

E. Taleisnik and R. S. Lavado (eds.), *Saline and Alkaline Soils in Latin America*, https://doi.org/10.1007/978-3-030-52592-7_19

Keywords *Chloris gayana* · *Prosopis* · Salt tolerance mechanisms · Ion compartmentalization · Reactive oxygen species (ROS) · Alkalinity tolerance

1 Soil Salinity in the Summit and Decline of Ancient Civilizations and as a Factor in Natural Vegetation Distribution

A large volume of literature illustrates about the negative effects of salinity, alkalinity and their combination on plant growth and crop yields, as well as on differences in salt-susceptibility among species, as exposed in one of the most cited references on this topic (Maas and Hoffman 1977). Salt-associated decreases in plant yield have had a significant impact on civilization since ancient times. Jacobsen and Adams (1958) studied agricultural evolution in the fertile plains of southern Iraq, in the period spanning from 3500 to 1700 BC and noticed a gradual replacement of salt-sensitive wheat for the more salt-tolerant barley until wheat was completely abandoned. Diminishing yields of both crops were associated with increasing soil salinization due to over-irrigation. This process was contemporary with a gradual migration of the population to the northern part of the alluvial plain. The authors conclude that increasing soil salinity played an important role in the decline of Sumerian civilization.

Today, soil salinization represents the second major cause of land degradation after soil erosion. It is estimated that nearly 2000 ha of arable land are being lost every day by salinization process (Shahid et al. 2018). Yet, salt-affected soils support the natural occurrence of halophytic plant species, adapted to such substrates on account of inherent salt tolerance traits (Flowers and Colmer 2015). Halophytic species are potential candidates for domestication and soil reclamation, as well as sources of salt tolerance mechanisms for increasing the adaptation of susceptible species (Flowers and Colmer 2015). Some crop wild relative species are natural halophytes, as has been well documented in tomato (Rick 1973) and the transfer of genes from wild relatives can contribute to improve salt tolerance in this species (Li et al. 2018). A similar approach has been successful in rice (Quan et al. 2018; Li et al. 2018).

2 Physiological Effects of Soil Salinity on Plants and Tolerance Mechanisms

The nature of salt tolerance mechanisms has recurrently been reviewed in the literature. Although there are older reviews on the responses of non-halophytic plants to salinity (e.g., Bernstein and Hayward 1958), the conceptual framework for this topic was outlined in the seminal review by Greenway and Munns (1980) setting the foundations for subsequent work. Earlier, Greenway and Osmond (1972) had shown that the in vitro activity of various cytosolic enzymes was affected more severely by inorganic salts than by similar concentrations of organic solutes, regardless of the origin of the enzymes, whether it was animals or plants, either halophytic or not. These findings contributed significantly to the development of the ideas on the responses of non-halophytes to salinity reviewed by Greenway and Munns (1980). They indicated that salinity imposes water and ion limitations on plants that in turn can cause nutritional imbalances and toxicity. Initially, plants respond mainly to low water potentials that characterize saline substrates, while ion-specific effects are observed later, as the concentration of potentially toxic ions that are in excess in saline substrates, such as Na^+ and Cl^-, builds up internally. High internal concentrations of these ions can result in alterations in the accumulation of essential elements such as K^+ (Wu et al. 2015), Ca^{2+} (Cramer 2002) and N (Grattan and Grieve 1998) or in water imbalance within cells generated by the accumulation of toxic ions in various cell compartments, including the apoplast (Oertli 1968).

According to this scheme, the central aspects of adaptation to salt stress are (a) access restriction of potentially toxic ions such as Na^+ and Cl^- to the sites of active metabolism in organs (Davenport et al. 2005) and within cells (Tester and Davenport 2003), (b) water balance rendered by accumulation of organic solutes in the cytosol and ions in vacuoles (Zhang et al. 1999), (c) the maintenance of intracellular concentrations of essential nutrients required for normal metabolism (Wu et al. 2015) and (d) presence of mechanisms for the detoxification of reactive oxygen species (ROS) that are normally originated in excess under stress situations (Miller et al. 2010). These responses are triggered and coordinated by complex networks of gene expression regulation and signaling whose components and function are being gradually identified (Hasegawa et al. 2000; Park et al. 2016).

Consequently, three types of mechanisms contribute to plant salinity tolerance (Rajendran et al. 2009; Roy et al. 2014), which are not excluding and can operate either simultaneously or not. The first type (Type I) confers tolerance to substrate low water potential caused by salinity, through the accumulation of organic and inorganic solutes (Zhang et al. 1999) that lower internal water potential and generate the necessary gradient to insure plant water uptake. The second type (Type II) involves the control of the uptake of potentially toxic ions such as Na^+ and Cl^- and the maintenance of intracellular concentrations of essential nutrients suitable for normal metabolism. These mechanisms also involve Na^+ and Cl^- compartmentalization at cellular and intracellular levels to avoid reaching toxic concentrations in the cytoplasm, especially in mesophyll cells of the youngest leaves. In general, these mechanisms contribute to the exclusion of toxic ions from sites of active metabolism, mainly in leaf blades, and help to prevent premature leaf death. The activity of these mechanisms explains apparent contradictions such as this one: the exclusion of Na^+ from expanded wheat leaves is correlated with salinity tolerance (Munns et al. 2006), yet, moderate concentrations of Na^+ can stimulate plant growth not only in halophytes but in other species as well, as in tomato (Taleisnik and Grunberg 1994). When Na^+ is compartmentalized in vacuoles, it becomes an osmotic agent of low energy cost, contributing to mobilize water essential for cell expansion. Finally, the third type of salinity tolerance mechanisms (Type III) confers tolerance to toxic ions in leaf tissues (James et al. 2008). This group of mechanisms involves processes that control oxidative stress by increasing the activity of ROS detoxification systems. The presence of high levels of ROS causes oxidative damage to membranes and macromolecules which can lead to cell death, visualized macroscopically as dry leaves. ROS levels depend on their rates of generation and detoxification by enzymatic and non-enzymatic mechanisms. ROS participate in stiffening but also in softening of cell walls, and the link between ROS and leaf growth in salinity conditions emerges in the context of mechanical constraints to cell expansion (Schmidt et al. 2016). Studies in maize showed that the decrease in ROS production could be linked to the decrease in leaf growth under saline stress condition (Rodríguez et al. 2004, 2009). Intraspecific variability can be observed in the three types of mechanisms (James et al. 2008) as it has been seen in wheat (Rajendran et al. 2009) and in sunflower (Céccoli et al. 2012).

Significant advances in the identification of the molecular bases of the salinity tolerance mechanisms mentioned in the review by Greenway and Munns (1980) have been registered since then (Hasegawa 2013; Roy et al. 2014; Arzani and Ashraf 2016). Up-to-date progress in these areas can be verified in recent reviews. Among many others, Liang et al. (2018) review salt tolerance mechanisms within the framework outlined above, while Yang and Guo (2018) focus on molecular components of salt tolerance. The comprehensive review by Morton et al. (2019) addresses definitions of salt tolerance and phenotyping technologies, the profitable tapping of genetic diversity for improving salt tolerance and discusses the use of newly available genomic tools for genomic selection. Therefore, it is beyond this chapter to provide such updates. It will address salt tolerance mechanisms in crops and woody species that grow in Argentina, highlighting the contribution of local research to such topics.

3 Agriculture and Livestock Production in Salt-Affected Areas in Argentina as Promoters of Local Research Efforts

Agriculture currently contributes roughly 6% to Argentina's GDP according the World Bank database (World Bank 2018). Beef and soybean are the main components of this figure. Other important crops in Argentina are maize, wheat, sunflower, barley, sorghum, cotton, sugarcane, beans and peanut.

The steady increase in Argentina's soybean production since the introduction of the genetically modified (GM) soybean varieties in 1996 (Choumert and Phélinas 2015) has displaced livestock production from traditional areas in the Pampa plains to other regions of the country, considered marginal for agriculture (Guevara and Grünwaldt 2012). Such marginal areas are characterized by environmental constraints to crop production, derived from climate and soil characteristics, salinity among them (Taleisnik and Lavado 2017). This trend has prompted active research efforts in forage plants salt tolerance. Work in salt-tolerant woody species and native plant resources in Argentina had a long tradition before the new developments in livestock production, and it has acquired new relevance in this context.

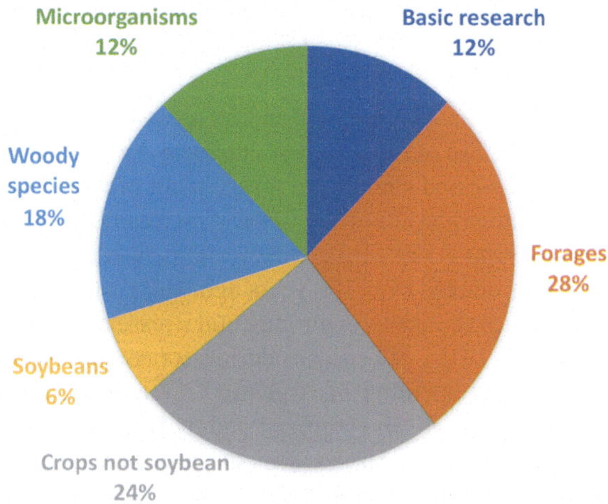

Fig. 1 Topical distribution of plant- and microorganisms-salinity papers published from Argentina. *Source* www.scopus.com

Figure 1 indicates the subject distribution of research addressing plant salinity issues, in papers published by Argentine authors in international journals. Soil and ecological studies have been excluded from this picture, and the analysis focuses only on plant and microorganism responses to salinity. Not surprisingly, work on forages (28%) and woody species (18%), together, represent 46% of the published research. Basic research on plant salinity takes up only 12% of this distribution, highlighting that it still does not attract sufficient consideration among researchers.

Despite the importance of soybean in the Argentine economy and the widespread occurrence of salt-affected soils in the country, relatively little research attention has been paid to soybean-salinity issues. This is the situation with most of the main crops in Argentina (the percentage of salinity research in all crops not soybean is similar to that devoted only to forages), possibly acknowledging the fact that the most important field crops have traditionally been grown in the best soils, where salinity is not an issue.

4 Forages to Sustain Livestock Production in Areas Considered Marginal for Agricultural Production

Areas considered marginal for agricultural production in Argentina, due to soil and climate constraints, support an array of unique natural plant resources (Marinoni et al. 2019; Pensiero and Zabala 2017), some with promising potential as forages. This is particularly relevant since the steady increase in soybean production has displaced cattle from traditional production areas in the Argentine Pampa plains to other regions of the country (Guevara and Grünwaldt 2012; Ruolo 2010). One of such areas is the Chaco region, where cattle production, traditionally based on natural pastures, gradually turned to introduced perennial subtropical grasses such as *Panicum coloratum* L., *Cenchrus ciliaris* L., *Chloris gayana* Kunth, *Digitaria eriantha* Steud. and *Urochloa brizantha* (Hochst. ex A. Rich.) R.D. Webster. The Chaco region is rich in native grass and leguminous forage resources which have been degraded by mismanagement and are a viable alternative to ecosystem restoration and sustainable use of rangeland in the Great American Chaco (see Chapter by Pensiero et al. "Native and naturalized forage plant genetic resources for saline environments of the southernmost portion of the American Chaco", in this book).

C. gayana is one of the most widely cultivated subtropical grasses in Argentina. It was introduced over 100 years ago, and it became widely accepted on account of its tolerance to various environmental stresses, heat, drought, light frosts and salinity (Toll Vera 2016). When considering forage plants for saline areas, *C. gayana* has concentrated attention of Argentine researchers. Early information about salinity tolerance in *C. gayana* (summarized by Bogdan 1969) stimulated interest to investigate its physiological bases. This species features several of the salt tolerance mechanisms described above. Salt glands are localized in the epidermis of its leaves (Liphschitz et al. 1974). These structures specialize in the extrusion of saline solution to the leaf surface, thus contributing to reducing internal salt concentrations (Céccoli et al. 2015) and therefore can be considered among Type II salt tolerance mechanisms (Fig. 2). Salt glands have been used to screen for tolerant lines *C. gayana* (de Luca et al. 2001) and as a tool to develop salt-tolerant cultivars of this species (Zorin and Loch 2007). The presence of salt glands distinguishes *C. gayana* from other moderately salt-tolerant subtropical forage grasses such as *C. ciliaris* and *P. coloratum*, as the concentration of Na^+ in leaves of both species is low (Ruiz and Taleisnik 2013; Pittaro et al. 2016), suggesting the occurrence of Type II mechanisms that compartmentalize Na^+ in organs located before the leaf blades.

C. gayana also presents Type I mechanisms because it accumulates organic compounds in the growing regions of the leaf blades (Fig. 2), enabling it to sustain

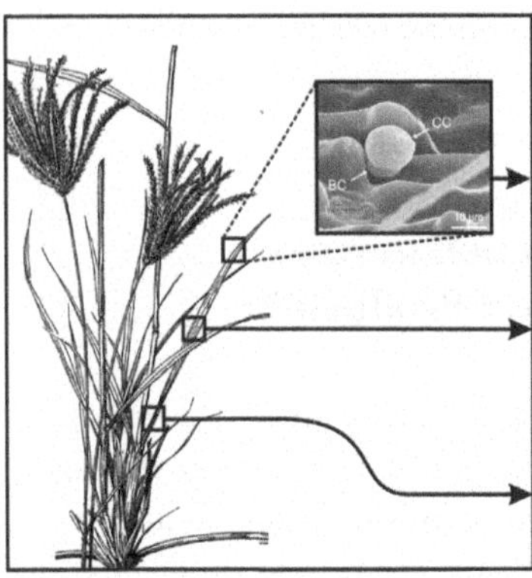

Salt-tolerance mechanism	Type	References
Salt glands	II	Liphschitz et al. 1974 Céccoli et al. 2015
Antioxidant enzymatic system stimulation	I	Luna et al. 2000 de Luca et al. 2001
Osmolyte accumulation in the leaf blade growth zone	III	de Luca et al. 2001

Fig. 2 *Chloris gayana* plants have three types of salt tolerance mechanisms. Line drawing of a *C. gayana* Kunth plant (adapted from www.plants.usda.gov). Inset: scanning electron image of a salt glands on the abaxial leaf surface (adapted from Oi et al. 2012; BC, basal cell, CC, cap cell of the salt glands)

water uptake and leaf growth (de Luca et al. 2001). *C. gayana* roots are concentrated in the upper soil layers, where they obtain the main fraction of the water taken up by the plant (Chiacchiera et al. 2016). Nevertheless, hydraulic conductivity decreases in *C. gayana* plants under salinity. This phenomenon has been associated with changes in xylem vessel diameters (Ortega et al. 2006) and root mass reduction (Córdoba et al. 2001).

In forage grasses, leaf blades are the main components of cattle intake. In two genetically close *C. gayana* genotypes with contrasting responses to salinity, the decrease in forage yield was associated with reduced leaf growth and increased leaf death (de Luca et al. 2001). Saline stress effects on the growth of monocotyledonous leaves have been summarized by Taleisnik et al. (2009). In *C. gayana* under salinity, final leaf length was more severely reduced than cell length, suggesting that cell division was affected by salinity (Ortega et al. 2006). Similar conclusions were reached in studies of *P. coloratum* plants under salinity (Pittaro et al. 2016).

Leaf death is, in many cases, restricted to the expanded zone of the blades, where a decrease in C fixation rates is usually registered under salinity condition (Shomer-Ilan et al. 1979). The decrease in photosynthetic surface and in C fixation rates result is less C supply to support growth and to sustain re-growth after defoliation caused by grazing, or after winter. On the other hand, a decrease in C fixation rates under saline condition can also contribute to ROS generation. Type III mechanisms aimed at reducing ROS presence were also described in *C. gayana* (Fig. 2). It was observed that oxidative damage in the expanded zone of leaf blades was lower in genotypes with high salinity tolerance (Luna et al. 2000; de Luca et al. 2001) and coincided with the stimulation of some antioxidant enzyme activities. One major process involved in ROS damage is the accumulation of lipid peroxidation-derived reactive aldehydes determined by malondialdehyde (MDA), generation and low MDA values are considered a tolerant response to salt stress (Sunkar et al. 2003). MDA accumulation was observed in *C. gayana* under salinity condition and was low in tolerant genotypes (Luna et al. 2000). The determination of MDA content was proposed as an auxiliary tool to identify genotypes with tolerance to salinity in this species (Luna et al. 2000) and also in *C. ciliaris* (Castelli et al. 2010).

Another cattle-production zone on marginal soils is the center-east of Buenos Aires province, in a vast depression known as the Flooding Pampas, which covers nearly 9,000,000 ha. It is characterized by recurrent periods of drought-flood, poor drainage, heterogeneous soils with low nutrient contents, high levels of sodic salts and also alkaline pH (Lavado and Alconada 1994, chapter by Imbellone et al. "Genesis, properties and management of salt affected soils in the Flooding Pampas, Argentina" in this book). Plants growing in saline-alkaline soils must deal with salinity effects, as well as with soil structural and nutritional limitations associated with alkalinity itself. Alkaline soils restrict plant growth and crop yield on account of their agrophysical, hydrological and physiological properties (Luna et al. 2017). Agrophysical properties are related to the low porosity and oxygen deficiency inherent of these soils (Lavado and Taboada 2017). Hydrological factors are a consequence of low porosity and lead to the low water permeability of these soils; a high content of unavailable moisture also characterizes them. Physiological factors are associated with high concentrations of water-soluble salts and nutritional imbalances, specifically, macronutrient (N and P) and micronutrient deficiencies (Fe and Zn) in plants are typical consequences of high soil pH (Marschner 1995).

In the Flooding Pampas, cattle production is supported mainly on native grasslands, as the growth of other fodder species such as red and white clover, and lucerne is limited (Escaray et al. 2012). Given these restrictive soils conditions, *Lotus tenuis* was introduced to confer sustainability and restoration of these soils, becoming rapidly naturalized, producing forage of high nutritional value (Dear et al. 2003; Antonelli et al. 2016). Local research efforts to elucidate the molecular and physiological basis behind saline-alkaline tolerance of this species are summarized in the chapter by Nieva and Ruiz "*Lotus* spp.—a foreigner that came to stay forever: economic and environmental changes caused by its naturalization in the Salado River Basin, Argentina" in this book.

Sorghum (*Sorghum bicolor* L. Moench), an annual or short-lived perennial species of the Poacea family, is also arising research interest for marginal areas. Sorghum is among the most important cereal crops along with rice, wheat, maize and barley, and it is used for food, feed, fodder and fuel production (Rao et al. 2016); its C4 metabolism supports high temperatures and water limited environments, and it is exceptionally efficient in the use of radiation, water and nitrogen (Mullet et al. 2014). Sorghum is considered a crop with moderate tolerance to salinity (Maas et al. 1986), though more tolerant than maize, and genotypic variability exists for this condition (Taylor 1975; Weimberg et al. 1982; Maiti et al. 1994). Research on salt and alkaline stress in this crop has recently been summarized (Huang 2017).

Currently, local research focusing on the evaluation and selection of sorghum genotypes with tolerance to salinity and water stress conditions, is being carried out by the National Institute of Agricultural Technology (INTA). Hybrids obtained from *Sorghum bicolor x sudangrass*, developed by the INTA Manfredi Experimental Station, exhibited higher photosynthetic performance than *S. bicolor* hybrids under alkaline conditions (Luna et al. 2018).

5 Woody Perennials for Environments Affected by Salinity

At the beginning of the exponential growth of research on salt tolerance mechanisms, attention was focused on halophytic species, and among the woody perennials, *Avicennia* spp. and *Atriplex* spp. were subject to intensive research (Flowers et al. 1977; Osmond et al. 1980). Those works set the basis for understanding the mechanisms that contribute to salt tolerance in halophytes and established the concept that those mechanisms are common across plant species (Glenn et al. 1999). The highly cited work by Maas and Hoffman (1977) indicates the existence of salt tolerance variability among woody perennials, and date palms were underscored as highly salt-tolerant species. Literature on genetic variability for salt tolerance in woody perennials prior to 1993 was reviewed by Allen et al. (1994) and later by Kozlowski

(1997). More recently, Qadir et al. (2008) made a critical assessment of salt tolerance and potential uses of woody perennials for land reclamation. A list of promising accessions of *Acacia*, *Eucalyptus* and *Melaleuca* for agroforestry purposes in salt-affected soils can be found in that review. Native woody resources for saline areas in Argentina are listed and discussed by Villagra et al. (2017), and in this book, Villagra et al. "Ecological restoration and productive recovery of saline environments from the Argentine Monte Desert using native plants".

Afforestation with salt-tolerant species is particularly attractive for the productive recovery of salt-affected soils (Gill and Abrol 1991; Hoy et al. 1994). A critical assessment on the effects of afforestation of pasture lands on water table level and salinity (Nosetto et al. 2008) highlights that such effects are tightly linked to rainfall, evapotranspiration, relative salt tolerance of stands, and that conclusions are influenced by plantation size (see also the chapter by Jobbaggy et al., in this book: "Salt accumulation and redistribution in the dry plains of Southern South America: Lessons from land use changes".

Salt tolerance in woody perennials in Argentina was reviewed by Taleisnik and López Launestein (2011). In general, *Prosopis* species in the Chaco region of Argentina are considered salt tolerant (Rhodes and Felker 1988; Villagra et al. 2017). Research on salt tolerance mechanisms in *Prosopis* species indicates the presence of Type II mechanisms since the ability to regulate and control Na^+ uptake and prevent its build-up in metabolically active tissues contributes to its adaptation (Meloni et al. 2004). Studies in *P. strombulifera* provide evidence that sulfate and chloride sodium salts exert different growth effects on this species (Reginato et al. 2014), due to specific effects on metabolism (Llanes et al. 2014, 2019).

Salt-tolerant shrubs and bushes are also important ecosystem components in arid and semi-arid regions, supplying forage and providing soil cover and organic matter. Salt tolerance is evident in local *Atriplex* spp. (Aiazzi et al. 2002, 2004, 2009).

Grapes and olives are grown in arid regions in Argentina, where the occurrence of saline soils is widespread. In *Vitis* spp., the search for salt-tolerant genotypes has underscored intraspecific variability for this trait (Cavagnaro et al. 2006), and shoot ion exclusion (Vila et al. 2016) may contribute to salt tolerance. Transgenic *Vitis* lines overexpressing the vacuolar Na^+/H^+ antiport AtNHX1 were more tolerant but without link to ion concentrations (Venier et al. 2014). Similarly, phenotyping in *Olea* has also shown the presence of intraspecific variability for salt tolerance (Ruiz et al. 2011). Overall, the volume of research in these species in Argentina is quite reduced, in comparison with the attention devoted to woody natural resources.

6 Upsurging Relevance for Salt Tolerance Research in Soybean

It was mentioned earlier that, as with the other main crops, relatively little research attention has been paid to soybean-salinity issues in Argentina, and it was suggested to be due to the fact that they are cultivated in non-saline soils. However, this is gradually changing. The situation is illustrated by recent agricultural developments in the Province of Córdoba, located in the central part of Argentina. There, 14% of the soils are affected by some degree of salinity, and 17% have high Na^+ concentration (Gorgas et al. 2003). These soils have traditionally been considered marginal for agriculture. Yet a recent study in the counties of Marcos Juárez and Unión (Aimetta et al., 2020) shows that between 17 and 29% of the surface sown with wheat, maize and soybean in these counties is in saline or alkaline soils, where about half are sown with soybean. This is a practical consequence of the natural plasticity of this crop.

Several reviews have summarized salt tolerance mechanisms in soybean, among them Phang et al. (2008) and Cao et al. (2018). Soybean are considered moderately sensitive to salinity (Katerji et al. 2003), though intraspecific variability exists (Abel and MacKenzie 1964; Velagaleti and Schweitzer 1993). Salt stress affects all stages in soybean development, from germination and vegetative growth (Abel and MacKenzie 1964; Wang and Shannon 1999) to seed yield and quality (Parker et al. 1983).

Various lines of evidence indicate that genetic diversity in the wild soybean ancestor (*Glycine soja* Sieb & Zucc.) is wider than in cultivated soybean *Glycine max* (L.) Merrill (Maughan et al. 1995; Xu and Gai 2003). Luo et al. (2005) reported the former is more salt tolerant than *G. max* in association with lower levels of leaf Na^+ and Cl^- accumulation, while higher levels of Na^+ were reported in roots (Zhang et al. 2011). Thus, *G. soja* is currently considered a valuable genetic resource to increase salt tolerance in cultivated soybean.

In Argentina, field studies have provided further evidence for the deleterious effects of Na^+ and Cl^- accumulation in soybean (Bustingorri and Lavado 2013; Muzlera Klappenbach et al. 2015). Physiological studies on the response of soybean to salinity carried out in the country often considered oxidative stress consequences of salinity (Comba et al. 1998; Balestrasse et al. 2008; Zilli et al. 2008, 2009, 2014; Campestre et al. 2011). Research efforts are currently focused on the effects of salinity on various stages of symbiosis establishment (Muñoz et al. 2012, 2014a, b; Robert et al. 2014, 2018). It is very encouraging that basic studies on drought tolerance have led to the release of the first stress-tolerant GM soybean, and it has been generated in Argentina (Waltz 2015).

7 Can Plant Physiology Contribute to Reduce Anthropic Salinization and Mitigate Agricultural Consequences? Some Final Considerations and Concluding Remarks

These issues were addressed by Ridley and Pannell (2006), who concluded that adequate policy responses to increasing dryland salinity include investments to improve salinity management technologies, and in research and development of new plant-based systems. Nevertheless, frequently, and with pessimism, it has been pointed out that a very small fraction of the knowledge about the physiological mechanisms and genes related with salinity tolerance has been transferred to breeding programs for the generation of salt-tolerant crops, a topic whose progress is considered slow. This has been attributed not only to the multigenic character of salinity tolerance (Flowers and Flowers 2005), but also, to the anarchy of the methods used to characterize it in plants. While the components of the tolerance mechanisms are usually identified under controlled conditions that allow to isolate variables and to accurately set the salinity of the medium (Munns and James 2003), yet even under controlled conditions, experiments using hydroponics setups or pots with soil do not necessarily yield coincident results (Tavakkoli et al. 2010;Luna et al. (2017). Passioura (2006) has highlighted the difficulties in transferring information from experiments performed in pots under controlled conditions to the field. The lack of agreement in the responses observed in mature *C. ciliaris* plants exposed to salinity in the field (Ruiz and Taleisnik 2013) and of seedlings under controlled conditions (Griffa et al. 2010) illustrates this problem. In addition, plants under controlled conditions are usually exposed to a single type of stress when, obviously, they are exposed simultaneously to multiple stresses in the field. These considerations must be taken into account in the design of tests to determine plant salinity tolerance and in the interpretation of their results.

Concerted actions are essential to meet the challenges posed for the productive incorporation of Argentine native, salt-tolerant germplasm, into breeding schemes. At this point, it is not redundant to emphasize that only transdisciplinary approaches, including professionals from various fields (agronomists, breeders, plant scientists, among others) along with farmers, will provide potential solutions that take into account the delicate equilibrium between rational exploitation and breached limits in the use of natural resources. Climate change is not a scenario of the future, and as we write this chapter, bushfires are ravaging Australia. It is our responsibility, as plant scientists, to keep this unprecedented tragedy in mind while we orient our work and propose research plans.

References

Abel GH, MacKenzie AJ (1964) Salt tolerance of soybean varieties (*Glycine max* L. Merrill) during germination and later growth. Crop Sci 4(2):157–161

Aiazzi MT, Carpane PD, Di Rienzo JA, Argüello JA (2002) Effects of salinity and temperature on the germination and early seedling growth of *Atriplex cordobensis* Gandoger et Stuckert (Chenopodiaceae). Seed Sci Technol 30(2):329–338

Aiazzi MT, Carpane PD, Argüello JA, Piotto B (2004) Salt tolerance at the germination stage of *Atriplex cordobensis* Gandoger et Stuckert (Chenopodiaceae) from different provenances. Seed Sci Technol 32(1):43–52

Aiazzi MT, Di Rienzio JA, Sosa L (2009) Effects of different salts on the germination and early seedling growth of *Atriplex cordobensis* Gandoger et Stuckert (Chenopodiaceae). Seed Sci Technol 37(1):17–24

Aimetta MB, Muñoz SA, Bustos D, Davidenco V, Cazorla CR, Galarza CM, Salvagiotti F (2020) Cuantificación del uso agrícola de suelos con limitantes salinas y sódicas mediante el índice NDVI. Cienc. del suelo 38 (1):174-186

Allen JA, Chambers JL, Stine M (1994) Prospects for increasing the salt tolerance of forest trees: a review. Tree Physiol 14:843–853

Antonelli CJ, Calzadilla PI, Escaray FJ, Babuin MF, Campestre MP, Rocco R, Bordenave CD, Perea García A, Nieva AS, Llames ME (2016) Lotus spp: biotechnological strategies to improve the bioeconomy of lowlands in the Salado River Basin (Argentina). AGROFOR Int J 1(2):43–53

Arzani A, Ashraf M (2016) Smart engineering of genetic resources for enhanced salinity tolerance in crop plants. Crit Rev Plant Sci 35(3):146–189

Balestrasse KB, Zilli CG, Tomaro ML (2008) Signal transduction pathways and haem oxygenase induction in soybean leaves subjected to salt stress. Redox Rep 13(6):255–262

Bernstein L, Hayward H (1958) Physiology of salt tolerance. Annu Rev Plant Physiol 9(1):25–46

Bogdan AV (1969) Rhodes grass. Herbage Abs 39:1–13

Bustingorri C, Lavado R (2013) Soybean response and ion accumulation under sprinkler irrigation with sodium-rich saline water. J Plant Nutr 36(11):1743–1753

Campestre MP, Bordenave CD, Origone AC, Menéndez AB, Ruiz OA, Rodríguez AA, Maiale SJ (2011) Polyamine catabolism is involved in response to salt stress in soybean hypocotyls. J Plant Physiol 168(11):1234–1240

Cao D, Li Y, Liu B, Kong F, Tran LSP (2018) Adaptive mechanisms of soybean grown on salt-affected soils. Land Degrad Dev 29(4):1054–1064

Castelli SL, Grunberg K, Muñoz N, Griffa S, Colomba EL, Ribotta A, Biderbost E, Luna C (2010) Oxidative damage and antioxidant defenses as potential indicators of salt-tolerant *Cenchrus ciliaris* L. genotypes. Flora 205(9):622–626

Cavagnaro JB, Ponce MT, Guzmán J, Cirrincione MA (2006) Argentinean cultivars of *Vitis vinifera* grow better than European ones when cultured in vitro under salinity. Biocell 30(1):1–7

Céccoli G, Senn ME, Bustos D, Ortega LI, Córdoba A, Vegetti A, Taleisnik E (2012) Genetic variability for responses to short- and long-term salt stress in vegetative sunflower plants. J Plant Nutr Soil Sci 175(6):882–890

Céccoli G, Ramos J, Pilatti V, Dellaferrera I, Tivano JC, Taleisnik E, Vegetti AC (2015) Salt glands in the *Poaceae* family and their relationship to salinity tolerance. Bot Rev 81(2):162–178

Chiacchiera S, Bertram N, Taleisnik E, Jobbágy E (2016) Effect of watertable depth and salinity on growth dynamics of Rhodes grass (*Chloris gayana*). Crop Pasture Sci 67(8):881–887

Choumert J, Phélinas P (2015) Determinants of agricultural land values in Argentina. Ecol Econ 110:134–140

Comba ME, Benavides MP, Tomaro ML (1998) Effect of salt stress on antioxidant defence system in soybean root nodules. Aust J Plant Physiol 25(6):665–671

Córdoba A, García Seffino L, Moreno H, Arias C, Grunberg K, Zenoff A, Taleisnik E (2001) Characterization of the effect of high salinity on roots of *Chloris gayana* Kunth: carbohydrate and lipid accumulation and growth. Grass Forage Sci 56(2):162–168

Cramer GR (2002) Sodium-calcium interactions under salinity stress. In: Läuchli A, Lüttge U (eds) Salinity: environment-plants-molecules. Springer, Dordrecht, pp 205–227

Davenport R, James RA, Zakrisson-Plogander A, Tester M, Munns R (2005) Control of sodium transport in durum wheat. Plant Physiol 137(3):807–818

de Luca M, García Seffino L, Grunberg K, Salgado M, Córdoba A, Luna C, Ortega L, Rodríguez AA, Castagnaro A, Taleisnik E (2001) Physiological causes for decreased productivity under high salinity in Boma, a tetraploid *Chloris gayana* cultivar. Aust J Agric Res 52(9):903–910

Dear B, Moore G, Hughes S (2003) Adaptation and potential contribution of temperate perennial legumes to the southern Australian wheatbelt: a review. Aust J Exp Agric 43(1):1–18

Escaray FJ, Menendez AB, Gárriz A, Pieckenstain FL, Estrella MJ, Castagno LN, Carrasco P, Sanjuán J, Ruiz OA (2012) Ecological and agronomic importance of the plant genus *Lotus*. Its application in grassland sustainability and the amelioration of constrained and contaminated soils. Plant Sci 182:121–133

Flowers TJ, Flowers SA (2005) Why does salinity pose such a difficult problem for plant breeders. Agric Water Manage 78(1–2):15–24

Flowers TJ, Colmer TD (2015) Plant salt tolerance: adaptations in halophytes. Ann Bot 115(3):327–331

Flowers TJ, Troke PF, Yeo AR (1977) The mechanism of salt tolerance in halophytes. Annu Rev Plant Physiol 28:89–125

Gill HS, Abrol IP (1991) Salt affected soils, their afforestation and its ameliorating influence. Int Tree Crops J 6(4):239–260

Glenn EP, Jed BJ, Blumwald E (1999) Salt tolerance and crop potential of halophytes. Crit Rev Plant Sci 18:227–255

Gorgas J, Tassile J, Jarsún B, Zamora E, Bosnero H, Lovera E, Ravelo A (2003) Recursos naturales de la provincia de Córdoba, los suelos: nivel de reconocimiento escala 1: 500.000. Agencia Córdoba DAC y TSEM Dirección Ambiente, Instituto Nacional de Tecnología Agropecuaria INTA-Manfredi

Grattan S, Grieve C (1998) Salinity–mineral nutrient relations in horticultural crops. Sci Hortic 78:127–157

Greenway H, Osmond CB (1972) Salt responses of enzymes from species differing in salt tolerance. Plant Physiol 49(2):256–259

Greenway H, Munns R (1980) Mechanisms of salt tolerance in nonhalophytes. Annu Rev Plant Physiol 31:149–190

Griffa S, Ribotta A, López Colomba E, Tommasino E, Carloni E, Luna C, Grunberg K (2010) Evaluation seedling biomass and its components as selection criteria for improving salt tolerance in Buffel grass genotypes. Grass Forage Sci 65(3):358–361

Guevara J, Grünwaldt E (2012) Status of beef cattle production in Argentina over the last decade and its prospects. In: Javed K (ed) Livestock production. BoD–Books on Demand, pp 117–134

Hasegawa PM (2013) Sodium (Na^+) homeostasis and salt tolerance of plants. Environ Exp Bot 92:19–31

Hasegawa PM, Bressan RA, Zhu JK, Bohnert HJ (2000) Plant cellular and molecular responses to high salinity. Annu Rev Plant Physiol 51:463–499

Hoy NT, Gale MJ, Walsh KB (1994) Revegetation of a scalded saline discharge zone in central Queensland. 1. Selection of tree species and evaluation of an establishment technique. Aust J Exp Agric 34(6):765–776

Huang RD (2017) Research progress on plant tolerance to soil salinity and alkalinity in sorghum. J Integr Agric 17(4):739–746

Jacobsen T, Adams RM (1958) Salt and silt in ancient Mesopotamian agriculture. Science 128(3334):1251–1258

James RA, von Caemmerer S, Condon AG, Zwart AB, Munns R (2008) Genetic variation in tolerance to the osmotic stress component of salinity stress in durum wheat. Funct Plant Biol 35(2):111–123

Katerji N, Van Hoorn J, Hamdy A, Mastrorilli M (2003) Salinity effect on crop development and yield, analysis of salt tolerance according to several classification methods. Agric Water Manage 62(1):37–66

Kozlowski TT (1997) Responses of woody plants to flooding and salinity. Tree Physiol 17(7):490. https://doi.org/10.1093/treephys/17.7.490

Lavado RS, Alconada M (1994) Soil properties behavior on grazed and ungrazed plots of a grassland sodic soil. Soil Technol 7(1):75–81

Lavado RS, Taboada MA (2017) Génesis y propiedades de los suelos halomórficos. In: Taleisnik E, Lavado R (eds) Ambientes salinos y alcalinos de la Argentina: recursos y aprovechamiento productivo. Orientación Gráfica Editora, Universidad Católica de Córdoba, Buenos Aires, pp 9–28

Li T, Yang X, Yu Y, Si X, Zhai X, Zhang H, Dong W, Gao C, Xu C (2018) Domestication of wild tomato is accelerated by genome editing. Nature Biotec 36(12):1160–1163

Liang W, Ma X, Wan P, Liu L (2018) Plant salt-tolerance mechanism: a review. Biochem Biophys Res Co 495(1):286–291

Liphschitz N, Shomer-Ilan A, Eshel A, Waisel Y (1974) Salt glands on leaves of Rhodes Grass (*Chloris gayana* Kth.). Ann Bot 38(2):459–462

Llanes A, Masciarelli O, Luna V (2014) Growth responses to sulfate and chloride are related to different phytohormone profiles in the halophyte *Prosopis strombulifera*. Emir J Food Agric:1097–1113

Llanes A, Pinamonti P, Iparraguirre J, Bertazza G, Luna V (2019) Abscisic acid alters carbohydrate accumulation induced by differential response to sodium salts in the halophyte *Prosopis strombulifera*. Plant Biosyst. https://doi.org/10.1080/11263504.2019.1610114

Luna C, García Seffino L, Arias C, Taleisnik E (2000) Oxidative stress indicators as selection tools for salt tolerance in *Chloris gayana*. Plant Breed 119(4):341–345

Luna DF, Aguirre A, Pittaro G, Bustos D, Ciacci B, Taleisnik E (2017) Nutrient deficiency and hypoxia as constraints to *Panicum coloratum* growth in alkaline soils. Grass Forage Sci 72(4):640–653

Luna DF, Saavedra Pons AB, Bustos D, Taleisnik E (2018) Early responses to Fe-deficiency distinguish *Sorghum bicolor* genotypes with contrasting alkalinity tolerance. Environ Exp Bot 155:165–176

Luo Q, Yu B, Liu Y (2005) Differential sensitivity to chloride and sodium ions in seedlings of *Glycine max* and *G. soja* under NaCl stress. J Plant Physiol 162(9):1003–1012

Maas EV, Hoffman GJ (1977) Crop salt tolerance. Current assessment. J Irrig Drain E-ASCE 102:115–134

Maas E, Poss J, Hoffman G (1986) Salinity sensitivity of sorghum at three growth stages. Irrigation Sci 7(1):1–11

Maiti R, de la Rosa-Ibarra M, Sandoval ND (1994) Genotypic variability in glossy sorghum lines for resistance to drought, salinity and temperature stress at the seedling stage. J Plant Physiol 143(2):241–244

Marinoni LDR, Zabala JM, Taleisnik EL, Schrauf GE, Richard GA, Tomas PA, Giavedoni JA, Pensiero JF (2019) Wild halophytic species as forage sources: key aspects for plant breeding. Grass Forage Sci 74(3):321–344

Marschner H (1995) Mineral nutrition of higher plants, 2nd edn. Academic Press, London

Maughan P, Maroof MS, Buss G (1995) Microsatellite and amplified sequence length polymorphisms in cultivated and wild soybean. Genome 38(4):715–723

Meloni DA, Gulotta MR, Martínez CA, Oliva MA (2004) The effects of salt stress on growth, nitrate reduction and proline and glycinebetaine accumulation in *Prosopis alba*. Braz J Plant Physiol 16(1):39–46

Miller GAD, Suzuki N, Ciftci-Yilmaz S, Mittler RON (2010) Reactive oxygen species homeostasis and signalling during drought and salinity stresses. Plant Cell Environ 33(4):453–467

Morton MJ, Awlia M, Al-Tamimi N, Saade S, Pailles Y, Negrão S, Tester M (2019) Salt stress under the scalpel–dissecting the genetics of salt tolerance. Plant J 97(1):148–163

Mullet J, Morishige D, McCormick R, Truong S, Hilley J, McKinley B, Anderson R, Olson SN, Rooney W (2014) Energy Sorghum—a genetic model for the design of C4 grass bioenergy crops. J Exp Bot 65(13):3479–3489

Munns R, James RA (2003) Screening methods for salinity tolerance: a case study with tetraploid wheat. Plant Soil 253(1):201–218

Munns R, James RA, Lauchli A (2006) Approaches to increasing the salt tolerance of wheat and other cereals. J Exp Bot 57(5):1025–1043

Muñoz N, Robert G, Melchiorre M, Racca R, Lascano R (2012) Saline and osmotic stress differentially affects apoplastic and intracellular reactive oxygen species production, curling and death of root hair during *Glycine max* L.-*Bradyrhizobium japonicum* interaction. Environ Exp Bot 78:76–83

Muñoz N, Rodríguez M, Robert G, Lascano R (2014a) Negative short-term salt effects on the soybean–*Bradyrhizobium japonicum* interaction and partial reversion by calcium addition. Funct Plant Biol 41(1):96–105

Muñoz N, Soria-Díaz ME, Manyani H, Sánchez-Matamoros RC, Serrano AG, Megías M, Lascano R (2014b) Structure and biological activities of lipochitooligosaccharide nodulation signals produced by *Bradyrhizobium japonicum* USDA 138 under saline and osmotic stress. Biol Fertil Soils 50(2):207–215

Muzlera Klappenbach A, Bustingorri C, Lavado RS (2015) Respuesta de la soja a elementos tóxicos (cloruros, arseniatos, fluoruros y vanadatos) presentes naturalmente en aguas y suelos. Agronomía & Ambiente 35(1):59–70

Nosetto MD, Jobbágy EG, Tóth T, Jackson RB (2008) Regional patterns and controls of ecosystem salinization with grassland afforestation along a rainfall gradient. Glo Biogeochem Cycles 22. https://doi.org/10.1029/2007GB003000

Oertli J (1968) Extracellular salt accumulation a possible mechanism of salt injury in plants. Agrochimica 12(5):461

Oi T, Taniguchi M, Miyake H (2012) Morphology and ultrastructure of the salt glands on the leaf surface of Rhodes grass (*Chloris gayana* Kunth). Int J Plant Sci 173(5):454–463

Ortega L, Fry SC, Taleisnik E (2006) Why are *Chloris gayana* leaves shorter in salt-affected plants? Analyses in the elongation zone. J Exp Bot 57(14):3945–3952

Osmond CB, Björkman O, Anderson DJ (1980) Physiological processes in plant ecology. Toward a synthesis with *Atriplex*. Ecological studies, vol 36. Springer, Berlin, Heidelberg, New York

Park HJ, Kim W-Y, Yun D-J (2016) A new insight of salt stress signaling in plant. Mol Cells 39(6):447

Parker MB, Gascho G, Gaines T (1983) Chloride toxicity of soybeans grown on atlantic coast flatwoods soils. Agron J 75(3):439–443

Passioura JB (2006) The perils of pot experiments. Funct Plant Biol 33:1075–1079

Pensiero JF, Zabala JM (2017) Recursos fitogenéticos forrajeros nativos y naturalizados para los bajos submeridionales: prospección y priorización de especies para planes de introducción a cultivo. Revista FAVE—Ciencias Agrarias 16(1):67–98

Phang TH, Shao G, Lam HM (2008) Salt tolerance in soybean. J Integr Plant Biol 50(10):1196–1212

Pittaro G, Cáceres L, Bruno C, Tomás A, Bustos D, Monteoliva M, Ortega L, Taleisnik E (2016) Salt tolerance variability among stress-selected Panicum coloratum cv Klein plants. Grass Forage Sci 71(4):683–698

Qadir M, Tubeileh A, Akhtar J, Larbi A, Minhas PS, Khan MA (2008) Productivity enhancement of salt-affected environments through crop diversification. Land Degrad Dev 19(4):429–453

Quan R, Wang J, Hui J, Bai H, Lyu X, Zhu Y, Zhang H, Zhang Z, Li S, Huang R (2018) Improvement of salt tolerance using wild rice genes. Front Plant Sci 8:2269

Rajendran K, Tester M, Roy SJ (2009) Quantifying the three main components of salinity tolerance in cereals. Plant Cell Environ 32(3):237–249

Rao PS, Vinutha KS, Kumar GSA, Chiranjeevi T, Uma A, Lal P, Prakasham RS, Singh HP, Rao RS, Chopra S, Jose S (2016) Sorghum: a multipurpose bioenergy crop. In: Ciampitti I, Prasad V (eds) Sorghum: state of the art and future perspectives. Agronomy Monographs, vol 58. American Society of Agronomy and Crop Science Society of America, Inc., Madison, pp 1–26

Reginato M, Sosa L, Llanes A, Hampp E, Vettorazzi N, Reinoso H, Luna V (2014) Na_2SO_4 and NaCl determine different growth responses and ion accumulation in the halophytic legume *Prosopis strombulifera*. Plant Biol 16:97–106

Rhodes D, Felker P (1988) Mass screening of *Prosopis* (mesquite) seedlings for growth at seawater salinity concentrations. Forest Ecol Manage 24(3):169–176

Rick CM (1973) Potential genetic resources in tomato species: clues from observations in native habitats. In: Srb AM (ed) Genes, enzymes, and populations. Plenus Press, New York, London, pp 255–269

Ridley AM, Pannell DJ (2006) The role of plants and plant-based research and development in managing dryland salinity in Australia. Aust J Exp Agric 45(11):1341–1355

Robert G, Muñoz N, Melchiorre M, Sánchez F, Lascano R (2014) Expression of animal anti-apoptotic gene Ced-9 enhances tolerance during *Glycine max* L.-*Bradyrhizobium japonicum* interaction under saline stress but reduces nodule formation. PLoS ONE 9(7):e101747
Robert G, Muñoz N, Alvarado-Affantranger X, Saavedra L, Davidenco V, Rodríguez-Kessler M, Estrada-Navarrete G, Sánchez F, Lascano R (2018) Phosphatidylinositol 3-kinase function at very early symbiont perception: a local nodulation control under stress conditions? J Exp Bot 69(8):2037–2048
Rodríguez AA, Córdoba AR, Ortega L, Taleisnik E (2004) Decreased reactive oxygen species concentration in the elongation zone contributes to the reduction in maize leaf growth under salinity. J Exp Bot 55(401):1383–1390
Rodríguez AA, Maiale SJ, Menéndez AB, Ruiz OA (2009) Polyamine oxidase activity contributes to sustain maize leaf elongation under saline stress. J Exp Bot 60(15):4249–4262
Roy SJ, Negrão S, Tester M (2014) Salt resistant crop plants. Curr Opin Biotec 26:115–124
Ruiz M, Olivieri G, Vita Serman F (2011) Effects of saline stress in two cultivars of *Olea europea* L: 'Arbequina' and 'Barnea'. Acta Hortic 924:117–124
Ruiz M, Taleisnik E (2013) Field hydroponics assessment of salt tolerance in *Cenchrus ciliaris* (L.): growth, yield, and maternal effect. Crop Pasture Sci 64(6):631–639
Ruolo MS (2010) Morfogénesis, estructura, producción y calidad de *Chloris gayana* Kunth bajo distintos regímenes de defoliación. Doctoral dissertation, Universidad de Buenos Aires, p 67
Schmidt R, Kunkowska AB, Schippers JH (2016) Role of reactive oxygen species during cell expansion in leaves. Plant Physiol 172(4):2098–2106
Shahid SA, Zaman M, Heng L (2018) Introduction to soil salinity, sodicity and diagnostics techniques. In: Zaman M, Shahid S, Heng L (eds) Guideline for salinity assessment, mitigation and adaptation using nuclear and related techniques. Springer International Publishing, pp 1–42
Shomer-Ilan ASYB, Kipnis T, Elmer D, Waisel Y (1979) Effects of salinity, N-nutrition and humidity on photosynthesis and protein metabolism of *Chloris gayana* Kunth. Plant Soil 53:477–486
Sunkar R, Bartels D, Kirch HH (2003) Overexpression of a stress-inducible aldehyde dehydrogenase gene from *Arabidopsis thaliana* in transgenic plants improves stress tolerance. Plant J 35(4):452–464
Taleisnik E, Grunberg K (1994) Ion balance in tomato cultivars differing in salt tolerance. I. Sodium and potassium accumulation and fluxes under moderate salinity. Physiol Plant 92(3):528–534
Taleisnik E, López Launestein D (2011) Leñosas perennes para ambientes afectados por salinidad: Una sinopsis de la contribución argentina a este tema. Ecol Austral 21(1):3–14
Taleisnik E, Lavado RS (eds) (2017) Ambientes salinos y alcalinos en la Argentina. Recursos y aprovechamiento productivo. Orientación Gráfica Editora, Universidad Católica de Córdoba, Buenos Aires
Taleisnik E, Rodríguez AA, Bustos D, Erdei L, Ortega L, Senn ME (2009) Leaf expansion in grasses under salt stress. J Plant Physiol 166(11):1123–1140
Tavakkoli E, Rengasamy P, McDonald GK (2010) The response of barley to salinity stress differs between hydroponic and soil systems. Funct Plant Biol 37(7):621–633
Taylor R, Young E Jr, Rivera R (1975) Salt tolerance in cultivars of grain sorghum. Crop Sci 15(5):734–735
Tester M, Davenport R (2003) Na^+ tolerance and Na^+ transport in higher plants. Ann Bot 91(5):503–527
Toll Vera JR (ed) (2016) Grama Rhodes: centenario de su liberación en Argentina. Publicación conmemorativa. Universidad Nacional de Tucumán. Facultad de Agronomía y Zootecnia, San Miguel de Tucumán
Velagaleti R, Schweitzer SM (1993) General effects of salt stress on growth and symbiotic nitrogen fixation in soybean. In: Pessarakli M (ed) Handbook of plant and crop stress. Marcel Dekker, New York, pp 461–471
Venier M, Bermejillo A, Filippini MF, Fernández Alonso S, Agüero CB, Blumwald E, Dandekar A (2014) Phenotypic evaluation of 'Thompson Seedless' grapes transformed with AtNHX1 growing in hydroponics and potted soils. Acta Hortic 1046:423–430

Vila HF, Di Filippo ML, Venier M, Filippini MF (2016) How rootstocks influence salt tolerance in grapevine? The roles of conferred vigor and ionic exclusion. Acta Hortic 1136:145–154

Villagra P, Passera C, Greco S, Sartor C, Naranibar J, Meglioli P, Alvarez J, Allegretti L, Fernández N, Cony M, Kozub P, Vega Riveros C (2017) Uso de plantas nativas en la restauración y recuperación productiva de ambientes salinos de las zonas áridas de la región del Monte, Argentina. In: Taleisnik E, Lavado RS (eds) Ambientes salinos y alcalinos de la Argentina. Recursos y aprovechamiento productivo. Orientación Gráfica Editora y Universidad Católica de Córdoba, Buenos Aires, pp 419–444

Waltz E (2015) First stress-tolerant soybean gets go-ahead in Argentina. Nat Biotechnol 33(7):682

Wang D, Shannon M (1999) Emergence and seedling growth of soybean cultivars and maturity groups under salinity. Plant Soil 214(1–2):117–124

Weimberg R, Lerner H, Poljakoff-Mayber A (1982) A relationship between potassium and proline accumulation in salt-stressed *Sorghum bicolor*. Physiol Plant 55(1):5–10

World Bank (2018) https://data.worldbank.org/indicator/NV.AGR.TOTL.ZS?end=2018&locations=AR&start=2003

Wu H, Zhu M, Shabala L, Zhou M, Shabala S (2015) K^+ retention in leaf mesophyll, an overlooked component of salinity tolerance mechanism: a case study for barley. J Integr Plant Biol 57(2):171–185

Xu D, Gai J (2003) Genetic diversity of wild and cultivated soybeans growing in China revealed by RAPD analysis. Plant Breed 122(6):503–506

Yang Y, Guo Y (2018) Elucidating the molecular mechanisms mediating plant salt-stress responses. New Phytol 217(2):523–539

Zhang J, Nguyen HT, Blum A (1999) Genetic analysis of osmotic adjustment in crop plants. J Exp Bot 50(332):291–302

Zhang X, Zhou Q, Cao J, Yu B (2011) Differential Cl^-/salt tolerance and NaCl-induced alternations of tissue and cellular ion fluxes in *Glycine max*, *Glycine soja* and their hybrid seedlings. J Agron Crop Sci 197(5):329–339

Zilli CG, Balestrasse KB, Yannarelli GG, Polizio AH, Santa-Cruz DM, Tomaro ML (2008) Heme oxygenase up-regulation under salt stress protects nitrogen metabolism in nodules of soybean plants. Environ Exp Bot 64(1):83–89

Zilli CG, Santa-Cruz DM, Balestrasse KB (2014) Heme oxygenase-independent endogenous production of carbon monoxide by soybean plants subjected to salt stress. Environ Exp Bot 102:11–16

Zilli CG, Santa-Cruz DM, Yannarelli GG, Noriega GO, Tomaro ML, Balestrasse KB (2009) Heme oxygenase contributes to alleviate salinity damage in *Glycine max* L. leaves. Int J Cell Biol. https://doi.org/10.1155/2009/848516

Zorin M, Loch DS (2007) Development of new *Chloris gayana* cultivars with improved salt tolerance from 'Finecut' and 'Topcut.' Proceedings sixth international Herbage seed conference. Gjennestad, Norway, pp 92–96

Genetic Improvement of Perennial Forage Plants for Salt Tolerance

Gustavo E. Schrauf, Flavia Alonso Nogara, Pablo Rush, Pablo Peralta Roa, Eduardo Musacchio, Sergio Ghio, Luciana Couso, Elena Ramos, Matías F. Schrauf, Lisandro Voda, Andrea Giordano, Julio Giavedoni, José F. Pensiero, Pablo Tomas, Juan M. Zabala, and Germán Spangenberg

Abstract The difficulties of genetic improvement of forage species are further complicated by the intricacies of salinity stress. Multiple evidence of the effects of salinity on germination and establishment highlight some of the limitations that must be overcome in order to carry out successful breeding programs for pastures. Different sources of variation feasible to be used in such breeding programs are

---

G. E. Schrauf (✉) · F. Alonso Nogara · P. Rush · P. P. Roa · E. Musacchio · S. Ghio · L. Couso · E. Ramos · L. Voda · A. Giordano
Facultad de Agronomía, Cátedra de Genética, Universidad de Buenos Aires, Buenos Aires, Argentina
e-mail: gschrauf@agro.uba.ar

F. Alonso Nogara
e-mail: alonso@agro.uba.ar

P. Rush
e-mail: rush@agro.uba.ar

P. P. Roa
e-mail: pperalta@agro.uba.ar

E. Musacchio
e-mail: musacchi@agro.uba.ar

S. Ghio
e-mail: sergioghio@gmail.com

L. Couso
e-mail: lcouso@agro.uba.ar

E. Ramos
e-mail: eleramos@agro.uba.ar

L. Voda
e-mail: lisandro.voda@basf.com

A. Giordano
e-mail: andy_gior@yahoo.com.ar

M. F. Schrauf
Facultad de Agronomía, Departamento de Métodos Cuantitativos, Universidad de Buenos Aires, Buenos Aires, Argentina
e-mail: matiasfschrauf@agro.uba.ar

E. Taleisnik and R. S. Lavado (eds.), *Saline and Alkaline Soils in Latin America*, https://doi.org/10.1007/978-3-030-52592-7_20

analyzed. The application of morpho-physiological selection criteria such as "salt glands" or "Na exclusion," simulation of the saline environment to assess germination behavior, initial growth or production before defoliation are considered. Methodological advances in sequencing and bioinformatics allow us to predict a prominent role in the application of "Genomic Selection". On the other hand, the advances in gene technologies have allowed direct the changes to specific sites by "Gene Edition" techniques, which are also very promising. The different methodologies of population management are largely dependent on reproductive systems, and it is a field where knowledge and "art" combine for successful results in plant breeding. Conclusions are drawn from the experiences carried out, and future perspectives for the improvement of perennial forage are analyzed. Both classical and molecular breeding come together not as alternatives but as complements.

Keywords Gene technologies · Genomic selection · Molecular markers · Na exclusion · Plant breeding · Salt glands

1 Genetic Improvement of Perennial Forage Species

One of the main disadvantages in the improvement of forage species is that they are sold through seeds, but their agronomic value is in the production of leaves, and both characteristics, forage and seed production, are frequently negatively correlated (Díaz et al. 2004). To the contradictions between forage and seed production, we must add the additional challenge of the decline in forage quality with flowering (Schrauf et al. 2017). When plants bloom, tissues become lignified, the leaf/stem ratio is reduced and this leads to a reduction of digestibility. Another problem is that the initial production is not correlated with the accumulated production over the years and also with perennity (Snaydon 1985), forcing evaluations longer than a year in perennial species to predict pasture yield over time (Berdahl and Barker 1978).

The success in the improvement of grain crops is largely due to alterations in photoassimilate allocation, favoring grains as sink. In forage species, changes in

J. Giavedoni · J. F. Pensiero · P. Tomas · J. M. Zabala
Facultad de Ciencias Agrarias, Universidad Nacional del Litoral, Esperanza, Santa Fe, Argentina
e-mail: jgiavedo@fca.unl.edu.ar

J. F. Pensiero
e-mail: jfpensi@fca.unl.edu.ar

P. Tomas
e-mail: patomas@fca.unl.edu.ar

J. M. Zabala
e-mail: jmzabala@fca.unl.edu.ar

A. Giordano · G. Spangenberg
Centre for AgriBioscience, AgriBio, La Trobe University, Bundoora, VIC, Australia
e-mail: German.Spangenberg@agriculture.vic.gov.au

sinks usually have a high cost, especially toward a lower perennity (Díaz et al. 2004). Another difficulty is that forage species are not usually sown in pure stands. Sowing in mixtures causes enormous evaluation difficulties. Among the most frequent, is that selection leads to single plants which are then evaluated in pure dense plots for production, despite the fact that they are usually grown in mixed populations in the field (Hill 1990).

Tolerance to saline stress may also be negatively associated with forage production and quality, so the evaluation is complex and, in turn, the heterogeneity of saline environments is often high. But, in addition, the final evaluation should be given by animal grazing (Berdahl and Barker 1978) and since animal production depends on the quantity and quality of the forage, it is necessary to estimate these two variables that are frequently negatively associated (Oba and Allen 1999; Oliver et al. 2005). Grazing efficiency depends not only on the management method but also on the architecture of the plants (da Silva et al. 2015), that is, the architecture of the plant that facilitates the efficiency of the bite should be a selection criterion. The impact of grazing on traction and compaction on pasture production and perennity should also be evaluated (Striker et al. 2006). The above illustrates some of the difficulties in genetic improvement of perennial forage plants. If the complexity of breeding for salinity tolerance is added, the challenge seems enormous. However, the generated knowledge at all scales (molecular, physiological, and ecological) comes to the aid of the breeder and will facilitate the success of programs for the development of forage species for salt-affected environments. Taking into account all these restrictions, this chapter deals with the application of different selection criteria, the difficulties of establishing pastures in saline environments, and classical improvement methods as well as molecular tools are analyzed, focusing on their current and future integration.

2 Selection Criteria

Part of the success in genetic improvement is based on the analysis of a high number of individuals and progenies. Finding criteria for easy measurement and simple evaluation environments, that allow predicting field behavior, are key for plant breeding. Salt glands, a feature associated with salinity tolerance, are among characters that are easy to measure, as salt crystals on the leaf surface indicate the presence of underlying salt glands. Salt glands were described by Taleisnik and Anton (1988) in *Pappophorum philippianum*. The authors observed that the adaxial surface of *P. philippianum* leaves had three times higher salt gland density than *P. pappiferum*, and higher salt tolerance correlated with that character Marcum et al. (1998) found a similar association in *Zoysia* species, Pérez et al. (2009) and Loch and Zorin (2010) in *Chloris gayana*.

In the *Trichloris genus, T. crinita* showed higher density of these glands in relation to *T. pluriflora* and also greater salinity tolerance (Fig. 1) (Zabala et al. 2011). Naz et al. (2009) found differences between species in their glands ability to discriminate against the Na^+ excretion from other ions such as Ca^{2+} and K^+ in grass salt glands and

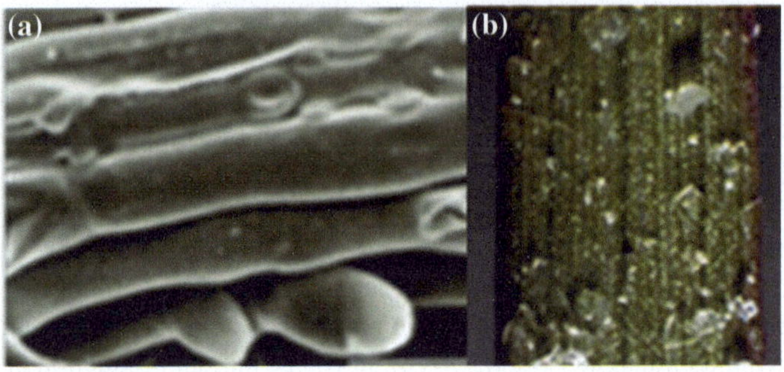

Fig. 1 Salt glands **a** and salt crystals **b** on the surface of *Trichloris crinita* leaves. *Photographs* J. M. Zabala

also plastic responses to environmental conditions (Shabala 2013). It is considered that until molecular mechanisms that mediate Na^+ excretion are elucidated, it may be practical to use salt glands characteristics (such as their number) as a selection criterion to improve the Na^+ balance in grass leaves.

Maas and Hoffman (1977) indicated that salinity sensitivity may change throughout ontogeny, yet a favorable association that has often been observed is that germination behavior and initial growth, taken together, are positively correlated with forage production in salinity conditions (Roundy 1985; Krishnamurthy et al. 2007; Yu et al. 2016). However, when germination behavior is analyzed, without considering the relation to initial growth, its association with salinity tolerance is controversial (Läuchli and Grattan 2007). Multiple evidences of salinity effects on germination and establishment underscore limitations that must be overcome for the success of degraded grasslands restoration programs or pasture implantation (Ungar 1995; Khan and Gulzar 2003; Lin and Tang 2005; Guan et al. 2009; Ruiz and Terenti 2012). It is difficult to find well-defined plant indicators for salinity tolerance that could practically be used by plant breeders, partly due the fact that the mechanisms of salt tolerance are so complex (Ashraf and Harris 2004). However, there are some proposed characteristics as: glycinebetaine (quaternary ammonium compound) and proline (amino acid) accumulation, or increasing activities of antioxidant enzymes superoxide dismutase (SOD), ascorbate peroxidase (APX) , and guaiacol peroxidase (GP), the maintenance of the K/Na ratio, membrane stability, and a reduction in chlorophyll loss (Ashraf and Harris 2004; Ahmed et al. 2013; Chunthaburee et al. 2016).

3 Molecular Tools Applied to Plant Breeding

Molecular techniques allow both the variability estimation and the possibility of assisted selection. Thanks to a significant reduction in costs, the use of anonymous markers has been changed to the possibility of sequencing massively and applying genomic selection. For example, in Alfalfa, full genome association studies (GWAS) and genotyping by sequencing (GBS) have been carried out on germination behavior in salinity conditions generating numerous markers to apply assisted selection (MAS) (Yu et al. 2016). On the other hand, gene technologies give the possibility of generating new variants through gene delivery or in the future through gene editing, the latter with the advantage of being able to restrict genome modification to a specific site (Liu et al. 2013). This chapter shows examples of studies using cytological-molecular, molecular , and gene technology techniques in two native species:

3.1 Molecular Cytology and Molecular Techniques Applied in **Elymus scabrifolius**

Argentine wheatgrass (*Elymus scabrifolius* (Doll) H.Z.) ($2n = 4x = 28$) is a forage grass native to Argentina and Uruguay, perennial, self-fertile and autumn–winter-spring production. Covas (1978) considered it "one of the most productive grasses in the wintering region of our country." Tomas et al. (2012) were able to identify each of the species' chromosomes and determine their allopolyploid origin (Fig. 2a, b). By applying markers (AFLPs) (Fig. 2c, d), it was possible to estimate the genetic distances between the accessions collected (Tomas et al. 2013) and to design crosses among genotypes that were most different at the molecular level in order to generate variability (Fig. 2e). F1 embryos were rescued and these plants showed normal meiosis (Fig. 2e). After seven generations following the single seed descent (SSD) method, transgressive combinations were found, both under salinity conditions and without saline stress (Fig. 2f), with respect to both parents who stood out for their biomass production.

Zabala (2016) determined the Na exclusion as the main mechanism of salinity tolerance in this species. He found a high heritability and a simple genetic control for this character, however he could not find associated sequence-related amplified polymorphism (SRAP) markers. In *Arabidopsis thaliana*, it was found that salinity tolerance was increased in a regulatory way operated by mRNA, and not due to differences in coding sequences (Chen et al. 2015). Since mRNAs are distributed in repetitive DNA regions (Zhang et al. 2009), it is possible to consider that the absence of association between SRAPs and sodium exclusion in Argentine wheatgrass is due to the fact that SRAPs are not located in repetitive DNA regions, but are located in regions encoding active transcription (Aneja et al. 2012). Alonso Nogara et al. (2016) obtained recombinant inbred lines (RILs) which show a wide phenotypic variation. Through their transcriptomic analysis, it will be tested whether it is the differential expression of Na transporters which explains the variation range in the response to salinity (Fig. 2f).

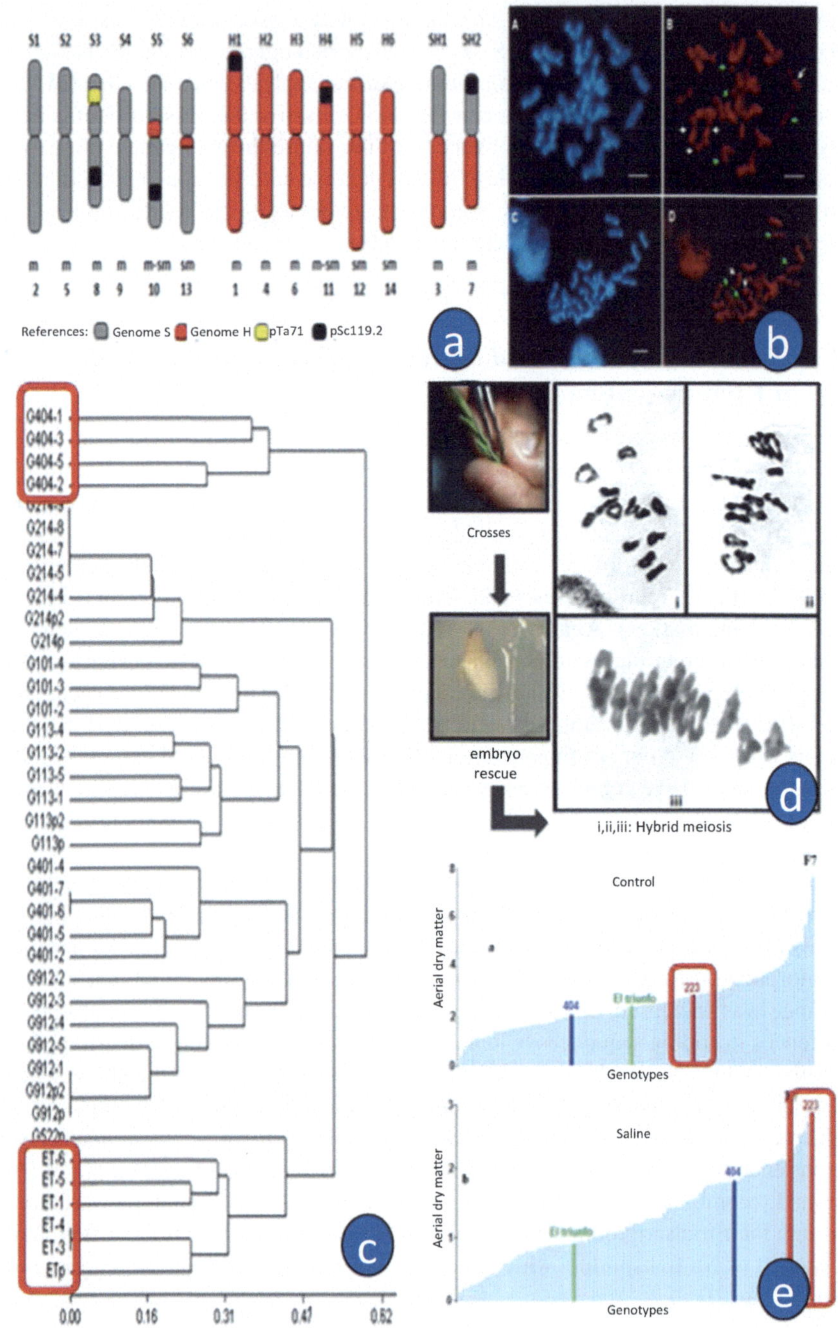
S1
S2
S3
S4
S5
S6
H1
H2
H3
H4
H5
H6
SH1
SH2
m m m m m-sm sm m m m m-sm sm sm m m
2 5 8 9 10 13 1 4 6 11 12 14 3 7
References: Genome S Genome H pTa71 pSc119.2
a
A
B
C
D
b
G404-1
G404-3
G404-5
G404-2
G214-8
G214-7
G214-5
G214-4
G214p2
G214p
G101-4
G101-3
G101-2
G113-4
G113-2
G113-5
G113-1
G113p2
G113p
G401-4
G401-7
G401-6
G401-5
G401-2
G912-2
G912-3
G912-4
G912-5
G912-1
G912p2
G912p
ET-6
ET-5
ET-1
ET-4
ET-3
ETp
0.00
0.16
0.31
0.47
0.62
c
Crosses
embryo rescue
i
ii
iii
d
i,ii,iii: Hybrid meiosis
Control
Aerial dry matter
404
El triunfo
223
Genotypes
Saline
Aerial dry matter
404
El triunfo
223
Genotypes
e

◀**Fig. 2** **a** Karyotype of *E. scabrifolius* showing the alopoliploid origin, **b** FISH-GISH techniques applied in chromosomal identification. *Photograph* P. Tomas. **c** Dendrogram showing the distances between genotypes estimated from molecular markers (AFLPs), the extreme genotypes used in the crosses are indicated with red. **d** Castration and cross-breeding technique, embryo rescue and meiotic analysis. *Photograph* P. Tomas. **e** Production of parental dry matter and F7 progenies obtained by applying the SSD methodology (a transgressive genotype in both conditions is indicated by red as an example)

3.2 Molecular Markers, Transcriptomic and Gene Technology in **Paspalum dilatatum**

Dallisgrass (*Paspalum dilatatum* Poir.) is a South American native C4 grass species, with spring–summer–autumn growth, and a strong candidate to cover the summer productivity deficit in pastoral systems. Within the species, the predominant reproductive system is apomixis (aposporic pseudogamic), and this reproductive mode implies a difficulty for improvement since it does not allow crossings and usually results in a population variability reduction. But within *P. dilatatum*, there are sexual forms that have allowed introgressions with other species and the generation of high variability. High molecular variation (AFLPs) observed within the progenies originated through crosses between *P. dilatatum* Biotype Virasoro and *Paspalum urvillei* (Schrauf et al. 2017). The use of microsatellite markers developed by Speranza and Malosetti (2007) allows to verify the hybridization of the inter- and intraspecific cross-breeding progenies. Within the forage species genetic improvement programs of the FAUBA Genetics Department, two cultivars have been obtained, "Relincho" (INaSe 2003---apomictic pentaploid genotype) with high capacity to establish and "Primo" (INase 2013---sexual tetraploid genotype) which incorporated the low susceptibility to *Claviceps paspali* of *P. urvillei* and shows a high productive potential.

AgriBio–Australia and FAUBA-Argentina, carried out transcriptomic studies in "Primo" cultivar (Giordano et al. 2014). The cDNA analysis allowed to obtain key basic information for the future improvement of the species, 324,695 readings were made with the average length of 338pb, when assembled, reached a number of 33,922 "singletons." Since both registered cultivars have a low tolerance to salinity and that the natural variability for this character is limited within the species, transgenesis becomes a useful tool to overcome this barrier.

One strategy to increase salinity tolerance is the introduction of *Arabidopsis thaliana* NHX genes that code for sodium/proton vacuolar antiporters (Blumwald et al. 2000). The objective of the work was to increase the salinity tolerance through the incorporation via transgenesis of the *Atnhx*1 and *Atnhx*5 genes. The *P. dilatatum* "Relincho" cultivar was genetically transformed with the Atnhx1 and Atnhx5 genes (Fig. 3a). The transformation of the plants was carried out by biolistics on embryogenic calli. These were induced from mature embryos and were proliferated and regenerated according to the technique described by Schrauf (2009). The genes presence in the plant genome was confirmed by molecular analysis. *At*nhx1 or *At*nhx5

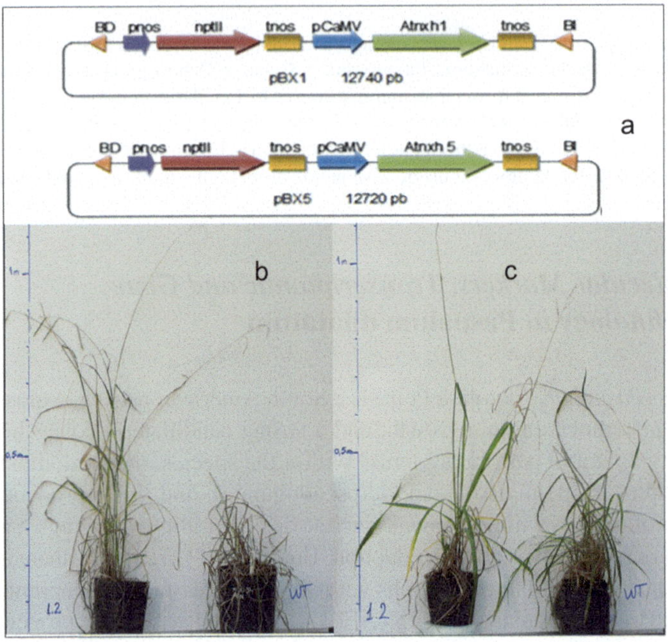

Fig. 3 Transformation of *Paspalum dilatatum* "Relincho" cultivar with the Atnhx1 and Atnhx5 genes. **a** Chimeric constructs used for the biolistic transformation, **b** transgenic (+) and wild-type (−) plants subject to salinity, **c.** (+) and (−) plants under non-saline conditions. *Photographs* (b) and (c): GE Schrauf

positive plants were vegetatively propagated and placed in terrines subjected to weekly 50 mM Na concentration increases until 250 mM concentration was achieved, control terrines were not salinized. The transgenic plants showed qualitative differences with respect to the wild-type plants (wt) under salinity (Fig. 3b) and did not differ under non-saline conditions (Fig. 3c).

4 Breeding Methodologies for Perennial Forage Species

Techniques such as crossings, mutagenesis, transgenesis, or gene editing are sources of variation, and usually the new variant must be included in a plant breeding program to generate a new cultivar. The methodology for the development of a cultivar depends mainly on the reproductive system of the species. In a species of agamic or apomictic reproduction, the variant can quickly be multiplied, evaluated, and converted into a new cultivar. Species of sexual reproduction require several generations until the cultivar is obtained. Furthermore, the breeding method can be very different depending on whether the species is autogamous or allogamous. Here, only some methodologies that were implemented by the FAUBA working group are described. Autogamous (self-compatible) populations are usually conducted in different ways, mass, genealogical, or SSD, trying to obtain a superior genotype.

The FAUBA Department of Genetics in its breeding program for *Bromus catharticus* developed a method in which instead of trying to obtain the superior genotype, it is sought to obtain the best combination of genotypes set so as to make efficient use of the resources (complementary genotype selection method—CGSM). "Quintun" and "Quidel" cultivars were obtained by applying the CGSM. Figure 4a shows the results of the trials network of the "Cámara de Semilleristas de la Bolsa de Cereales-Argentina", where both "Quintun" and "Quidel" cultivars obtained the first two positions and significantly differentiating themselves from the rest of the cultivars. One of the circles shows a mixture of "Ñandú" and "Copetona" cultivars as Exp 287, which shows a lower yield compared to pure cultivars. That is, in the same trial, not only was the value of the complementary genotype selection method demonstrated, but it is not a matter of mixing genotypes but also of mixing complementary genotypes (Garyulo et al. 2006). The genotypes included in the "Quintun" and "Quidel" cultivars were selected because they used resources such as light, water, and nutrients in a different and complementary way. More recently, Zabala et al. (2018) applied CGSM in Sweet clover (*Melilotusalbus)*, an annual legume of saline environments, by mixing two populations that complemented their temporary production they found a greater production in mixtures of 50% compared to pure populations. This species was introduced in the 1950s and since then it was not the target of planned breeding. Instead a greater adaptation to saline and semi-arid environments and a greater seed production were increased, probably through unconscious selection, and surely in detriment of the production of forage. Together with the Faculty of Agronomy of the UBA and the Faculty of Agricultural Sciences of the UNL, a plant breeding program of *M. albus* was generated trying to capture the adaptation and reverse the reduction in the leaf/stem ratio; it was initially selected by late flowering, but a negative correlation was found with the production of fodder, so the criterion of selection of number of basal ramifications was incorporated. The progress in the selection led to the obtaining of highly productive materials and also with high leaf/stem ratios, that is, to say with greater response to defoliation and higher forage quality that was called "Faraón" (INaSe 2007a; Zabala et al. 2012) (Fig. 4).

In allogamous species, populations are usually improved through Phenotypic Recurrent Selection Methods or Progeny Tests. Although the second methodology requires two years for each selection cycle, it has proved advantageous with respect to phenotypic recurrent selection or mass selection when the heritability is not high (Palmieri 2009). Obtaining hybrids can be done in different ways, exploiting androsterility (cytoplasmic, nuclear or barnase/barstar system) or self-incompatibility. This last route made it possible to produce hybrids from vegetatively propagated self-incompatible clones. Having generated knowledge about the genetic control of self-incompatibility allows currently planning crossings and using molecular information to generate hybrids (Aghili et al. 2015). Synthetic varieties and hybrids are generated by combinatorial aptitude tests. But for both the recurrent selection and for the aptitude tests, the evaluation environments are generally in an isolated plant or pure plot, far from real production situations. In white clover (*Trifolium repens*),

CULTIVAR	YEAR 1	YEAR 2	YEAR 3	TOTAL
"Quidel"	8.34 a	13.21 ab	8.53 a	30.08a
"Quintun"	8.22 ab	13.88 a	7.96 ab	30.06a
Rosabel	7,31 cd	13,03 ab	7.92 ab	28,26 b
Copetona	7,42 cd	12,28 bcd	8.42 a	28,12 b
Don Enrique (B. parodii)	7,03 cd	12,15 bcd	8.57 a	27,74 b
Fierro Plus INTA (T)	7.49 bcd	12.67 bc	7.17 b	27.32 bc
Ñandú	7,50 bcd	11,85 cd	7,80 ab	27,15 bc
Tijereta (T)	7,79 abc	12,60 bc	6,22 c	26,60 bc
Exp. 287	7,24 cd	11,20 d	7,39 b	25,83 cd
Maitén	6,75 d	10,11 e	7,50 b	24,35 d
L.S.D. (0,05)	0,800	1,030	0,820	1,714
C.V.%	7.33	5.75	7.28	4.29

Fig. 4 Results of the trials network of the "Cámara de Semilleristas de la Bolsa de Cereales-Argentina"

a synthetic variety was generated from clones selected under competition from a *Festuca arundinacea* canopy, the progeny of these clones generated progeny with a greater general combinatorial ability for forage production with respect to those selected clones in conditions of isolated plant, generating the cultivar "Junin" (INaSe 2007b). This methodology is feasible to apply in a similar way by selecting *Lotus* spp. materials in wheatgrass or Argentine wheatgrass canopies, that is, generating evaluation environments more similar to the production environment (grass/legume mixtures). Hill (1990) raised the need for a double program by jointly selecting both components of the pasture both legume and grass.

To incorporate variation generated by mutations, transgenes or edited genes backcrossing is the most applied methodology although it may require special effort for self-incompatible allogamous plants such as many of the perennial forage grasses. Currently, it is not the elite materials that are transformed but the events obtained must be used as donor genotype and the elite materials used as recurrent genotype in the backcross method. Here, molecular markers can be useful not only in the detection of the gene to be transferred by backcrossing, but especially to recover in a few generations the "background" of the elite material. In the near future, it may be possible to transform directly to elite materials by increasing the totipotence of their tissues as stated by Lowe et al. (2018).

5 Genomic Selection

Genomic selection is a modified version of marker-assisted selection (MAS) as a way of integrating whole genome panels of molecular markers into the prediction of genetic values. The approach as a concept was introduced in Meuwissen et al. (2001) and differs from traditional MAS in that it does not use previously identified regions or QTLs, but rather uses all markers available to inform breeding choices. In genomic selection, a reference population, which has both genotypic and phenotypic records, is used to train a predictive model. This model is then used to predict the so-called genomic estimated breeding value (GEBV) for each genotype, even for those without direct phenotypic record. Then, breeding choices can be assisted by this GEBV. Figure 5 shows the basics of a genomic selection implementation.

The novelty of genomic selection is the realization that markers do not need to form part of genes or regulatory segments for predictions to be effective. As long as the marker panel covers the genome with a high enough density, many markers will be physically close regions of DNA responsible for the relevant variations. Nevertheless, proximity is not enough: in order to possess predictive value, the markers must be in linkage disequilibrium with the causal loci. That is, some marker alleles must be statistically associated with the polymorphisms affecting the trait. This statistical linkage is shorter in range in outbred populations (it rapidly decays with increasing physical linkage), an issue which needs consideration when applying genomic selection to such populations as ryegrass (Pembleton et al. 2018). It is also important that, during selection in a population, recombination and changes in allele frequencies

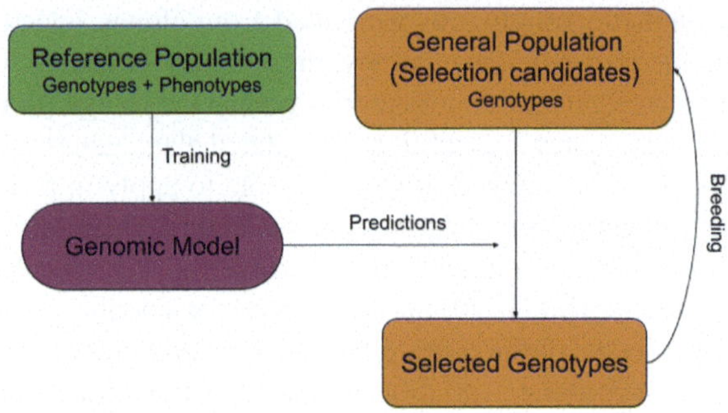

Fig. 5 Basics of a genomic selection implementation

occur with the consequence that genomic models need updating to remain accurate through generation. Some of the first empirical evidence that genomic selection works was obtained for dairy cattle (VanRaden et al. 2009; Hayes et al. 2009) and similar evidence is now available for many crops (Lorenzana and Bernardo 2009; de los Campos et al. 2009; Lorenz et al. 2012).

Genomic selection is a versatile technique which does not require knowledge of the genetic basis for the traits of interest. This allows its use in emerging crops, or for selection in novel and poorly understood traits. When there is physiological knowledge of the trait, or even identified QTLs, this can be used to enhance the prediction models. For an example of the first, integrating crop ecophysiology has been proposed to better select for drought adaptation in *Sorghum* (Hammer et al. 2016). For an example of the second, inclusion of genotypes for known rust resistance genes increased accuracy of genomic models in wheat landraces (Daetwyler et al. 2014). More generally, the contribution of genomic selection is of higher impact when: (a) the traits of interest are of costly measurement (e.g., require long field trials) or can only be measured late in the crop development (e.g., in perennial crops); (b) the generational interval is lengthy (time to reproductive maturity); (c) early selection (e.g., at a seedling stage) is valuable. In consequence, genomic selection can be a valuable approach to breed perennial forage adapted to salinity.

6 Future Perspectives

The omics (genomics, transcriptomics, proteomics, metabolomics, and phenomics) will have a high impact and will become indispensable tools of any plant breeding program. Genomics provides the basic information and raw material for improvement, that is why Spangenberg et al. (2001) developed a program to search for salinity resistance genes in *Agrostis adamsonii* as the Australian grassland grass representative. In Argentina, numerous species that successfully colonize saline environments have been described (Marinoni et al. 2019) that present a forage value per se but can also be considered as a source of genes for different salinity tolerance mechanisms, potential candidates for genomic studies.

Pastures composed of mixtures of legumes and grasses offer a unique opportunity in genomic investigations to study plant-pathogen, legume-nitrogen-fixing bacteria, legume-mycorrhizae, grasses-endophyte symbionts interactions. These studies will generate knowledge applied both for the development of resistance to pathogens and to improve beneficial associations in forages (Spangenberg 2005).

The study of the metabolites of various salinity-resistant plants allows discovering and proposing new strategies to provide tolerance to pastures of interest through their genetic modification. The metabolomics is the most transversal of all the "omics," since the metabolites reflect the integration of genetic expression, protein interaction, and other different regulatory processes and therefore are closer to the phenotype than mRNA transcripts (transcriptomic) or proteins alone (proteomics). Satisfactory results can be obtained by analyzing the molecular phenotypes of plants in responses to abiotic stresses in order to find patterns associated with stress tolerance and use them to introduce them into non-resistant forage plants (Arbona et al. 2013).

The association of different types of molecular markers (especially SSRs and SNPs) with the expression of these metabolites allows to predict an acceleration of progress in the improvement. While in the gene edition technique it is where the highest expectations are deposited, since it adds to the transgenesis technique the ability to direct the gene modification at specific sites and promptly change regulatory sequences that are expected to involve overexpression of the genes associated with salinity tolerance without these modifications involving transgenesis. Having described the genes responsible for excluding Na and conferring significant increases in salinity performance (Munns et al. 2012) makes it possible to propose that gene editing constitute a successful alternative in improving salinity tolerance (Schroeder et al. 2013).

References

Aghili H, Shinozuka H, Baillie R, Drayton M, Shinozuka M, Bain M, Ran Y, Wijesinghe N, Spangenberg G, Slater A, Cogan N, Forster J (2015) Genetics and genomics of self-incompatibility in forage grasses. In: International symposium of forage breeding. Buenos Aires, Argentina

Ahmed IM, Cao F, Zhang M, Chen X, Zhang G, Wu F (2013) Difference in yield and physiological features in response to drought and salinity combined stress during anthesis in Tibetan wild and cultivated barleys. PLoS ONE 8:e77869

Alonso Nogara F, Zabala JM, Rush P, Schrauf GE (2016) Comparación de caracteres productivos relacionados a la tolerancia a la salinidad en dos filiales (F2 y F7) conducidas mediante SSD. In: IV Reunión de la RAS (Red Argentina de Salinidad). Reconquista, Santa Fe, Argentina

Aneja B, Yadav NR, Chawla V, Yadav RC (2012) Sequence-related amplified polymorphism (SRAP) molecular marker system and its applications in crop improvement. Mol Breed 30:1635–1648

Arbona V, Manzi M, de Ollas C, Gómez-Cadenas A (2013) Metabolomics as a tool to investigate abiotic stress tolerance in plants. Int J Mol Sci 14:4885–4911

Ashraf MP, Harris PJ (2004) Potential biochemical indicators of salinity tolerance in plants. Plant Sci 166(1):3–16

Berdahl JD, Barker RE (1978) Genetic improvement of forage grasses. North Dakota Farm Res 36:31–34

Blumwald E, Aharon GS, Apse MP (2000) Sodium transport in plant cells. Biochim Biophys Acta Biomembr 1465:140–151
Chen Z, Hu L, Han N, Hu J, Yang Y, Xiang T, Zhang X, Wang L (2015) Overexpression of a miR393-resistant form of *Transport Inhibitor Response Protein 1 ($MTIR_1$)* enhances salt tolerance by increased osmoregulation and Na^+ exclusion in *Arabidopsis thaliana*. Plant Cell Physiol 56:73–83
Chunthaburee S, Dongsansuk A, Sanitchon J, Pattanagul W, Theerakulpisut P (2016) Physiological and biochemical parameters for evaluation and clustering of rice cultivars differing in salt tolerance at seedling stage. Saudi J Biol Sci 23:467–477
Covas G (1978) Forrajeras indígenas. Especies que requieren un plan de conservación de germoplasma. Cienc Invest 34:209–213
da Silva S, Sbrissia A, Pereira L (2015) Ecophysiology of C4 forage grasses—understanding plant growth for optimising their use and management. Agriculture 5:598–625
Daetwyler HD, Bansal UK, Bariana HS, Hayden MJ, Hayes BJ (2014) Genomic prediction for rust resistance in diverse wheat landraces. Theor Appl Genet 127:1795–1803. https://doi.org/10.1007/s00122-014-2341-8
de Los Campos G, Naya H, Gianola D, Crossa J, Legarra A, Manfredi E, Weigel K, Cotes JM (2009) Predicting quantitative traits with regression models for dense molecular markers and pedigree. Genetics 182:375–385
Díaz ML, Echenique CV, Schrauf G, Cardone S, Lutz EE, Spangenberg G (2004) Biotecnología y mejoramiento genético de especies forrajeras. RIA 33(3):77–104
Garyulo M, Marcantoni F, Figueras I, Rush P, Capurro P, Schrauf GE (2006) Mejora del rendimiento en Cebadilla criolla (*Bromus catharticus*) a través de la selección de genotipos complementarios. In: III Congreso Científico y Cultural. Ciudad del Beni –Bolivia, pp 1–15
Giordano A, Cogan NOI, Kaur S, Drayton M, Mouradov A, Panter S, Schrauf GE, Mason JE; Spangenberg G (2014) Gene discovery and molecular marker development, based on high-throughput transcript sequencing of *Paspalum dilatatum* Poir. PLOS ONE 9(2):e85050
Guan B, Zhou D, Zhang H, Tian Y, Japhet W, Wang P (2009) Germination responses of *Medicago ruthenica* seeds to salinity, alkalinity, and temperature. J Arid Environ 73:135–138
Hammer G, Messina C, van Oosterom E, Chapman S, Singh V, Borrell A, Jordan D, Cooper M (2016) Molecular breeding for complex adaptive traits: how integrating crop ecophysiology and modelling can enhance efficiency. In: Yin X, Struik P (eds) Crop systems biology. Springer, Cham
Hayes BJ, Bowman PJ, Chamberlain AJ, Goddard ME (2009) Invited review: genomic selection in dairy cattle: progress and challenges. J Dairy Sci 92:433–443
Hill J (1990) The three C's—competition, coexistence and coevolution—and their impact on the breeding of forage crop mixtures. Theor Appl Genet 79:168–176
INaSe (2003) Inscripción del cultivar "RELINCHO" de pasto miel (*Paspalum dilatatum*) en los Registros de la Propiedad de Cultivares y Nacional de Cultivares del INaSe. Record number 7774
INaSe (2007a) Inscripción del cultivar "FARAÓN" de trébol de olor blanco (*Melilotus albus*) en los Registros de la Propiedad de Cultivares y Nacional de Cultivares del INaSe. Record number 9085
INaSe (2007b) Inscripción del cultivar "JUNÍN" de trébol blanco (*Trifolium repens*) en los Registros de la Propiedad de Cultivares y Nacional de Cultivares del INaSe. Record number 9547
INaSe (2013) Inscripción del cultivar "PRIMO" de pasto miel (*Paspalum dilatatum*), en los Registros de la Propiedad de Cultivares y Nacional de Cultivares del INaSe. Record number 12252
Khan MA, Gulzar S (2003) Light, salinity, and temperature effects on the seed germination of perennial grasses. Am J Bot 90:131–134
Krishnamurthy L, Serraj R, Hash CT, Dakheel AJ, Reddy BVS (2007) Screening sorghum genotypes for salinity tolerant biomass production. Euphytica 156:15–24
Läuchli A, Grattan SR (2007) Plant growth and development under salinity stress. In: Jenks MA, Hasegawa P, Mohan Jain S (eds) Advances in molecular breeding toward drought and salt tolerant crops. Springer, pp 1–32
Lin N, Tang JL (2005) Study on the environment evolution and the analysis of causes to land salinization and desertification in Songnen plain. Quater Sci 25:474–483

Liu W, Yuan JS, Stewart CN (2013) Advanced genetic tools for plant biotechnology. Nat Rev Genet 14:781–793

Loch DS, Zorin M (2010) Development of new tetraploid *Chloris gayana* cultivars with improved salt tolerance from 'Callide' and 'Samford'. Int Herb Seed Conf 190–194

Lorenz AJ, Smith KP, Jannink J-L (2012) Potential and optimization of genomic selection for *Fusarium* head blight resistance in six-row barley. Crop Sci 52:1609–1621

Lorenzana RE, Bernardo R (2009) Accuracy of genotypic value predictions for marker-based selection in biparental plant populations. Theor Appl Genet 120:151–161

Lowe K, La Rota M, Hoerster G, Hastings C, Wang N, Chamberlin M, Wu E, Jones T, Gordon-Kamm W (2018) Rapid genotype "independent" *Zea mays* L. (maize) transformation via direct somatic embryogenesis. In Vitro Cell Dev-Pl 54:240–252

Maas EV, Hoffman GJ (1977) Crop salt tolerance–current assessment. J Irrig Drain Div 103:115–134

Marcum KB, Anderson SJ, Engelke MC (1998) Salt gland ion secretion: a salinity tolerance mechanism among five zoysiagrass species. Crop Sci 38:806–810

Marinoni LDR, Zabala JM, Taleisnik EL, Schrauf GE, Richard GA, Tomas PA, Giavedoni JA, Pensiero JF (2019) Wild halophytic species as forage sources: key aspects for plant breeding. Grass Forage Sci 74:321–344

Meuwissen THE, Hayes BJ, Goddard ME (2001) Prediction of total genetic value using genome-wide dense marker maps. Genetics 157:1819–1829

Munns R, James RA, Xu B, Athman A, Conn SJ, Jordans C, Byrt CS, Hare RA, Tyerman SD, Tester M, Plett D, Gilliham M (2012) Wheat grain yield on saline soils is improved by an ancestral Na^+ transporter gene. Nat Biotechnol 30:360–364

Naz N, Hameed M, Wahid A, Arshad M, Aqeel Ahmad MS (2009) Patterns of ion excretion and survival in two stoloniferous arid zone grasses. Physiol Plant 135:185–195

Oba M, Allen MS (1999) Evaluation of the importance of the digestibility of neutral detergent fiber from forage: effects on dry matter intake and milk yield of dairy cows. J Dairy Sci 82:589–596

Oliver AL, Pedersen JF, Grant RJ, Klopfenstein TJ (2005) Comparative effects of the sorghum bmr-6 and bmr-12 genes: I. Forage sorghum yield and quality. Crop Sci 45:2234–2239

Palmieri JL (2009) Evaluación de progenies de clones selectos de agropiro alargado (*Thinopyrum ponticum*) bajo condiciones de salinidad. Undergraduate Thesis, Facultad de Agronomía, Universidad de Buenos Aires

Pembleton LW, Inch C, Baillie RC, Drayton MC, Thakur P, Ogaji YO, Spangenberg GC, Forster JW, Daetwyler HD, Cogan NOI (2018) Exploitation of data from breeding programs supports rapid implementation of genomic selection for key agronomic traits in perennial ryegrass. Theor Appl Genet 131:1891–1902

Pérez H, Taleisnik E, Pemán R (2009) Development of a tetraploid salt-tolerant *Chloris gayana* cultivar. In: II Simpósio Internacional sobre Melhoramento de Forrageiras. Embrapa Gado de Corte, Campo Grande, Brazil

Roundy BA (1985) Root penetration and shoot elongation of tall wheatgrass and basin wildrye in relation to salinity. Can J Plant Sci 65:335–343

Ruiz M, Terenti O (2012) Germination of four grasses under salt stress. Phyton (Buenos Aires) 81:169–176

Schrauf G, Alonso Nogara F, Rush P, Peralta Roa P, Musacchio E, Ghio S, Giordano A, Giavedoni JA, Pensiero JF, Tomas PA, Zabala JM, Spangenberg G (2017) Mejoramiento de forrajeras para tolerancia a la salinidad. In: Taleisnik E, Lavado R (eds) Ambientes salinos y alcalinos de la Argentina: Recursos y aprovechamiento productivo. Orientación Gráfica Editora and Universidad Católica de Córdoba, Buenos Aires, pp 467–484

Schrauf GE (2009) Aplicación de métodos convencionales y biotecnológicos en el mejoramiento de *Paspalum dilatatum* Poir. Doctoral Thesis, Universidad de Buenos Aires

Schroeder JI, Delhaize E, Frommer WB, Lou GM, Harrison MJ, Herrera-Estrella L, Horie T, Kochian LV, Munns R, Nishizawa NK, Tsay YF, Sanders D (2013) Using membrane transporters to improve crops for sustainable food production. Nature 497:60–66

Shabala S (2013) Learning from halophytes: physiological basis and strategies to improve abiotic stress tolerance in crops. Ann Bot 112:1209–1221

Snaydon R (1985) Aspects of the ecological genetics of pasture species. In: Haeck J, Woldendorp JW, Holl N (eds) Structure and functioning of plant populations. North-Holland Pub. Co., Amsterdam

Spangenberg G (2005) Transgenesis and genomics in molecular breeding of pasture grasses and legumes for forage quality and other traits. In: Makkar HP, Viljoen GJ (eds) Applications of gene-based technologies for improving animal production and health in developing countries. Springer, Berlin, pp 357–372

Spangenberg G, Kalla R, Lidgett A, Sawbridge T, Ong EK, John U (2001) Breeding forage plants in the genome era. In: Spangenberg GC (ed) Molecular breeding of forage crops. Developments in plant breeding. Springer, Dordrecht, pp 1–39

Speranza P, Malosetti M (2007) Nuclear and cytoplasmic microsatellite markers for the species of the Dilatata group of *Paspalum* (Poaceae). Plant Genet Resour C 5:14–26

Striker GG, Insausti P, Grimoldi AA, Leon RJC (2006) Root strength and trampling tolerance in the grass *Paspalum dilatatum* and the dicot *Lotus glaber* in flooded soil. Funct Ecol 20:4–10

Taleisnik EL, Anton AM (1988) Salt glands in *Pappophorum* (Poaceae). Ann Bot 62:383–388

Tomas PA, González GE, Schrauf GE, Poggio L (2012) Chromosomal characterization in native populations of *Elymus scabrifolius* from Argentina through classical and molecular cytogenetics (FISH–GISH). Genome 55:591–598

Tomas PA, Gottlieb AM, Schrauf GE, Poggio L (2013) Utilization of morphological and AFLP molecular markers in the identification of native and cultivated germplasm of *Elymus scabrifolius* (Poaceae) | Utilización de marcadores morfológicos y moleculares AFLP en la identificación de germoplasma nativo y cult. Rev Fac Cienc Agrar 45:85–100

Ungar IA (1995) Seed germination and seed-bank ecology in halophytes. In: Kigel J (ed) Seed development and germination. Routledge, New York, p 30

VanRaden PM, Van Tassell CP, Wiggans GR, Sonstegard TS, Schnabel RD, Taylor JF, Schenkel FS (2009) Invited review: reliability of genomic predictions for North American Holstein bulls. J Dairy Sci 92:16–24

Yu L-X, Liu X, Boge W, Liu X-P (2016) Genome-wide association study identifies loci for salt tolerance during germination in autotetraploid Alfalfa (*Medicago sativa* L.) using genotyping-by-sequencing. Front Plant Sci 7:956

Zabala JM (2016) Bases fisiológicas y control genético de la exclusión de sodio en agropiro criollo *Elymus scabrifolius* (Doll) Hunz. Doctoral thesis. Universidad Nacional del Litoral, Argentina

Zabala JM, Delbino M, Giavedoni JA, Pensiero JF (2011) Glándulas de sal como criterio de selección y evaluación de la tolerancia a la salinidad en *Trichloris crinita* y *Trichloris pluriflora*. In: 2da Reunión de la Red Argentina de Salinidad (RAS). San Miguel de Tucumán, Argentina

Zabala JM, Marinoni L, Giavedoni JA, Schrauf GE (2018) Breeding strategies in *Melilotus albus* Desr., a salt-tolerant forage legume. Euphytica 214:1–15

Zabala JM, Schrauf G, Baudracco J, Giavedoni J, Quaino O, Rush P (2012) Selection for late flowering and greater number of basal branches increases the leaf dry matter yield in *Melilotus albus* Desr. Crop Pasture Sci 63:370–376

Zhang L, Chia J-M, Kumari S, Stein JC, Liu Z, Narechania A, Maher CA, Guill K, McMullen MD, Ware D (2009) A genome-wide characterization of microRNA genes in maize. PLoS Genet 5:1000716

Antioxidant Mechanisms Involved in the Control of Cowpea Root Growth Under Salinity

Josemir Moura Maia, Cristiane E. C. Macedo, Ivanice da Silva Santos, Yuri Lima Melo, and Joaquim A. G. Silveira

Abstract Salinity affects plant productivity in all agricultural crops. In the Brazilian semiarid region, considered one of the most salinized areas in Brazil, the cultivation of cowpea beans (*Vigna unguiculata*) is of paramount importance for the region's economy and food security. However, the development of the plant roots is affected by the inherent salinity of the soils. In addition to its direct effects, salt stress can cause osmotic stress and oxidative stress. The latter occurs when there is an imbalance between the production and removal of reactive oxygen species (ROS), which in general is a sign of stress in the plant, as in excess ROS can cause oxidative damage in cell membranes. Saline stress in the Brazilian semiarid region can cause a reduction in the size of cowpea roots but without apparent peroxidation of the membranes, suggesting that stress caused by salinity does not induce oxidative damage in root membrane lipids. Thus, this work presents evidence that the balance between the production and removal of ROS is crucial for responses related to the vegetative growth of cowpea roots.

Keywords Antioxidant enzymes · NADPH oxidase · Reactive oxygen species

J. M. Maia (✉) · I. da Silva Santos
Laboratory of Plant Production Technologies, Universidade Estadual da Paraíba, Campus IV, Catolé do Rocha, Paraíba, Brazil
e-mail: jmouram@uepb.edu.br

I. da Silva Santos
e-mail: ivaniceagrarias@gmail.com

C. E. C. Macedo
Laboratory of Plant Biotechnology Studies, Universidade Federal do Rio Grande do Norte, Natal, Rio Grande do Norte, Brazil
e-mail: macedocristianee@gmail.com

Y. L. Melo
Laboratory of Ecophysiology of Cultivated Plants, Universidade Estadual da Paraíba, Campina Grande, Paraíba, Brazil
e-mail: yurimelo86@gmail.com

J. A. G. Silveira
Laboratory of Plant Metabolism, Universidade Federal do Ceará, Fortaleza, Ceará, Brazil
e-mail: silveira@ufc.br

E. Taleisnik and R. S. Lavado (eds.), *Saline and Alkaline Soils in Latin America*, https://doi.org/10.1007/978-3-030-52592-7_21

1 Introduction

Soil salinity is one of the main problems facing farmers in Latin America and the Caribbean, affecting plant productivity in all agricultural crops. According to Sá et al. (2010), the main salinized areas in Brazil are found in the Brazilian semiarid region. Considered one of the largest semiarid regions on the planet, it is located almost entirely in the northeast of Brazil and has remarkable climatic conditions with variable rainfall associated with high temperatures during much of the year (Souza et al. 2016). In addition to these characteristics, the relief is irregular and the soils are shallow with low fertility and little organic matter content (Souza et al. 2016). Even so, according to the National Supply Company (CONAB 2018), the grain harvest in the 2018/2019 biennium in northeast Brazil was estimated to be 19 million tons, representing 8.3% of the national productivity.

In Brazil, cowpea stands out as an economically and nutritionally important crop because it is considered one of the main sources of vegetable proteins. In the Northeast Region of Brazil, it was estimated that for the 2018/2019 biennium, it was cultivated on up to 396.5 thousand hectares (CONAB 2018). However, as in other plants, the agricultural productivity of this species is drastically affected by edaphoclimatic factors, one of the main ones being soil salinity.

2 Vegetative Root Growth Responses to Abiotic Stresses

The main response of cowpea roots to salinity is a reduction in the rate of root growth (Maia et al. 2012). Maia (2008) evaluated the effects of salt stress on seedlings of two cowpea cultivars previously described as having different degrees of sensitivity to salt stress at the germination stage (Freitas 2006). The increase in the concentration of NaCl in the cultivation substrate promoted a proportional reduction in the size of the main root.

The question of what is intimately involved in the control of root growth under salt stress arises, given that this is one of the biggest problems faced by farmers in the Brazilian semiarid region and the roots are the main organs responsible for plant nutrition.

Saline stress causes a cascade of reactions and effects on plant metabolism, which generally helps plants to respond effectively to this deficit, but which in more severe cases can cause death. It is known that associated with salt stress is ionic stress, which occurs when the concentration of Na^+ ions reaches a level higher than the concentration of K^+ ions in the cytosol (Lucena et al. 2012). K^+ is considered the main cationic inorganic nutrient in plants and is associated with charge balancing in the cytoplasm, activation of enzymatic reactions, and regulation of turgor pressure. The exchange of K^+ for Na^+ causes disruption in many metabolic processes and is initiated by competition from Na^+ for K^+ transport sites (Fig. 1). The competition is motivated by the physicochemical similarities between these two ions and culminates

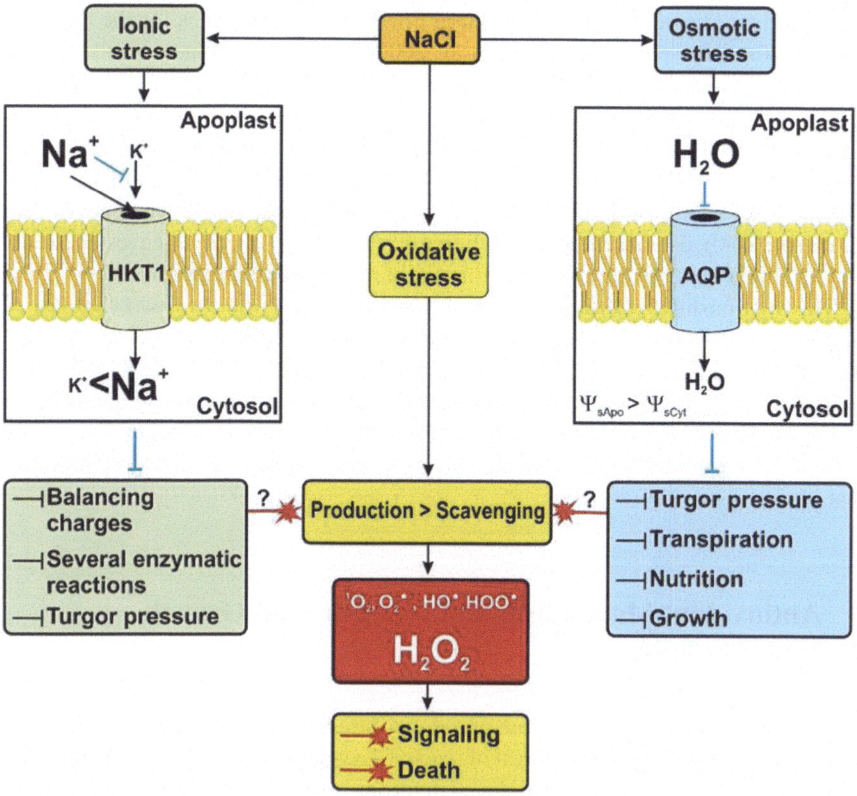

Fig. 1 Components of salt stress. Saline treatment can cause ionic, osmotic, and oxidative stress. Ionic stress is caused by an increase in the concentration of Na^+ in the cytosol, mainly by competition with K^+ for transport channels such as HKT1. In the cytosol,Na^+—in concentrations higher than that of K^+—causes load imbalance, inhibition of several enzymatic reactions dependent on K^+, and interferes with turgor pressure. Osmotic stress is caused by changes in the solute potential (ψs) of the apoplast (Apo) and cytosol (Cit). This effect causes a reduction in the rate of translocation of water to the cytosol through aquaporins (AQP) and consequently interferes in turgor pressure, decreases the transpiration rate, and interferes in nutritional processes and ultimately in growth. Both ionic and osmotic stress can cause oxidative stress by interfering with the balance between the production and removal of reactive oxygen species (ROS). The production of ROS can be used to signal and/or culminate in the death of the tissue or the entire plant

in an intracellular K^+ deficiency. In this compartment, Na^+ competes for K^+ binding sites and thus inhibits crucial metabolic processes dependent on that ion (Maathuis and Amtmann 1999).

In addition to ionic stress, osmotic stress can also be caused by saline stress (Fig. 1). This is caused by decreased availability of water to the plant through changes in the solute potential (ψs) of the apoplast (Apo) and cytosol (Cit), and, it can result in a decrease in the turgor pressure of the tissues. This could be caused by a reduction in the rate of translocation of water to the cytosol through aquaporins (AQP), in addition the reduction in the transpiration flow with consequences for the nutrition and growth of the plant (Hasegawa et al. 2000; Zhu 2002; Willadino and Camara 2010).

Reductions in dry matter content and relative growth in cowpea roots treated with NaCl had been reported by Cavalcanti et al. (2007). The effects could not be related to leaf relative water content (RWC), which estimates leaf water accumulation, as it did not differ significantly between the treatments and their respective controls. Lin and Kao (2000) also observed an inhibition of the growth of cereal leaves and roots under saline stress but did not relate this effect to a decrease in turgor.

The relationship between K^+ and Na^+ was evaluated in roots of two varieties of cowpea (*Vigna unguiculata*) under saline stress by Maia (2008). As expected, the Na^+ concentration increased 77.23% on average in both cultivars while the K^+/Na^+ ratio decreased by 72.5%. These results suggested that the ionic effect was significant in the reduction of root length. Additionally, the high Na^+ concentration in the roots (about 1400 μmol g^{-1}DM), would generate Na^+ toxicity (Coelho et al. 2014), reinforcing that idea.

3 Antioxidant Metabolism and Cowpea Root Growth Under Salinity

Both ion and osmotic stress can cause oxidative stress by interfering with the balance between the production and removal of reactive oxygen species (ROS) such as superoxide ($O_2^{\bullet -}$), hydrogen peroxide (H_2O_2), hydroxyl radicals ($HO^{\bullet -}$), peroxyl radicals ($ROO^{\bullet}$), and singlet oxygen ($^1O_2$) (Møller et al. 2007). Although the two components of salt stress (ionic and osmotic) can cause oxidative stress, little is known about the real contribution of these components in the production of ROS in roots. However, it is widely known that an imbalance between the production and removal of these radicals is the main cause of ROS production and that all processes involve highly regulated signaling cascades (Møller et al. 2007).

Since the production of ROS is a process common to several metabolic pathways during aerobic respiration in healthy plants (Møller et al. 2007), several cellular compartments are responsible for their production, and the increase in the pool is mainly due to the mismatch between production and removal of these radicals (Møller et al. 2007) (Table 1). Of the known metabolic pathways, electron transport chains (ETC) in the chloroplast and mitochondria produce the most ROS during stressful situations (Foyer and Noctor 2003; Barbosa et al. 2010). The chloroplast ETC stands out as the main ROS production site. However, in the case of non-photosynthetic tissues such as roots, mitochondria become the main site of ROS production. Other cellular compartments also produce ROS, such as the apoplast and peroxisomes, which in roots can be compartments as active in the production of ROS as mitochondria (Møller et al. 2007) (Table 1).

Table 1 Production of reactive oxygen species in roots

Mechanism	Location	ROS
Oxalate oxidase	Apo	H_2O_2
Amine oxidase	Apo	H_2O_2
Peroxidases	Apo	$HO^{\bullet}$, $HOO^{\bullet}$
Ascorbate degradation	Apo	$HO^{\bullet}$, $HOO^{\bullet}$
Peroxidases, Mn^{+2} and NADH	CW	$O_2^{\bullet-}$, H_2O_2
Complex I and III at CTE	Mit	$O_2^{\bullet-}$
Mn-SOD	Mit	H_2O_2
Glycolate oxidase	Per	H_2O_2
β-oxidation of fatty acids	Per	H_2O_2
Xanthine oxidase	Per	$O_2^{\bullet-}$
Urea oxidase	Per	$O_2^{\bullet-}$
NADH oxidase	Per	$O_2^{\bullet-}$
Cu/Zn-SOD	Per, Cit, Apo	H_2O_2
NADPH oxidase	Pm	$O_2^{\bullet-}$

Acronyms: *Apo* apoplast, *Cit* cytosol, *ETC* electron transport chain, *CW* cell wall, *ROS* reactive oxygen species, *HO*• hydroxyl radical, *HOO*• peroxyl radical, *Mit* mitochondria, *Per* peroxisome, *MP* plasma membrane, *SOD* superoxide dismutase
Updated from Alscher et al. (2002); Mittler 2002; Vranová et al. 2002; Passardi et al. 2004; Green and Fry 2005

The plant cell and its organelles—peroxisomes, chloroplasts (Barbosa et al. 2014; Del Río et al. 2006), and mitochondria (Møller et al. 2007; Navrot et al. 2007; Barbosa et al. 2010)—contain several enzymatic and non-enzymatic systems for the removal of ROS (Møller et al. 2007; Apel and Hirt 2004) (Table 2). The balance between these two systems under stress conditions can be regulated by the concentration of O_2 in the system (Blokhina et al. 2003). The distinct subcellular location and biochemical properties of antioxidant enzymes, their different patterns of induction and gene expression, and the large number of non-enzymatic removers make the antioxidant system a flexible and versatile unit that can control the accumulation and detoxification of temporal and spatial ROS (Silveira et al. 2010; Barbosa et al. 2014; Vranová et al. 2002).

Reduced growth rate under stress conditions has been related with processes involving antioxidant mechanisms (Foreman et al. 2003; De Cnodder et al. 2005; Foyer and Noctor 2005; Carol and Dolan 2006). Foyer and Noctor (2005) explain that the decrease in growth precedes an oxidative explosion, but according to Qin et al. (2004), the cell compartment in which the oxidative explosion first occurs is controversial. One hypothesis is that it is the apoplast, involving NADPH oxidase (NOX), which transfers electrons from the cytoplasmic NADPH to apoplastic O_2, forming $O_2^{\bullet-}$, which is then dismutated to H_2O_2 (Ogawa et al. 1996; Qin et al. 2004; Passardi et al. 2004; Foyer and Noctor 2005).

Table 2 Mechanisms for the elimination of reactive oxygen species in roots, their cellular compartments, and the respective species removed

Mechanism	Scavenged (product)	Cellular site
Ascorbate/glutathione cycle	H_2O_2 (H_2O)	Cytosol, mitochondria, peroxisomes
Catalase	H_2O_2 (H_2O)	Mitochondria, peroxisomes
Carotenoids and tocopherols	$^1O_2$ (O_2)	Plastids
Alternative oxidase	$O_2^{\bullet-}$	Mitochondria
Ascorbate peroxidase	H_2O_2 (H_2O)	Cytosol, mitochondria, plasma membrane, microbodies, glyoxysomes, peroxisomes
Phenols peroxidase	H_2O_2 (H_2O)	Cytosol, vacuoles, apoplast, cell wall
Glutathione peroxidase	H_2O_2 (H_2O)	Cytosol, endoplasmic reticulum, mitochondria
	Lipid hydroperoxide	Cytosol, endoplasmic reticulum, mitochondria
	Other hydroperoxides	Cytosol, endoplasmic reticulum, mitochondria
Peroxyredoxin system	H_2O_2 (H_2O)	Cytosol, mitochondria, nucleus
	Hydroperoxide alkil	Cytosol, mitochondria, nucleus
	Peroxynitrite	Cytosol, mitochondria, nucleus
Thioredoxin system	H_2O_2 (H_2O)	Cytosol, mitochondria, secretory pathway
Glutaredoxin system	H_2O_2 (H_2O)	
Superoxide dismutase	$O_2^{\bullet-}$ (H_2O_2)	Cytosol, mitochondria, peroxisomes, microsomes, glyoxysomes
Thioredoxin	H_2O_2	

Update from Dat et al. (2000), Møller (2001), Shigeoka et al. (2002), Mittler et al. (2004), D'arcy-Lameta et al. (2006), Møller et al. (2007)

Neves et al. (2010) suggest that in soybeans, for example, growth reduction is related to an increase in the lignin content in the cell wall which is promoted by the increased activity of cell wall peroxidase, stiffening this structure and, consequently, restricting growth. Other evidence points out that antioxidants are cellular signalers and that a differential distribution between superoxides and hydrogen peroxide (H_2O_2) is required for the growth and development of the root system (Noctor et al. 2018).

Some evidence indicates that peroxidases located in the apoplast are the main components responsible for the reduction in the growth rate of several vegetative organs through the synthesis of lignin in a process corresponding to an early maturation of the tissues during abiotic stresses (Passardi et al. 2004). These apoplastic peroxidases can participate in both restriction and growth stimulation, depending on the physiological conditions of the organism (Passardi et al. 2004). Stimuli for growth are provided by a metabolic pathway in which apoplastic peroxidases catalyze the

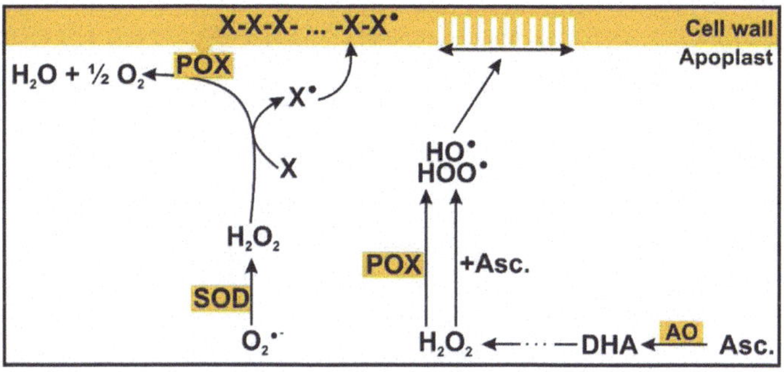

Fig. 2 Mechanisms of growth control by the activity of apoplastic peroxidases and by the metabolism of ascorbate. The $O_2^{\bullet-}$ produced in the apoplast is converted to H_2O_2 by an apoplastic SOD. A phenol peroxidase (POX) bound to the cell wall degrades H_2O_2 to form H_2O and O_2 using apoplastic phenols (X) as reducing substrates. Phenol radicals ($X^{\bullet}$) are adjuvants in the formation of lignin. This polymer stiffens the cell wall, restricting cell growth. On the other hand, POX can also induce growth through the production of hydroxyl ($HO^{\bullet}$) and peroxyl ($HOO^{\bullet}$) radicals. These radicals degrade components of the cell wall, allowing the wall to loosen and allowing growth by stretching the cells. Ascorbate also participates in growth control. This molecule can react with H_2O_2 to form $HO^{\bullet}$ and $HOO^{\bullet}$ or even produce H_2O_2 through the degradation pathway

formation of hydroxyl ($HO^{\bullet}$ e) and peroxyl ($HOO^{\bullet}$) radicals. The produced radicals degrade components of the cell wall, inducing loosening of the wall and subsequently growth by cell stretching (Passardi et al. 2004) (Fig. 2).

Ascorbate also participates in growth regulation and can react with H_2O_2, producing oxygen radicals that induce growth, or even contributing to the increase of the H_2O_2 pool in the apoplast through the degradation pathway (Passardi et al. 2004).

Maia (2008) studied fluctuations in H_2O_2 and lipid peroxidation in roots of cowpea plants submitted to salt stress. After 120 min of treatment, the concentration of H_2O_2 increased sharply, followed by a gradual reduction up to 180 min. In 30 h of treatment, there was an increase in concentration by 57.8%, with a decrease in H_2O_2 again in 48 h. In three days of treatment, the concentrations again tended to increase, but decreased again in the four days of treatment at levels of 64.3%. There were no significant changes in lipid peroxidation in all observed time series (Fig. 3). In contrast, Cavalcanti et al. (2004) identified that the increase in the lipid peroxidation index in cowpea leaves treated with NaCl may be related to the osmotic effect of salinity.

The drastic increase in the concentration of H_2O_2 after 120 min of treatment may suggest the occurrence of a phenomenon called oxidative explosion (Minibaeva and Gordon 2003; Foyer and Noctor 2005). According to Bolwell et al. (1998) and Mahalingam and Fedorof (2003), this process is transient and occurs very quickly—within seconds—in systems such as cell cultures of common beans (*Phaseolus vulgaris*) and soybeans (*Glycine max*). After the detection of stress, a hypersensitive response is triggered, responsible for the early maturation of tissues, which is caused by the activation of the apoplastic antioxidative metabolism (Passardi et al.

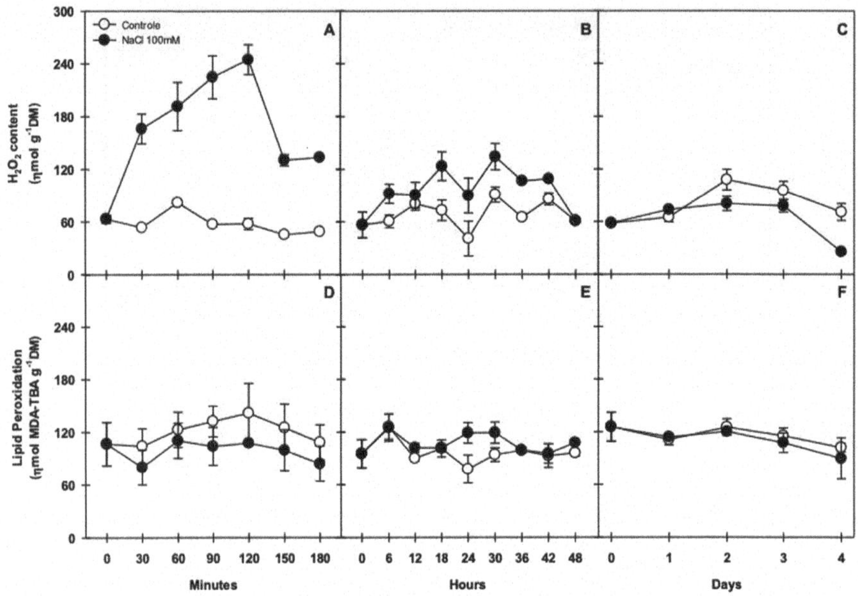

Fig. 3 H_2O_2 concentration (**a**; **b** and **c**) and lipid peroxidation (**d**; **e**; **f**) in cowpea seedling roots submitted to 0 (control) and 100 mM NaCl for 180 min (**a**; **d**), 48 h (**b**; **e**) and four days (**c**; **f**)

2004; Foyer and Noctor 2005). This hypersensitive response is believed to occur as a result of the release of $O_2^{\bullet-}$ and H_2O_2 in the apoplast, as well as causing the synthesis of defensive barriers such as the cell wall (Sgherri et al. 2007).

To observe the interaction of saline treatment with the intracellular antioxidant response, the activities of total APX and CAT were also evaluated (Fig. 4). APX activity showed an average reduction of 42.5% in treatments S-24 and S-48 in relation to their respective controls. CAT activity showed no significant changes in roots treated for 24 h with NaCl. However, saline treatment for 48 h caused a drastic (68%) reduction in the activity of this enzyme.

Total and apoplastic SOD activities, as well as the content of H_2O_2 in these fractions, were analyzed to understand the interactions between the pools of apoplastic and total H_2O_2 (Maia 2008). Total SOD activity did not vary significantly in the presence of salt during 48 h of treatment (Fig. 5a). However, the activity of this enzyme in the apoplastic extract increased significantly in 48 h of treatment with NaCl (37%) (Fig. 5b). The saline treatment for 48 h still increased the total H_2O_2 content (Fig. 5c); however, this decreased again by 50% in 48 h.

According to Passardi et al. (2004), POX can regulate growth by two different metabolic cycles. They favor elongation by the generation of oxygen radicals or inhibit growth by controlling the local concentration of H_2O_2; it is believed that this, in turn, is involved as a cellular signal of stress and as an adjunct to wall lignification. In cowpea, it is possible that POX activity is related to the reduction in the length of roots treated with NaCl. To corroborate the hypothesis that saline treatment may

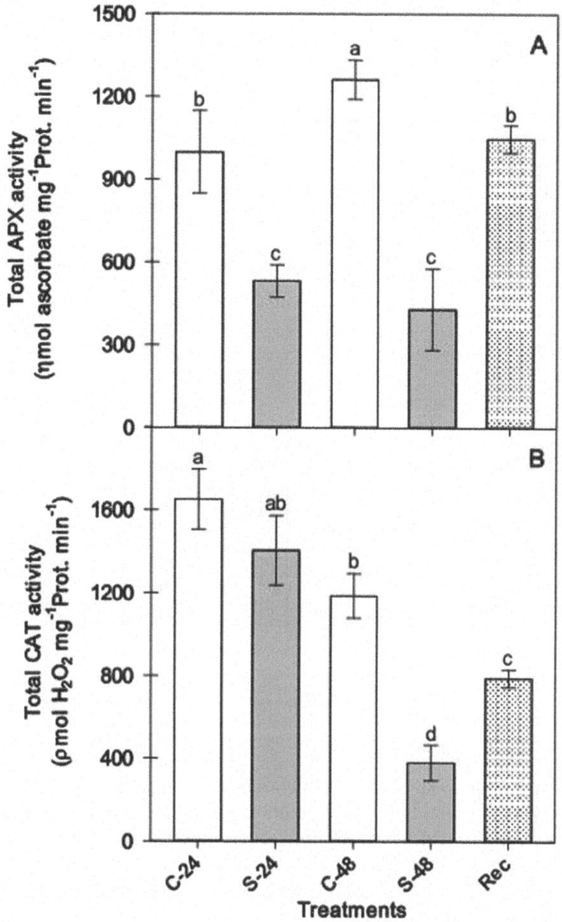

Fig. 4 Total ascorbate peroxidase (APX) **a** and total catalase (CAT) activity **b** in cowpea seedling roots, grown under control conditions (0 mM NaCl) and treated with 100 mM NaCl for 24 and 48 h. Treatments: 24 h control (C-24); 100 mM NaCl-24 h (S-24); control-48 h (C-48); 100 mM NaCl-48 h (S-48); 100 mM NaCl-24 h H_2O-24 h (Rec). The letters on the bars indicate significant differences tested by Tukey ($p \leq 0.05$)

cause the activation of apoplastic oxidative metabolism, Maia (2008) evaluated the activities of membrane NOX, apoplast SOD, and cell wall POX (Fig. 6). The activities of these three enzymes increased after 24 h of treatment. According to Foyer and Noctor (2005), NOX actively participates in the hypersensitive response through the production of $O_2^{\bullet-}$ radicals. The ROS produced by NOX are spontaneously converted to H_2O_2 in a reaction that can be catalyzed by a Cu/Zn-SOD detected in the apoplast (Ogawa et al. 1996; Sgherri et al. 2007). In turn, a POX ionically bound to the cell wall can use these ROS as oxidants for the formation of lignin monomers (Passardi et al. 2004).

Along with the increase in NOX, SOD, and POX activity, there was a reduction in apoplastic H_2O_2 content and a significant increase in tissue lignin content (Fig. 7). The decrease in the H_2O_2 content of this compartment can be confirmed by the decrease in the content of free phenols in the apoplast. The consumption of these two molecules is possibly due to the POX activity directed to the deposition of lignin

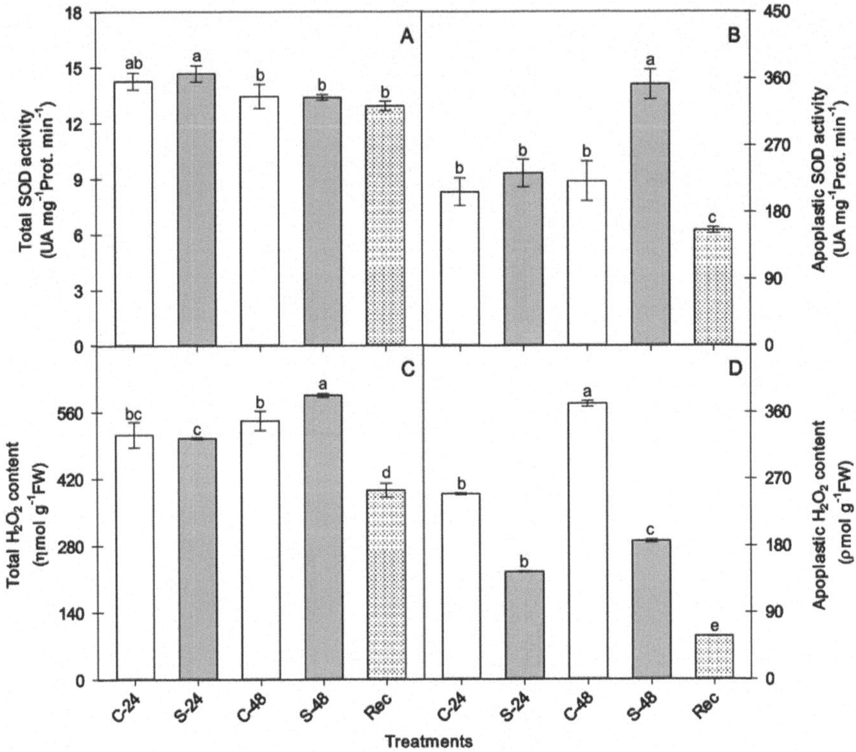

Fig. 5 Total superoxide dismutase (SOD) activity (**a**), apoplastic superoxide dismutase (SOD) activity (**b**), total H_2O_2 concentration (**c**), and apoplastic H_2O_2 concentration (**d**) in cowpea seedling roots, grown under control conditions (0 mM NaCl) and treated with 100 mM NaCl for 24 and 48 h. Treatments: 24 h control (C-24); 100 mM NaCl-24 h (S-24); control-48 h (C-48); 100 mM NaCl-48 h (S-48); 100 mM NaCl-24 h H2O-24 h (Rec). The letters on the bars indicate significant differences tested by Tukey ($p \leq 0.05$)

in the cell wall. It can be inferred, therefore, that the decrease in root length may be associated to the intense activity of lignin production by a POX isoform associated with the cell wall.

Exposure to salt caused an average reduction of 16.5% in root length over time, concomitantly with total and apoplastic POX activity, providing further support to its involvement in reducing root growth (Fig. 8).

Descriptions of reduced growth rates, related to an increase in POX activity, are relatively common in the literature (Cavalcanti et al. 2004; Passardi et al. 2004; Foyer and Noctor 2005). The study by Maia (2008) provides evidence that the mechanisms of growth regulation during saline stress can be regulated by the apoplastic H_2O_2 pool and, possibly, by the ascorbate of the medium. Additionally, maintaining the H_2O_2 concentration in the tissue may be a reflection of the decreased activity of intracellular antioxidant enzymes.

Fig. 6 NADPH oxidase (NOX) activity of the plasma membrane (**a**), apoplastic superoxide dismutase (SOD) (**b**), and cell wall phenol peroxidase (POX) (**c**) in cowpea seedling roots, grown under control conditions and treated with 100 mM NaCl for 24 and 48 h. The letters on the bars indicate significant differences tested by Tukey ($p \leq 0.05$)

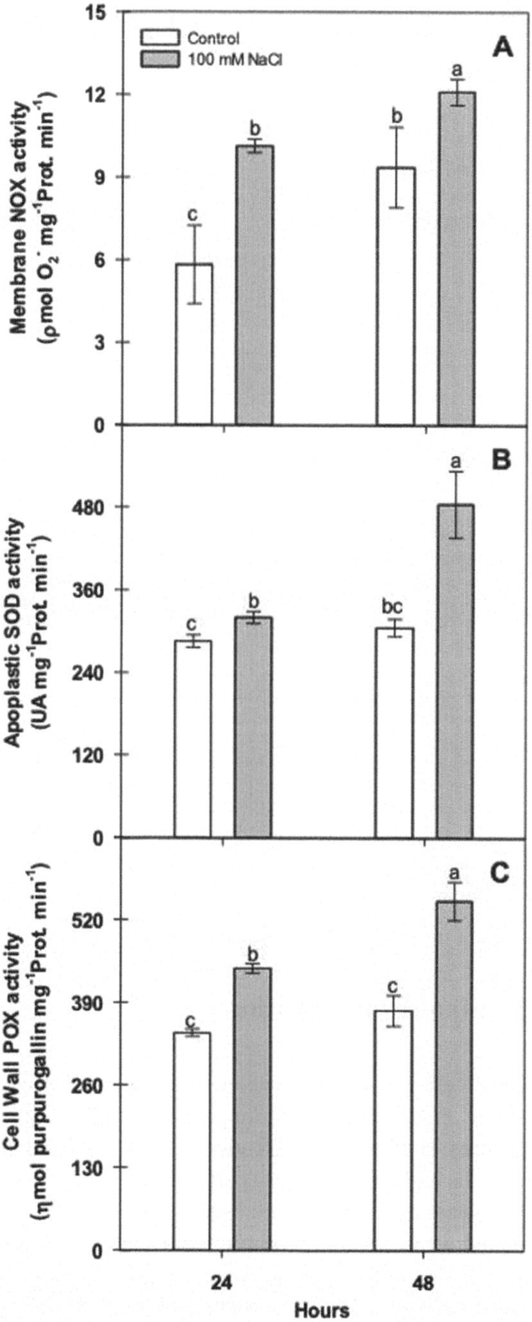

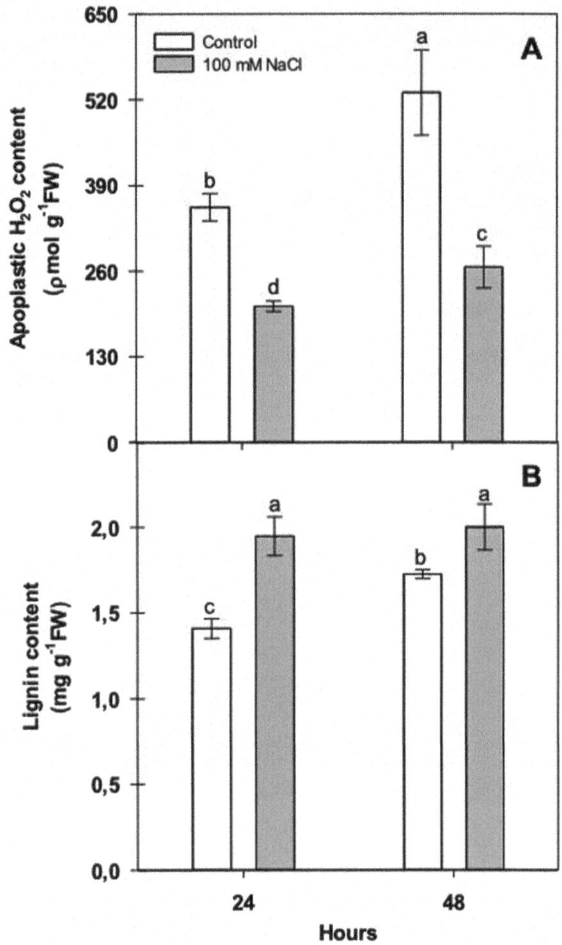

Fig. 7 Apoplastic H_2O_2 concentration (**a**) and lignin content (**b**) in cowpea seedling roots grown under control conditions and treated with 100 mM NaCl for 24 and 48 h. The letters on the bars indicate significant differences tested by Tukey ($p \leq 0.05$)

4 Final Considerations

The Brazilian semiarid region is composed of edaphoclimatic characteristics that enable an environment of unique and sensitive aspects for the cultivation of different cultures. However, it is notorious for the great adaptability that its native vegetation presents in the face of these particularities and how much they manage to circumvent the negative effects that are usually caused by the unique climate of the region, such as high irradiance, low rainfall, and high evaporation, among others. In addition to these factors, the inherent salinity of the semiarid soils is today one of the greatest difficulties in producing agricultural crops in the region.

However, crops like cowpea and several others make use of cellular defense mechanisms to circumvent the possible damage caused by abiotic stresses. The production of ROS can cause oxidative stress, as well as being part of an intricate signaling system

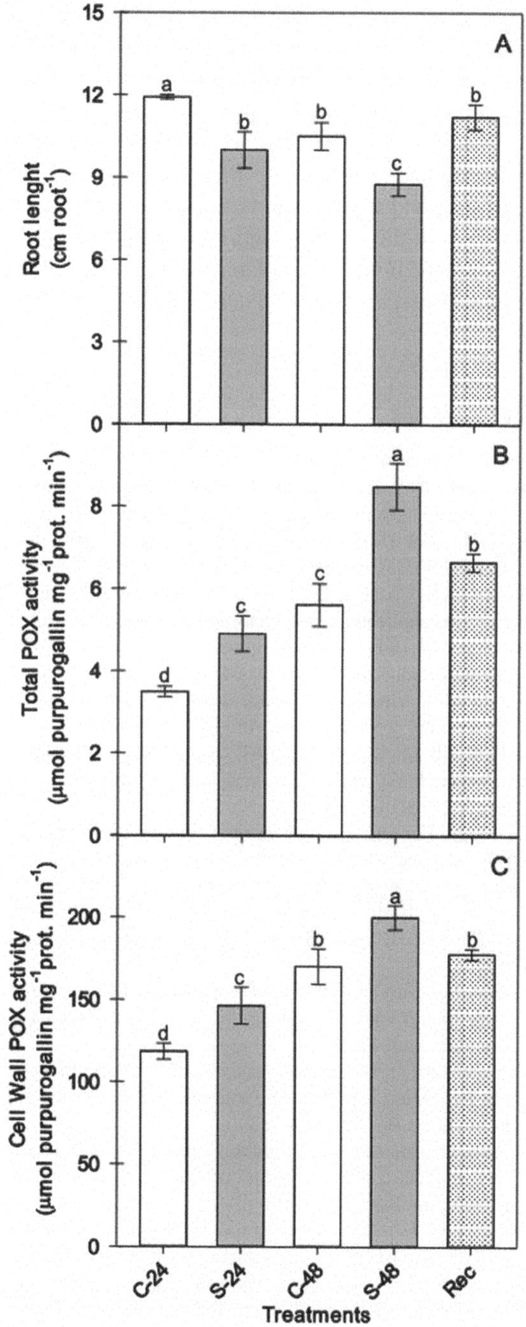

Fig. 8 Root length (**a**), total phenol peroxidase activity (POX) (**b**), and apoplastic phenol peroxidase activity (POX) (**c**) in cowpea seedlings roots, grown under control conditions (0 mM NaCl) and treated with 100 mM NaCl for 24 and 48 h. Treatments: 24 h control (C-24); 100 mM NaCl-24 h (S-24); control-48 h (C-48); 100 mM NaCl-48 h (S-48); 100 mM NaCl-24 h H_2O-24 h (Rec). The letters on the bars indicate significant differences tested by Tukey ($p \leq 0.05$)

of the plant organism, allowing antioxidant enzymes such as POX, SOD, APX, and many others to come into action and make use of mitigating mechanisms of possible damage such as, for example, the production of lignin in the cell wall of the cells of the plant root system, thus preventing its growth.

In addition, the understanding of the mitigating mechanisms of abiotic stresses in agricultural crops in the Brazilian semiarid region, as well as the conditions of their soils and water resources, is important. This is one of the regions that produces the most in the country, and a large portion of the production comes from family farming, the main sector of agribusiness that suffers negative impacts in coping with abiotic stresses, especially saline stress.

References

Alscher RG, Erturk N, Heath LS (2002) Role of superoxide dismutases (SODs) in controlling oxidative stress in plants. J Exp Bot 53:1331–1341

Apel K, Hirt H (2004) Reactive oxygen species: metabolism, oxidative stress, and signal transduction. Annu Rev Plant Biol 55:373–399

Barbosa KB, Costa NM, Alfenas RD, De Paula SO, Minim VP, Bressan J (2010) Oxidative stress: concept, implications and modulating factors. Rev Nutr 23:629–643

Barbosa MR, Silva MMA, Ulisses LWC, Rangel Camara T (2014) Geração e desintoxicação enzimática de espécies reativas de oxigênio em plantas. Ciência Rural 44:453–460

Blokhina O, Virolainen E, Fagerstedt KV (2003) Antioxidants, oxidative damage and oxygen deprivation stress: a review. Ann Bot 91:179–194

Bolwell GP, Davies DR, Gerrish C, Auh C-K, Murphy TM (1998) Comparative biochemistry of the oxidative burst produced by rose and French bean cells reveals two distinct mechanisms. Plant Physiol 116:1379–1385

Cavalcanti FR, Oliveira JT, Martins-Miranda AS, Viégas RA, Silveira JA (2004) Superoxide dismutase, catalase and peroxidase activities do not confer protection against oxidative damage in salt-stressed cowpea leaves. New Phytol 163:563–571

Cavalcanti FR, Lima JP, Ferreira-Silva SL, Viégas RA, Silveira JA (2007) Roots and leaves display contrasting oxidative response during salt stress and recovery in cowpea. J Plant Physiol 164:591–600

Coelho DS, Simões WL, Mendes A, Dantas BF, Rodrigues JA, Souza MA (2014) Germinação e crescimento inicial de variedades de sorgo forrageiro submetidas ao estresse salino. Rev Bras Eng Agr Amb 18:25–30

CONAB (2018) Acompanhamento da safra brasileira de grãos. CONAB, Brasília

Carol RJ, Dolan L (2006) The role of reactive oxygen species in cell growth: lessons from root hairs. J Exp Bot 57:1829–1834

D'Arcy-Lameta A, Ferrari-Iliou R, Contour-Ansel D, Pham-Thi AT, Zuily-Fodil Y (2006) Isolation and characterization of four ascorbate peroxidase cDNAs responsive to water deficit in cowpea leaves. Ann Bot 97:133–140

Dat J, Vandenabeele S, Vranová EV, Van Montagu M, Inzé D, Van Breusegem F (2000) Dual action of the active oxygen species during plant stress responses. CLMS 57:779–795

De Cnodder T, Vissenberg K, Van Der Straeten D, Verbelen JP (2005) Regulation of cell length in the *Arabidopsis thaliana* root by the ethylene precursor 1-aminocyclopropane-1-carboxylic acid: a matter of apoplastic reactions. New Phytol 168:541–550

Del Río LA, Sandalio LM, Corpas FJ, Palma JM, Barroso JB (2006) Reactive oxygen species and reactive nitrogen species in peroxisomes: production, scavenging and role in cell signaling. Plant Physiol 141:330–335

Foreman J, Demidchik V, Bothwell JHF Mylona P, Miedema H, Torres MA, Linstead P, Costa S, Brownlee C, Jones JD, Davies JM (2003) Reactive oxygen species produced by NADPH oxidase regulate plant cell growth. Nature 422:442–446

Foyer CH, Noctor G (2003) Redox sensing and signalling associated with reactive oxygen in chloroplasts, peroxisomes and mitochondria. Physiol Plant 119:355–364

Foyer CH, Noctor G (2005) Redox homeostasis and antioxidant signaling: a metabolic interface between stress perception and physiological responses. Plant Cell 17:1866–1875

Freitas JBS (2006) Respostas fisiológicas ao estresse salino de duas cultivares de feijão caupi. Universidade Federal do Ceará, Thesis

Green MA, Fry SC (2005) Vitamin C degradation in plant cells via enzymatic hydrolysis of 4-O-oxalyl-L-threonate. Nature 433:83–87

Hasegawa PM, Bressan RA, Pardo JM (2000) The dawn of plant salt to tolerance genetics. Trends Plant Sci 5:317–319

Lin CC, Kao CH (2000) Effect of NaCl stress on H_2O_2 metabolism in rice leaves. Plant Growth Regul 30:151–155

Lucena CC, Siqueira DL, Martinez HE, Cecon PR (2012) Efeito do estresse salino na absorção de nutrientes em mangueira. Rev Bras Frutic 34:297–308

Maathuis FJM, Amtmann A (1999) K^+ nutrition and Na^+ toxicity: the basis of cellular K^+/Na^+ ratios. Ann Bot 84:123–133

Mahalingam R, Fedoroff N (2003) Stress response, cell death and signalling: the many faces of reactive oxygen species. Physiol Plant 119:56–68

Maia JM (2008) Restrição de crescimento induzida por estresse salino como uma estratégia de defesa oxidativa em raízes de feijão-caupi. Universidade Federal do Ceará, Thesis

Maia JM, Ferreira-Silva SL, Voigt EL, Macêdo CE, Ponte LF, Silveira JA (2012) Atividade de enzimas antioxidantes e inibição do crescimento radicular de feijão caupi sob diferentes níveis de salinidade. Acta Bot Bras 26:342–349

Minibaeva FV, Gordon LK (2003) Superoxide production and the activity of extracellular peroxidase in plant tissues under stress conditions. Russ J Plant Physiol 50:411–416

Mittler R (2002) Oxidative stress, antioxidants and stress tolerance. Trends Plant Sci 7:405–410

Mittler R, Vanderauwera S, Gollery M, Van Breusegem F (2004) Reactive oxygen gene network of plants. Trends Plant Sci 9:490–498

Møller IM (2001) Plant mitochondria and oxidative stress: electron transport, NADPH turnover, and metabolism of reactive oxygen species. Annu Rev Plant Physiol Plant Mol Biol 52:561–591

Møller IM, Jensen PE, Hansson A (2007) Oxidative modifications to cellular components in plants. Annu Rev Plant Biol 58:459–481

Navrot N, Rouhier N, Gelhaye E, Jacquot JP (2007) Reactive oxygen species generation and antioxidant systems in plant mitochondria. Physiol Plant 129:185–195

Neves GY, Marchiosi R, Ferrarese ML, Siqueira-Soares RC, Ferrarese-Filho O (2010) Root growth inhibition and lignification induced by salt stress in soybean. J Agr Crop Sci 196:467–473

Noctor G, Reichheld JP, Foyer CH (2018) ROS-related redox regulation and signaling in plants. Sem Cell Dev Biol 80:3–12

Ogawa K, Kanematsu S, Asada K (1996) Intra- and extra-cellular localization of "cytosolic" Cu/Zn-superoxide dismutase in spinach leaf and hypocotyl. Plant Cell Physiol 37:790–799

Passardi F, Penel C, Dunand C (2004) Performing the paradoxical: how plant peroxidases modify the cell wall. Trends Plant Sci 9:534–540

Qin WM, Lan WZ, Yang X (2004) Involvement of NADPH oxidase in hydrogen peroxide accumulation by *Aspergillus niger* elicitor-induced *Taxus chinensis* cell cultures. J Plant Physiol 161:355–361

Sá IB, Cunha TJ, Teixeira AD, Angelotti F, Drumond MA (2010) Processos de desertificação no Semiárido brasileiro. In: Sá IB, Silva PCG (eds) Semiárido brasileiro: pesquisa, desenvolvimento e inovação, 1st edn. Embrapa Semiárido, Petrolina, pp 126–158

Sgherri C, Quartacci MF, Navari-Izzo F (2007) Early production of activated oxygen species in root apoplast of wheat following copper excess. J Plant Physiol 164(9):1152–1160

Shigeoka S, Ishikawa T, Tamoi M, Miyagawa Y, Takeda T, Yabuta Y, Yoshimura K (2002) Regulation and function of ascorbate peroxidase isoenzymes. J Exp Bot 53(372):1305–1319

Silveira JAG, Silva SLF, Silva EN, Viégas RA (2010) Mecanismos biomoleculares envolvidos com a resistência ao estresse salino em plantas. In: Gheyi HR, Dias NS, Lacerda CF et al (eds) (2016) Manejo da salinidade na agricultura: Estudos básicos e aplicados, 2nd edn. INCTSal, Fortaleza, pp 181–197

Souza NGM, Silva JA, Maia JM Silva JB, Júnior ED, Meneses CH (2016) Tecnologias sociais voltadas para o desenvolvimento do Semiárido Brasileiro. Biofarm 12(3):1–12

Vranová E, Inzé D, Van Breusegem F (2002) Signal transduction during oxidative stress. J Exp Bot 53:1227–1236

Willadino L, Camara TR (2010) Tolerância das plantas à salinidade: aspectos fisiológicos e bioquímicos. Encicl Biosfera 6:1–23

Zhu JK (2002) Salt and drought stress signal transduction in plants. Annu Rev Plant Biol 53:247–273

Lotus spp.: A Foreigner that Came to Stay Forever: Economic and Environmental Changes Caused by Its Naturalization in the Salado River Basin (Argentina)

Amira Susana Nieva and Oscar Adolfo Ruiz

Abstract The Flooding Pampas in Buenos Aires Province is one of Argentina's cattle-raising areas. Climatic, topographic and edaphic conditions limit its potential in this area for growing crops such as soybean and wheat. The introduction of *L. tenuis* in the Flooding Pampas area triggered research based on its ability to tolerate the abiotic stresses that characterize the region and on its role in the improvement of the quality of forage resources. Along with research on abiotic stress tolerance, productive strategies have been developed to enhance the establishment of *L. tenuis* grassland and beef production. Research on *Lotus* spp. in the Flooding Pampas has therefore studied not only the biotechnological development and evaluation of new plant resources, but also the accompanying plant diversity, soil microorganisms and symbionts and their impact on environmental dynamics and sustainability. Based on this research, productive strategies have been designed, including continuous evaluation of the impact of cattle production on vulnerable ecosystems. In addition, basic and applied research on grasslands have been combined in order to respond to the environmental impact of the introduction and use of *Lotus* in these particular ecosystems.

Keywords Legume · Cattle raising · Tannins · Plant naturalization · Mycorrhizal fungi

A. S. Nieva · O. A. Ruiz (✉)
Instituto Tecnológico de Chascomús (INTECH), Chascomús, Argentina
e-mail: ruiz@intech.gov.ar

A. S. Nieva
e-mail: susanna_nieva@hotmail.com

A. S. Nieva
Max Planck Institute of Molecular Plant Physiology, Potsdam, Germany

E. Taleisnik and R. S. Lavado (eds.), *Saline and Alkaline Soils in Latin America*, https://doi.org/10.1007/978-3-030-52592-7_22

1 Introduction

The Salado River Basin is the central and most representative area of the Flooding Pampas. A combination of factors such as weather, topography, edaphic restrictions and anthropic activity have made it inappropriate for extensive agricultural activities. The Salado River Basin lowlands are periodically affected by recurrent flooding periods. The major edaphic limiting factors are salinity and alkalinity (see the chapter by Imbellone et al., in this book: "Genesis, properties and management of salt affected soils in the Flooding Pampas, Argentina").

In recent years the "agricultural frontier" has advanced, reducing the "traditional" pasture areas used for livestock, and the areas with soil limitations have experienced an increase in grazing animals stocking rate. It has therefore been essential to increase the supply of fodder in this lowland system, in order to support its economic profitability.

In the absence of native legumes tolerant environmental constraints, the introduction of *Lotus* spp. in this area has improved grassland resources. *Lotus* species are legumes of the *Fabaceae* family which are able to grow and develop fully under diverse abiotic stresses (Escaray et al. 2012a, 2019b). Research on this legume in recent years has reinforced the available information on its tolerance to multiple abiotic stress factors, and has enabled the design of novel tools for improving grassland yield and quality. Some ecologists define *L. tenuis* as a "keystone species" because it can modify the characteristics of the soils where it has been planted, thereby enabling the progressive development of other plant species with economic importance as forage but which lack the tolerance that characterizes *L. tenuis* (Quinos et al. 1998).

2 Introduction and Naturalization of *Lotus* spp. in Argentina: Stress Tolerance as the Main Factor in the Survival and Dispersion of *L. Tenuis* in the Flooding Pampas

Lotus species are native to the European Mediterranean region and other parts of Europe and Western Asia (Degtjareva et al. 2006). Their distinctive characteristic is their tolerance to different abiotic stresses. Furthermore, their different adaptations enable them to be used in remediation, restoration and conservation of environments, and for grazing management and productive protocols for soil conservation.

Lotus species were introduced in Argentina from Europe in the 1950s (Vignolio et al. 1995) and easily became distributed in Natraquoll soils. The two most widely used Lotus species for forage in Argentina are *L. tenuis* and *L. corniculatus*. Their distribution has been mainly determined by soil characteristics. In the Argentinean Northeast, *L. corniculatus* was easily incorporated into cattle-raising systems (Real et al. 2013), while in the Flooding Pampas, *L. tenuis* has demonstrated high adaptability to the predominant environmental stresses: salinity, alkalinity and contrasting periods of "available soil water" (drought periods in summer and flood periods in autumn–winter) (Escaray et al. 2012a). Research on *L. tenuis* to date has demonstrated its ability to tolerate the stresses caused by flooding periods and salinity better than *L. corniculatus* (Vignolio et al. 1994). Therefore, commercial crops of tetraploid *L. corniculatus* have been destined to soils of better quality than those where *L. tenuis* is promoted through different agricultural management practices. The tolerance of *L. tenuis* to a combination of abiotic stresses is an interesting trait that accounts for its successful establishment and dispersion in the Flooding Pampas fields, and has been the object of intensive studies (Escaray et al. 2012a, b, 2014; Babuin et al. 2014; Antonelli et al. 2016, 2019; Campestre et al. 2016a, b; Bordenave et al. 2019; Calzadilla et al. 2016a, b, 2019; Sanchez et al. 2005, 2011; Paz et al. 2012). *L. tenuis* seeds and seedlings have the ability to remain in water for a long time without damage to teguments or cotyledons (Vignolio et al. 1995; García and Mendoza 2014), maintaining their viability and normal growth (Clua et al. 2009). Based in these results, the ability of *L. tenuis* to survive in submersion has also been intensively studied. Evidence shows that *L. tenuis* growth and development cycle are unaffected by the seed having been under waterlogged conditions. Its adaptation to flooding periods has been attributed to a combination of morphological and physiological features (Striker et al. 2005; Antonelli et al. 2019). Its tolerance to the flooding is an important determining factor in promoting the dispersion of the germplasm (Strittmatter et al. 1992) due to the limitation imposed by water in other plant species. The heterogeneity and seasonal character of flooding events have determined the establishment of "*L. tenuis* patch communities", following the heterogeneity of edaphic characteristics. It is currently easy to find *L. tenuis* communities growing spontaneously throughout the region (León 2000).

In addition to tolerating waterlogging stress, *L. tenuis* is tolerant to other stresses in the Flooding Pampas (Fig. 1). Salt stress tolerance is a feature shared by most *Lotus* species. In *L. tenuis*, a direct mechanism to ameliorate salt stress involves

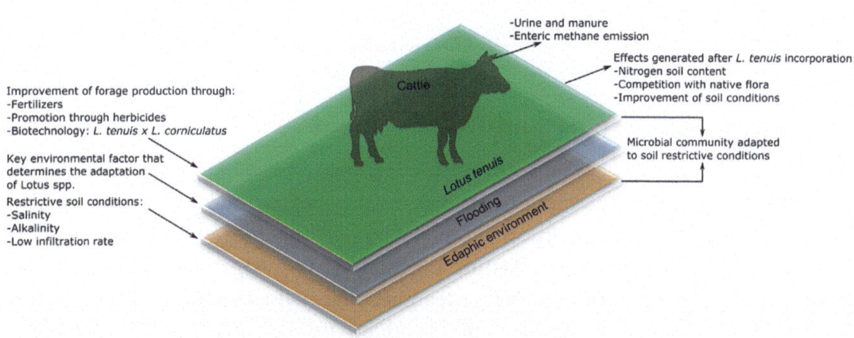

Fig. 1 Main factors involved in the *L. tenuis* naturalization in the Flooding Pampas lowlands, including the environmental components and system management

the extrusion of chlorides and sodium from the xylem (Teakle et al. 2007). Other responses of *Lotus* species subjected to salt stress, such as polyamine accumulation and root length reduction, have also been reported (Sanchez et al. 2005; Paz et al. 2012). *L. tenuis*, in particular, withstands osmotic stress caused by salt in the soil by means of a metabolic pre-adaptation that involves the accumulation of proline, myo-inositol, serine and some organic acids (Sanchez et al. 2011). Different *Lotus* species respond in different ways to the combination of salt and waterlogging stress, suggesting that each of them has specific physiological stress response traits.

The responses of *Lotus* spp. to salinity, alkalinity and low temperature stresses have been widely studied in the *L. japonicus* model (Paz et al. 2012; Babuin et al. 2014; Campestre et al. 2016a; Calzadilla et al. 2016a, b). This research made major progress regarding understanding physiological traits, providing an opportunity to extrapolate results to the crop species (Fig. 2). In addition, the *L. japonicus* model has been used in research to determine the response to the abiotic stress conditions involving photosynthesis and photosynthesis-derived components, in a study that reported carbon accumulation as a mechanism to adapt to and mitigate the stress caused by sub-optimal temperatures (Calzadilla et al. 2019). Chlorophyll fluorescence has become important in the evaluation of the stress caused by salinity, alkalinity, and the combination of both (Bordenave et al. 2019). These results promote the design of innovative strategies regarding the use of radiation efficiency in breeding programs.

Although the results obtained in the *L. japonicus* model may be extrapolated to crop species such as *L. tenuis* and *L. corniculatus*, it is the direct study of the mechanisms involved in the stress tolerance of the latter two that has provided useful technological information. *In vitro* examination of stress conditions has demonstrated the ability of *L. tenuis* to grow and fully develop under different abiotic stresses (Escaray et al. 2012a) and even under combined stresses (Antonelli et al. 2019). These results suggest the possibility of exploring outcomes under similar conditions for other *Lotus* forage species by combining the information obtained from the *L. japonicus* model, specific environmental limitations, and economic interest in improving *Lotus* forage species plantation and establishment in order to manage a sustainable, productive pasture ecosystem.

3 Pasture Productivity and Management

In addition to providing germplasm resources, the introduction of *L. tenuis* in the Flooding Pampas has also incorporated management strategies widely used in other grassland systems, for example, pasture promotion. The Argentine National Institute of Agricultural Technology (INTA) designed and evaluated different approaches for improving *L. tenuis* conservation and productivity. Fertilizer- and/or herbicide-mediated promotion has been used in order to improve its yield and successful establishment (Bailleres and Sarena 2011). Rodríguez et al. (2007) demonstrated

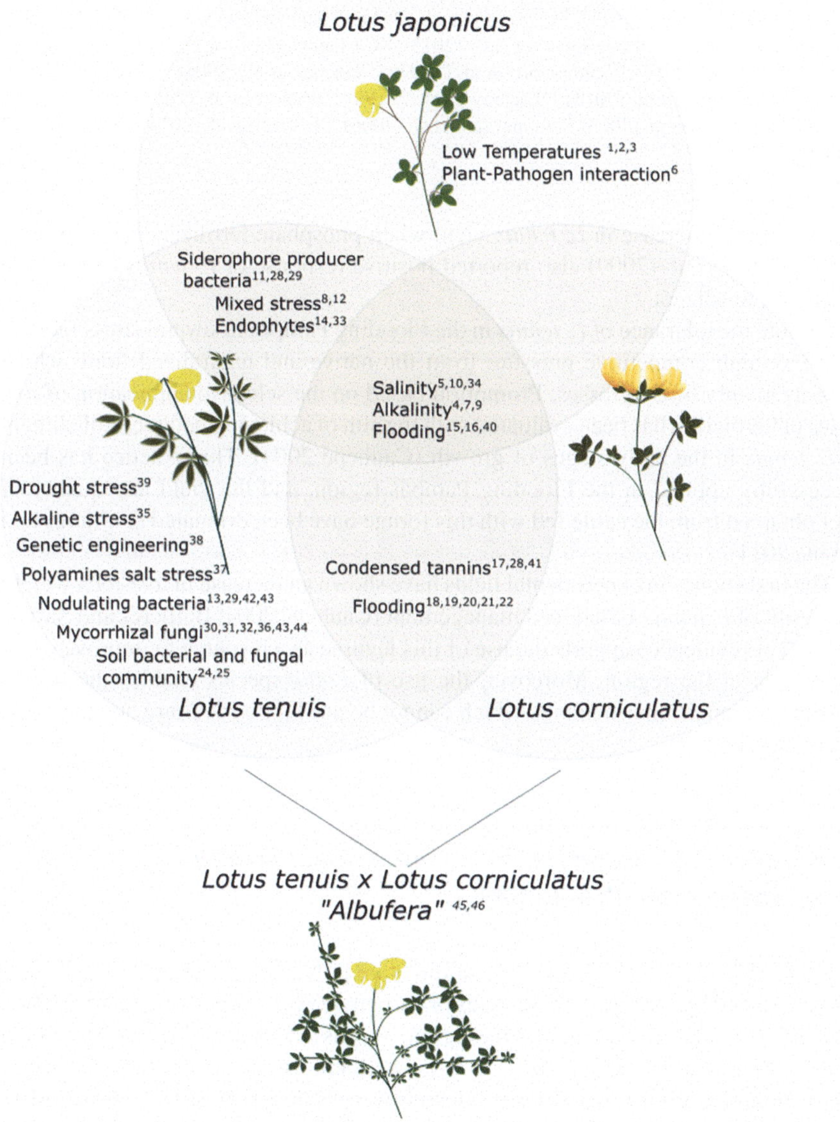

Fig. 2 Main topics analysed in recent years during research on *Lotus* species after their introduction into the lowlands of the Flooding Pampas Area (Buenos Aires Province, Argentina). The information gathered prompted the development of the hybrid *Lotustenuis* x *Lotus corniculatus* ("Albufera") [1]Calzadilla et al. (2016a),[2]Calzadilla et al. (2016b),[3]Calzadilla et al. (2019),[4]Babuin et al. (2014),[5]Espasandin et al. (2018),[6]Bordenave et al. (2013),[7]Bordenave et al. (2017),[8]Bordenave et al. (2019),[9]Campestre et al. (2016a),[10]Sanchez et al. (2008),[11]Campestre et al. (2016b),[12]Paz et al. (2012),[13]Sannazzaro et al. (2011),[14]Nieva et al. (2019),[15]Striker (2012),[16]Escaray et al.

Fig. 2 (continued) (2012a),[17]Escaray et al. (2012b),[18]Vignolio et al. (1994),[19]Vignolio et al. (1995),[20]García and Mendoza (2014),[21]Striker et al. (2005),[22]Antonelli et al. (2019),[23]Sanchez et al. (2011),[24]Nieva et al. (2016),[25]Nieva et al. (2018),[26]Acuña et al. 2008,[27]Estrella et al. (2009),[28]Castagno et al. (2011),[29]Castagno et al. (2014),[30]Sannazzaro et al. (2004),[31]Escudero and Mendoza (2005),[32]Echeverria et al. (2013),[33]Castagno et al. (2011),[34]Uchiya et al. (2016),[35]Paz et al. (2014),[36]Sannazzaro et al. (2007),[37]Sanchez et al. (2005),[38]Espasandin et al. (2010),[39]Espasandin et al. (2014),[40]Escaray et al (2019a),[41]Escaray et al. (2017),[42]Solans et al. (2015),[43]Echeverria et al. (2008),[44]Sannazzaro et al. (2006),[45]Escaray et al. (2014),[46]Escaray et al. (2019b).

a 5- to 20-fold increase in *L. tenuis* yield when phosphate fertilizers were applied, and Mendoza et al. (2009) also reported positive response of *L. tenuis* to different phosphate fertilizers.

Despite the tolerance of *L. tenuis* in the Flooding Pampas to several stress factors, it suffers high competitive pressure from the native and naturalized flora, which consists mainly of C4 grasses. Promotion based on the selective application of low doses of herbicides has been evaluated with the aim of achieving better establishment of *L. tenuis* in the early stages of growth (Cauhépé 2004). This practice has been successfully applied in the Flooding Pampas region, and the yield and quality of beef obtained from the cattle fed with this forage have been evaluated (Bailleres and Sarena 2011).

The first studies in experimental fields have shown an increase in livestock weight as a result of *L. tenuis*-based feed management (Cauhépé 2004; Bailleres and Sarena 2011). This evidence supports the use of this legume as a resource to improve cattle production in the region. Moreover, the use of *Lotus* species enables the use of restrictive edaphic environments which cannot be used for agriculture and have low forage production.

4 Condensed Tannins (CT): From Animal Health to Sustainable Production

Condensed Tannins (*syn.* Proanthocyanidins) are polymers of flavan-3-ol or flavan-3,4-diol linked by carbon–carbon bonds or carbon-oxygen-carbon bonds (McMahon et al. 2000), which are present in some legume organs (Jackson et al. 1996; McMahon et al. 2000; Escaray et al. 2012b). It is well known that some *Lotus* species can accumulate CT in their leaves (Gebrehiwot et al. 2002). CT are involved in several aspects of the cattle-raising ecosystem, including nutrition and animal health. Regarding nutrition, tannins are able to bind proteins and mitigate rumen fermentation (McMahon et al. 2000). At the same time, they provide protection against gastrointestinal parasites (Aerts et al. 1999).

Condensed tannins are also involved in the improvement of ecological and environmental functions. Tannins in the rumen can reduce ammonia and methane emission (Mueller-Harvey et al. 2019). These traits increase the nutritional value of

Lotus spp. as a forage resource and provide a strategy to be used in the sustainable management of cattle productive systems.

As noted above, legumes can accumulate condensed tannins. In some species, such as *Medicago* and *Trifolium,* CT are located mostly in the seeds (Lees et al. 1995; Bagci 2011). However, the beneficial effects of CT in animal feed are relevant when CT are located in the leaves. *L. corniculatus* tetraploid varieties accumulate higher levels of CT in leaves than *L. tenuis* (Acuña et al. 2008). In an attempt to leverage the well-known attributes of forage legumes with adequate levels of CT in livestock production systems, Escaray et al. (2014, 2017) have designed an interspecific hybrid between *L. tenuis* and *L. corniculatus.* This new plant material combines adequate concentrations of CT in leaves of commercial varieties of *L. corniculatus* with the tolerance to abiotic stresses of the accessions of *L. tenuis* naturalized in the lowlands of the Salado River Basin.

Recent evaluations of the *L. tenuis* x *L. corniculatus* hybrid have demonstrated the tolerance of this new, improved plant material to salt stress by accumulation of chlorides and anthocyanins (Escaray et al. 2019a, b). The hybrid has also performed well during partial submergence, displaying similar tolerance to flooding as that observed in *L. tenuis* (Antonelli et al. 2019). These studies suggest that the *L. tenuis* stress tolerance trait remains intact in the hybrid.

Current research is evaluating the productivity and stress tolerance of this new material in different environments and locations, even outside the Flooding Pampas region. The ultimate aim of the studies on the hybrid is to extend the use of *Lotus* species as forage sources to new environments. Accordingly, this interspecific hybrid has been registered in the National Seed Institute (INASE-Argentina) under the name of "Albufera", and is currently being analysed in other landscapes with different edaphic and environmental conditions.

5 Microorganisms Associated to Naturalization of *L. tenuis* in the Flooding Pampas Environments

Bacteria and fungi have been found to influence the adaptation of *L. tenuis* and improve its tolerance to abiotic stresses through their dynamic interaction. Like most legumes, members of the *Lotus* genus establish mutualistic interactions with rhizobial bacteria. There is a wide range of nodulating bacteria that can interact with *L. tenuis*, the main genera in the Flooding Pampas are *Mesorhizobium* and *Rhizobium* (Estrella et al. 2009; Sannazzaro et al. 2011).

The nutritional traits related to microorganisms, such as nitrogen fixation and their adaptation to low nutrients levels, have been the main focus in the research and selection of microbes associated with *Lotus* in the Flooding Pampas. In the context of the shortage of available phosphorus and iron in this ecosystem, phosphate-solubilising and siderophore-producing bacteria play important roles. *L. tenuis* is able to establish interactions with both phosphate-solubilising and siderophore-producing bacteria such as *Pantoea eucalypti* (Castagno et al. 2011, 2014; Campestre et al. 2016b). In addition, endophytic fungi of *L. tenuis* are able to solubilise phosphate and tolerate high salt and pH levels (Nieva et al. 2019).

The role that these microorganisms play in the mitigation of abiotic stress in *L. tenuis* has been exhaustively evaluated and reported. In the Flooding Pampas soils, *L. tenuis* is able to interact with *Rhizophagus* spp. (ex. *Glomus*) and *Acaulospora* sp. (Sannazzaro et al. 2004; Escudero and Mendoza 2005). In addition to the nutritional benefits that they provide through symbiotic interactions, these organisms have other important functional traits, such as the interaction between mycorrhizal fungi and *L. tenuis* and its relationship with the amelioration of stress due to salinity (Sannazzaro et al. 2006) and flooding (Mendoza et al. 2005).

Rhizobial bacteria have also been shown to be associated with the CT content in roots, because the concentration and composition of these secondary metabolites can modulate the symbiotic compatibility between *Lotus* spp. and their rhizobial partners (Lorite et al. 2018). Moreover, other plant metabolites involved in the plant stress response, (e.g.: polyamines and proline) have also been related to the microbial interactions mediated by *Glomus intraradices* and *Mesorhizobium thianshaense* (Echeverria et al. 2013). These studies highlight the importance of the microbiome in the ecosystem and in the establishment of a non-native plant, and of its potential participation in the selection and establishment of the plant community related to *L. tenuis* naturalization.

6 Impact of the Introduction of *Lotus* spp. on Soil Properties

The introduction of legumes in pasture systems can improve physical properties of the soil by reducing bulk density and improving water conductance (Armstrong et al. 1999). *Lotus* species can improve fertility and soil characteristics by improving drainage and further reducing soil salinity (Teakle et al. 2007).

The root systems of legumes in the Flooding Pampas can ameliorate flooding effects in some areas by increasing the infiltration rate (Sentís 2014). In this regard, the use of *L. tenuis* in the region has demonstrated its ability to improve the physical properties of the soil. Furthermore, as a consequence of the modification of the water-related properties in the soil, *L. tenuis* can also help to manage salt lixiviation. *L. tenuis* provides an important source of nitrogen in the cattle-raising systems of the Flooding Pampas. As a legume, it can interact with nitrogen-fixing bacteria and improve soil fertility by increasing nitrogen concentration. *Lotus* spp. can fix approximately 40 kg N/ha/year (Refi et al. 1989; García et al. 2008). This capacity constitutes an advantageous ability in comparison to other plant genera, determining its role as "pioneer plant" in degraded, contaminated or burned ecosystems, with direct effect on microbial activity, organic matter cycling and the dynamics of the C/N ratio.

This improvement in the soil can directly or indirectly affect all members of the plant community, leading to better yield and quality of the grassland ecosystems. The balance of observations made in the Salado River Basin suggests that the introduction of *L. tenuis* constitutes a "positive biological invasion" with potential benefits for other members of the plant community.

7 Impact of the Introduction of *Lotus* spp. on Soil Microbial Diversity

The introduction of exotic species can shift the balance in the soil ecosystem, and the functional attributes of the microbial community could be affected by the modification of the physical and chemical characteristics. As mentioned above, the introduction of *L. tenuis* in the Flooding Pampas was followed by the use of herbicides and fertilizers, which improved its productivity and establishment.

Analysis of the effect of 8 years of herbicide-mediated promotion of *L. tenuis* on the soil microbial community showed that the change in the composition of the flora did not affect the diversity of the soil bacterial community (Nieva et al. 2016), and that soil fungal diversity increased (Nieva et al. 2018). These results suggest that the fungal community may be more sensitive to changes in the soil physicochemical features than the bacterial community is. It has been also demonstrated that there was a 40% reduction in the number of mycorrhizal arbuscules in *L. tenuis* as consequence of glyphosate-mediated promotion (Druille et al. 2017). Recent observations provide evidence of a complex interrelation between *Fusarium solani* and *Lotus* spp. (Nieva et al. 2019), which has given rise to the hypothesis of a differential regulation between plants and microorganisms, with potential impact on stress response and biodiversity. The interactions between microorganisms and *L. tenuis* in the framework of abiotic stress conditions are an important topic in which further research is needed in order to achieve better understanding of the outcome of the interactions between soil, epiphytic microorganisms, endophytes and *L. tenuis.*

8 Relationships Between *L. tenuis* and its Partners in the Plant Community in the Flooding Pampas

Lotus spp. share the habitat with other introduced species and native species in the Flooding Pampas ecosystem, which creates competition. As mentioned above, the presence of *L. tenuis* in the Flooding Pampas ecosystems caused changes in the grassland composition (Perelman et al. 2001; Cid et al. 2011). However, despite competition with other members of the plant community, *Lotus* spp. were able to prevail thanks to their ability to grow and develop under restrictive soil conditions, and over several years, *L. tenuis* became successfully established and naturalized in the Flooding Pampas.

Lotus spp. have been used in mixed forage systems in combination with *Festuca arundinacea*, *Paspalum* spp., *Agropyron sp.* and *Trifolium repens.* The grass *F. arundinacea* and the legume *L. tenuis* have both been introduced and naturalized in the Flooding Pampas, and despite competing with each other, they both survive the same environment (Sevilla et al. 1996). This is an advantage that has been leveraged in high-quality, management systems using mixed pastures as a productive strategy to improve cattle feed. Moreover, *L. tenuis* is not poisonous to cattle, in contrast to *Schedonorus arundinaceus* (tall fescue), in which the endophytic fungus *Epichloë coenophiala,* which produces fungal alkaloids, is often observed growing in its apoplast and is transmitted vertically through the seeds to successive plant generations.

In addition, there is strong evidence of the facilitation effect of *L. tenuis* on other plant species in the community, such as *Paspalum dilatatum* (Quinos et al. 1998). The use of *L. tenuis* may generate new benefits related to pasture resource enrichment in association with other valuable forage species and the possibility of taking advantage of this quality in the management of grassland systems.

9 Greenhouse Gas Emission: Relevance and Perspectives

Within the framework of global climate change and the search for sustainable agriculture, the use of new strategies for improving crop production and at the same time maintaining the balance of natural resources is becoming relevant.

Greenhouse gas emissions (GHG) are strongly related to the global warming phenomenon. Although the burning of fossil fuels is the largest source of GHG production, agricultural activities are also significantly involved in global warming (Khandekar et al. 2005). Cattle-raising systems are of particular importance in sustainable agriculture management and involve points of discussion regarding the mitigation of GHG that contribute directly to global warming. Cattle production contributes to GHG through the methane gas produced by livestock during digestion, which is directly released into the atmosphere. Furthermore, urine and nitrogen-based fertilizers, in combination with microbial activity, contribute to the production of nitroxide gases.

Strategies to mitigate GHG in cattle-raising ecosystems depend on pasture conservation, cattle management and soil type (Sudmeyer et al. 2016). The amount of GHG produced by each component of cattle-raising in the Flooding Pampas is currently under study.

The use of *L. tenuis* in the Flooding Pampas environment may contribute to mitigating GHG. The effects of *L. tenuis* colonization in recent years has improved water movement through the soil. Better water infiltration and circulation of oxygen in the soil may in the long term contribute to the reduction of the *Archaea*, a domain that includes organisms responsible for soil methane production.

In spite of the normal GHG production by cattle-raising systems, the nutritional benefits provided by *L. tenuis* could ameliorate enteric methane emissions. MacAdam et al. (2016) reported the reduction of enteric methane emission by *L. corniculatus* compared to *Bromus riparus*. Furthermore, Naumann et al. (2018) demonstrated the correlation between CT content and reduction in methane emission. This recent evidence provides a starting point for new research perspectives on the sustainable management of grasslands and further improvement in productivity. The introduction of *L. tenuis* could be useful to improve cattle production in the Flooding Pampas. Similarly, the progressive introduction of plants with improved agricultural traits, such as the "Albufera" hybrid could also be relevant to the nutritional improvement of grasslands in the saline-alkaline lowlands, at the same time contributing to the reduction of GHG emissions.

10 Conclusions

The introduction of exotic legumes in the livestock system improves the quality of forage resources. The introduction of *L. tenuis* in Argentina has provided an important tool for improving livestock production. It is also an ecological access point for improving carbon balance in the Flooding Pampas. The ability of *L. tenuis* to tolerate abiotic stress conditions means that its establishment could improve soil conditions, thereby benefiting the rest of the plant community and biome, potentially enabling these soils to be used for different purposes. In addition, the research on the *L. japonicus* has provided further information on the mechanisms involved in the tolerance of *Lotus* species to abiotic and biotic stresses. At the same time, research on *L. tenuis,* the development of biotechnological approaches, and the environmental effects derived from its successful naturalization, provide an important source of information that has accumulated progressively.

In summary, the use of *L. tenuis* in the Flooding Pampas has turned an area restricted for agriculture into a sustainable region used for cattle production. The results obtained to date show that the promotion of this legume in the region benefits animal nutrition and soil fertility, and constitutes an environmental-friendly improvement in livestock production systems.

References

Acuña H, Concha A, Figueroa M (2008) Condensed tannin concentrations of three Lotus species grown in different environments. Chil J Agr Res 68(1):31–41

Aerts RJ, Barry TN, McNabb WC (1999) Polyphenols and agriculture: beneficial effects of proanthocyanidins in forages. Agric Ecosyst Environ 75(1–2):1–12

Antonelli CJ, Calzadilla PI, Escaray FJ, Babuin MF, Campestre MP, Rocco R, Bordenave C, Perea-García A, Nieva AS, Maguire V, Melani G, Sarena D, Bailleres C, Carrasco P, Paolocci F, Gárriz

A, Menéndez A, Ruiz O (2016) *Lotus spp*: Biotechnological strategies to improve the bioeconomy of lowlands in the Salado River Basin (Argentina). AGROFOR Int J 1:43–53

Antonelli CJ, Calzadilla PI, Vilas JM, Campestre MP, Escaray FJ, Ruiz OA (2019) Physiological and anatomical traits associated with tolerance to long-term partial submergence stress in the *Lotus* genus: responses of forage species, a model and an interspecific hybrid. J Agron Crop Sci 205(1):65–76

Armstrong RD, Kuskopf BJ, Millar G, Whitbread AM, Standley J (1999) Changes in soil chemical and physical properties following legumes and opportunity cropping on a cracking clay soil. Aust J Exp Agric 39(4):445–456

Babuin MF, Campestre MP, Rocco R, Bordenave CD, Escaray FJ, Antonelli C, Calzadilla P, Gárriz A, Serna E, Carrasco P, Menéndez A, Ruiz OA (2014) Response to long-term NaHCO3-derived alkalinity in model *Lotus japonicus* Ecotypes Gifu B-129 and Miyakojima MG-20: transcriptomic profiling and physiological characterization. PLoS ONE 9(5):e97106

Bagci E (2011) Tannin, protein contents and fatty acid composition of the seeds of some *Trifolium L.* species from Turkey. Asian J Anim Vet Adv 6(1):88–95

Bailleres M, Sarena D (2011) Promoción Estival de *Lotus tenuis*. Ida & Vuelta Rural 9:29

Bordenave CD, Escaray FJ, Menéndez AB, Serna E, Carrasco P, Ruiz OA, Gárriz A (2013) Defense responses in two ecotypes of *Lotus japonicus* against non-pathogenic *Pseudomonas syringae*. PLoS ONE 8(12):e83199

Bordenave CD, Rocco R, Babuin MF, Campestre MP, Escaray FJ, Gárriz A, Antonelli C, Carrasco P, Ruiz OA, Menéndez AB (2017) Characterization of the primary metabolome during the long-term response to $NaHCO_3$-derived alkalinity in *Lotus japonicus* ecotypes Gifu B-129 and Miyakojima MG-20. Acta Physiol Plant 39(3):76

Bordenave CD, Rocco R, Maiale SJ, Campestre MP, Ruiz OA, Rodríguez AA, Menéndez AB (2019) Chlorophyll a fluorescence analysis reveals divergent photosystem II responses to saline, alkaline and saline–alkaline stresses in the two *Lotus japonicus* model ecotypes MG20 and Gifu-129. Acta Physiol Plant 41(9):167

Calzadilla PI, Maiale SJ, Ruiz OA, Escaray FJ (2016a) Transcriptome response mediated by cold stress in *Lotus japonicus*. Front Plant Sci 7:374

Calzadilla PI, Signorelli S, Escaray FJ, Menéndez AB, Monza J, Ruiz OA, Maiale SJ (2016b) Photosynthetic responses mediate the adaptation of two *Lotus japonicus* ecotypes to low temperature. Plant Sci 250:59–68

Calzadilla PI, Vilas JM, Escaray FJ, Unrein F, Carrasco P, Ruiz OA (2019) The increase of photosynthetic carbon assimilation as a mechanism of adaptation to low temperature in *Lotus japonicus*. Sci Rep 9(1):863

Campestre MP, Antonelli C, Calzadilla PI, Maiale SJ, Rodríguez AA, Ruiz OA (2016a) The alkaline tolerance in *Lotus japonicus* is associated with mechanisms of iron acquisition and modification of the architectural pattern of the root. J Plant Physiol 206:40–48

Campestre MP, Castagno LN, Estrella MJ, Ruiz OA (2016b) *Lotus japonicus* plants of the Gifu B-129 ecotype subjected to alkaline stress improve their Fe^{2+} bio-availability through inoculation with *Pantoea eucalypti* M91. J Plant Physiol 192:47–55

Castagno LN, Estrella MJ, Sannazzaro AI, Grassano AE, Ruiz OA (2011) Phosphate -solubilization mechanism and in vitro plant growth promotion activity mediated by *Pantoea eucalypti* isolated from *Lotus tenuis* rhizosphere in the Salado River Basin (Argentina). J Appl Microbiol 110(5):1151–1165

Castagno LN, García IV, Sannazzaro AI, Bailleres M, Ruiz OA, Mendoza RE, Estrella MJ (2014) Growth, nutrient uptake and symbiosis with rhizobia and arbuscular mycorrhizal fungi in *Lotus tenuis* plants fertilized with different phosphate sources and inoculated with the phosphate-solubilizing bacterium *Pantoea eucalypti* M91. Plant Soil 385(1–2):357–371

Cauhépé M (2004) Does *Lotus glaber* improve beef production at the Flooding Pampas? Lotus Newsletter 34:38–43

Cid MS, Grecco RF, Oesterheld M, Paruelo JM, Cibils AF, Brizuela MA (2011) Grass-fed beef production systems of Argentina's Flooding Pampas: Understanding ecosystem heterogeneity to improve livestock production. Outlook Agriculture 40(2):181–189

Clua A, Orsini H, Beltrano J (2009) Incidence of variable flooding period on *Lotus tenuis* biomass production and leaf senescence. Lotus Newsletter 39(1):13–20

Degtjareva GV, Kramina TE, Sokoloff DD, Samigullin TH, Valiejo-Roman CM, Antonov AS (2006) Phylogeny of the genus *Lotus* (Leguminosae, Loteae): evidence from nrITS sequences and morphology. Botany 84(5):813–830

Druille M, Acosta AP, Acosta GL, Rossi JL, Golluscio RÁ, Bailleres M (2017) Response to glyphosate application of beneficial soil fungi associated with *Lotus tenuis*. RIA Rev Investig Agropecu 43(3):297–302

Echeverria M, Scambato AA, Sannazzaro AI, Maiale S, Ruiz OA, Menéndez AB (2008) Phenotypic plasticity with respect to salt stress response by *Lotus glaber*: The role of its AM fungal and rhizobial symbionts. Mycorrhiza 18(6–7):317–329

Echeverria M, Sannazzaro AI, Ruiz OA, Menéndez AB (2013) Modulatory effects of *Mesorhizobium tianshanense* and *Glomus intraradices* on plant proline and polyamine levels during early plant response of *Lotus tenuis* to salinity. Plant Soil 364(1–2):69–79

Escaray FJ, Menéndez AB, Gárriz A, Pieckenstain FL, Estrella MJ, Castagno LN, Carrasco P, Sanjuan J, Ruiz OA (2012a) Ecological and agronomic importance of the plant genus *Lotus*. Its application in grassland sustainability and the amelioration of constrained and contaminated soils. Plant Sci 182:121–133

Escaray FJ, Passeri V, Perea García A, Antonelli CJ, Damiani F, Ruiz OA, Paolocci F (2017) The R2R3- MYB TT2b and the bHLH TT8 genes are the major regulators of proanthocyanidin biosynthesis in *Lotus* species. Planta 246(2):243–261

Escaray FJ, Passeri V, Babuin FM, Marco F, Carrasco P, Damiani F, Pieckenstain FL, Paolocci F, Ruiz OA (2014) *Lotus tenuis x L. corniculatus* interspecific hybridization as a means to breed bloat-safe pastures and gain insight into the genetic control of proanthocyanidin biosynthesis in legumes. BMC Plant Biol 14(1):40

Escaray FJ, Rosato M, Pieckenstain FL, Menéndez AB, Rosselló JA, Carrasco P, Ruiz OA (2012b) The proanthocyanidin content as a tool to differentiate between *Lotus tenuis* and *L. corniculatus* individuals. Phytochemistry Lett 5(1):37–40

Escaray FJ, Antonelli CJ, Copello GJ, Puig S, Peñarrubia L, Ruiz OA, Perea-García A (2019a) Characterization of the Copper Transporters from *Lotus* spp. and their Involvement under Flooding Conditions. Int J Mol Sci 20(13):3136

Escaray FJ, Antonelli CJ, Carrasco P, Ruiz OA (2019b) Interspecific hybridization improves the performance of *Lotus* spp. under saline stress. Plant Sci 283:202–210

Escudero V, Mendoza R (2005) Seasonal variation of arbuscular mycorrhizal fungi in temperate grasslands along a wide hydrologic gradient. Mycorrhiza 15(4):291–299

Espasandin FD, Collavino MM, Luna CV, Paz RC, Tarragó JR, Ruiz OA, Mroginski LA, Sansberro PA (2010) *Agrobacterium tumefaciens*-mediated transformation of *Lotus tenuis* and regeneration of transgenic lines. Plant Cell Tiss Org Cult 102(2):181–189

Espasandin FD, Maiale SJ, Calzadilla P, Ruiz OA, Sansberro PA (2014) Transcriptional regulation of 9-cis-epoxycarotenoid dioxygenase (NCED) gene by putrescine accumulation positively modulates ABA synthesis and drought tolerance in *Lotus tenuis* plants. Plant Physiol Biochem 76:29–35

Espasandin FD, Calzadilla PI, Maiale SJ, Ruiz OA, Sansberro PA (2018) Overexpression of the arginine decarboxylase gene improves tolerance to salt stress in *Lotus tenuis* plants. J Plant Growth Regul 37(1):156–165

Estrella MJ, Munoz S, Soto MJ, Ruiz O, Sanjuan J (2009) Genetic diversity and host range of rhizobia nodulating *Lotus tenuis* in typical soils of the Salado River Basin (Argentina). Appl Environ Microbiol 75(4):1088–1098

García I, Mendoza R (2014) *Lotus tenuis* seedlings subjected to drought or waterlogging in a saline sodic soil. Environ Exp Bot 98:47–55

García I, Mendoza R, Pomar MC (2008) Deficit and excess of soil water impact on plant growth of *Lotus tenuis* by affecting nutrient uptake and arbuscular mycorrhizal symbiosis. Plant Soil 304(1–2):117–131

Gebrehiwot L, Beuselinck PR, Roberts CA (2002) Seasonal variations in condensed tannin concentration of three *Lotus* species. Agron J 94(5):1059–1065

Jackson FS, Barry TN, Lascano C, Palmer B (1996) The extractable and bound condensed tannin content of leaves from tropical tree, shrub and forage legumes. J Sci Food Agric 71(1):103–110

Khandekar ML, Murty TS, Chittibabu P (2005) The global warming debate: A review of the state of science. Pure Appl Geophys 162(8–9):1557–1586

Lees GL, Suttill NH, Wall KM, Beveridge TH (1995) Localization of condensed tannins in apple fruit peel, pulp, and seeds. Can J Bot 73(12):1897–1904

León RJ (2000) Heterogeneidad espacial y temporal de la vegetación pampeana. Anales de la Acad. Nac. de Agron. y Vet. Tomo LIV pp. 336-379

Lorite MJ, Estrella MJ, Escaray FJ, Sannazzaro A, Videira e Castro IM, Monza J, Sanjuán J, León-Barrios M (2018) The Rhizobia-Lotus Symbioses: deeply specific and widely diverse. Front Microbiol 9:2055

MacAdam JW, Beauchemin KA, Bolletta AI, Pitcher LR (2016) Reduced enteric methane emissions on legume versus grass irrigated pastures. J Anim Sci 94:315–316

McMahon LR, McAllister TA, Berg BP, Majak W, Acharya SN, Popp JD, Coulman BE, Wang Y, Cheng KJ (2000) A review of the effects of forage condensed tannins on ruminal fermentation and bloat in grazing cattle. Can J Plant Sci 80(3):469–485

Mendoza R, Escudero V, García I (2005) Plant growth, nutrient acquisition and mycorrhizal symbioses of a waterlogging tolerant legume (*Lotus glaber Mill.*) in a saline-sodic soil. Plant Soil 275(1–2):305–315

Mendoza R, Lamas MC, García I (2009) How do soil P tests, plant yield and P acquisition by *Lotus tenuis* plants reflect the availability of added P from different phosphate sources. Nutr Cycling Agroecosyst 85(1):17–29

Mueller-Harvey I, Bee G, Dohme-Meier F, Hoste H, Karonen M, Kölliker R, Lüscher A, Niderkorn V, Pellikaan WF, Salminen JP, Skøt L, Smith LMJ, Totterdell P, Wilkinson I, William AR, Azuhnwi B, Baert N, Brinkhaus AG, Copani G, Desrues O, Drake C, Engström M, Fryganas C, Girard M, Huyen NT, Kempf K, Malisch C, Mora-Ortizi M, Quijada J, Ramsay A, Kopiak HM, Waghrn GC (2019) Benefits of condensed tannins in forage legumes fed to ruminants: Importance of structure, concentration, and diet composition. Crop Sci 59(3):861–885

Naumann H, Sepela R, Rezaire A, Masih S, Zeller W, Reinhardt L, Robe J, Sullivan M, Hagerman A (2018) Relationships between structures of condensed tannins from Texas legumes and methane production during in vitro rumen digestion. Molecules 23(9):2123

Nieva AS, Bailleres MA, Corriale MJ, Llames ME, Menéndez AB, Ruiz OA (2016) Herbicide-mediated promotion of *Lotus tenuis* (Waldst. & Kit. ex Wild.) did not influence soil bacterial communities, in soils of the Flooding Pampa, Argentina. Appl Soil Ecol 98:83–91

Nieva AS, Bailleres MA, Llames ME, Taboada MA, Ruiz OA, Menéndez A (2018) Promotion of *Lotus tenuis* in the Flooding Pampa (Argentina) increases the soil fungal diversity. Fungal Ecol 33:80–91

Nieva AS, Vilas JM, Gárriz A, Maiale SJ, Menéndez AB, Erban A, Kopka J, Ruiz OA (2019) The fungal endophyte *Fusarium solani* provokes differential effects on the fitness of two *Lotus* species. Plant Physiol Biochem 144:100–109

Paz RC, Rocco RA, Reinoso H, Menéndez AB, Pieckenstain FL, Ruiz OA (2012) Comparative study of alkaline, saline, and mixed saline–alkaline stresses with regard to their effects on growth, nutrient accumulation, and root morphology of *Lotus tenuis*. J Plant Growth Regul 31(3):448–459

Paz RC, Reinoso H, Espasandin FD, González Antivilo FA, Sansberro PA, Rocco RA, Ruiz OA, Menéndez AB (2014) Alkaline, saline and mixed saline–alkaline stresses induce physiological and morpho-anatomical changes in *Lotus tenuis* shoots. Plant Biol 16(6):1042–1049

Perelman SB, León RJC, Oesterheld M (2001) Cross-scale vegetation patterns of Flooding Pampa grasslands. J Ecol 89(4):562–577

Quinos PM, Insausti P, Soriano A (1998) Facilitative effect of *Lotus tenuis* on *Paspalum dilatatum* in a lowland grassland of Argentina. Oecologia 114(3):427–431

Real D, Sandral GA, Rebuffo M, Hughes SJ, Kelman WM, Mieres J M, Dods K, Crossa J (2013). Breeding of an early-flowering and drought-tolerant *Lotus corniculatus L.* variety for the high-rainfall zone of southern Australia. Crop Pasture Sci 63(9):848–857

Refi RO, Navarro CA, Escudero CJ (1989) Section 2: Nitrogen fixation, cycling and utilization. Nitrogen fixation in mixed pastures under two grazing intensities. Presented in 16. International Grassland Congress, Nice (France), 4–11 Oct 1989

Rodríguez AM, Jacobo EJ, Scardaoni P, Deregibus VA (2007) Effect of phosphate fertilization on Flooding Pampa grasslands (Argentina). Rangeland Ecol Manag 60(5):471–478

Sanchez DH, Cuevas JC, Chiesa MA, Ruiz OA (2005) Free spermidine and spermine content in *Lotus glaber* under long-term salt stress. Plant Sci 168(2):541–546

Sanchez DH, Lippold F, Redestig H, Hannah MA, Erban A, Krämer U, Kopka J, Udvardi MK (2008) Integrative functional genomics of salt acclimatization in the model legume *Lotus japonicus*. Plant J 53(6):973–987

Sanchez DH, Pieckenstain FL, Escaray F, Erban A, Kraemer UTE, Udvardi MK, Kopka J (2011) Comparative ionomics and metabolomics in extremophile and glycophytic *Lotus* species under salt stress challenge the metabolic pre-adaptation hypothesis. Plant Cell Environ 34(4):605–617

Sannazzaro AI, Ruiz OA, Albertó E, Menéndez AB (2004) Presence of different arbuscular mycorrhizal infection patterns in roots of *Lotus glaber* plants growing in the Salado River basin. Mycorrhiza 14(2):139–142

Sannazzaro AI, Ruiz OA, Alberto EO, Menéndez AB (2006) Alleviation of salt stress in *Lotus glaber* by *Glomus intraradices*. Plant Soil 285(1–2):279–287

Sannazzaro AI, Echeverría M, Albertó EO, Ruiz OA, Menéndez AB (2007) Modulation of polyamine balance in *Lotus glaber* by salinity and arbuscular mycorrhiza. Plant Physiol Biochem 45(1):39–46

Sannazzaro AI, Bergottini VM, Paz RC, Castagno LN, Menéndez AB, Ruiz OA, Pieckenstain LF, Estrella MJ (2011) Comparative symbiotic performance of native rhizobia of the Flooding Pampa and strains currently used for inoculating *Lotus tenuis* in this region. Antonie Leeuwenhoek 99(2):371–379

Sentís IP (2014) Advances in the prognosis of soil sodicity under dryland irrigated conditions. Int Soil Water Conserv Res 2(4):50–63

Sevilla GH, Fernández ON, Miñón DP, Montes L (1996) Emergence and seedling survival of *Lotus tenuis* in *Festuca arundinacea* pastures. Rangeland Ecol Manag 49(6):509–511

Solans M, Ruiz OA, Wall LG (2015) Effect of actinobacteria on *Lotus tenuis–Mesorhizobium loti* symbiosis: preliminary study. Symbiosis 65(1):33–37

Striker GG, Insausti P, Grimoldi AA, Ploschuk EL, Vasellati V (2005) Physiological and anatomical basis of differential tolerance to soil flooding of *Lotus corniculatus* L. and *Lotus glaber* Mill. Plant Soil 276(1–2):301–311

Striker GG, Izaguirre RF, Manzur ME, Grimoldi AA (2012) Different strategies of *Lotus japonicus, L. corniculatus* and *L. tenuis* to deal with complete submergence at seedling stage. Plant Biol 14(1):50–55

Strittmatter CD, Wagner ML, Kade M, Gurni AA (1992) Identification of *Lotus tenuis* (Waldst et Kit.) flavonoids. Biochem System Ecol 20(7):685–687

Sudmeyer RA, Bennett A, Strawbridge M (2016) Climate-ready agriculture: a situation statement for Western Australia. Department of Agriculture and Food, Western Australia, Perth. Bulletin 4876

Teakle NL, Flowers TJ, Real D, Colmer TD (2007) *Lotus tenuis* tolerates the interactive effects of salinity and waterlogging by 'excluding' Na^+ and Cl^- from the xylem. J Exp Bot 58(8):2169–2180

Uchiya P, Escaray FJ, Bilenca D, Pieckenstain F, Ruiz OA, Menéndez AB (2016) Salt effects on functional traits in model and in economically important *Lotus* species. Plant Biol 18(4):703–709

Vignolio OR, Maceira NO, Fernández ON (1994) Efectos del anegamiento en invierno y verano sobre el crecimiento y la supervivencia de *Lotus tenuis* y *Lotus corniculatus*. Ecol Austral 4(01):019–028

Vignolio OR, Maceira NO, Fernández ON (1995) Efectos del anegamiento sobre el poder germinativo de las semillas de *Lotus tenuis* y *Lotus corniculatus*. implicancias para la propagación diferencial en la Depresión del Salado (Buenos Aires, Argentina). Ecol Austral 5(157):63

Future Perspectives

Climate Change and Salinity-Vulnerable Ecosystems in Latin America

Ernesto F. Viglizzo and M. Florencia Ricard

Abstract Soil salinization is the cause of soil degradation affecting the functional health of ecosystems, the habitat and the local biodiversity. No climatic zone in Latin America (LA) is completely free from salinization, although the most vulnerable areas are located in arid and semi-arid regions. Specific local cases in Argentina, Chile, Colombia, Paraguay and Mexico are described. Considering that global food production will need to increase by 38% by 2025 and by 57% by 2050, careful attention must be devoted to all degraded lands, especially those affected by salinity or sodicity. The threat increases when such lands trigger a feedback effect on the climate system. In turn, climate can affect ecosystem functions and food security. The combined result is a greater sensitivity of crops and pastures to drought. Adaptation requires a technological control of land-use/land-cover change, cropping types, cropping periods, agronomic practices, and water management. A new paradigm is growing in recent years that aims at remediating, restoring and converting degraded lands into protected areas to preserve the habitat as well as the local flora and fauna. Protection is usually accompanied by processes of carbon sequestration in plants and soils that mitigate the emission of greenhouse gases to the atmosphere.

Keywords Salinization · Climate change · Ecosystem functions · Food security

E. F. Viglizzo (✉) · M. F. Ricard
INCITAP-CONICET, Santa Rosa, La Pampa, Argentina
e-mail: eviglizzo@gmail.com

M. F. Ricard
e-mail: flor.ricard@gmail.com

E. F. Viglizzo
Facultad de Ciencias Empresariales, Universidad Austral, Buenos Aires, Argentina

M. F. Ricard
Facultad de Ciencias Exactas y Naturales, UNLPam, Santa Rosa, La Pampa, Argentina

E. Taleisnik and R. S. Lavado (eds.), *Saline and Alkaline Soils in Latin America*, https://doi.org/10.1007/978-3-030-52592-7_23

1 Rationale

Soil salinization is the cause of soil degradation in agricultural lands that are exposed to rain-fed, irrigation or flood conditions. In turn, it also affects the functional integrity of ecosystems by threatening the habitat and the local biodiversity.

While rainfall, aeolian deposits, mineral weathering, and stored salts are the sources of salts, surface and groundwater can redistribute the accumulated salts and may also provide additional amounts. In general, sodium salts dominate in many saline soils, but salts of other cations such as calcium, magnesium, and iron are also found in specific locations (Rengasamy 2006).

An increasing concern rises in areas that are menaced by drought and aridity expansion due to climate change (FAO 2011). In response to this, the scientific community is urged to detect early-warning signals of salinization in fragile areas that are exposed to rapid temperature, precipitation and evapotranspiration changes.

The salt-water balance and thus the water quality of arid and semiarid lands are particularly sensitive to temperature increase and precipitation decrease, depending on the lithology of the area (Pankova and Konyushkova 2013). Signals of water quality change and trends can be detected in rivers, streams and water bodies, as well as underground aquifers. It also should be noted that such changes can directly affect the above- and below-ground biodiversity of soils, which in turn will indirectly affect the development of plants that are necessary to provide food for humans and animals (Várallyay 2010).

2 The Identification of Vulnerable Ecosystems

The identification of ecosystems that are vulnerable to salinization is necessary to undertake adaptation policies in order to reduce the negative effects of salinization on food production and security (Battle Sales 2011). Beyond the well-known biophysical impacts, novel insights are needed to assess the social and economic vulnerability of regions to salts accumulation. Vulnerability should start by recognizing, both in time and space, the response of plants, animals and ecosystems to salinization episodes caused by climate change and variability (Jobbágy et al. 2017).

In general, flat sedimentary regions with low topographic gradients and poorly developed drainage networks impede draining the excess of water, determining stagnant hydrological systems that favor shallow water tables and long-lasting floods (Fan et al. 2013). Extended flat landscapes show low horizontal water transport as a result of low surface runoff and slow groundwater fluxes (Tóth 1963).

2.1 *General Picture of Salinization-Vulnerable Areas in Latin America*

Salinization affects the soils at different latitudes and times in Latin America (Fig. 1). No climatic zone in Latin America is completely free from salinization, although the general perception is focused on arid and semi-arid regions (Rengasamy 2006). According to different information summarized in Table 1, the actual total global area of salt-affected soils including saline and sodic soils is around 17,000,000 km^2, doubling those registered almost 30 years ago by Szabolcs (1989). That represents a little more than 12% of the total global land area. In the case of Latin America, about 1,730,000 km^2 of are affected by salinization, the third most important region in the world. Paraguay and Argentina have almost 30% of their territory with some type of salinity restriction, leading the Top 10 of countries with the greatest restrictions in Latin America (Fig. 2). In Brazil, due to the dimensions of the country, the area affected by salts does not seem to be significantly important in relative terms. However, in absolute terms, the affected area -especially in the Northeastern states- is critical for the region.

Salinization affects the soils at different latitudes and times in Latin America (Fig. 1).

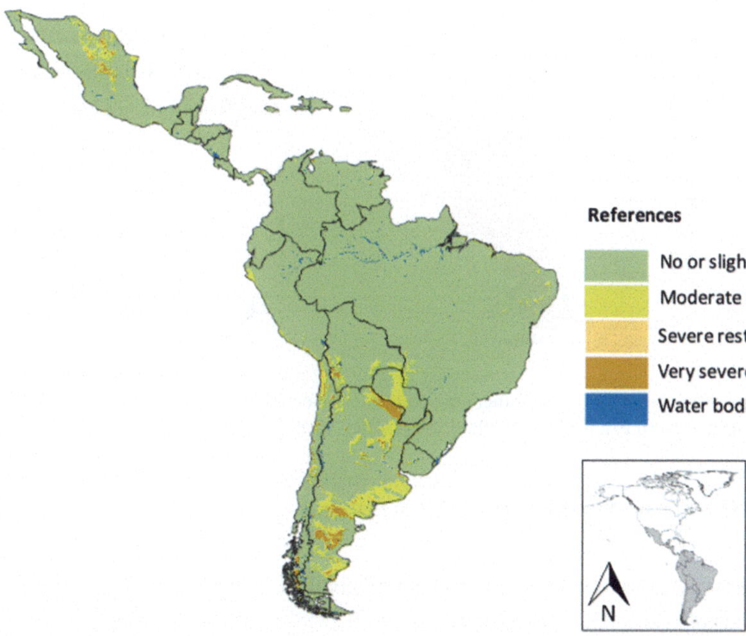

Fig. 1 Salts excess in Latin American. Modified from FAO/IIASA/ISRIC/ISSCAS/JRC (2012)

Table 1 Global distributions of saline and sodic soils

Continent	Area (million hectares)		
	Saline	Sodic	Total
Africa	412,2	208,0	620,2
Asia	378,6	236,8	615,3
Europe	19,6	57,7	77,3
Latin America	94,5	78,9	173,4
North America	36,6	56,7	93,4
Oceania	5,5	106,7	112,2

Source FAO/IIASA/ISRIC/ISSCAS/JRC (2012) and Szabolcs (1989).

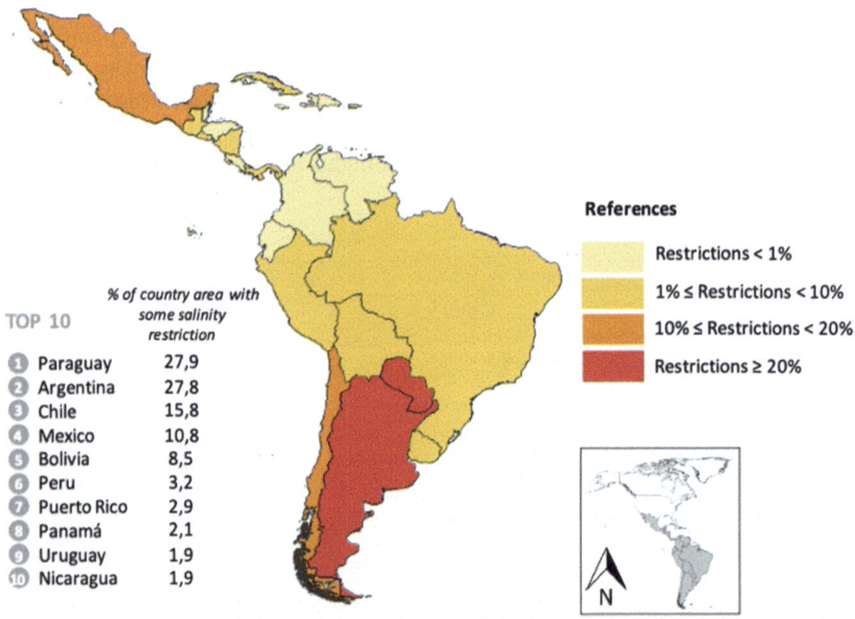

Fig. 2 Surface proportion of Latin American countries affected by some degree of salinity restriction. Modified from FAO/IIASA/ISRIC/ISSCAS/JRC (2012)

2.2 Specific Case Studies in Latin America

Selected cases correspond to countries that show greatest restrictions for saline soils in Latin America.

To assess the evolution of salinity in the dry Paraguayan Chaco, Glatzle et al. (2020) designed an experimental study on saline groundwater dynamics to better understand dryland salinity as a natural phenomenon, which is a potential consequence of land use change. Results show a relation among topography, land cover, precipitation, groundwater salinity and surface salinization. These findings provide information for land-use planning in the region, which is prone to dryland salinity due to a generally high saline water table.

To assess the evolution of salinity in the Cauca River Valley (Colombia) within the period 1980–2009, a study (Jiménez et al. 2011) demonstrated that of the 41,200 ha monitored, 20,830 ha have some degree of salinity, which remained constant throughout many years. The study also showed that the valley is very sensitive to climate variation events, especially to those associated with the El Niño Southern Oscillation (ENSO). The assessment of that sensitivity is essential to understand and predict the possible long-term impact of climate change and design adaptation strategies.

A study on the Yaqui Valley, Sonora, Mexico (González et al. 2011) aimed at analyzing and determining the groundwater salinization in the valley. The results showed that in some years the rate of water removed from the aquifer was larger than the rate of recharge. This situation increases the risk of saline intrusion of the coastal aquifer. On the contrary, in periods of high precipitation, infiltration from agricultural land may reach the aquifer thus raising the phreatic levels towards the surface. The salinization is aggravated because the potential evaporation of the region is about 2000 mm/year causing direct phreatic evapotranspiration and increased salt accumulation on soils and water. Adaptation strategies include the construction of 2350 km of agricultural drainage. Complementary actions enabled the recovery of 28,000 ha affected by salts.

Two significant flood episodes occurred in 2000–2003 and 2012–2013 in two of the most flood-prone areas of the Argentine Pampas (Kuppel el al. 2015), the so-called Western and Flooding Pampas with a coverage of 60,000 km^2 each. The surface water cover reached 31 and 19% during the first flooding episode in each subregion, while up to 22 and 10% were respectively covered during the second episode. Because of the different hydrological responses, precipitation and evapotranspiration were strongly linked in the Lower Pampas only, while the connection between water fluxes and storage was limited to the Western Pampas. In both regions, evapotranspiration losses were strongly linked to flooded conditions as a regulatory feedback, while liquid water outflows remained negligible. The two distinctive hydrological behaviors support the designing of efficient flood risk anticipation systems: in addition to the real-time rainfall and surface water monitoring in the entire region, a more extensive network of phreatic sensors would help accomplish that purpose.

Sandoval et al. (2013) reported an *in vitro* experiment to restore the soil affected by the tsunami of February 2010 that struck the Coliumo District, Bio-Bio region, Chile. The agricultural productivity of many coastal lands was severely affected, rendering them unfit for crop production. For this reason, the researchers analyzed multiple options to favor the growth of plants in that area.

3 The Effect of Climate Change-Salinization Relations on Ecosystem Functions and Food Security

Global food production will need to increase by 38% by 2025 and by 57% by 2050 if food supply to the growing world population is to be maintained at current levels. But on the other hand, more efforts will be required to guarantee food security as more lands are becoming degraded, especially those affected by salinization (Wild 2003). Nearly 6% of the total world land surface is affected by either salinity or sodicity events (Igbal et al. 2011).

Beyond land degradation, a changing climate will cause considerable changes in soil properties, natural vegetation, land use practices and food production security (Várallyay 2010). Of significant importance is that these changes in turn will result in a feedback effect on climate: modified albedo, surface roughness, micro-circulation processes and heat and energy balance of the near-surface atmosphere. This complex picture will demand a careful analysis to foresee uncertain consequences on food security. New analytical tools will be necessary to face this challenge.

What we know today is that change in the climate conditions can affect the ecosystems functions and food security in two ways: (i) the rise of temperature increase the potential evaporation and transpiration, and at the same time decreases runoff, infiltration, groundwater flow and water storage, especially if accompanied by low precipitation; (ii) The decrease in atmospheric precipitation will result in a decrease in water runoff, infiltration, water recharge and storage in soils, and will increase the sensitivity of crops and pastures to drought.

To some degree, humans can manipulate the conditions through land-use and land-cover change, cropping patterns, cultivation technologies, and water management through irrigation techniques, drainage developments and so on. Jointly, those human-dependent factors can help adaptation to increase food security under a likely scenario of increasing salinization risk.

Not always those effects seem to be clear. For example, in the so-called Flooding Pampas of Argentina, Chanetton and Lavado (1996) tested the hypothesis that long-term, human driven grazing exclusion in flooded lands would affect the impact of livestock on soil chemical properties and thus on salinization. They found that soil-vegetation changes in response to grazing appeared to be loosely coupled to changes in soil chemical properties in those rangeland ecosystems.

4 Concluding Remarks

The effect of climate change, with the expected changes in water flows and temperature and, consequently, in salt flows, will undoubtedly modify soil management practices or at least modify their application on the field. The techniques to increase soil cover and to improve the biological, physical and chemical soil condition will allow the control of the capillary rise from the phreatic water and the water flux toward depth.

On the other hand, regardless of global changes, soil management undergoes a changing paradigm that will re-shape the future. Since ancient times humans have drastically and irreversibly modified natural lands to convert them into croplands and grazing lands. Different marshlands, wetlands and lagoons have been transformed by Europeans since the middle ages. From the end of the nineteenth century, several attempts to colonize saline wetlands took place in Latin America. Practices such as drainage, surface water management, deep tillage, introduction of salt-tolerant crops and pastures, etc. were extensively applied (Shahid et al. 2017). In these cases the underlying idea was that those modifications were necessary and convenient, but should not have undesirable consequences. Thus, beyond ideologies and socio-political contexts, ecosystems in various areas of the American continent were modified with more or less successful results. On the other hand, both the habitat and biodiversity have been affected in variable degree.

In recent years a new growing paradigm drives the conversion of those lands in protected areas to preserve the local flora and fauna (e.g., Ramsar sites, national parks, natural reserves, etc.). The process is complemented by restoration efforts that aim at recovering, as much as possible, valuable functions from the original ecosystems. Remediation technologies for soil and vegetation are necessary to deal with such restoration efforts.

Another aspect in which changes will be evidenced is in the case of contamination. This will happen in all production systems, but in the case of soils affected by salts, chemical amendments are recommended. However, soils remediation with some minerals, or industrial and urban waste should be restricted, for example, when the metal load of the amendment is too heavy (Sharma and Nagpal 2018).

Following the model of urban soils, the rebuilding of highly disturbed soils is another challenging field to develop in the near future. The availability of new plant varieties, more tolerant to salinity and alkalinity, and the introduction of halophyte species to agriculture (biosaline agriculture) are expected to be instrumental for this challenge. Biotechnology can substantially contribute to address these objectives in the future, and along with technologies that favor increased carbon sequestration in soils, should provide ways to mitigate the accumulation of greenhouse gases in the atmosphere.

References

Batlle-Sales J (2011) Salinization: an environmental concern under climate change scenarios. In: Proceedings of the Global Forum on Salinization and Climate Change (GFSCC2010)

Chaneton EJ, Lavado RS (1996) Soil nutrients and salinity after long-term grazing exclusion in a Flooding Pampa grassland. J Range Manage 49:182–187

Fan Y, Li H, Miguez-Macho G (2013) Global patterns of groundwater table depth. Science 339:940–943. https://doi.org/10.1126/science.1229881

FAO (2011) Global forum on salinization and climate change (GFSCC2010), Valencia, 25–29 Oct 2010, World Soil Resources Reports No. 105. FAO, Rome

FAO/IIASA/ISRIC/ISSCAS/JRC (2012) Base de Datos Armonizada de los Suelos del Mundo v1.2. https://www.fao.org/soils-portal/soil-survey/mapas-historicos-de-suelos-y-bases-de-datos/base-de-datos-armonizada-de-los-suelos-del-mundo-v12/es/

Glatzle A, Reimer L, Núñez Cobo J, Smeenk A, Musálem K, Laino R (2020) Groundwater dynamics, land cover and salinization in the dry Chaco in Paraguay. Ecohydrol Hydrobiol 20:175-182. https://doi.org/10.1016/j.ecohyd.2019.10.003

Gonzalez A, Canales AG, Eduardo Devora E (2011) Salinization of soils and aquifers: the case of the Yaqui Valley, Sonora, Mexico. In: Proceedings of the global forum on salinization and climate change (GFSCC2010)

Iqbal MM, Goheer MA, Khan AM (2011) Facing the food challenge under climate change threats to land resources through increased salinization. In: Proceedings of the Global Forum on Salinization and Climate Change (GFSCC2010)

Jiménez H, Carvajal Y, Calero A, Romero G (2011) Salt-affected soils and climate change in the Valle del Cauca, Colombia. Salt-affected soils and climate change in the Valle del Cauca, Colombia. In: Proceedings of the global forum on salinization and climate change (GFSCC2010)

Jobbágy EG, Tóth T, Nosetto MD, Earman S (2017) On the fundamental causes of high environmental alkalinity (pH $\geq$ 9): an assessment of its drivers and global distribution. Land Degrad Dev 28:1973–1981. https://doi.org/10.1002/ldr.2718

Kuppel S, Houspanossian J, Nosetto MD, Jobbagy EG (2015) What does it take to flood the Pampas? Lessons from a decade of strong hydrological fluctuations. Water Resour Res 51:2937–2950. https://doi.org/10.1002/2015WR016966

Pankova YI, Konyushkova MV (2013) Effect of global warming on soil salinity of the arid regions. Russ Agric Sci 39:464–467

Rengasamy P (2006) World salinization with emphasis on Australia. J Exp Bot 57:1017–1023

Shahid SA, Zaman M, Zaman M, Lee H (2017) Soil salinity: historical perspectives and a world overview of the problem. In: Guideline for salinity assessment, mitigation and adaptation using nuclear and related techniques. International Atomic Energy Agency, https://doi.org/10.1007/978-3-319-96190-3_2

Sharma A, Nagpal AK (2018) Soil amendments: a tool to reduce heavy metal uptake in crops for production of safe food. Rev Environ Sci Bio 17:187–203

Sandoval M, Celis J, Pedreros L, Capulín J (2013) Chemical remediation of an agricultural soil: a case Study of the tsunami-affected area of Chile. Water Air Soil Pollut 224(1590). https://doi.org/10.1007/s11270-013-1590-5

Szabolcs I (1989) Salt-affected soils. CRC Press, Boca Raton, Fl

Tóth J (1963) A theoretical analysis of groundwater flow in small drainage basins. J Geophys Res 68:4795–4812

Várallyay G (2010) The impact of climate change on soils and on their water management. Agr Res 8(Special Issue II):385–396

Wild A (2003) Soils, land and food: managing the land during the twenty-first century. Cambridge University Press, Cambridge, UK

Index

E. Taleisnik and R. S. Lavado (eds.), *Saline and Alkaline Soils in Latin America*, https://doi.org/10.1007/978-3-030-52592-7

The manufacturer's authorised representative in the EU is Springer Nature Customer Service Centre GmbH, Europaplatz 3, 69115 Heidelberg, Germany. If you have any concerns regarding our products, please contact ProductSafety@springernature.com

Printed and bound by CPI Group (UK) Ltd, Croydon, CR0 4YY
30/09/2026
02225503-0017